Elementary Number Theory in Nine Chapters

Elementary Number Theory in Nine Chapters

Second Edition

JAMES J. TATTERSALL

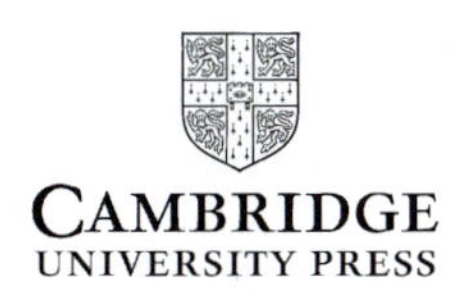

CAMBRIDGE UNIVERSITY PRESS
Cambridge, New York, Melbourne, Madrid,Cape Town, Singapore, São Paulo

CAMBRIDGE UNIVERSITY PRESS
The Edinburgh Building, Cambridge CB2 2RU, UK

Published in the United States of America by
Cambridge University Press, New York

www.cambridge.org
Information on this title: www.cambridge.org/9780521850148

First published 2005

Printed in the United Kingdom at the University Press, Cambridge

Typeset in Times 10/13pt, in 3B2 [KT]

A catalogue record for this book is available from the British Library

Library of Congress Cataloguing in publication data

Tattersall, James J. (James Joseph), 1941–
Elementary number theory in nine chapters/James J. Tattersall.
p. cm.
Includes bibliographical references.
ISBN 0 521 58503 1 (hb).–ISBN 0 521 58531 7 (pb)
1. Number theory. I. Title.
QA241.T35 1999
512'.72–dc21 98–4541 CIP

ISBN-13 978-0-521-85014-8 hardback
ISBN-10 0-521-85014-2 hardback
ISBN-13 978-0-521-61524-2 paperback
ISBN-10 0-521-61524-0 paperback

To Terry

Contents

Preface

Elementary Number Theory in Nine Chapters is primarily intended for a one-semester course for upper-level students of mathematics, in particular, for prospective secondary school teachers. The basic concepts illustrated in the text can be readily grasped if the reader has a good background in high school mathematics and an inquiring mind. Earlier versions of the text have been used in undergraduate classes at Providence College and at the United States Military Academy at West Point.

The exercises contain a number of elementary as well as challenging problems. It is intended that the book should be read with pencil in hand and an honest attempt made to solve the exercises. The exercises are not just there to assure readers that they have mastered the material, but to make them think and grow in mathematical maturity.

While this is not intended to be a history of number theory text, a genuine attempt is made to give the reader some insight into the origin and evolution of many of the results mentioned in the text. A number of historical vignettes are included to humanize the mathematics involved. An algorithm devised by Nicholas Saunderson the blind Cambridge mathematician is highlighted. The exercises are intended to complement the historical component of the course.

Using the integers as the primary universe of discourse, the goals of the text are to introduce the student to:

the basics of pattern recognition,
the rigor of proving theorems,
the applications of number theory,
the basic results of elementary number theory.

Students are encouraged to use the material, in particular the exercises, to generate conjectures, research the literature, and derive results either individually or in small groups. In many instances, knowledge of a programming language can be an effective tool enabling readers to see patterns and generate conjectures.

The basic concepts of elementary number theory are included in the first six chapters: finite differences, mathematical induction, the Euclidean Algorithm, factoring, and congruence. It is in these chapters that the number theory rendered by the masters such as Euclid, Fermat, Euler, Lagrange, Legendre, and Gauss is presented. In the last three chapters we discuss various applications of number theory. Some of the results in Chapter 7 and Chapter 8 rely on mathematical machinery developed in the first six chapters. Chapter 7 contains an overview of cryptography from the Greeks to exponential ciphers. Chapter 8 deals with the problem of representing positive integers as sums of powers, as continued fractions, and p-adically. Chapter 9 discusses the theory of partitions, that is, various ways to represent a positive integer as a sum of positive integers.

A note of acknowledgment is in order to my students for their persistence, inquisitiveness, enthusiasm, and for their genuine interest in the subject. The idea for this book originated when they suggested that I organize my class notes into a more structured form. To the many excellent teachers I was fortunate to have had in and out of the classroom, in particular, Mary Emma Stine, Irby Cauthen, Esayas Kundert, and David C. Kay, I owe a special debt of gratitude. I am indebted to Bela Bollobas, Jim McGovern, Mark Rerick, Carol Hartley, Chris Arney and Shawnee McMurran for their encouragement and advice. I wish to thank Barbara Meyer, Liam Donohoe, Gary Krahn, Jeff Hoag, Mike Jones, and Peter Jackson who read and made valuable suggestions to earlier versions of the text. Thanks to Richard Connelly, Frank Ford, Mary Russell, Richard Lavoie, and Dick Jardine for their help solving numerous computer software and hardware problems that I encountered. Thanks to Mike Spiegler, Matthew Carreiro, and Lynn Briganti at Providence College for their assistance. Thanks to Roger Astley and the staff at Cambridge University Press for their first class support. I owe an enormous debt of gratitude to my wife, Terry, and daughters Virginia and Alexandra, for their infinite patience, support, and understanding without which this project would never have been completed.

Preface to the Second Edition

The organization and content of this edition is basically the same as the previous edition. Information on several conjectures and open questions noted in the earlier edition have been updated. To meet the demand for more problems, over 375 supplementary exercises have been added to the text. The author is indebted to his students at Providence College and colleagues at other schools who have used the text. They have pointed out small errors and helped clarify parts that were obscure or diffuse. The advice of the following colleages was particularly useful: Joe Albree, Auburn University at Montgomery; Ed Burger, Williams College; Underwood Dudley, DePauw University; Stan Izen, the Latin School of Chicago; John Jaroma, Austin College; Shawnee McMurran, California State University at San Bernardino; Keith Matthews, University of Queensland; Thomas Moore, Bridgewater State College; Victor Pambuccian, Arizona State University; Tim Priden, Boulder, Colarado; Aldo Scimone, Italy; Jeff Stopple, University of California at Santa Barbara; Robert Vidal, Narbonne, France; and Thomas Weisbach, San Jose, California. I am also particularly indebted to the helpful suggestions from Mary Buckwalter, Portsmouth, Rhode Island, John Butler of North Kingston, Rhode Island, and Lynne DeMasi of Providence College. The text reads much better as a result of their help. I remain solely responsible for any errors or shortcomings that remain.

1

The intriguing natural numbers

'The time has come,' the Walrus said, 'To talk of many things.'
Lewis Carroll

1.1 Polygonal numbers

We begin the study of elementary number theory by considering a few basic properties of the set of natural or counting numbers, $\{1, 2, 3, \ldots\}$. The natural numbers are closed under the binary operations of addition and multiplication. That is, the sum and product of two natural numbers are also natural numbers. In addition, the natural numbers are commutative, associative, and distributive under addition and multiplication. That is, for any natural numbers, a, b, c:

$$\begin{aligned} a + (b + c) &= (a + b) + c, & a(bc) &= (ab)c & &\text{(associativity);} \\ a + b &= b + a, & ab &= ba & &\text{(commutativity);} \\ a(b + c) &= ab + ac, & (a + b)c &= ac + bc & &\text{(distributivity).} \end{aligned}$$

We use juxtaposition, xy, a convention introduced by the English mathematician Thomas Harriot in the early seventeenth century, to denote the product of the two numbers x and y. Harriot was also the first to employ the symbols '$>$' and '$<$' to represent, respectively, 'is greater than' and 'is less than'. He is one of the more interesting characters in the history of mathematics. Harriot traveled with Sir Walter Raleigh to North Carolina in 1585 and was imprisoned in 1605 with Raleigh in the Tower of London after the Gunpowder Plot. In 1609, he made telescopic observations and drawings of the Moon a month before Galileo sketched the lunar image in its various phases.

One of the earliest subsets of natural numbers recognized by ancient mathematicians was the set of polygonal numbers. Such numbers represent an ancient link between geometry and number theory. Their origin can be traced back to the Greeks, where properties of oblong, triangular, and square numbers were investigated and discussed by the sixth century BC, pre-Socratic philosopher Pythagoras of Samos and his followers. The

Greeks established the deductive method of reasoning whereby conclusions are derived using previously established results.

At age 18, Pythagoras won a prize for wrestling at the Olympic games. He studied with Thales, father of Greek mathematics, traveled extensively in Egypt and was well acquainted with Babylonian mathematics. At age 40, after teaching in Elis and Sparta, he migrated to Magna Graecia, where the Pythagorean School flourished at Croton in what is now Southern Italy. The Pythagoreans are best known for their theory of the transmigration of souls and their belief that numbers constitute the nature of all things. The Pythagoreans occupied much of their time with mysticism and numerology and were among the first to depict polygonal numbers as arrangements of points in regular geometric patterns. In practice, they probably used pebbles to illustrate the patterns and in doing so derived several fundamental properties of polygonal numbers. Unfortunately, it was their obsession with the deification of numbers and collusion with astrologers that later prompted Saint Augustine to equate mathematicans with those full of empty prophecies who would willfully sell their souls to the Devil to gain the advantage.

The most elementary class of polygonal numbers described by the early Pythagoreans was that of the oblong numbers. The nth oblong number, denoted by o_n, is given by $n(n+1)$ and represents the number of points in a rectangular array having n columns and $n+1$ rows. Diagrams for the first four oblong numbers, 2, 6, 12, and 20, are illustrated in Figure 1.1.

The triangular numbers, 1, 3, 6, 10, 15, ..., t_n, ..., where t_n denotes the nth triangular number, represent the numbers of points used to portray equilateral triangular patterns as shown in Figure 1.2. In general, from the sequence of dots in the rows of the triangles in Figure 1.2, it follows that t_n, for $n \geqslant 1$, represents successive partial sums of the first n natural numbers. For example, $t_4 = 1 + 2 + 3 + 4 = 10$. Since the natural numbers are commutative and associative,

$$t_n = 1 + 2 + \cdots + (n-1) + n$$

and

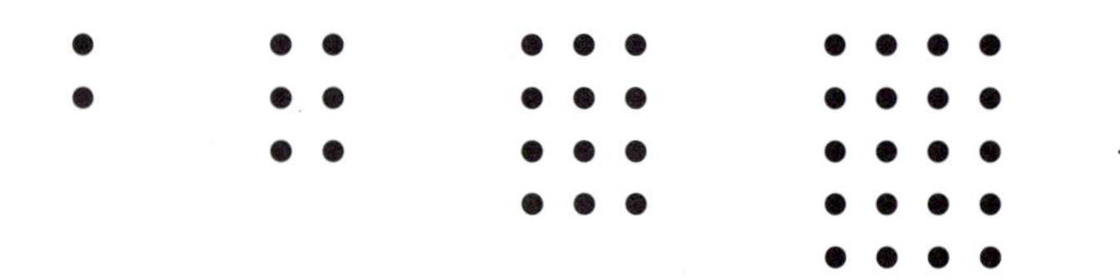

Figure 1.1

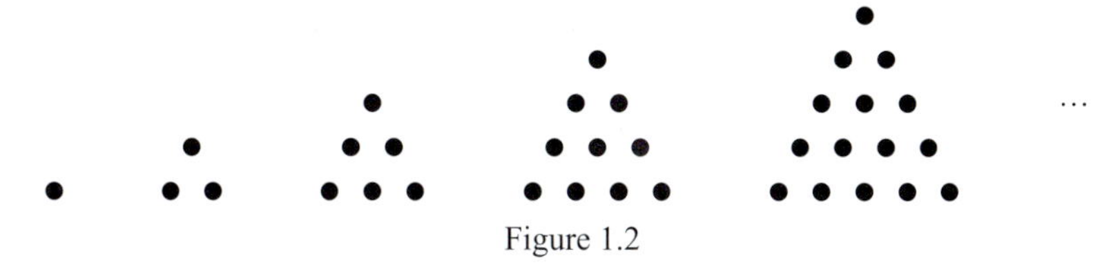

Figure 1.2

$$t_n = n + (n - 1) + \cdots + 2 + 1;$$

adding columnwise, it follows that $2t_n = (n + 1) + (n + 1) + \cdots + (n + 1) = n(n + 1)$. Hence, $t_n = n(n + 1)/2$. Multiplying both sides of the latter equation by 2, we find that twice a triangular number is an oblong number. That is, $2t_n = o_n$, for any positive integer n. This result is illustrated in Figure 1.3 for the case when $n = 6$. Since $2 + 4 + \cdots + 2n = 2(1 + 2 + \cdots + n) = 2 \cdot n(n + 1)/2 = n(n + 1) = o_n$, the sum of the first n even numbers equals the nth oblong number.

The square numbers, 1, 4, 9, 16, ..., were represented geometrically by the Pythagoreans as square arrays of points, as shown in Figure 1.4. In 1225, Leonardo of Pisa, more commonly known as Fibonacci, remarked, in *Liber quadratorum* (*The Book of Squares*) that the nth square number, denoted by s_n, exceeded its predecessor, s_{n-1}, by the sum of the two roots. That is $s_n = s_{n-1} + \sqrt{s_n} + \sqrt{s_{n-1}}$ or, equivalently, $n^2 = (n - 1)^2 + n + (n - 1)$. Fibonacci, later associated with the court of Frederick II, Emperor of the Holy Roman Empire, learned to calculate with Hindu–Arabic

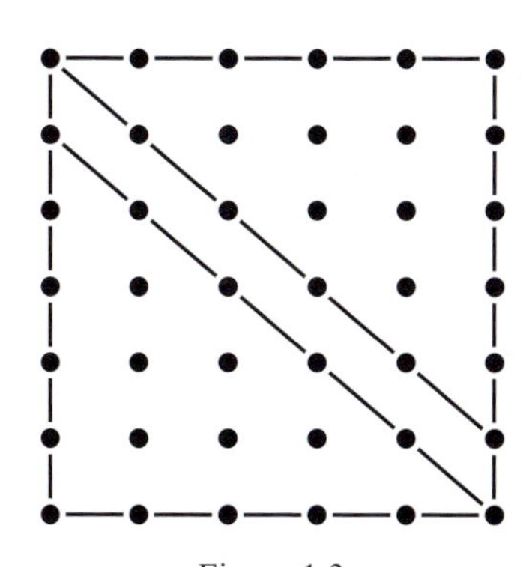

Figure 1.3

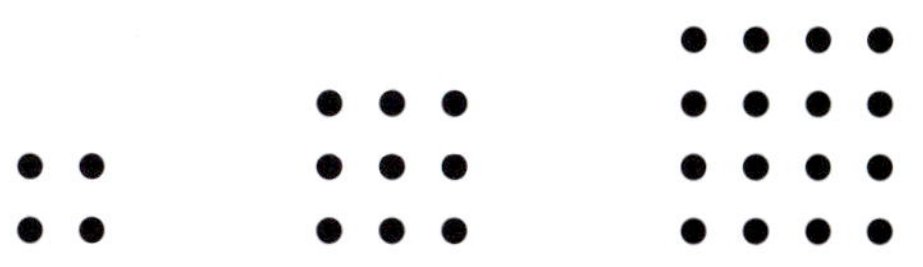

Figure 1.4

numerals while in Bougie, Algeria, where his father was a customs officer. He was a direct successor to the Arabic mathematical school and his work helped popularize the Hindu–Arabic numeral system in Europe. The origin of Leonardo of Pisa's sobriquet is a mystery, but according to some sources, Leonardo was figlio de (son of) Bonacci and thus known to us patronymically as Fibonacci.

The Pythagoreans realized that the nth square number is the sum of the first n odd numbers. That is, $n^2 = 1 + 3 + 5 + \cdots + (2n - 1)$, for any positive integer n. This property of the natural numbers first appears in Europe in Fibonacci's *Liber quadratorum* and is illustrated in Figure 1.5, for the case when $n = 6$.

Another interesting property, known to the early Pythagoreans, appears in Plutarch's *Platonic Questions*. Plutarch, a second century Greek biographer of noble Greeks and Romans, states that eight times a triangular number plus one is square. Using modern notation, we have $8t_n + 1 = 8[n(n+1)/2] + 1 = (2n+1)^2 = s_{2n+1}$. In Figure 1.6, the result is illustrated for the case $n = 3$. It is in Plutarch's biography of Marcellus that we find one of the few accounts of the death of Archimedes during the siege of Syracuse, in 212 BC.

Around the second century BC, Hypsicles [HIP sih cleez], author of

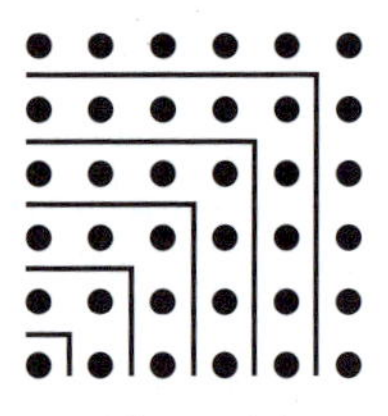

Figure 1.5

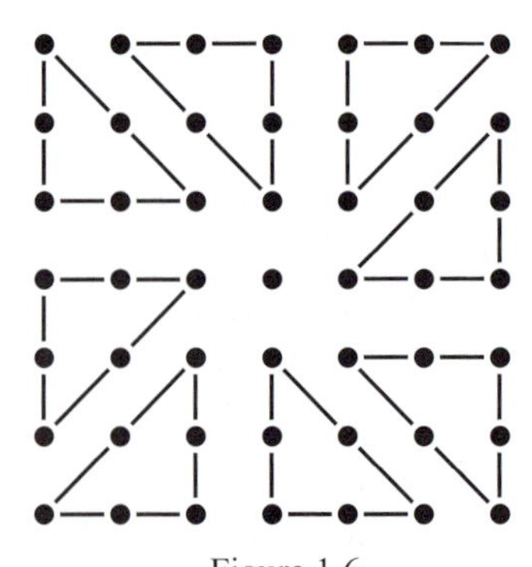

Figure 1.6

Book XIV, a supplement to Book XIII of Euclid's *Elements* on regular polyhedra, introduced the term polygonal number to denote those natural numbers that were oblong, triangular, square, and so forth. Earlier, the fourth century BC philosopher Plato, continuing the Pythagorean tradition, founded a school of philosophy near Athens in an area that had been dedicated to the mythical hero Academus. Plato's Academy was not primarily a place for instruction or research, but a center for inquiry, dialogue, and the pursuit of intellectual pleasure. Plato's writings contain numerous mathematical references and classification schemes for numbers. He firmly believed that a country's leaders should be well-grounded in Greek arithmetic, that is, in the abstract properties of numbers rather than in numerical calculations. Prominently displayed at the Academy was a maxim to the effect that none should enter (and presumably leave) the school ignorant of mathematics. The epigram appears on the logo of the American Mathematical Society. Plato's Academy lasted for nine centuries until, along with other pagan schools, it was closed by the Byzantine Emperor Justinian in 529.

Two significant number theoretic works survive from the early second century, *On Mathematical Matters Useful for Reading Plato* by Theon of Smyrna and *Introduction to Arithmetic* by Nicomachus [nih COM uh kus] of Gerasa. Smyrna in Asia Minor, now Izmir in Turkey, is located about 75 kilometers northeast of Samos. Gerasa, now Jerash in Jordan, is situated about 25 kilometers north of Amman. Both works are philosophical in nature and were written chiefly to clarify the mathematical principles found in Plato's works. In the process, both authors attempt to summarize the accumulated knowledge of Greek arithmetic and, as a consequence, neither work is very original. Both treatises contain numerous observations concerning polygonal numbers; however, each is devoid of any form of rigorous proofs as found in Euclid. Theon's goal was to describe the beauty of the interrelationships between mathematics, music, and astronomy. Theon's work contains more topics and was a far superior work mathematically than the *Introduction*, but it was not as popular. Both authors note that any square number is the sum of two consecutive triangular numbers, that is, in modern notation, $s_n = t_n + t_{n-1}$, for any natural number $n > 1$. Theon demonstrates the result geometrically by drawing a line just above and parallel to the main diagonal of a square array. For example, the case where $n = 5$ is illustrated in Figure 1.7. Nicomachus notes that if the square and oblong numbers are written alternately, as shown in Figure 1.8, and combined in pairs, the triangular numbers are produced. That is, using modern notation, $t_{2n} = s_n + o_n$ and $t_{2n+1} = s_{n+1} + o_n$, for any natural

Table 1.1.

	1	2	3	4	5	6	7	8	9	10
1	1	2	3	4	5	6	7	8	9	10
2	2	4	6	8	10	12	14	16	18	20
3	3	6	9	12	15	18	21	24	27	30
4	4	8	12	16	20	24	28	32	36	40
5	5	10	15	20	25	30	35	40	45	50
6	6	12	18	24	30	36	42	48	54	60
7	7	14	21	28	35	42	49	56	63	70
8	8	16	24	32	40	48	56	64	72	80
9	9	18	27	36	45	54	63	72	81	90
10	10	20	30	40	50	60	70	80	90	100

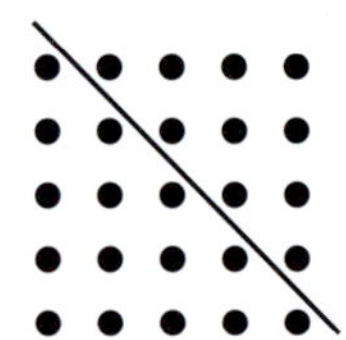

Figure 1.7

s_1 o_1 s_2 o_2 s_3 o_3 s_4 o_4 s_5 o_5

1 2 4 6 9 12 16 20 25 30

3 6 10 15 21 28 36 45 55

t_2 t_3 t_4 t_5 t_6 t_7 t_8 t_9 t_{10}

Figure 1.8

number n. From a standard multiplication table of the first ten natural numbers, shown in Table 1.1, Nicomachus notices that the major diagonal is composed of the square numbers and the successive squares s_n and s_{n+1} are flanked by the oblong numbers o_n. From this, he deduces two properties that we express in modern notation as $s_n + s_{n+1} + 2o_n = s_{2n+1}$ and $o_{n-1} + o_n + 2s_n = s_{2n}$.

Nicomachus extends his discussion of square numbers to the higher dimensional cubic numbers, 1, 8, 27, 64, ..., and notes, but does not establish, a remarkable property of the odd natural numbers and the cubic numbers illustrated in the triangular array shown in Figure 1.9, namely, that the sum of the nth row of the array is n^3. It may well have been Nicomachus's only original contribution to mathematics.

1	1
3 5	8
7 9 11	27
13 15 17 19	64
21 23 25 27 29	125
....	

Figure 1.9

In the *Introduction*, Nicomachus discusses properties of arithmetic, geometric, and harmonic progressions. With respect to the arithmetic progression of three natural numbers, he observes that the product of the extremes differs from the square of the mean by the square of the common difference. According to this property, known as the *Regula Nicomachi*, if the three terms in the progression are given by $a-k$, a, $a+k$, then $(a-k)(a+k)+k^2=a^2$. In the Middle Ages, rules for multiplying two numbers were rather complex. The Rule of Nicomachus was useful in squaring numbers. For example, applying the rule for the case when $a=98$, we obtain $98^2=(98-2)(98+2)+2^2=96\cdot 100+4=9604$.

After listing several properties of oblong, triangular, and square numbers, Nicomachus and Theon discuss properties of pentagonal and hexagonal numbers. Pentagonal numbers, 1, 5, 12, 22, $\ldots$, $p^5{}_n$, $\ldots$, where $p^5{}_n$ denotes the nth pentagonal number, represent the number of points used to construct the regular geometric patterns shown in Figure 1.10. Nicomachus generalizes to heptagonal and octagonal numbers, and remarks on the patterns that arise from taking differences of successive triangular, square, pentagonal, heptagonal, and octagonal numbers. From this knowledge, a general formula for polygonal numbers can be derived. A practical technique for accomplishing this involving successive differences appeared in a late thirteenth century Chinese text *Works and Days Calendar* by Wang Xun (SHUN) and Guo Shoujing (GOW SHOE GIN). The method was mentioned in greater detail in 1302 in *Precious Mirror of the Four*

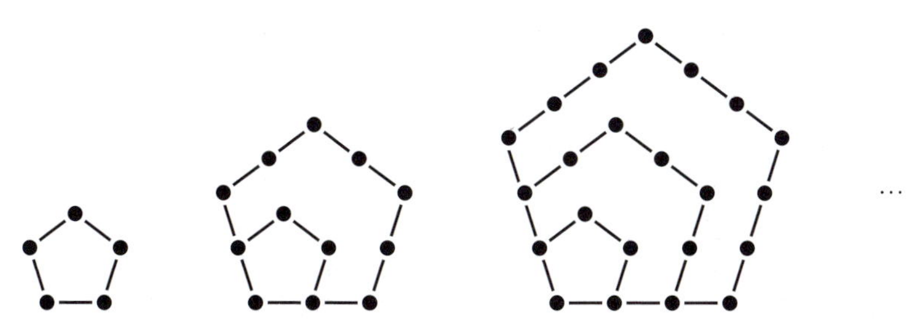

Figure 1.10

Elements by Zhu Shijie (ZOO SHE GEE), a wandering scholar who earned his living teaching mathematics. The method of finite differences was rediscovered independently in the seventeenth century by the British mathematicians Thomas Harriot, James Gregory, and Isaac Newton.

Given a sequence, $a_k, a_{k+1}, a_{k+2}, \ldots,$ of natural numbers whose rth differences are constant, the method yields a polynomial of degree $r-1$ representing the general term of the given sequence. Consider the binomial coefficients

$$\binom{n}{k} = \frac{n!}{k!(n-k)!}, \quad \text{for } 0 \leqslant k \leqslant n, \quad \binom{n}{0} = 1, \quad \text{and otherwise } \binom{n}{k} = 0,$$

where for any natural number n, n factorial, written $n!$, represents the product $n(n-1)(n-2)\cdots 3\cdot 2\cdot 1$ and, for consistency, $0! = 1$. The exclamation point used to represent factorials was introduced by Christian Kramp in 1802. The numbers, $\binom{n}{k}$, are called the binomial coefficients because of the role they play in the expansion of $(a+b)^n = \sum_{k=0}^{n}\binom{n}{k}a^{n-k}b^k$. For example,

$$\begin{aligned}(a+b)^3 &= \binom{3}{0}a^3b^0 + \binom{3}{1}a^2b^1 + \binom{3}{2}a^1b^2 + \binom{3}{3}a^0b^3\\ &= a^3 + 3a^2b + 3ab^2 + b^3.\end{aligned}$$

Denote the ith differences, Δ_i, of the sequence $a_k, a_{k+1}, a_{k+2}, \ldots$ by $d_{i1}, d_{i2}, d_{i3}, \ldots,$ and generate the following finite difference array:

$$\begin{array}{lccccccccccccc}
n & k & & k+1 & & k+2 & & k+3 & & k+4 & & k+5 & & k+6\\
a_n & a_k & & a_{k+1} & & a_{k+2} & & a_{k+3} & & a_{k+4} & & a_{k+5} & & a_{k+6}\\
\Delta_1 & & d_{11} & & d_{12} & & d_{13} & & d_{14} & & d_{15} & & d_{16} & \\
\Delta_2 & & & d_{21} & & d_{22} & & d_{23} & & d_{24} & & d_{25} & & \\
\cdots & & & & & & & & & & & & & \\
\Delta_r & & & & d_{r1} & & d_{r2} & & d_{r3} & & d_{r4} & & &
\end{array}$$

If the rth differences $d_{r1}, d_{r2}, d_{r3}, \ldots$ are equal, then working backwards and using terms in the leading diagonal each term of the sequence a_k, $a_{k+1}, a_{k+2}, \ldots$ can be determined. More precisely, the finite difference array for the sequence $b_n = \binom{n-k}{m}$, for $m = 0, 1, 2, 3, \ldots, r$, $n = k, k+1, k+2, \ldots,$ and a fixed value of k, has the property that the mth differences, Δ_m, consist of all ones and, except for $d_{m1} = 1$ for $1 \leqslant m \leqslant r$, the leading diagonal is all zeros. For example, if $m = 0$, the finite difference array for $a_n = \binom{n-k}{0}$ is given by

$$\begin{array}{lccccccccccccc}
n & k & & k+1 & & k+2 & & k+3 & & k+4 & & k+5 & & k+6\\
b_n & 1 & & 1 & & 1 & & 1 & & 1 & & 1 & & 1\\
\Delta_1 & & 0 & & 0 & & 0 & & 0 & & 0 & & 0 &
\end{array}$$

If $m = 1$, the finite difference array for $a_n = \binom{n-k}{1}$ is given by

n	k		$k+1$		$k+2$		$k+3$		$k+4$		$k+5$		$k+6$
b_n	0		1		2		3		4		5		6
Δ_1		1		1		1		1		1		1	
Δ_2			0		0		0		0		0		0

If $m = 2$, the finite difference array for $a_n = \binom{n-k}{2}$ is given by

n	k		$k+1$		$k+2$		$k+3$		$k+4$		$k+5$		$k+6$
b_n	0		0		1		3		6		10		15
Δ_1		0		1		2		3		4		5	
Δ_2			1		1		1		1		1		1
Δ_3				0		0		0		0		0	

The leading diagonals of the finite difference array for the sequence a_k, $a_{k+1}, a_{k+2}, \ldots$, and the array defined by

$$a_k\binom{n-k}{0} + d_{11}\binom{n-k}{1} + d_{21}\binom{n-k}{2} + \cdots + d_{r1}\binom{n-k}{r}$$

are identical. Therefore,

$$a_n = a_k\binom{n-k}{0} + d_{11}\binom{n-k}{1} + d_{21}\binom{n-k}{2} + \cdots + d_{r1}\binom{n-k}{r},$$

for $n = k, k+1, k+2, \ldots$.

Example 1.1 The finite difference array for the pentagonal numbers, 1, 5, 12, 22, 35, ..., $p^5{}_n$, ... is given by

n	1		2		3		4		5		6		...
$p^5{}_n$	1		5		12		22		35		51		...
Δ_1		4		7		10		13		16		...	
Δ_2			3		3		3		3		...		

Our indexing begins with $k = 1$. Therefore

$$p^5{}_n = 1\cdot\binom{n-1}{0} + 4\cdot\binom{n-1}{1} + 3\cdot\binom{n-1}{2} = 1 + 4(n-1) + 3\frac{(n-1)(n-2)}{2}$$
$$= \frac{3n^2 - n}{2}.$$

A more convenient way to determine the general term of sequences with finite differences is the following. Since the second differences of the pentagonal numbers sequence is constant, consider the sequence whose general term is $f(n) = an^2 + bn + c$, whose first few terms are given by $f(1) = a + b + c$, $f(2) = 4a + 2b + c$, $f(3) = 9a + 3b + c$, $f(4) = 16a + 4b + c$, and whose finite differences are given by

$a+b+c$	$4a+2b+c$	$9a+3b+c$	$16a+4b+c\ldots$
	$3a+b$	$5a+b$	$7a+b\ldots$
	$2a$	$2a$	$\ldots$

Matching terms on the first diagonal of the pentagonal differences with those of $f(n)$ yields

$$2a = 3$$

$$3a+b = 4$$

$$a+b+c = 1.$$

Hence, $a=\frac{3}{2}$, $b=-\frac{1}{2}$, $c=0$, and $f(n)=\frac{3}{2}n^2-\frac{1}{2}n$.

From Table 1.2, Nicomachus infers that the sum of the nth square and the $(n-1)$st triangular number equals the nth pentagonal number, that is, for any positive integer n, $p^5{}_n = s_n + t_{n-1}$. For example, if $n=6$, $s_6+t_5 = 36+15 = 51 = p^5{}_6$. He also deduces from Table 1.2 that three times the $(n-1)$st triangular number plus n equals the nth pentagonal number. For example, for $n=9$, $3\cdot t_8+9 = 3\cdot 36+9 = 117 = p^5{}_9$.

In general, the m-gonal numbers, for $m = 3, 4, 5, \ldots$, where m refers to the number of sides or angles of the polygon in question, are given by the sequence of numbers whose first two terms are 1 and m and whose second common differences equal $m-2$. Using the finite difference method outlined previously we find that $p^m{}_n = (m-2)n^2/2 - (m-4)n/2$, where $p^m{}_n$ denotes the nth m-gonal number. Triangular numbers correspond to 3-gonal numbers, squares to 4-gonal numbers, and so forth. Using Table 1.2, Nicomachus generalizes one of his previous observations and claims that $p^m{}_n + p^3{}_{n-1} = p^{m+1}{}_n$, where $p^3{}_n$ represents the nth triangular number.

The first translation of the *Introduction* into Latin was done by Apuleius of Madaura shortly after Nicomachus's death, but it did not survive. However, there were a number of commentaries written on the *Introduction*. The most influential, *On Nicomachus's Introduction to Arithmetic*, was written by the fourth century mystic philosopher Iamblichus of Chalcis in Syria. The Islamic world learned of Nicomachus through Thabit ibn Qurra's *Extracts from the Two Books of Nicomachus*. Thabit, a ninth century mathematician, physician, and philosopher, worked at the House of Wisdom in Baghdad and devised an ingenious method to find amicable numbers that we discuss in Chapter 4. A version of the *Introduction* was written by Boethius [beau EE thee us], a Roman philosopher and statesman who was imprisoned by Theodoric King of the Ostrogoths on a charge of conspiracy and put to death in 524. It would be hard to overestimate the influence of Boethius on the cultured and scientific medieval mind. His *De*

Table 1.2.

n	1	2	3	4	5	6	7	8	9	10
Triangular	1	3	6	10	15	21	28	36	45	55
Square	1	4	9	16	25	36	49	64	81	100
Pentagonal	1	5	12	22	35	51	70	92	117	145
Hexagonal	1	6	15	28	45	66	91	120	153	190
Heptagonal	1	7	18	34	55	81	112	148	189	235
Octagonal	1	8	21	40	65	96	133	176	225	280
Enneagonal	1	9	24	46	75	111	154	204	261	325
Decagonal	1	10	27	52	85	126	175	232	297	370

institutione arithmetica libri duo was the chief source of elementary mathematics taught in schools and universities for over a thousand years. He coined the term *quadrivium* referring to the disciplines of arithmetic, geometry, music, and astronomy. These subjects together with the *trivium* of rhetoric, grammar, and logic formed the seven liberal arts popularized in the fifth century in Martianus Capella's book *The Marriage of Mercury and Philology*. Boethius's edition of Nicomachus's *Introduction* was the main medium through which the Romans and people of the Middle Ages learned of formal Greek arithmetic, as opposed to the computational arithmetic popularized in the thirteenth and fourteenth centuries with the introduction of Hindu–Arabic numerals. Boethius wrote *The Consolation of Philosophy* while in prison where he reflected on the past and on his outlook on life in general. The *Consolation* was translated from Latin into Anglo-Saxon by Alfred the Great and into English by Chaucer and Elizabeth I.

In the fourth century BC Philip of Opus and Speusippus wrote treatises on polygonal numbers that did not survive. They were, however, among the first to extend polygonal numbers to pyramidal numbers. Speusippus [spew SIP us], a nephew of Plato, succeeded his uncle as head of the Academy. Philip, a mathematician–astronomer, investigated the connection between the rainbow and refraction. His native home Opus, the modern town of Atalandi, on the Euboean Gulf, was a capital of one of the regions of Locris in Ancient Greece.

Each class of pyramidal number is formed from successive partial sums of a specific type of polygonal number. For example, the nth tetrahedral number, $P^3{}_n$, can be obtained from successive partial sums of triangular numbers, that is, $P^3{}_n = p^3{}_1 + p^3{}_2 + \cdots + p^3{}_n$. For example, $P^3{}_4 = 1 + 3 + 6 + 10 = 20$. Accordingly, the first four tetrahedral numbers are 1, 4,

10, and 20. An Egyptian papyrus written about 300 BC gives $\frac{1}{2}(n^2 + n)$ as the sum of the first n natural numbers and $\frac{1}{3}(n+2)\frac{1}{2}(n^2+n)$ as the sum of the first n triangular numbers. That is, $t_n = p^3{}_n = n(n+1)/2$ and $P^3{}_n = n(n+1)(n+2)/6$. The formula for $P^3{}_n$ was derived by the sixth century Indian mathematician–astronomer Aryabhata who calculated one of the earliest tables of trigonometric sines using 3.146 as an estimate for π.

Example 1.2 Each triangle on the left hand side of the equality in Figure 1.11 gives a different representation of the first four triangular numbers, 1, 3 $(1+2)$, 6 $(1+2+3)$, and 10 $(1+2+3+4)$. Hence, $3\cdot(1+3+6+10) = 1\cdot 6 + 2\cdot 6 + 3\cdot 6 + 4\cdot 6 = (1+2+3+4)\cdot 6 = t_4(4+2)$. In general, $3(t_1 + t_2 + t_3 + \cdots + t_n) = t_n(n+2) = n(n+1)(n+2)/2$. Therefore, $P^3{}_n = n(n+1)(n+2)/6$.

In Figure 1.11, the sum of the numbers in the third triangle is the fourth tetrahedral number. That is, $1\cdot 4 + 2\cdot 3 + 3\cdot 2 + 4\cdot 1 = 20$. Thus, in general, $1\cdot n + 2\cdot(n-1) + \cdots + (n-1)\cdot 2 + n\cdot 1 = P^3{}_n$. Hence, we can generate the tetrahedral numbers by summing the terms in the SW–NE diagonals of a standard multiplication table as shown in Table 1.3. For example, $P^3{}_6 = 6 + 10 + 12 + 12 + 10 + 6 = 56$.

Pyramidal numbers with a square base are generated by successive

$$\begin{matrix} & & & 1 \\ & & 1 & & 2 \\ & 1 & & 2 & & 3 \\ 1 & & 2 & & 3 & & 4 \end{matrix} \quad + \quad \begin{matrix} & & & 1 \\ & & 2 & & 1 \\ & 3 & & 2 & & 1 \\ 4 & & 3 & & 2 & & 1 \end{matrix} \quad + \quad \begin{matrix} & & & 4 \\ & & 3 & & 3 \\ & 2 & & 2 & & 2 \\ 1 & & 1 & & 1 & & 1 \end{matrix} \quad = \quad \begin{matrix} & & & 6 \\ & & 6 & & 6 \\ & 6 & & 6 & & 6 \\ 6 & & 6 & & 6 & & 6 \end{matrix}$$

Figure 1.11

Table 1.3.

	$P^3{}_1$	$P^3{}_2$	$P^3{}_3$	$P^3{}_4$	$P^3{}_5$	$P^3{}_6$	$P^3{}_7$	$P^3{}_8$	$P^3{}_9$
	1	4	10	20	35	56	84	120	165
1	2	3	4	5	6 ↗	7	8	9	
2	4	6	8	10	12	14	16		
3	6	9	12	15	18	21			
4	8	12	16	20	24				
5	10	15	20	25					
6	12	18	24						
7	14	21							
8	16								
9									

Table 1.4.

n	1	2	3	4	5	6	7	8	9	10
$f^0{}_n$	1	1	1	1	1	1	1	1	1	1
$f^1{}_n$	1	2	3	4	5	6	7	8	9	10
$f^2{}_n$	1	3	6	10	15	21	28	36	45	55
$f^3{}_n$	1	4	10	20	35	56	84	120	165	220
$f^4{}_n$	1	5	15	35	70	126	210	330	495	715
$f^5{}_n$	1	6	21	56	126	252	462	792	1287	2002
$f^6{}_n$	1	7	28	84	210	462	924	1716	3003	5005

partial sums of square numbers. Hence, the nth pyramidal number, denoted by $P^4{}_n$, is given by $1^2 + 2^2 + 3^2 + \cdots + n^2 = n(n+1)(2n+1)/6$. For example, $P^4{}_4 = 1 + 4 + 9 + 16 = 30$. The total number of cannonballs in a natural stacking with a square base is a pyramidal number.

Slicing a pyramid through a vertex and the diagonal of the opposite base results in two tetrahedrons. Hence, it should be no surprise to find that the sum of two consecutive tetrahedral numbers is a pyramidal number, that is, $P^4{}_n = P^3{}_{n-1} + P^3{}_n$.

In the tenth century, Gerbert of Aurillac in Auvergne included a number of identities concerning polygonal and pyramidal numbers in his correspondence with his pupil Adalbold, Bishop of Utrecht. Much of Gerbert's *Geometry* was gleaned from the work of Boethius. One of the more difficult problems in the book asks the reader to find the legs of a right triangle given the length of its hypotenuse and its area. Gerbert was one of the first to teach the use of Hindu–Arabic numerals. He was elected Pope Sylvester II in 999, but his reign was short.

Triangular and tetrahedral numbers form a subclass of the figurate numbers. In the 1544 edition of *Arithmetica Integra*, Michael Stifel defined the nth mth-order figurate number, denoted by $f^m{}_n$, as follows: $f^m{}_n = f^m{}_{n-1} + f^{m-1}{}_n$, $f^m{}_1 = f^0{}_n = f^0{}_1 = 1$, for $n = 2, 3, \ldots,$ and $m = 1, 2, 3, \ldots.$ An array of figurate numbers is illustrated in Table 1.4, where the nth triangular number corresponds to $f^2{}_n$ and the nth tetrahedral number to $f^3{}_n$. In 1656, John Wallis, the English mathematician who served as a cryptanalyst for several Kings and Queens of England, and introduced the symbol ∞ to represent infinity, showed that, for positive integers n and r, $f^r{}_{n+1} = f^0{}_n + f^1{}_n + f^2{}_n + \cdots + f^r{}_n$.

Stifel was the first to realize a connection existed between figurate numbers and binomial coefficients, namely that $f^m{}_n = \binom{n+m-1}{m}$. In particular, $f^2{}_n = t_n = \binom{n+1}{2}$ and $f^3{}_n = P^3{}_n = \binom{n+2}{3}$. Stifel earned a Master's degree at Wittenberg University. He was an avid follower of Martin Luther, an ardent biblical scholar, and a millenarian. Stifel must have thought he was standing in the foothills of immortality when, through his reading, he inferred that the world was going to end at 8 o'clock on the morning of October 18, 1533. He led a band of followers to the top of a nearby hill to witness the event, a nonoccurrence that did little to enhance his credibility.

Nicomachus's *Introduction to Arithmetic* was one of the most significant ancient works on number theory. However, besides Books VII–IX of Euclid's *Elements*, whose contents we will discuss in the next chapter, the most influential number theoretic work of ancient times was the *Arithmetica* of Diophantus, one of the oldest algebra treatises in existence. Diophantus, a mathematician who made good use of Babylonian and Greek sources, discussed properties of polygonal numbers and included a rule to determine the nth m-gonal number which he attributed to Hypsicles. Unfortunately, a complete copy of the *Arithmetica* was lost when the Library of Alexandria was vandalized in 391 by Christians acting under the aegis of Theophilus, Bishop of Alexandria, and a decree by Emperor Theodosius concerning pagan monuments. Portions of the treatise were rediscovered in the fifteenth century. As a consequence, the *Arithmetica* was one of the last Greek mathematical works to be translated into Latin.

There were a number of women who were Pythagoreans, but Hypatia, the daughter of the mathematician Theon of Alexandria, was the only notable female scholar in the ancient scientific world. She was one of the last representatives of the Neo-platonic School at Alexandria, where she taught science, art, philosophy, and mathematics in the early fifth century. She wrote a commentary, now lost, on the first six books of the *Arithmetica* and may very well have been responsible for editing the version of Ptolemy's *Almagest* that has survived. Some knowledge of her can be gleaned from the correspondence between her and her student Synesius, Bishop of Cyrene. As a result of her friendship with Alexandria's pagan Prefect, Orestes, she incurred the wrath of Cyril, Theophilus's nephew who succeeded him in 412 as Bishop of Alexandria. In 415, Hypatia was murdered by a mob of Cyril's followers. During the millennium following her death no woman distinguished herself in science or mathematics.

In the introduction to the *Arithmetica*, Diophantus refers to his work as consisting of thirteen books, where a book consisted of a single scroll representing material covered in approximately twenty to fifty pages of ordinary type. The first six books of the *Arithmetica* survived in Greek and four books, which may have a Hypatian rather than a Diophantine origin, survived in Arabic. In addition, a fragment on polygonal numbers by

Diophantus survives in Greek. The *Arithmetica* was not a textbook, but an innovative handbook involving computations necessary to solve practical problems. The *Arithmetica* was the first book to introduce consistent algebraic notation and systematically use algebraic procedures to solve equations. Diophantus employed symbols for squares and cubes but limited himself to expressing each unknown quantity in terms of a single variable. Diophantus is one the most intriguing and least known characters in the history of mathematics.

Much of the *Arithmetica* consists of cleverly constructed positive rational solutions to more than 185 problems in indeterminate analysis. Negative solutions were not acceptable in Diophantus's time or for the next 1500 years. By a rational solution, we mean a number of the form p/q, where p and q are integers and $q \neq 0$. In one example, Diophantus constructed three rational numbers with the property that the product of any two of the numbers added to their sum or added to the remaining number is square. That is, in modern notation, he determined numbers x, y, z such that $xy + x + y$, $xz + x + z$, $yz + y + z$, $xy + z$, $xz + y$, and $yz + x$ are all square. In another problem, Diophantus found right triangles with sides of rational length such that the length of the hypotenuse minus the length of either side is a cube. In the eleventh century, in Baghdad, the Islamic mathematician al-Karaji and his followers expanded on the methods of Diophantus and in doing so undertook a systematic study of the algebra of exponents.

Problems similar to those found in the *Arithmetica* first appear in Europe in 1202 in Fibonacci's *Liber abaci* (*Book of Calculations*). The book introduced Hindu–Arabic numerals to European readers. It was revised by the author in 1228 and first printed in 1857. However, the first reference to Diophantus's works in Europe is found in a work by Johannes Müller who, in his day, was called Joannes de Regio monte (John of Königsberg). However, Müller is perhaps best known today by his Latinized name Regiomontanus, which was popularized long after his death. Regiomontanus, the first publisher of mathematical and astronomical literature, studied under the astronomer Georges Peurbach at the University of Vienna. He wrote a book on triangles and finished Peurbach's translation of Ptolemy's *Almagest*. Both Christopher Columbus and Amerigo Vespucci used his *Ephemerides* on their voyages. Columbus, facing starvation in Jamaica, used a total eclipse of the Moon on February 29, 1504, predicted in the *Ephemerides*, to encourage the natives to supply him and his men with food. A similar idea, albeit using a total solar eclipse, was incorporated by Samuel Clemens (Mark Twain) in *A Connecticut Yankee in King Arthur's*

Court. Regiomontanus built a mechanical fly and a 'flying' eagle, regarded as the marvel of the age, which could flap its wings and saluted when Emperor Maximilian I visited Nuremberg. Domenico Novarra, Copernicus's teacher at Bologna, regarded himself as a pupil of Regiomontanus who, for a short period, lectured at Padua.

Regiomontanus wrote to the Italian mathematician Giovanni Bianchini in February 1464 that while in Venice he had discovered Greek manuscripts containing the first six books of *Arithmetica*. In 1471, Regiomontanus was summoned to Rome by Pope Sixtus IV to reform the ecclesiastical calendar. However, in 1476, before he could complete his mission, he died either a victim of the plague or poisoned for his harsh criticism of a mediocre translation of the *Almagest*.

In 1572, an Italian engineer and architect, Rafael Bombelli, published *Algebra*, a book containing the first description and use of complex numbers. The book included 271 problems in indeterminate analysis, 147 of which were borrowed from the first four books of Diophantus's *Arithmetica*. Gottfried Leibniz used Bombelli's text as a guide in his study of cubic equations. In 1573, based on manuscripts found in the Vatican Library, Wilhelm Holtzman, who wrote under the name Xylander, published the first complete Latin translation of the first six books of the *Arithmetica*. The Dutch mathematician, Simon Stevin, who introduced a decimal notation to European readers, published a French translation of the first four books of the *Arithmetica*, based on Xylander's work.

In 1593, François Viète [VYET], a lawyer and cryptanalyst at the Court of Henry IV, published *Introduction to the Analytic Art*, one of the first texts to champion the use of Latin letters to represent numbers to solve problems algebraically. In an effort to show the power of algebra, Viète included algebraic solutions to a number of interesting problems that were mentioned but not solved by Diophantus in the *Arithmetica*.

A first-rate translation, *Diophanti Alexandrini arithmeticorum libri sex*, by Claude-Gaspard Bachet de Méziriac, appeared in 1621. Bachet, a French mathematician, theologian, and mythologist of independent means, included a detailed commentary with his work. Among the number theoretic results Bachet established were

(a) ${p^m}_{n+r} = {p^m}_n + {p^m}_r + nr(m-2)$,
(b) ${p^m}_n = {p^3}_n + (m-3){p^3}_{n-1}$, and
(c) $1^3 + 2^3 + 3^3 + \cdots + n^3 = ({p^3}_n)^2$,

where ${p^m}_n$ denotes the nth m-gonal number. The third result is usually expressed as $1^3 + 2^3 + 3^3 + \cdots + n^3 = (1 + 2 + 3 + \cdots + n)^2$ and referred to as Lagrange's identity.

In the fourth book of the *Arithmetica* Diophantus found three rational numbers, $\frac{153}{81}$, $\frac{6400}{81}$, and $\frac{8}{81}$, which if multiplied in turn by their sum yield a triangular number, a square number, and a cube, respectively. Bachet extended the problem to one of finding five numbers which if multiplied in turn by their sum yield a triangular number, a square, a cube, a pentagonal number, and a fourth power, respectively.

Bachet was an early contributor to the field of recreational mathematics. His *Problèmes plaisants et délectables qui se font par les nombres*, first published in 1612, is replete with intriguing problems including a precursor to the cannibals and missionaries problem, the Christians and Turks problem, and a discussion on how to create magic squares. At age 40, Bachet married, retired to his country estate, sired seven children, and gave up his mathematical activity forever. Except for recurring bouts with gout and rheumatism, he lived happily ever after.

The rediscovery of Diophantus's work, in particular through Bachet's edition which relied heavily on Bombelli's and Xylander's work, greatly aided the renaissance of mathematics in Western Europe. One of the greatest contributors to that renaissance was Pierre de Fermat [fair MAH], a lawyer by profession who served as a royal councillor at the Chamber of Petitions at the Parlement of Toulouse. Fermat was an outstanding amateur mathematician. He had a first-class mathematical mind and, before Newton was born, discovered a method for finding maxima and minima and general power rules for integration and differentiation of polynomial functions of one variable. He rarely, however, published any of his results. In 1636, he wrote, in a burst of enthusiasm, that he had just discovered the very beautiful theorem that every positive integer is the sum of at most three triangular numbers, every positive integer is the sum of at most four squares, every positive integer is the sum of at most five pentagonal numbers, and so on *ad infinitum*, but added, however, that he could not give the proof, since it depended on 'numerous and abstruse mysteries of numbers'. Fermat planned to devote an entire book to these mysteries and to 'effect in this part of arithmetic astonishing advances over the previously known limits'. Unfortunately, he never published such a book.

In 1798, in *Théorie des nombres*, the Italian mathematician and astronomer, Joseph-Louis Lagrange, used an identity discovered by the Swiss mathematician Leonhard Euler [OILER] to prove Fermat's claim for the case of square numbers. Karl Friedrich Gauss proved the result for triangular numbers when he was nineteen and wrote in his mathematical diary for 10 July 1796: '$\varepsilon\upsilon\rho\eta\kappa\alpha$! num $= \blacktriangle + \blacktriangle + \blacktriangle$.' Two years later, Gauss's result was proved independently by the French mathematician, Adrien Marie Legendre. In the introduction to *Disquisitiones arithmeticae* (*Arithmetical Investigations*) Gauss explains his indebtedness to Diophantus's *Arithmetica.* In Chapters 5, 6, and 8, we discuss the contents of Gauss's *Disquisitiones.* In 1808, Legendre included a number of quite remarkable number theoretic results in his *Théorie des nombres*; in particular, the property that every odd number not of the form $8k + 7$, where k is a positive integer, can be expressed as the sum of three or fewer square numbers. In 1815, Augustin-Louis Cauchy proved that every positive integer is the sum of m m-gonal numbers of which all but four are equal to 0 or 1. Cauchy's *Cours d'analyse*, published in 1821, advocated a rigorous approach to mathematical analysis, in particular to the calculus. Unfortunately, Cauchy was very careless with his correspondence. Evariste Galois and Niels Henrik Abel sent brilliant manuscripts to Cauchy for his examination and evaluation, but they were lost.

One of the first results Fermat established was that nine times any triangular number plus one always yielded another triangular number. Fermat later showed that no triangular number greater than 1 could be a cube or a fourth power. Fermat, always the avid number theorist, once challenged Lord Brouncker, first President of the Royal Society, and John Wallis, the best mathematician in England at the time, to prove there is no triangular number other than unity that is a cube or a fourth power. Neither was able to answer his query.

Fermat often used the margins of texts to record his latest discoveries. In 1670, Fermat's son, Clément-Samuel, published a reprint of Bachet's *Diophantus* together with his father's marginal notes and an essay by the Jesuit, Jacques de Billy, on Fermat's methods for solving certain types of Diophantine-type equations. His most famous marginal note, the statement of his 'last' theorem, appears in his copy of Bachet's edition of the *Arithmetica.* Fermat wrote to the effect that it was impossible to separate a cube into two cubes, or a biquadratic into two biquadratics, or generally any power except a square into two powers with the same exponent. Fermat added that he had discovered a truly marvelous proof of this result; however, the margin was not large enough to contain it. Fermat's Last Theorem was 'last' in the sense that it was the last major conjecture by Fermat that remained unproven. Fermat's Last Theorem has proven to be a veritable fountainhead of mathematical research and until recently its proof eluded the greatest mathematicians. In 'The Devil and Simon Flagg'

Arthur Porges relates a delightful tale in which the Devil attempts to prove Fermat's Last Theorem.

The Swiss mathematician, Leonhard Euler, perused a copy of Bachet's Diophantus with Fermat's notes and was intrigued by Fermat's emphasis on integer, rather than rational, solutions. At the University of Basel, Euler was a student of Johann Bernoulli. Bernoulli won the mathematical prize offered by the Paris Academy twice. His son Daniel Bernoulli won it ten times. Euler, who won the prize twelve times, began a lifelong study of number theory at age 18. Euler's papers are remarkably readable. He has a good historical sense and often informs the reader of things that have impressed him and of ideas that led him to his discoveries. Even though over half of Euler's 866 publications were written when he was blind, he laid the foundation of the theory of numbers as a valid branch of mathematics. His works were still appearing in the *Memoirs* of the St Petersburg Academy fifty years after his death. It is estimated that he was responsible for one-third of all the mathematical work published in Europe from 1726 to 1800. He had a phenomenal memory and knew Vergil's *Aeneid* by heart. At age 70, given any page number from the edition he owned as a youth, he could recall the top and bottom lines. In addition, he kept a table of the first six powers of the first hundred positive integers in his head.

Before proceeding further, it is important in what follows for the reader to be able to distinguish between a conjecture and an open question. By a conjecture we mean a statement which is thought to be true by many, but has not been proven yet. By an open question we mean a statement for which the evidence is not very convincing one way or the other. For example, it was conjectured for many years that Fermat's Last Theorem was true. It is an open question, however, whether $4! + 1 = 5^2$, $5! + 1 = 11^2$, and $7! + 1 = 71^2$ are the only solutions to the equation $n! + 1 = m^2$.

Exercises 1.1

1. An even number can be expressed as $2n$ and an odd number as $2n + 1$, where n is a natural number. Two natural numbers are said to be of the same parity if they are either both even or both odd, otherwise they are said to be of opposite parity. Given any two natural numbers of the same parity, show that their sum and difference are even. Given two numbers of opposite parity, show that their sum and difference are odd.

2. Nicomachus generalized oblong numbers to rectangular numbers, which are numbers of the form $n(n+k)$, denoted by $r_{n,k}$, where $k \geqslant 1$ and $n > 1$. Determine the first ten rectangular numbers that are not oblong.
3. Prove algebraically that the sum of two consecutive triangular numbers is always a square number.
4. Show that $9t_n + 1$ [Fermat], $25t_n + 3$ [Euler], and $49t_n + 6$ [Euler] are triangular.
5. Show that the difference between the squares of any two consecutive triangular numbers is always a cube.
6. In 1991, S.P. Mohanty showed that there are exactly six triangular numbers that are the product of three consecutive integers. For example, $t_{20} = 210 = 5 \cdot 6 \cdot 7$. Show that t_{608} is the product of three consecutive positive integers.
7. Show that the product of any four consecutive natural numbers plus one is square. That is, show that for any natural number n, $n(n+1)(n+2)(n+3)+1 = k^2$, for some natural number k.
8. The nth star number, denoted by $*_n$, represents the sum of the nth square number and four times the $(n-1)$st triangular number, where $*_1 = 1$. One geometric interpretation of star numbers is as points arranged in a square with equilateral triangles on each side. For example $*_2$ is illustrated in Figure 1.12. Derive a general formula for the nth star number.
9. Show that Gauss's discovery that every number is the sum of three or fewer triangular numbers implies that every number of the form $8k+3$ can be expressed as the sum of three odd squares.
10. Verify Nicomachus's claim that the sum of the odd numbers on any row in Figure 1.9 is a cube.
11. For any natural number n prove that
    (a) $s_{2n+1} = s_n + s_{n+1} + 2o_n$. [Nicomachus]
    (b) $s_{2n} = o_{n-1} + o_n + 2s_n$. [Nicomachus]

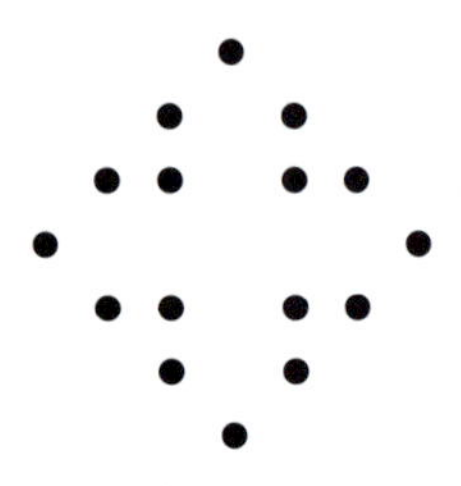

Figure 1.12

12. Show that $s_n + t_{n-1} = p^5{}_n$, for any natural number n. [Nicomachus]
13. Prove that $p^5{}_n = 3t_{n-1} + n$, for any natural number n. [Nicomachus]
14. Show that every pentagonal number is one-third of a triangular number.
15. Find a positive integer $n > 1$ such that $1^2 + 2^2 + 3^2 + \cdots + n^2$ is a square number. [*Ladies' Diary*, 1792] This problem was posed by Edouard Lucas in 1875 in *Annales de Mathématique Nouvelles*. In 1918, G. N. Watson proved that the problem has a unique solution.
16. Prove the square of an odd multiple of 3 is the difference of two triangular numbers, in particular show that for any natural number n, $[3(2n+1)]^2 = t_{9n+4} - t_{3n+1}$.
17. Show that there are an infinite number of triangular numbers that are the sum of two triangular numbers by establishing the identity $t_{[n(n+3)+1]/2} = t_{n+1} + t_{n(n+3)/2}$.
18. Prove that $t_{2mn+m} = 4m^2 t_n + t_m + mn$, for any positive integers m and n.
19. Paul Haggard and Bonnie Sadler define the nth m-triangular number, $T^m{}_n$, by $T^m{}_n = n(n+1)\cdots(n+m+1)/(m+2)$. When $m = 0$, we obtain the triangular numbers. Generate the first ten $T^1{}_n$ numbers.
20. Derive a formula for the nth hexagonal number. The first four hexagonal numbers 1, 6, 15, 28 are illustrated geometrically in Figure 1.13.
21. Show that 40 755 is triangular, pentagonal, and hexagonal. [*Ladies' Diary*, 1828]
22. Use the method of finite differences to derive a formula for the nth m-gonal number $p^m{}_n$. [Diophantus]
23. Prove that for any natural numbers m and n, $p^{m+1}{}_n = p^m{}_n + p^3{}_{n-1}$. [Nicomachus]
24. Prove that $p^m{}_{n+r} = p^m{}_n + p^m{}_r + nr(m-2)$, where n, m, and r, are natural numbers and $m > 2$. [Bachet]

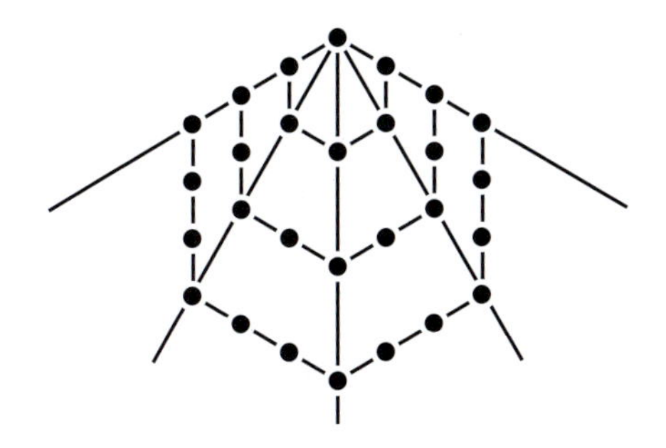

Figure 1.13

25. Prove that $p^m{}_n = p^3{}_n + (m-3)p^3{}_{n-1}$. [Bachet]
26. In 1897, G. Wertheim devised a method to determine in how many ways a number r appears as a polygonal number. He used the fact that $p^m{}_n = \frac{1}{2}n(2+(m-2)(n-1))$, let $2r = n(2+(m-2)(n-1)) = n \cdot s$, and concentrated on such factorizations of $2r$ where $2 < n < s$ and $n-1$ divides $s-2$. For example, $72 = 3 \cdot 24 = 6 \cdot 12 = 8 \cdot 9 = n \cdot s$. Hence, $36 = p^{13}{}_3 = p^4{}_6 = p^3{}_8$. Using Wertheim's method determine how many ways 120 appears as a polygonal number.
27. In the 1803 edition of *Recreations in Mathematics and Natural Philosophy*, a revision of a text first published by Ozanam in 1692 and revised by Jean Etienne Montucla in 1778, it is stated that a number n is m-gonal if $8n(m-2)+(m-4)^2$ is a square number. Use Ozanam's rule to show that 225 is octagonal.
28. Derive Ozanam's rule.
29. Use the method of finite differences to show that the nth tetrahedral number, $P^3{}_n$, is given by $n(n+1)(n+2)/6$. [Aryabhata]
30. There are only five numbers less than 10^9 which are both triangular and tetrahedral, namely, 1, 10, 120, 1540, and 7140. Show that 1540 and 7140 are both triangular and tetrahedral.
31. Show that $P^4{}_n = P^3{}_{n-1} + P^3{}_n$, for any natural number n.
32. Show that $P^5{}_n = \frac{1}{3}n(2n^2+1)$, for any natural number n.
33. Show

$$P^m{}_n = \frac{n+1}{6}(2p^m{}_n + n),$$

for any natural numbers m and n, where $m \geqslant 3$. The relation between pyramidal and polygonal numbers appears in a fifth century Roman codex.
34. The nth octahedral number, denoted by O_n, is defined as the sum of the nth and $(n-1)$st pyramidal numbers. Determine the first 10 octahedral numbers.
35. Use the binomial representation of figurate numbers to show that $f^2{}_n$ represents the nth triangular number and $f^3{}_n$ represents the nth tetrahedral number.
36. Justify the formula, $f^3{}_{n-1} + f^3{}_n = n(n+1)(2n+1)/6$, found in an ancient Hindu manuscript.
37. In the fall of 1636, Fermat wrote to Marin Mersenne and Gilles Persone de Roberval that he had discovered that $n \cdot f^r{}_{n+1} = (n+r) \cdot f^{r+1}{}_n$, where n and r are natural numbers. Justify Fermat's formula.

38. Show that a general solution to Problem 17 in Book III of Diophanus's *Arithmetica*, find x, y, z such that $xy + x + y$, $yz + y + z$, $zx + z + x$, $xy + z$, $xz + y$, and $yz + x$ are square, is given by $x = n^2$, $y = (n + 1)^2$, and $z = 4(n^2 + n + 1)$.
39. Use algebra to solve Gerbert's problem: given the area and length of the hypotenuse of a right triangle, find the lengths of the sides of the triangle.
40. The nth central trinomial coefficient, denoted by a_n, is defined as the coefficient of x^n in $(1 + x + x^2)^n$. Determine a^n for $0 \leqslant n \leqslant 10$.

1.2 Sequences of natural numbers

A sequence is a finite or infinite ordered linear array of numbers. For example, 2, 4, 6, 8, ... represents the infinite sequence of even positive integers. Analytically, an infinite sequence can be thought of as the range of a function whose domain is the set of natural numbers. For example, polygonal, oblong, pyramidal, and figurate numbers are examples of infinite sequences of natural numbers. In this section, we investigate a number of patterns that arise from imposing various conditions on the terms of a sequence. The construction of some sequences can seem to be almost diabolical. For example, each successive term in the sequence 1, 5, 9, 31, 53, 75, 97, ... is obtained by adding 4 to the previous term and reversing the digits. Properties of look and say sequences were developed by John H. Conway at Cambridge University. For example, each successive term in the look and say sequence 1, 11, 21, 1 211, 111 221, 312 211, ... is generated from the previous term as follows: the first term is 1, the second term indicates that the first term consists of one one, the third term indicates that the second term consists of two ones, the fourth term indicates that the third term consists of one two and one one, the fifth term indicates that the fourth term consists of one one, one two, and two ones, and so forth. A look and say sequence will never contain a digit greater than 3 unless that digit appears in the first or second term.

In 1615, Galileo remarked that

$$\frac{1}{3} = \frac{1+3}{5+7} = \frac{1+3+5}{7+9+11} = \cdots.$$

Hence, we call a sequence a_1, a_2, a_3, ... a Galileo sequence with ratio k, for k a positive integer, if it has the property that $S_{2n}/S_n = k + 1$ or, equivalently, $S_{2n} - S_n = kS_n$, where S_n denotes the nth partial sum, $a_1 + a_2 + a_3 + \cdots + a_n$. Thus, the increasing sequence of odd positive

Table 1.5.

n	1	2	3	4	5	6	7	8	9	10	...
a_n	0	3	6	12	24	48	96	192	384	768	...
$a_n + 4$	4	7	10	16	28	52	100	196	388	772	...
$(a_n + 4)/10$	0.4	0.7	1.0	1.6	2.8	5.2	10.0	19.6	38.8	77.2	...
Actual distance (AU)	0.387	0.723	1	1.52		5.2	9.59	19.2	30.1	39.5	
	Mercury	Venus	Earth	Mars		Jupiter	Saturn	Uranus	Neptune	Pluto	

natural numbers is a Galileo sequence with ratio 3. If $a_1, a_2, a_3, \ldots$ is a Galileo sequence with ratio k, then, for r a positive integer, $ra_1, ra_2, ra_3, \ldots$ is also a Galileo sequence with ratio k. A strictly increasing Galileo sequence $a_1, a_2, a_3, \ldots,$ with ratio $k \geqslant 3$, can be generated by the recursive formulas

$$a_{2n-1} = \left[\!\left[\frac{(k+1)a_n - 1}{2} \right]\!\right]$$

and

$$a_{2n} = \left[\!\left[\frac{(k+1)a_n}{2} \right]\!\right] + 1,$$

for $n \geqslant 2$, where $a_1 = 1$, $a_2 = k$, for $k \geqslant 2$, and $[\![x]\!]$ denotes the greatest integer not greater than x. For example, when $k = 3$, the formula generates the sequence of odd natural numbers. For $k = 4$, the Galileo sequence generated is $1, 4, 9, 11, 22, 23, 27, 28, 54, 56, \ldots$.

One of the most intriguing sequences historically is generated by Bode's law. The relation was discovered in 1766 by Johann Daniel Titius, a mathematician at Wittenberg University, and was popularized by Johann Bode [BO duh], director of the Berlin Observatory. According to Bode's law, the distances from the Sun to the planets in the solar system in astronomical units, where one astronomical unit equals the Earth–Sun distance or approximately 93 million miles, can be obtained by taking the sequence which begins with 0, then 3, then each succeeding term is twice the previous term. Then 4 is added to each term and the result is divided by 10, as shown in Table 1.5.

Initially, Bode's law is a fairly accurate predictor of the distances to the planets from the Sun in astronomical units. The penultimate row in Table 1.5 gives the actual average distance from the planets to the Sun in astronomical units. Bode became an astronomical evangelist for the law and formed a group called the celestial police to search for a missing planet 2.8 AU from the Sun. On January 1, 1801, the first day of the nineteenth century, Father Giuseppe Piazzi at the Palermo Observatory found what he thought was a new star in the constellation Taurus and informed Bode of his discovery. Bode asked the 23-year-old Gauss to calculate the object's orbit. It took Gauss two months to devise a technique, the method of least squares, that would take an observer a few hours to calculate the orbit of a body in 3-space. The previous method, due to Euler, took numerous observations and several weeks of calculation. Using Gauss's method the object was rediscovered December 7, 1801 and named Ceres, after the Roman goddess of vegetation and protector of Sicily. Three years later another minor planet was discovered. A few years later another sun object was discovered, then another. Today the orbits of about 80 000 minor planets are known. Almost all minor planets ply orbits between those of Mars and Jupiter, called the asteroid belt. Their average distance from the Sun is amazingly close to 2.8 AU.

Superincreasing sequences of positive integers have the property that each term is greater than the sum of all the preceding terms. For example, $2, 4, 8, 16, 32, \ldots, 2^n, \ldots$ is an infinite superincreasing sequence and 3, 9, 14, 30, 58, 120, 250, 701 is a finite superincreasing sequence with eight terms. We will use superincreasing sequences in Chapter 7 to create knapsack ciphers.

Consider the sequence of positive integers where each succeeding term is the sum of the decimal digits of the previous term. More formally, if $S_r(n)$ denotes the sum of the rth powers of the decimal digits of the positive integer n the general term to the sequences will be $a_k = S_r^k(n) = S_r(S_r^{k-1}(n))$ where $r = 2$. In particular, since $1^2 + 2^2 = 5$, $5^2 = 25$, $2^2 + 5^2 = 29$, and $2^2 + 9^2 = 85$, the sequence generated by 12 is given by 12, 5, 25, 26, 85, 89, 145, 42, 20, 4, 16, 37, 58, 89, 145, ... Numbers whose sequences eventually reach the cycle 4, 16, 37, 58, 89, 145, 42, 20 of period 8 as 12 does are called sad numbers. A positive integer n is called happy if $S_2^m(n) = 1$, for some positive integer m. The height of a happy number is the number of iterations necessary to reach unity. For example, 31 is a happy number of height two and 7 is a happy number of height five. For any positive integer 10^n is happy and $2(10)^n$ is sad, hence there are an infinite numbers of both happy and sad numbers. About 1/7 of all positive integers are happy. In 2002, E. El-Sedy and S. Siksek showed the existence of sequences of consecutive happy integers of arbitrary length. In 1945, Arthur Porges of the Western Military Academy in Alton, Illinois proved that every positive integer is either happy or sad.

A natural generalization of happy and sad numbers is to sequences formed where each succeeding term is the sum of the rth powers of the digits of the previous term. That is, when the general term of the sequence is $a_k = S_r^k(n)$ with $r > 2$ a positive integer. For example when $r = 3$, eight distinct cycles arise. In particular, $3^3 + 7^3 + 1^3 = 371$. Hence, 371 self-replicates. In 1965, Y. Matsuoka proved that if n is a multiple of 3 then there exists a positive integer m such that $S_3^m(n) = 153$, another self-replecate. A positive integer n is called a cubic happy number is $S_3^m(n) = 1$, for some positive integer m.

Sidney sequences, $a_1, a_2, \ldots, a_n$, named for their 15-year-old discoverer Sidney Larison of Ceres, California, are defined as follows: given any

m-digit natural number $a_1a_2 \cdots a_m$, let the first m terms of the Sidney sequence be $a_1, a_2, \ldots, a_m$; then, for $k > m$, a_k is defined to be the units digit of $a_{k-m} + \cdots + a_{k-2} + a_{k-1}$, the sum of the previous m terms of the sequence. A Sidney sequence terminates when the last m terms of the sequence match the first m terms of the sequence. For example, with $m = 2$ the Sidney sequence for 76 is given by 7, 6, 3, 9, 2, 1, 3, 4, 7, 1, 8, 9, 7, 6.

For the case when $m = 2$, Larison showed there are six different repeating patterns generated by Sidney sequences. One of the cycles has period 60, a property noted by Lagrange in 1744 when he discovered that the units digits of the Fibonacci numbers form a sequence with period 60. When $m = 3$, there are 20 patterns, and 11 exist if $m = 4$. Similar results occur if we are given an m-digit natural number and proceed to construct a product instead of a sum.

Undoubtedly, the most famous sequence of natural numbers is the Fibonacci sequence, 1, 1, 2, 3, 5, 8, 13, 21, 34, 55, 89, ..., u_n ..., where $u_1 = 1$, $u_2 = 1$, and $u_{n+1} = u_n + u_{n-1}$. The sequence first appeared in Europe in 1202 in *Liber abaci* by Leonardo of Pisa, more commonly known as Fibonacci. Albert Girand, a mathematician from the Netherlands and a disciple of Viète, first defined the sequence recursively in 1634 in a posthumous publication. Fibonacci numbers were used prior to the eighth century to describe meters in Sanskrit poetry.

Fibonacci first mentions the sequence in connection with the number of pairs of rabbits produced in n months, beginning with a single pair, assuming that each pair, from the second month on, begets a new pair, and no rabbits die. The number of pairs of rabbits after n months is the sum of the number of pairs which existed in the previous month and the number of pairs which existed two months earlier, because the latter pairs are now mature and each of them now produces another pair. In Figure 1.14, A_n represents the nth pair of rabbits in their first month and B_n the nth pair of rabbits in succeeding months.

The sequence never gained much notoriety until the late nineteenth century when Edouard Lucas popularized the sequence in *Théorie des nombres* and attached the name Fibonacci to it. Lucas was a French artillery officer during the Franco-Prussian War and later taught at the Lycée Saint-Louis and at the Lycée Charlemagne in Paris. In *Mathematical Recreations*, he introduced the Tower of Hanoi puzzle where, according to Lucas, three monks of Benares in northeastern India (not Vietnam) maintained a device consisting of three pegs onto which 64 different sized disks were placed. Initially, all the disks were on one peg and formed a pyramid. The monks' task was to move the pyramid from one peg to another peg.

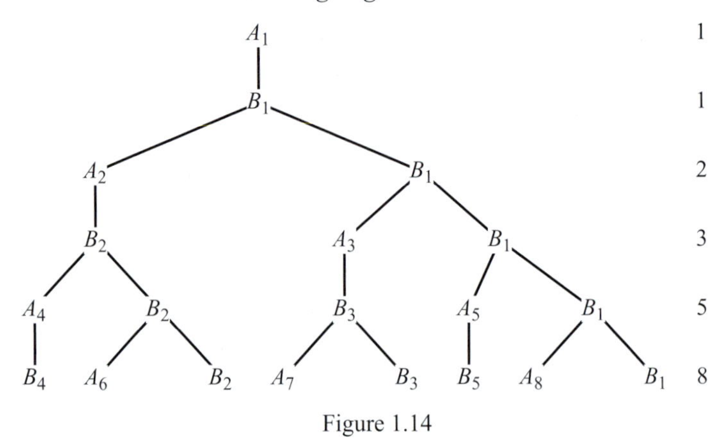

Figure 1.14

The rules were simple. Only one disk could be moved at a time from one peg to another peg, and no larger disk could be placed on a smaller disk. According to legend, when the monks finished their task the world would end. Lucas explained how it would take at least $2^{64} - 1$ moves to complete the task. At the rate of one move a second, the monks would take almost 6×10^9 centuries to complete their task. Unfortunately, Lucas died of erysipelas after a freak accident in a restaurant where a waiter dropped a tray of dishes and a shard gashed his cheek.

Lucas numbers, denoted by v_n, are defined recursively as follows: $v_{n+1} = v_n + v_{n-1}$, $v_1 = 1$, and $v_2 = 3$. Lucas originally defined v_n to be u_{2n}/u_n. He derived many relationships between Fibonacci and Lucas numbers. For example, $u_{n-1} + u_{n+1} = v_n$, $u_n + v_n = 2u_{n+1}$, and $v_{n-1} + v_{n-1} = 5u_n$. The sequence of Lucas numbers is an example of a Fibonacci-type sequence, that is, a sequence $a_1, a_2, \ldots,$ with $a_1 = a$, $a_2 = b$, and $a_{n+2} = a_{n+1} + a_n$, for $n \geqslant 2$.

Fibonacci numbers seem to be ubiquitous in nature. There are abundant references to Fibonacci numbers in phyllotaxis, the botanical study of the arrangement or distribution of leaves, branches, and seeds. The numbers of petals on many flowers are Fibonacci numbers. For example, lilies have 3, buttercups 5, delphiniums 8, marigolds 13, asters 21, daisies 21 and 34. In addition, poison ivy is trifoliate and Virginia creeper is quinquefoliate.

The fraction 10000/9899 has an interesting connection with Fibonacci numbers for its decimal representation equals 1.010 203 050 813 213 455 .... There are only four positive integers which are both Fibonacci numbers and triangular numbers, namely, 1, 3, 21, and 55. There are only three number which are Lucas and triangular numbers, namely, 1, 3, and 5778. In 1963, J. H. E. Cohn showed that except for unity, the only square Fibonacci number is 144.

Geometrically, we say that a point C divides a line segment AB, whose length we denote by $|AB|$, in the golden ratio when $|AB|/|AC| = |AC|/|CB|$, as shown in Figure 1.15. Algebraically, let $|AC| = a$ and $|AB| = b$; then $b/a = a/(b-a)$, hence, $b^2 - ab = a^2$. Dividing both sides of the equation by a^2 and setting $x = b/a$, we obtain $x^2 = x + 1$, whose roots are $\tau = (1+\sqrt{5})/2$, the golden ratio, and $\sigma = (1-\sqrt{5})/2$, its reciprocal. It is thought by many who search for human perfection that the height of a human body of divine proportion divided by the height of its navel is the golden ratio. One of the most remarkable connections between the Fibonacci sequence and the golden ratio, first discovered by Johannes Kepler the quintessential number cruncher, is that as n approaches infinity the limit of the sequence of ratios of consecutive Fibonacci numbers, u_{n+1}/u_n, approaches τ, the golden ratio.

Using only Euclidean tools, a compass and a straightedge, a line segment AB may be divided in the golden ratio. We construct DB perpendicular to AB, where $|DB| = \frac{1}{2}|AB|$, as shown in Figure 1.16. Using a compass, mark off E on AD such that $|DE| = |DB|$ and C on AB so that $|AC| = |AE|$. From the construction, it follows that $|AB|/|AC| = \tau$.

Golden right triangles have their sides in the proportion $1{:}\ \sqrt{\tau}{:}\ \tau$. In 1992, Duane DeTemple showed that there is a golden right triangle

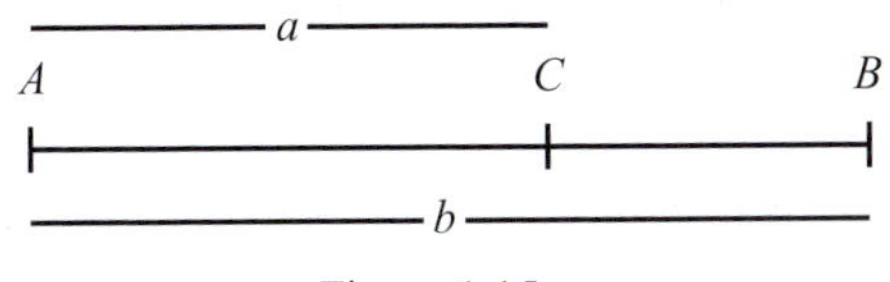

Figure 1.15

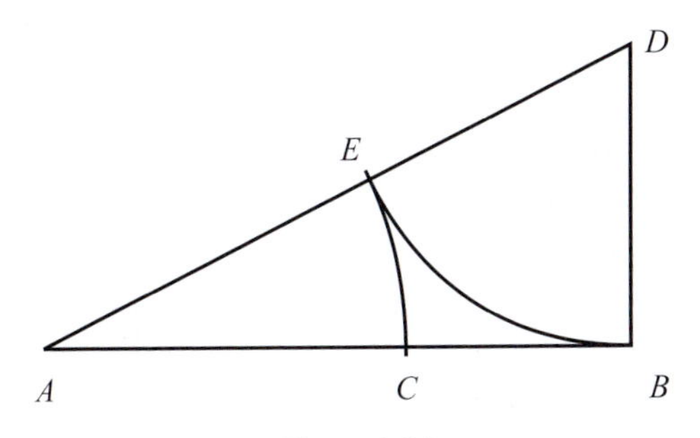

Figure 1.16

associated with the isosceles triangle of smallest perimeter circumscribing a given semicircle. Rectangles whose sides are of length a and b, with $b/a = \tau$, are called golden rectangles. In the late nineteenth century, a series of psychological experiments performed by Gustav Fechner and Wilhelm Wundt indicated that golden rectangles were the quadrilaterals which were, aesthetically, most pleasing to the eye. Such rectangles can be found in 3×5 file cards, 5×8 photographs, and in Greek architecture, in particular, in the design of the Parthenon. A golden rectangle can be constructed from a square. In particular, given a square $ABCD$, let E be the midpoint of side DC, as shown in Figure 1.17. Use a compass to mark off F on DC extended such that $|EF| = |EB|$. Mark off G on AB such that $|AG| = |DF|$, and join GF, CF, and BG. From the construction, it follows that $|AG|/|AD| = \tau$. Hence, the quadrilateral $AGFD$ is a golden rectangle.

In 1718, Abraham de Moivre, a French mathematician who migrated to England when Louis XIV revoked the Edict of Nantes in 1685, claimed that $u_n = (\tau^n - \sigma^n)/(\tau - \sigma)$, where $\tau = (1 + \sqrt{5})/2$ and $\sigma = (1 - \sqrt{5})/2$. The first proof was given in 1728 by Johann Bernoulli's nephew Nicolas. Independently, the formula was established by Jacques-Philippe-Marie Binet in 1843 and by Gabriel Lamé a year later. It is better known today as Binet's or Lamé's formula.

Since $\tau + \sigma = 1$, $\tau - \sigma = \sqrt{5}$, multiplying both sides of the identity $\tau^2 = \tau + 1$ by τ^n, where n is any positive integer, we obtain $\tau^{n+2} = \tau^{n+1} + \tau^n$. Similarly, $\sigma^{n+2} = \sigma^{n+1} + \sigma^n$. Thus, $\tau^{n+2} - \sigma^{n+2} = (\tau^{n+1} + \tau^n) - (\sigma^{n+1} + \sigma^n) = (\tau^{n+1} - \sigma^{n+1}) + (\tau^n - \sigma^n)$. Dividing both sides by $\tau - \sigma$ and letting $a_n = (\tau^n - \sigma^n)/(\tau - \sigma)$, we find that

$$a_{n+2} = \frac{\tau^{n+2} - \sigma^{n+2}}{\tau - \sigma} = \frac{\tau^{n+1} - \sigma^{n+1}}{\tau - \sigma} + \frac{\tau^n - \sigma^n}{\tau - \sigma} = a_{n+1} + a_n,$$

with $a_1 = a_2 = 1$. Hence,

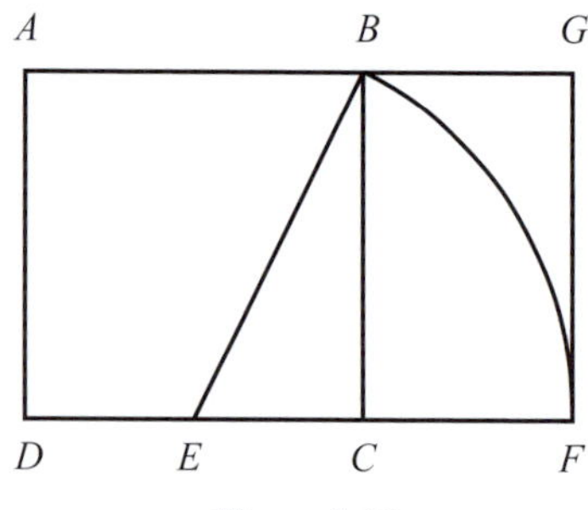

Figure 1.17

$$a_n = \frac{\tau^n - \sigma^n}{\tau - \sigma} = u_n,$$

the nth term in the Fibonacci sequence.

Another intriguing array of natural numbers appears in Blaise Pascal's *Treatise on the Arithmetic Triangle*. The tract, written in 1653, was published posthumously in 1665. Pascal was a geometer and one of the founders of probability theory. He has been credited with the invention of the syringe, the hydraulic press, the wheelbarrow, and a calculating machine. Pascal left mathematics to become a religious fanatic, but returned when a severe toothache convinced him that God wanted him to resume the study of mathematics.

Pascal exhibited the triangular pattern of natural numbers, known as Pascal's triangle, in order to solve a problem posed by a noted gamester, Chevalier de Méré. The problem was how to divide the stakes of a dice game if the players were interrupted in the midst of their game. For further details, see [Katz]. Each row of the triangle begins and ends with the number 1, and every other term is the sum of the two terms immediately above it, as shown in Figure 1.18. Pascal remarked that the nth row of the triangle yields the binomial coefficients found in the expansion of $(x + y)^n$.

The triangular array, however, did not originate with Pascal. It was known in India around 200 BC and appears in several medieval Islamic mathematical texts. The frontispiece of Zhu Shijie's *Precious Mirror of the Four Elements* contains a diagram of the triangle (Figure 1.19). In 1261, the triangular array appeared in Yang Hui's (CHANG WAY) *A Detailed Analysis of the Mathematical Methods in the 'Nine Chapters'*. Yang Hui noted that his source for the diagram was *The Key to Mathematics* by Jia Xian (GEE AH SHE ANN), an eleventh century work which has been lost. Yang Hui's method of extracting square roots uses the formula $(a + b)^2 = a^2 + (2a + b)b$, with a as an initial value. Cubic roots were extracted using the formula $(a + b)^3 = a^3 + (3a^2 + 3ab + b^2)b$. Higher

```
            1
          1   1
        1   2   1
      1   3   3   1
    1   4   6   4   1
  1   5  10  10   5   1
1   6  15  20  15   6   1
.........................
```

Figure 1.18

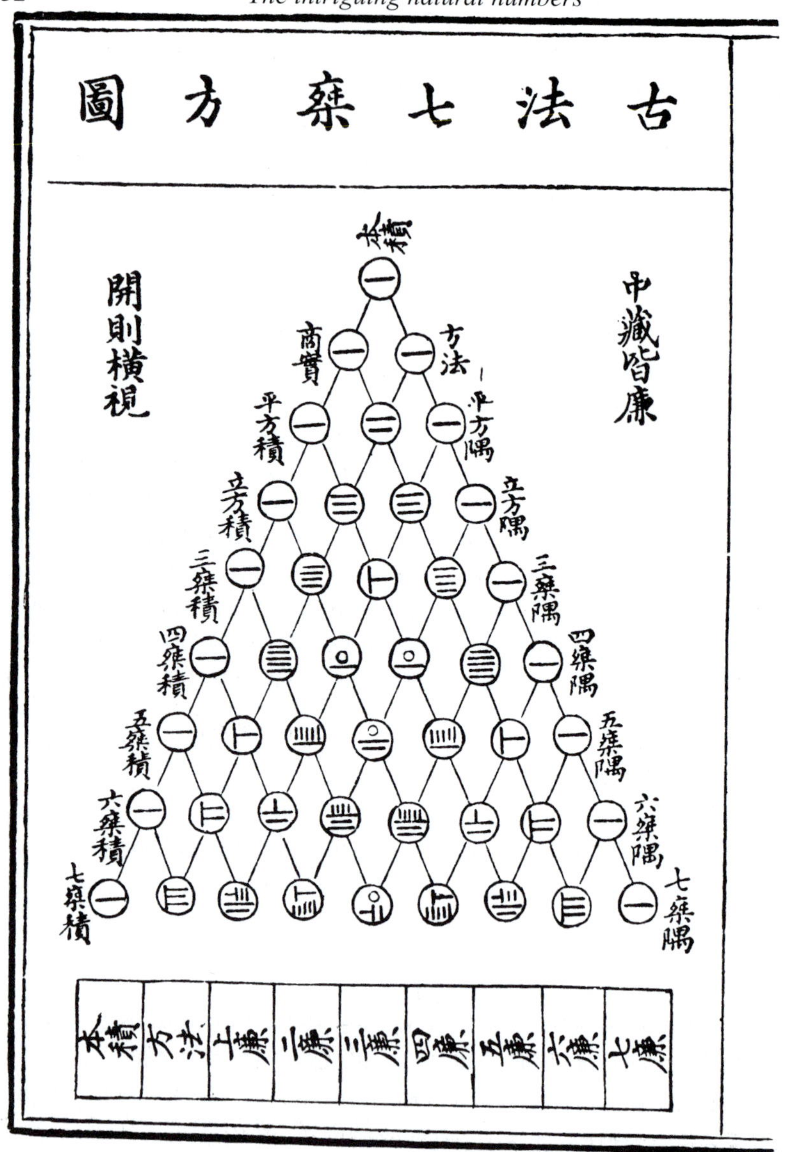

Figure 1.19

roots can be extracted by generalizing the formula using higher-order binomial coefficients. Prior to the introduction of the hand calculator such methods were sometimes taught in schools. Similar arrangements of numbers can be found in the works of Persian mathematicians Al-Karaji and Omar Khayyam. Pascal's triangle first appeared in Europe in 1225 in

Jordanus de Nemore's *On Arithmetic* and was conspicuously displayed on the title page of the 1527 edition of Peter Apian's *Arithmetic*. In 1524 Apian published a popular but very laborious method to calculate longitude using the Moon. In the eighteenth century, John Harrison constructed a reliable chronometer that enabled navigators to determine their longitude more accurately and with fewer calculations. In 1544, the triangle used as a tool in root extraction played a prominent role in Stifel's *Complete Arithmetic*. In 1556, the array appeared in Niccoló of Brescia's *General Treatise*. Niccoló was commonly known as Tartaglia, the stammerer, owing to an injury received as a boy. In Italy, the triangular array is known as Tartaglia's triangle.

The figurate-binomial relationship first observed by Stifel was rediscovered in 1631 by Henry Briggs, inventor of common logarithms, and William Oughtred [AWE tread], inventor of the slide rule. Oughtred worked at mathematics at a country vicarage in Albury, Surrey, where he served as rector and gave mathematical instruction to any who came to him provided they could write clearly. Oughtred believed that mathematics improved reasoning power and was a pathway to the understanding of God. Oughtred complained that many a good notion was lost and many a problem went unsolved because his wife took away his candles right after dinner. He was ecstatic when one of his pupils, perhaps John Wallis, brought him a box of candles.

Pascal's name was first attached to the array in 1708 by Pierre Rémond de Montmort. Pascal's original arrangement, shown in Table 1.6, is

Table 1.6. $f^m{}_n$

	n										
m	0	1	2	3	4	5	6	7	8	9	10
0	1	1	1	1	1	1	1	1	1	1	1
1	1	2	3	4	5	6	7	8	9	10	...
2	1	3	6	10	15	21	28	36	45	...	
3	1	4	10	20	35	56	84	120	...		
4	1	5	15	35	70	126	210	...			
5	1	6	21	56	126	252	...				
6	1	7	28	84	210	...					
7	1	8	36	120	...						
8	1	9	45	...							
10	1	...									

fundamentally a table of figurate numbers. Even though the array did not originate with Pascal, the conclusions that he drew from it with respect to solving problems in probability went far beyond any of his predecessors.

In the seventeenth century, René François de Sluse remarked that the sums of the slant ENE–WSW diagonals of Pascal's triangle in Figure 1.20 yield the Fibonacci numbers, a result rediscovered by Edouard Lucas in 1896.

In this text, you will encounter a number of mechanical computational procedures or algorithms. An algorithm is a specific set of rules used to obtain a given result from a specific input. The word is a Latin corruption of al-Khwarizmi, a ninth century mathematician–astronomer, member of the House of Wisdom in Baghdad, and author of a very influential work, *al-Kitab al-muhtasar fi hisab al-jabar wa-l-muqabala* (*The Condensed Book on Comparing and Restoring*), the text from which our word 'algebra' derives. One must wonder if students would be even more reticent about high school mathematics if they were required to take two years of 'muqabala'. In many cases, as we shall see, algorithms can generate very interesting sequences of natural numbers. For example, the Collatz algorithm, named for Lothar O. Collatz of the University of Hamburg who devised it in the 1930s, is as follows: given any positive integer a_1, let

$$a_{n+1} = \begin{cases} \dfrac{a_n}{2} & \text{if } a_n \text{ is even, and} \\ 3a_n + 1 & \text{if } a_n \text{ is odd.} \end{cases}$$

Collatz conjectured that for any natural number the sequence generated eventually reached unity. John Selfridge, of Northern Illinois University, has shown this to be the case for all natural numbers less than 7×10^{11}. The conjecture is one of the more well-known unsolved problems in number theory, as is the question of whether there is an upper limit to the number of iterations in the Collatz algorithm necessary to reach unity. Any

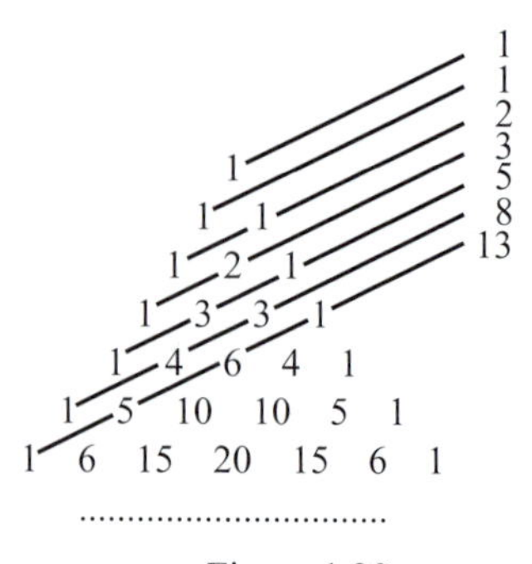

Figure 1.20

slight adjustment of the algorithm may change the outcome. For example, if $3a_n + 1$ is replaced by $3a_n - 1$, when a_n is odd, three distinct cycles are generated.

An interesting procedure, albeit not as intriguing as the Collatz algorithm, is the Kaprekar algorithm devised in 1949 by the Indian mathematician D.R. Kaprekar (kuh PREE kur). Kaprekar's sort–reverse–subtract routine goes as follows: given a four-digit natural number larger than 1000 for which not all digits are equal, arrange the digits in descending order, subtract the result from its reverse (the number with the digits in ascending order). Successive applications of this algorithm result in the four-digit Kaprekar constant, the self-replicating number 6174. For example, for 1979, we have

$$\begin{array}{r} 9971 \\ -\underline{1799} \\ 8172 \end{array} \quad \begin{array}{r} 8721 \\ -\underline{1278} \\ 7443 \end{array} \quad \begin{array}{r} 7443 \\ -\underline{3447} \\ 3996 \end{array} \quad \begin{array}{r} 9963 \\ -\underline{3699} \\ 6264 \end{array} \quad \begin{array}{r} 6642 \\ -\underline{2466} \\ 4176 \end{array} \quad \begin{array}{r} 7641 \\ -\underline{1467} \\ 6174 \end{array}$$

Given any m-digit number n, with not all digits the same and $m > 2$, let M_n and m_n denote the largest and smallest positive integers obtainable from permuting the digits of n and let $K(n) = M_n - m_n$. The m-digit Kaprekar constant, denoted by k_m, is the integer such that successive iterations of K on any m-digit positive integer generate k_m and $K(k_m) = k_m$. For four-digit numbers the Kaprekar constant is 6174.

The digital root of a positive integer n, denoted by $\rho(n)$, is the single digit obtained by adding the digits of a number. If the sum obtained has more than one digit, then the process is repeated until a single digit is obtained. For example, since $7 + 4 + 3 + 2 + 8 = 24$ and $2 + 4 = 6$, the digital root of 74 328 is 6, that is, $\rho(74\,328) = 6$. For natural numbers m and n, $\rho(\rho(n)) = \rho(n)$, $\rho(n + 9) = \rho(n)$, and the pairs $\rho(mn)$ and $\rho(m)\rho(n)$, and $\rho(n \pm m)$ and $\rho(n) \pm \rho(m)$, have the same remainder when divided by 9. For any positive integer k we may construct the auxiliary sequence $a_1, a_2, \ldots, a_n, \ldots$, where $a_1 = k$ and $a_{n+1} = a_n + \rho(a_n)$. From this sequence, the digital root sequence $\rho(a_1), \rho(a_2), \ldots$ can be generated. For example, the auxiliary sequence for 12 is given by 12, 15, 21, 24, 30, 33, 39. Hence, the digital root sequence for 12 is given by 3, 6, 3, 6, 3, 6, 3, ... In 1979, V. Sasi Kumar showed that there are only three basic digital root sequences.

We end this section with two sequences generated by the digits of a number, one constructed additively, the other multiplicatively. The digital sum sequence is defined as follows: let a_1 be any natural number. For $k \geqslant 2$, define $a_k = a_{k-1} + s_{\mathrm{d}}(a_{k-1})$, where $s_{\mathrm{d}}(n)$ denotes the sum of the digits of n. In 1906, A. Géradin showed that the 19th term of the digital sum sequence whose 1st term is 220 and the 10th term of the digital sum sequence whose 1st term is 284 both equal 418.

In 1973, Neil Sloane of AT&T Bell Labs, author of *A Handbook of Integer Sequences*, devised a sequence of natural numbers by defining each successive term in the sequence as the product of the digits of the preceding term. Sloane defined the persistence of a natural number as the number of steps required to obtain a single digit number. For example, the persistence of 74 is 3 since its persistence sequence is 74, 28, 16, 6. The smallest number with persistence 2 is 25. The smallest number with persistence 1 is 10. Sloane showed that no number less than 10^{50} has a persistence greater than 11. He conjectured that there is a natural number N such that every natural number has persistence less than N. Sloane's online encyclopedia of integer sequences contains an interactive database of information on known sequences. Currently, the website averages 30,000 hits a day and adds approximately 40 new sequences a day to the database.

Given any two positive integers m and n, S. Ulam defined the sequence of $u(m, n)$-numbers, $a_1, a_2, a_3, \ldots,$ such that $a_1 = m$, $a_2 = n$ and for $k > 2$, a_k is the least integer greater than a_{k-1} uniquely expressible as $a_i + a_j$ for $1 \leqslant i < j \leqslant k - 1$, that is, as the sum of two distinct previous terms of the sequence. For example, if $m = 1$ and $n = 2$, then the first $u(1, 2)$-numbers are 1, 2, 3, 4, 6, 8, 11, 13, 16, 18, 26, 28, 36, 38, 47, 48, 53, 57, 62, 69, 72, 77, 82, 87, 97, 99, and so forth. Note that $3 = 1 + 2$, $4 = 1 + 3$, however, $5 = 2 + 3 = 4 + 1$. Thus, 5 does not have a unique representation as a sum of previous terms and, hence, does not belong in the sequence. There are a number of open questions concerning $u(1, 2)$-numbers. For example, are there infinitely many numbers which are not the sum of $u(1, 2)$-numbers? Are there infinitely many pairs of consecutive $u(1, 2)$-numbers. Are there arbitrarily large gaps in the sequence of $u(1, 2)$-numbers? Ulam worked as a mathematician on the Manhattan Project in Los Alamos which led to the development of the first atomic bomb.

Similarly, given any two positive integers m and n, we define the sequence of $v(m, n)$-numbers $b_1, b_2, b_3, \ldots,$ such that $b_1 = m$, $b_2 = n$, and for $k > 2$, b_k is the least integer greater than b_{k-1} that is not of the form $b_i + b_j$, for $1 \leqslant i < j \leqslant k - 1$. That is, each succeeding term in the sequence is the next positive integer that cannot be written as a sum of two previous terms of the sequence. For example, the first ten $v(2, 5)$-numbers are 2, 5, 6, 9, 10, 13, 17, 20, 21, 24.

These concepts can be generalised. For example, the sequence of

$u(a_1, a_2, a_3, \ldots, a_n)$-numbers, $a_1, a_2, a_3, \ldots$, has the property that for $k > n$ a_k is the least integer greater than a_{k-1} uniquely expressible as $a_i + a_j$, for $1 \leqslant i < j \leqslant k - 1$. The sequence of $v(b_1, b_2, b_3, \ldots, b_n)$-numbers, $b_1, b_2, b_3, \ldots$, has the property that for $k > n$, b_k is the least integer greater than b_{k-1} that can not be represented as $b_i + b_j$, for $1 \leqslant i < j \leqslant k$.

Exercises 1.2

1. Determine the next three terms in the look and say sequence 1, 11, 21, 1 211, 111 221, 312 211, ...
2. Explain why a look and say sequence cannot contain a digit greater than 3 unless that digit appears in the first or second term.
3. Generate the first ten terms of a Galileo sequence with ratio 5 and first term 1.
4. Which of the following are superincreasing sequences?
   (a) 2, 3, 6, 12, 25, 50, 99, 199, 397,
   (b) 3, 5, 9, 18, 35, 72, 190, 1009,
   (c) 4, 7, 12, 24, 48, 96, 192, 384, 766.
5. Determine the next three terms of the sequence 1, 5, 9, 31, 53, 75, 97, ..., and the rule that generates the sequence.
6. Determine the next three terms of the sequence 5, 8, 21, 62, 86, 39, 74, 38, ..., and the rule that generates the sequence.
7. Are the following natural numbers happy or sad?
   (a) 392, (b) 193, (c) 269, (d) 285, (e) 521.
8. Determine the nine cycles that occur in sequences of natural numbers where each succeeding term is the sum of the cubes of the digits of the previous number.
9. Determine the six cycles that occur if succeeding terms of a sequence are the sum of the fourth powers of the digits of the previous term.
10. Determine the six different cycles that result from applying Sidney's algorithm to two-digit numbers. What is the sum of the periods of the six cycles?
11. Given any m-digit natural number $a_1 a_2 \cdots a_m$, let the first m terms of the sequence be $a_1, a_2, \ldots, a_m$; then, for $k > m$, a_{k+1} is defined to be the units digit of the product of the previous nonzero m terms of the sequence. The sequence terminates when a repeating pattern of digits occurs. What repeating patterns result from this Sidney product sequence algorithm for $m = 2$? (for $m = 3$?)
12. For what values of n is u_n, the nth Fibonacci number, even.

13. Show that the sum of any 10 consecutive Fibonacci-type numbers is always equal to 11 times the seventh term in the sequence.
14. Show that $\tau = (1 + (1 + (1+ \cdots)^{1/2})^{1/2})^{1/2}$ (hint: square both sides of the equation).
15. In Figure 1.16, show that $|AB|/|AC| = \tau$.
16. In Figure 1.17, show that $|AG|/|AD| = \tau$.
17. A golden box is a parallelepiped whose height, width, and length are in the geometric proportion σ:1:τ. Show that a golden box may also be defined as a parallelepiped whose height, width, and length are in the geometric proportion 1:τ:τ^2.
18. Determine the first ten Lucas numbers.
19. Show that 5778 is a triangular–Lucas number.
20. If $\tau = (1 + \sqrt{5})/2$ and $\sigma = (1 - \sqrt{5})/2$, show that $\upsilon_n = \tau^n + \sigma^n$.
21. The tribonacci numbers a_n are defined recursively as follows: $a_1 = a_2 = 1$, $a_3 = 2$, and $a_n = a_{n-1} + a_{n-2} + a_{n-3}$, for $n \geqslant 4$. Generate the first 20 tribonacci numbers.
22. The tetranacci numbers b_n are defined as follows: $b_1 = b_2 = 1$, $b_3 = 2$, $b_4 = 4$, and $b_n = b_{n-1} + b_{n-2} + b_{n-3} + b_{n-4}$, for $n \geqslant 5$. Generate the first 20 tetranacci numbers.
23. Verify the Collatz conjecture for the following numbers:
(a) 9, (b) 50, (c) 121.
24. Determine the three cycles that occur when $3a_n - 1$ is substituted for $3a_n + 1$ in the Collatz algorithm.
25. Perform the Kaprekar routine on the following natural numbers until you obtain the Kaprekar constant:
(a) 3996, (b) 1492, (c) your birth year, (d) the current calendar year.
26. Use the Kaprekar algorithm to determine the three-digit Kaprekar constant for three-digit numbers.
27. The reverse–subtract–reverse–add algorithm is stated as follows: given a three-digit natural number with the outer two digits differing by at least 2, reverse the digits of the number and subtract the smaller from the larger of the two numbers to obtain the number A, take A, reverse its digits to obtain the number B, add A and B. The sum, $A + B$, will always be 1089. Verify this algorithm for the following numbers: (a) 639, (b) 199, (c) 468.
28. Given a four-digit number n for which not all the digits are equal, let $abcd$ represent the largest integer possible from permuting the digits a, b, c, d of n, that is, so $a \geqslant b \geqslant c \geqslant d$. The Trigg operator, $T(n)$, is defined such that $T(n) = badc - cdab$. The Trigg constant is the

integer m such that iterations of T always lead to m and $T(m) = m$. Determine the Trigg constant.

29. Determine the three basic digital root sequences.
30. For any natural number n prove that $\rho(n+9) = \rho(n)$, where $\rho(n)$ denotes the digital root of n.
31. Show that the 19th term of the digital sum sequence whose 1st term is 220 and the 10th term of the digital sum sequence whose 1st term is 284 both equal 418.
32. Determine the sum of the digits of the first million positive integers.
33. The sequence $a_1, a_2, \ldots$ is called a Kaprekar sequence, denoted by K_{a_1}, if a_1 is a positive integer and $a_{k+1} = a_k + s_{\mathrm{d}}(a_k)$, for $k > 1$, where $s_{\mathrm{d}}(n)$ denotes the sum of the digits of n. For example, if $a_1 = 1$, we obtain the Kaprekar sequence $K_1 = 1, 2, 4, 8, 16, 23, 28, \ldots$. In 1959, Kaprekar showed that there are three types of Kaprekar sequence: (I) each term is not divisible by 3, (II) each term is divisible by 3 but not by 9, and (III) each term is divisible by 9. For example, K_1 is type I. Determine the Kaprekar type for K_{a_1}, when $a_1 = k$, for $2 \leqslant k \leqslant 10$.
34. Kaprekar called a positive integer a self number if it does not appear in a Kaprekar sequence except as the first term. That is, a natural number n is called a self, or Columbian, number if it cannot be written as $m + s_{\mathrm{d}}(m)$, where m is a natural number less than n. For example, 1 and 3 are self numbers. Determine all the self numbers less that 100.
35. In *The Educational Times* for 1884, Margaret Meyer of Girton College, Cambridge, discovered a set of conditions under which a number n is such that $s_{\mathrm{d}}(n) = 10$ and $s_{\mathrm{d}}(2n) = 11$. Find such a set of conditions.
36. Determine the persistence of the following natural numbers: (a) 543, (b) 6989, (c) 86 898, (d) 68 889 789, (e) 3 778 888 999.
37. Determine the smallest natural numbers with persistence 3, with persistence 4, with persistence 5.
38. Determine the first 15 $u(1, 3)$-numbers.
39. Determine the first 15 $u(2, 3)$-numbers.
40. Determine the first 30 $u(2, 5)$-numbers.
41. Determine the first 15 $u(2, 3, 5)$-numbers.
42. Determine the first 15 $v(1, 2)$-numbers.
43. Determine the first 15 $v(1, 3)$-numbers.
44. Determine the first 15 $v(3, 4, 6, 9, 10, 17)$-numbers.
45. Define the sequence $a_1, a_2, \ldots$ of $w(m, n)$-numbers as follows. Let $a_1 = m$, $a_2 = n$, and a_k, for $k > 2$, be the unique smallest number

greater than a_{k-1} equal to a product $a_i a_j$, where $i < j < k$. Determine the first eight $w(2, 3)$-numbers.

1.3 The principle of mathematical induction

Most students of mathematics realize that a theorem is a statement for which a proof exists, a lemma is a subordinate theorem useful in proving other theorems, and a corollary is a result whose validity follows directly from a theorem. Proofs of theorems and lemmas may be constructive or nonconstructive, that is, in general, practical or elegant. It should also be evident that in mathematical problems, ‘establish’, ‘show’ and ‘prove’ are the same commands.

One of the most important techniques in establishing number theoretic results is the principle of mathematical induction. The method was first employed by Pascal in 1665 and named as such by Augustus De Morgan in 1838. It is a technique that is not very satisfying to students since it is usually nonconstructive and does not give any clue as to the origin of the formula that it verifies. Induction is not an instrument for discovery. Nevertheless, it is a very important and powerful tool. The principle of mathematical induction follows from the well-ordering principle which states that every nonempty set of natural numbers has a least element.

Theorem 1.1 (Principle of mathematical induction) *Any set of natural numbers that contains the natural number m, and contains the natural number $n + 1$ whenever it contains the natural number n, where $n \geqslant m$, contains all the natural numbers greater than m.*

Proof Let S be a set containing the natural number m and the natural number $n + 1$ whenever it contains the natural number n, where $n \geqslant m$. Denote by T the set of all natural numbers greater than m that are not in S. Suppose that T is not empty. By the well-ordering principle T has a least element, say r. Now, $r - 1$ is a natural number greater than or equal to m and must lie in S. By the induction assumption, $(r - 1) + 1 = r$ must also lie in S, a contradiction. Hence, the assumption that T is not empty must be false. We conclude that T is empty. Therefore, S contains all the natural numbers greater than m. ■

In most applications of the principle of mathematical induction, we are interested in establishing results that hold for all natural numbers, that is, when $m = 1$. There is an alternate principle of mathematical induction, equivalent to the principle of mathematical induction stated in Theorem 1.1, in which, for a given natural number m, we require the set in question to contain the natural number $n+1$ whenever it contains all the natural numbers between m and n, where $n \geqslant m$. The alternate principle of mathematical induction is very useful and is stated in Theorem 1.2 without proof.

Theorem 1.2 (Alternate principle of mathematical induction) *Any set of natural numbers that contains the natural number m, and contains $n+1$ whenever it contains all the natural numbers between m and n, where $n \geqslant m$, contains all the natural numbers greater than m.*

The alternate principle of mathematical induction implies the well-ordering principle. In order to see this, let S be a nonempty set of natural numbers with no least element. For $n > 1$, suppose $1, 2, \ldots, n$ are elements $\overline{S}$, the complement of S. A contradiction arises if $n+1$ is in S for it would then be the least positive natural number in S. Hence, $n+1$ must be in $\overline{S}$. From the alternate principle of mathematical induction, with $m = 1$, $\overline{S}$ must contain all natural numbers. Hence, S is empty, a contradiction.

Establishing results using induction is not as difficult as it seems and it should be in every mathematician's repertoire of proof techniques. In Example 1.3, we use induction to establish a result known to the early Pythagoreans.

Example 1.3 The sum of consecutive odd natural numbers beginning with 1 is always a square. This result first appeared in Europe in 1225 in Fibonacci's *Liber quadratorum*. The statement of the problem can be expressed in the form of a variable proposition $P(n)$, a statement whose truth or falsity varies with the natural number n, namely $P(n)$: $1 + 3 + 5 + \cdots + (2n-1) = n^2$. In order to establish the truth of $P(n)$, for all natural numbers n, using induction, we first show that $P(1)$ is true. This follows since $1 = 1^2$. We now assume that proposition $P(n)$ holds for an arbitrary value of n, say k, and show that $P(k+1)$ follows from $P(k)$. Since we are assuming that $P(n)$ holds true for $n = k$, our assumption is

$$P(k)\text{: } 1 + 3 + 5 + 7 + \cdots + (2k-1) = k^2.$$

Adding $2k+1$ to both sides yields

$$1 + 3 + 5 + 7 + \cdots + (2k-1) + (2k+1) = k^2 + (2k+1),$$

or

$$1 + 3 + 5 + 7 + \cdots + (2k - 1) + (2k + 1) = (k + 1)^2,$$

establishing the truth of $P(k + 1)$. Hence, by the principle of mathematical induction, $P(n)$ is true for all natural numbers n.

Example 1.4 We show that $u_1 + u_2 + \cdots + u_n = u_{n+2} - 1$, where n is any natural number and u_n represents the nth Fibonacci number. We have $u_1 = 1 = 2 - 1 = u_3 - 1$, hence $P(1)$ is true. Assume that $P(n)$ is true for an arbitrary natural number k, hence, we assume that $u_1 + u_2 + u_3 + \cdots + u_k = u_{k+2} - 1$. Adding u_{k+1} to both sides of the equation we obtain $u_1 + u_2 + u_3 + \cdots + u_k + u_{k+1} = (u_{k+2} - 1) + u_{k+1} = (u_{k+1} + u_{k+2}) - 1 = u_{k+3} - 1$. Thus, $P(k + 1)$ follows from $P(k)$, and the result is established for all natural numbers by the principle of mathematical induction.

It is important to note that verifying both conditions of the principle of mathematical induction is crucial. For example, the proposition $P(n)$: $1 + 3 + 5 + \cdots + (2n - 1) = n^3 - 5n^2 + 11n - 6$ is only true when $n = 1, 2,$ or 3. Further, $P(n)$: $1 + 3 + 5 + \cdots + (2n - 1) = n^2 + n(n - 1)(n - 2) \cdots (n - 1000)$ is true for $n = 1, 2, 3, \ldots, 1000$ and only those natural numbers. Algebraically, the proposition $P(n)$: $1 + 3 + 5 + \cdots + (2n - 1) = n^2 + 5$ implies the proposition $P(n + 1)$: $1 + 3 + 5 + \cdots + (2n - 1) + (2n + 1) = (n + 1)^2 + 5$. However, $P(n)$ is not true for any value of n.

In the exercises the reader is asked to establish formulas for the natural numbers, many of which were known to the ancient mathematicians.

Exercises 1.3

Establish the following identities for all natural numbers n, unless otherwise noted, where u_n and v_n represent the general terms of the Fibonacci and Lucas sequences, respectively.

1. $1^2 + 2^2 + 3^2 + \cdots + n^2 = \dfrac{n(n + 1)(2n + 1)}{6}$.
2. $1^2 + 3^2 + 5^2 + \cdots + (2n - 1)^2 = \dfrac{n(4n^2 - 1)}{3}$. [Fibonacci]
3. $\dfrac{1}{1 \cdot 2} + \dfrac{1}{2 \cdot 3} + \dfrac{1}{3 \cdot 4} + \cdots + \dfrac{1}{n(n + 1)} = \dfrac{n}{n + 1}$.
4. $t_1 + t_2 + \cdots + t_n = \dfrac{n(n + 1)(n + 2)}{6}$. [Aryabhata]
5. $1^3 + 2^3 + 3^3 + \cdots + n^3 = (1 + 2 + \cdots n)^2$. [Aryabhata Bachet]

6. $(1+a)^n \geqslant 1+na$, where a is any real number greater than -1. [Jakob Bernoulli]
7. $n! > n^2$ for all natural numbers $n > 3$.
8. $u_1 + u_3 + u_5 + \cdots + u_{2n-1} = u_{2n}$.
9. $u_1^2 + u_2^2 + u_3^2 + \cdots + u_n^2 = u_n u_{n+1}$.
10. $u_2 + u_4 + u_6 + \cdots + u_{2n} = u_{2n+1} - 1$. [Lucas]
11. $u_n \geqslant \tau^{n-2}$.
12. $u_{n+1}^2 - u_n^2 = u_{n-1}u_{n+2}$, if $n \geqslant 1$. [Lucas]
13. $u_n + v_n = 2u_{n+1}$. [Lucas]
14. $v_{n-1} + v_{n+1} = 5u_n$, for $n \geqslant 2$. [Lucas]
15. $v_n = u_{n-1} + u_{n+1}$ if $n \geqslant 2$. [Lucas]
16. $u_{2n} = u_n v_n$. [Lucas]
17. $u_{n+2}^2 - u_n^2 = u_{2n+2}$. [Lucas]
18. In 1753 Robert Simson proved that $u_{n+1}u_{n-1} + (-1)^{n+1} = u_n^2$, for $n > 1$. Use induction to establish the formula.
19. If $S = \{a_1, a_2, a_3, \ldots\}$ is a set of natural numbers with $a_1 > a_2 > a_3 > \ldots$, then show S is finite.
20. Show that there are no natural numbers between 0 and 1.
21. Prove Wallis's result concerning figurate numbers, namely for natural numbers n and r, $f^r{}_{n+1} = f^1{}_n + f^2{}_n + \cdots + f^r{}_n$.

1.4 Miscellaneous exercises

1. Show that 1 533 776 805 is a triangular, pentagonal, and hexagonal number.
2. A natural number is called palindromic if it reads the same backwards as forwards. For example, 3 245 423 is palindromic. Determine all two- and three-digit palindromic triangular numbers.
3. Show that the squares of 1 270 869 and 798 644 are palindromic.
4. Show that the squares of the numbers 54 918 and 84 648 are pandigital, that is they contain all the digits.
5. In 1727, John Hill, of Staffordshire, England, claimed that the smallest pandigital square was $(11\,826)^2$. Was he correct?
6. The number 16 583 742 contains all the digits except 9 and 0. Show that $90 \cdot 16\,583\,742$ is pandigital.
7. A positive integer n is called k-transposable if, when its leftmost digit is moved to the unit's place, the result is $k \cdot n$. For example, 285 714 is 3-transposable since $857\,142 = 3 \cdot 285\,714$. Show that 142 857 is 3-transposable.
8. A number n is called automorphic if its square ends in n. For example,

25 is automorphic since $25^2 = 625$. Show that 76 and 625 are automorphic.

9. A number is called trimorphic if it is the nth triangular number and its last digits match n. For example, 15 is trimorphic since it is the fifth triangular number and it ends in 5. Show that 325, 195 625, and 43 959 376 are trimorphic.
10. A number is called a Kaprekar number if its square can be partitioned in 'half' such that the sum of the first half and the second half equals the given number. For example, 45 is a Kaprekar number since $45^2 = 2025$ and $20 + 25 = 45$. Show that 297, 142 857 and 1 111 111 111 are Kaprekar numbers.
11. A number is called an Armstrong number if it can be expressed as a sum of a power of its digits. For example 407 is Armstrong since $407 = 4^3 + 0^3 + 7^3$. Show that 153 and 371 are Armstrong numbers.
12. A number is called narcissistic if its digits can be partitioned in sequence so that it can be expressed as a power of the partitions. For example, 101 is narcissistic since $101 = 10^2 + 1^2$. All Armstrong numbers are narcissistic. Show that 165 033 is narcissistic.
13. A number $a_1a_2 \ldots a_n$ is called handsome if there exist natural numbers $x_1, x_2, \ldots, x_n$ such that $a_1a_2 \ldots a_n = a_1^{x_1} + a_2^{x_2} + \cdots + a_n^{x_n}$. For example 24 is handsome since $24 = 2^3 + 4^2$. Show that 43, 63, 89 and 132 are handsome.
14. A number $abcd$ is called extraordinary if $abcd = a^b c^d$. Show that 2592 is extraordinary.
15. A number is called curious if it can be expressed as the sum of the factorials of its digits. For example 1, 2, and 145 are curious since $1 = 1!$, $2 = 2!$, $145 = 1! + 4! + 5!$ Show that 40 585 is curious.
16. Multiplying 142 857 by 2, 3, 4, 5 or 6 permutes the digits of 142 857 cyclically. In addition, 4 times 2178 reverses the digits of 2178. Show that multiplying by 4 reverses the digits of 21 978 and 219 978 and multiplying by 9 reverses the digits of 10 989.
17. Determine how long it will take to return all the gifts mentioned in the song 'The twelve days of Christmas' if the gifts are returned at the rate of one gift per day.
18. Take the month that you were born, January = 1, December = 12, etc., multiply by 5, and add 6. Then multiply the result by 4 and add 9. Then take that result, multiply by 5, and add the day of the month that you were born. Now from the last result subtract 165. What does the answer represent?
19. Given any integer between 1 and 999, multiply it by 143. Take the

number represented by the last three digits of the result and multiply it by 7. The number represented by the last three digits of this result is the original number. Explain why.

20. Note that $\frac{2666}{6665} = \frac{266}{665} = \frac{26}{65} = \frac{2}{5}$, and $\frac{16}{64} = \frac{1}{4}$. Find all pairs of two-digit numbers ab and bc with the property that $ab/bc = a/c$.
21. The following puzzle was devised by William Whewell [YOU ell], Master of Trinity College, Cambridge. Represent each of the first 25 natural numbers using exactly four nines, any of the four basic operations (addition, subtraction, multiplication, and division), parentheses and, if absolutely necessary, allowing $\sqrt{9} = 3$ and $.\overline{9} = 1$. Whewell was a philosopher of science and historian who, in his correspondence with Michael Faraday, coined the terms anode, cathode, and ion. He also introduced the terms "physicist" and "scientist". Obtain a solution to Whewell's puzzle.
22. Exhibit 25 representations for zero using Whewell's conditions.
23. Solve Whewell's puzzle using four fours, the four basic operations and, if necessary, $\sqrt{4}$ and/or 4!
24. Prove that
$$\tau = \lim_{n\to\infty} \frac{u_{n+1}}{u_n}.$$
25. Consider integer solutions to the equation $x_1 + x_2 + \cdots + x_n = x_1 \cdot x_2 \cdots x_n$, where $x_1 \leqslant x_2 \leqslant \cdots \leqslant x_n$. For example, when $n = 2$, we have $2 + 2 = 2 \cdot 2$, hence, $x_1 = x_2 = 2$ is a solution. Find a general solution to the equation.
26. Gottfried Leibniz and Pietro Mengoli determined the sum of the reciprocals of the triangular numbers,
$$\sum_{n=1}^{\infty} \frac{1}{t_n} = 1 + \tfrac{1}{3} + \tfrac{1}{6} + \tfrac{1}{10} + \cdots.$$
What does the sum equal?
27. A lone reference to Diophantus in the form of an epitaph appears in the *Greek Anthology* of Metrodorus, a sixth century grammarian. According to the translation by W.R. Paton, 'This tomb holds Diophantus. Ah, how great a marvel! the tomb tells scientifically the measure of his life. God granted him to be a boy for the sixth part of his life, and adding a twelfth part to this, he clothed his cheeks with down; he lit him the light of wedlock after a seventh part, and five years after his marriage he granted him a son. Alas! late-born wretched child; after attaining the measure of half his father's life, chill Fate took him. After consoling his grief by this science of numbers for four

Table 1.7.

	0	1	2	3	4	5	6
Century	1500 1900 2300		1800 2200		1700 2100		1600* 2000* 2400*
Year	00 06 17 23 28* 34 45 51 56* 62 73 79 84* 90	01 07 12* 18 29 35 40* 46 57 63 68* 74 85 91 96*	02 13 19 24* 30 41 47 52* 58 69 75 80* 86 97	03 08* 14 25 31 36* 42 53 59 64* 70 81 87 92* 98	09 15 20* 26 37 43 48* 54 65 71 76* 82 93 99	04* 10 21 27 32* 38 49 55 60* 66 77 83 88* 94	05 11 16* 22 33 39 44* 50 61 67 72* 78 89 95
Month	January October	May	February* August	February March November	June	September December	January* April July
Day	7 14 21 28	1 8 15 22 29	2 9 16 23 30	3 10 17 24 31	4 11 18 25	5 12 19 26	6 13 20 27
Day of the week	Sunday 0 7 14 21	Monday 1 8 15 22	Tuesday 2 9 16 23	Wednesday 3 10 17 24	Thursday 4 11 18	Friday 5 12 19	Saturday 6 13 20

years he ended his life.' How old was Diophantus when he died? (Hint: if n denotes his age at his death then, according to the epitaph, $n/6 + n/12 + n/7 + n/2 + 9 = n$.)

28. De Morgan and Whewell once challenged each other to see who could come the closest to constructing sentences using each letter in the alphabet exactly once, precursors to 'the quick brown fox jumps over the lazy dog' and 'pack my box with five dozen liquor jugs'. It was decided that De Morgan's 'I, quartz pyx, who flung muck beds' just edged out Whewell's 'phiz, styx, wrong, buck, flame, quid'. Try your hand at the equally hard puzzle of trying to come up with a 26-word abecedarian phrase such that each word begins with a different letter of the alphabet in lexicographical order.
29. Table 1.7 is based on the Gregorian calendar that began replacing the Julian calendar in 1582. The table may be used to find the day of the week, given the date, by adding the figures at the top of each column and noting what column contains the sum. Asterisks denote leap years. For example, consider December 7, 1941.

Century	19	0
Year	41	2
Month	December	5
Day	7	0
SUM		7

Therefore, from Table 1.7, we find that December 7, 1941 was a Sunday. What day of the week was July 4, 1776?
30. On what day of the week were you born?
31. What was the date of the fourth Tuesday in June 1963?
32. What was the date of the first Tuesday of October 1917?
33. What day of the week was August 31, 1943?
34. Even though weekday names were not common until the fourth century, use the fact that in most Catholic countries Thursday October 4, 1582 in the Julian calendar was followed by Friday October 15, 1582 in the Gregorian and that all century years, prior to 1700, were leap years to determine the day of the week that each of these events occurred:
    (a) the Battle of Hastings (October 14, 1066),
    (b) the signing of Magna Carta (June 15, 1215), and
    (c) the marriage of Henry VIII and Ann Boleyn (January 25, 1533).
35. At a square dance each of the 18 dancers on the floor is identified with a distinct natural number from 1 to 18 prominently displayed on their

back. Suppose the sum of the numbers on the back of each of the 9 couples is a square number. Who is dancing with number 6?

36. An even natural number n is called a square dance number if the numbers from 1 to n can be paired in such a way that the sum of each pair is square. Show that 48 is a square dance number.
37. Determine all the square dance numbers.
38. John H. Conway and Richard K. Guy have defined an nth order zigzag number to be an arrangement of the numbers $1, 2, 3, \ldots, n$ in such a manner that the numbers alternately rise and fall. For example, the only first and second order zigzag numbers are 1 and 12, respectively. There are two third order zigzag numbers, namely, 231 and 132. There are five fourth order zigzag numbers, namely, 3412, 1423, 2413, 1324, and 2314. Determine all fifth order zigzag numbers.
39. In 1631, Johann Faulhaber of Ulm discovered that

$$1^{k-1} + 2^{k-1} + \cdots + n^{k-1} =$$
$$\frac{1}{k}\left[n^k + \binom{k}{1}n^{k-1}\cdot\frac{-1}{2} + \binom{k}{2}n^{k-2}\cdot\frac{1}{6} + \binom{k}{3}n^{k-3}\cdot 0 + \binom{k}{4}n^{k-4}\cdot\frac{-1}{30} + \cdots\right].$$

The coefficients, $1, -\frac{1}{2}, \frac{1}{6}, 0, -\frac{1}{30}, 0, \ldots$, are called Bernoulli numbers and appear in the 1713 edition of Jakob Bernoulli's *Ars conjectandi*. In general,

$$\binom{n+1}{1}B_n + \binom{n+1}{2}B_{n-1} + \cdots + \binom{n+1}{n}B_1 + B_0 = 0.$$

Hence, $B_0 = 1$, $B_1 = -\frac{1}{2}$, $B_2 = \frac{1}{6}$, $B_3 = 0$, $B_4 = -\frac{1}{30}$, $B_5 = 0$, and so forth. For example, $5B_4 + 10B_3 + 10B_2 + 5B_1 + B_0 = 0$. Hence, $B_4 = -\frac{1}{30}$. In addition, if $n > 1$, then $B_{2n+1} = 0$. Find the Bernoulli numbers B_6, B_8, and B_{10}.

40. Lucas defined the general term of a sequence to be $w_n = u_{3n}/u_n$. Determine the first six terms of the sequence. Is the sequence generated a Fibonacci-type sequence?
41. Given the 2×2 matrix

$$A = \begin{pmatrix} 1 & 1 \\ 1 & 0 \end{pmatrix},$$

use induction to show that, for $n \geqslant 1$,

$$A^n = \begin{pmatrix} u_{n+1} & u_n \\ u_n & u_{n-1} \end{pmatrix},$$

where u_n represents the nth Fibonacci number with the convention that $u_0 = 0$.

42. If

$$A = \begin{pmatrix} 1 & 1 \\ 1 & 0 \end{pmatrix},$$

find a numerical value for the determinant of A^n.

43. Evaluate

$$\frac{3^2 + 4^2 + 5^2 + 6^2 + 7^2 + 8^2 + 9^2}{1^2 + 2^2 + 3^2 + 4^2 + 5^2 + 6^2 + 7^2}.$$

44. Establish the following algebraic identity attributed to the Indian mathematician Srinivasa Ramanujan:

$$(a + 1)(b + 1)(c + 1) + (a - 1)(b - 1)(c - 1) = 2(a + b + c + abc).$$

45. Ramanujan stated a number of formulas for fourth power sums. Establish his assertion that $a^4 + b^4 + c^4 = 2(ab + bc + ca)^2$ provided $a + b + c = 0$.

46. Prove or disprove that $3a_n - a_{n+1} = u_{n-1}(u_{n-1} + 1)$, for $n \geqslant 1$, where a_n denotes the coefficient of x_n in $(1 + x + x^2)^n$, for $n = 0, 1, 2, \ldots,$ and u_n represents the nth Fibonacci number.

47. The curriculum of universities in the Middle Ages consisted of the seven liberal arts, seven flags flew over Texas, Rome and Providence, Rhode Island, were built on seven hills. Determine the following septets:
    (a) the seven wonders of the ancient world;
    (b) the seven sages of antiquity;
    (c) the seven wise women of antiquity.

48. In 1939, Dov Juzuk established the following extension to Nicomachus's method of generating the cubes from an arithmetic triangle. Show that if the even rows of the arithmetic triangle shown below are deleted, the sum of the natural numbers on the first n remaining rows is given by n^4. For example, $1 + (4 + 5 + 6) + (11 + 12 + 13 + 14 + 15) = 81 = 3^4$.

1
2 3
4 5 6
7 8 9 10
11 12 13 14 15
16 17 18 19 20 21
22 23 24 25 26 27 28
. . . .

49. In 1998, Ed Barbeau of the University of Toronto generalized Nicoma-

chus's cubic result to hexagonal numbers. Show that if the even rows of the arithmetic triangle shown below are deleted, the sum of the natural numbers on the first n remaining rows is given by $({p^6}_n)^2$. For example, $1 + (5 + 6 + 7 + 8 + 9) + (17 + 18 + \cdots + 25) = 225 = ({p^6}_3)^2$.

1

2 3 4

5 6 7 8 9

10 11 12 13 14 15 16

17 18 19 20 21 22 23 24 25

26 27 28 29 30 31 32 33 34 35 36

37 38 39 40 41 42 43 44 45 46 47 48 49

. . . .

50. If every other row in the following triangle is deleted, beginning with the second row, identify the partial sums of the first n remaining rows. Hint: there are $3n - 2$ numbers in the nth row.

1

2 3 4 5

6 7 8 9 10 11 12

13 14 15 16 17 18 19 20 21 22

1.5 Supplementary exercises

1. Use the method of finite differences to determine a formula for the general term of the sequence 2, 12, 28, 50, 78, 112, ...
2. Use the method of finite differences to determine a formula for the general term of the sequence whose first seven terms are: 14, 25, 40, 59, 82, 109, 149, ... Hint: Match the first diagonal of finite differences with that of $f(n) = an^3 + bn^2 + cn + d$.
3. Use the method of finite differences to determine a formula for the general term of the sequence whose first seven terms are: 5, 19, 49, 101, 181, 295, 449, ...
4. Given the following array

2

2 1

2 3 1

2 5 4 1

2 7 9 5 1

2 9 16 14 6 1

2 11 25 30 20 7 1

2 13 26 55 50 27 8 1

. . . .

where $a(i, j)$, the element in the ith row and jth column, is obtained by successive additions using the recursive formula $a(i, j) = a(i-1, j-1) + a(i-1, j)$, the second column consists of the odd positive integers and the third column squares of the positive integers, determine a formula for the positive integers 1, 5, 14, 30, 50, ... in the fourth column.

5. Determine a formula for the general term of the sequence 1, 6, 20, 50, ... in the fifth column of the previous exercise.
6. Determine a general formula for $a(i, j)$, in Exercise 5, where $i \geqslant j$.
7. Generalize the result in Exercise 4 in Section 1.1, by finding functions $f(m)$, $g(m)$ $h(m, n)$ such that $f(m)t_n + g(m) = t_{h(m,n)}$.
8. Show that $n^3 + 6t_n + 1 = (n+1)^3$. [Bachet]
9. Show that $t_{m+n} = t_m + t_n + mn$. [Bachet]
10. Which triangular number is $2^{n-1}(2^n - 1)$?
11. Can every positive integer be expressed as a difference of two nonconsecutive triangular numbers? If so, in how may ways?
12. Show that 3 divides t_{3k} and t_{3k+2} but not t_{3k+1}.
13. Let $r_n = t_n/3$. Show that r_{12k+5} and r_{12k+6} are both odd, r_{12k} and r_{12k+11} are both even.
14. Show that both r_{12k+2} and r_{12k+3} have opposite parity, as do r_{12k+8} and r_{12k+9}.
15. Show that each of the first fifty positive integers can be expressed as a sum of pentagonal numbers.
16. Find two hexagonal numbers such that their sum plus one is square and their difference minus one is square.
17. A positive integer is called pheptagonal if it is palindromic and heptagonal. Determine the first six pheptagonal numbers.
18. According to Nicomachus, oddly-even numbers are positive integers of the form $2n(2n+1)$ and oddly-odd numbers are numbers of the form $(2m+1)(2n+1)$. Determine the first fifteen oddly-even numbers and the first fifteen oddly-odd numbers.
19. Bantu numbers are positive integers that do not contain a "2" when expressed decimally. List the first twenty-five bantu numbers.
20. Eban numbers are positive integers that can be written in English without using the letter 'e'. Determine the first twenty eban numbers.
21. Curvaceous numbers are those positive integers that can be written with curves only, for example 3, 6, 8, and 9. Determine the first fifteen curvaceous numbers.
22. Cheap numbers are positive integers formed using 1, 2 and 3, but not two of them. The first ten cheap numbers are 1, 2, 3, 11, 22, 33, 111,

222, 333, 1111. Cheaper numbers are positive integers formed using any two of 1, 2, or 3. Cheapest numbers are positive integers formed using only 1, 2, and 3. Determine the first twenty-five cheaper numbers and the first fifteen cheapest numbers.

23. A Stern number a_n, first described by M.A. Stern in 1838, is a positive integer such that $a_1 = 1$ and a_n is the sum of the previous $n - 1$ numbers. Determine the first fifteen Stern numbers.
24. A positive integer is called a neve number if it is nonpalindromic, even, and remains even when the digits are reversed. Determine the first fifteen neve numbers.
25. A positive integer is called strobogrammatic if it is vertically palindromic, for example, 8 and 96. Determine the first fifteen strobogrammatic numbers.
26. Generate the first six terms of a Galileo sequence with first term 1 and ratio 6.
27. Generate the first six terms of a Galileo sequence with first term 2 and ratio 6.
28. A positive integer is called phappy if it is palindromic and happy. Determine the first five phappy numbers.
29. Happy couple numbers are consecutive positive integers that are both happy. Determine the first twenty-five happy couples.
30. What is the value of the smallest happy number of height 3, of height 4, of height 5, and of height 6.
31. Investigate the outcome of successive iterations of S_3 on positive integers of the form $3k + 1$ and of the form $3k + 2$, for k a positive integer.
32. Let $S_{r,b}(n)$ denote the sum of the rth powers of the digit in base b and $S^k_{r,b}(n) = S_{r,b}(S^{k-1}_{r,b}(n))$. For example, in base 3 with $r = 3$, $2 \rightarrow 22 \rightarrow 121 \rightarrow 101 \rightarrow 2$. Investigate the action of $S_{3,3}$ on the first 25 positive integers.
33. Find the height of the cubic happy number 112.
34. Find the cubic heights of 189 and 778.
35. Determine the least positive integer with persistence 6.
36. Investigate properties of the Pell sequence, determined by $P_0 = 0$, $P_1 = 1$, and $P_{n+1} = 2P_{n+1} + P_n$.
37. A MacMahon sequence is generated by excluding all the sums of two or more earlier members of the sequences. For example, if $a_1 = 1$ and $a_2 = 2$, the sequence formed is 1, 2, 4, 5, 8, 10, 14, 15, 16, 21. Determine the next ten terms of the sequence.
38. The general term of a Hofstader sum sequence has the property that it

is the least integer greater than the preceding term that can be expressed as a sum of two or more consecutive terms of the sequence. For example, if $a_1 = 1$ and $a_2 = 2$, the first twelve terms of the resulting Hofstadter sum sequence are 1, 2, 3, 5, 6, 8, 10, 11, 14, 16, 17, 18. Determine the next twelve terms of the sequence.

39. The general term of a Hofstater product sequence is defined as follows, given $a_1, a_2, \ldots, a_n$ form all possible products $a_i \cdot a_j - 1$ with $1 \leqslant i < j \leqslant n$ and append them to the sequence. For example, the first ten terms of the Hofstader sequence with $a_1 = 2$ and $a_2 = 3$ are 2, 3, 5, 9, 14, 17, 26, 27, 33, 41. Determine the next ten terms of the sequence.
40. Define $f_r(n)$ to be the least number of rs that can be used to represent n using rs and any number of plus and times signs and parentheses. For example, $f_1(2) = 2$ since $2 = 1 + 1$. Determine $f_1(3^k)$ for k a positive integer.
41. In 1972, D.C. Kay investigated properties of the following generalized Collatz algorithm given by $a_{n+1} = a_n/p$ if $p|n$ and $a_{n+1} = a_n \cdot q + r$ if $p \nmid n$. Investigate the actions of the generalized algorithm on the positive integers when $(p, q, r) = (2, 5, 1)$.
42. Determine the next two rows of the Lambda Triangle

$$\begin{array}{ccccccc} & & & 1 & & & \\ & & 2 & & 3 & & \\ & 4 & & 6 & & 9 & \\ 8 & & 12 & & 18 & & 27 \end{array}$$

43. Determine the next three terms of the sequence 71, 42, 12, 83, 54, ...
44. A postive integer is called pfibonacci if it is palindromic and Fibonacci. Determine the first six pfibonacci numbers.
45. If a_n denotes the nth tribonacci number determine $\lim_{n\to\infty}(a_{n+1}/a_n)$.
46. If b_n denotes the nth tetrancci number determine $\lim_{n\to\infty}(b_{n+1}/b_n)$.
47. If d_n denotes the nth deacci number (add the previous ten terms) where the first ten terms are 1, 1, 2, 4, 8, 16, 32, 64, 128, 256, determine $\lim_{n\to\infty}(d_{n+1}/d_n)$.
48. Determine the first fifteen $u(3, 4)$ numbers.
49. Determine the first fifteen $v(3, 4)$ numbers.
50. Show for any positive integer n: $1^3 + 3^3 + 5^3 + \cdots + (2n - 1)^3 = n^2(2n^2 - 1)$.
51. For any positive integer n show that

$$1 \cdot 2 + 2 \cdot 3 + \cdots + n(n + 1) = \frac{n(n + 1)(n + 2)}{3}.$$

52. For any positive integer n show that

$$\frac{1}{3}+\frac{1}{15}+\frac{1}{35}+\cdots+\frac{1}{4n^2-1}=\frac{n}{2n+1}.$$

53. For any positive integer n show that
$$2+5+8+\cdots+(3n-1)=\frac{n(3n+1)}{2}.$$
54. If u_n and v_n represent the nth Fibonacci and Lucas numbers, respectively, show that $u_n + v_n = 2u_{n+1}$.
55. Show that for any positive integer n, $(v_{n+1})^2 - (v_n)^2 = v_{n-1}v_{n+2}$.
56. Show that for any positive integer n, $(v_{n+2})^2 - (v_n)^2 = 5u_{2n+2}$.
57. Show that for any positive integer n, $u_{n+1}^2 + u_n^2 = u_{2n+1}$.
58. Show that for any positive integer n, $u_{n+1}^3 + u_n^3 - u_{n-1}^3 = u_{3n}$.
59. Show that for any positive integer n, $u_{n+2}^3 - 3u_n^3 + u_{n-2}^3 = 3u_{3n}$.
60. For any positive integer n, establish the Gelin-Cesàro identity $u_{n-2}\cdot u_{n-1}\cdot u_{n+1}\cdot u_{n+2} - u_n^4 = -1$.
61. Given a positive integer n an expression of the form $n=\sum_{k=0}^{m}\varepsilon_k u_k$, where the ε_k equal 0 or 1, $\varepsilon_k\varepsilon_{k+1}=0$ for $1\leqslant k\leqslant m$, and u_n denotes the nth Fibonacci number is called a Zeckendorf representation of n. We may denote the Zeckendorf representation of n using the function $Z(n)$, which represents binarily the Fibonacci numbers in the Zeckendorf representation of n. For example, $30 = 21 + 8 + 1$, hence $Z(30) = 101001$. Determine Zeckendorf representations for the first fifty positive integers and the associated function $Z(n)$.
62. Show that every positive integer has a Zeckendorf representation.
63. Compute the Zeckendorf representation for u_n^2, where $1\leqslant n\leqslant 25$.
64. Compute the Zeckendorf representation for v_n^2, where $1\leqslant n\leqslant 25$.
65. Compute the Zeckendorf representation for $u_n u_{n+1}$, where $1\leqslant n\leqslant 25$.
66. Let $u_n! = u_n\cdot u_{n-1}\cdots u_1$. Find the values of $u_n!$ for $1\leqslant n10$.
67. Let $\binom{n}{k}_u = u_n!/u_k!u_{n-k}!$ if $n\geqslant k$ and 0 otherwise. Determine the values of $\binom{n}{k}_u$ for $1\leqslant n\leqslant 10$ and $0\leqslant k\leqslant n$.
68. Consider the sum $\sum_{n=1}^{\infty}1/(u_n)^r$, where r is a positive integer. Estimate the sum when $r=1$.
69. Show that $(5/32)u_{6n}^2$ is an integer, whenever n is a positive integer.
70. Show that
$$\sum_{n=1}^{\infty}\frac{1}{u_n\cdot u_{n+2}}=1.$$

2

Divisibility

If you are going to play the game, you'd better know all the rules.

Barbara Jordan

2.1 The division algorithm

We extend our universe of discourse from the set of natural numbers to the set of integers, ..., $-3, -2, -1, 0, 1, 2, 3, \ldots$, by adjoining zero and the negatives of the natural numbers. The integers are closed under addition, subtraction, and multiplication. We use the additive inverse to define subtraction. That is, by the expression $a - b$, we mean $a + (-b)$. In order to work with integers efficiently we rely heavily on the following basic properties of addition and multiplication of integers.

Properties of the integers

Associativity	$a + (b + c) = (a + b) + c$	$a(bc) = (ab)c$
Commutativity	$a + b = b + a$	$ab = ba$
Distributivity	$a(b + c) = ab + ac$	$(a + b)c = ac + bc$
Identity	$a + 0 = 0 + a = a$	$a \cdot 1 = 1 \cdot a = a$
Inverse	$a + (-a) = (-a) + a = 0$	
Transitivity	$a > b$ and $b > c$ implies $a > c$	
Trichotomy	Either $a > b$, $a < b$, or $a = b$	
Cancellation law	If $a \cdot c = b \cdot c$ and $c \neq 0$, then $a = b$.	

The set of rational numbers, a superset of the integers, consists of numbers of the form m/n, where m and n are integers and $n \neq 0$. We employ multiplicative inverses to define division on the rationals, that is by $r \div s$ we mean $r \cdot (1/s)$. Since

$$\frac{a}{b} \pm \frac{c}{d} = \frac{ad \pm bc}{bd},$$

$$\frac{a}{b} \cdot \frac{c}{d} = \frac{ac}{bd},$$

and

$$\frac{a}{b} \div \frac{c}{d} = \frac{ad}{bc},$$

the rationals are closed under the binary operations of addition, subtraction, multiplication, and division (except by zero). Furthermore, every rational number can be expressed as a repeating decimal and vice versa. For example, if $n = 0.\overline{63}$, then $100n = 63.\overline{63}$. Thus, $99n = 100n - n = 63$. Therefore, $n = \frac{63}{99} = \frac{7}{11}$. Conversely, since there are only n possible remainders when dividing by the integer n, every rational number can be expressed as a repeating decimal.

The rationals are not closed under the unary operation of taking the square root of a positive number. However, if we adjoin nonrepeating decimal expansions, called irrational numbers, to the rationals we obtain the real numbers. The reals are closed under the four basic binary operations (except division by zero) and the unary operation of taking the square root of a positive number. By extending the reals to the complex numbers, numbers of the form $a + b\mathrm{i}$, where a and b are real and $\mathrm{i}^2 = -1$, we obtain a set closed under the four basic binary operations (except division by zero) and the unary operation of taking the square root.

A function is a rule or correspondence between two sets that assigns to each element of the first set a unique element of the second set. For example, the absolute value function, denoted by $|\cdot|$, is defined such that $|x|$ equals x when x is nonnegative and $-x$ when x is negative. It follows immediately from the definition that if $|x| < k$, then $-k < x < k$, and if $|x| > k$, then $x > k$ or $x < -k$. Two vertical bars, the notation used for the absolute value, were introduced by Karl Weierstrass in 1841. Weierstrass, a German mathematician, who taught at the University of Berlin, was advocate for mathematical rigor. An important property of the absolute value function is expressed in the following result.

Theorem 2.1 (Triangle inequality) *For any two real numbers a and b, $|a| + |b| \geqslant |a + b|$.*

Proof Since $-|a| \leqslant a \leqslant |a|$ and $-|b| \leqslant b \leqslant |b|$ it follows that $-|a| - |b| \leqslant a + b \leqslant |a| + |b|$. Therefore, $|a + b| \leqslant |a| + |b|$. ∎

We define the binary relation 'divides' on the integers as follows: if a and b are integers, with $a \neq 0$, and c is an integer such that $ac = b$, then we say that a divides b and write $a|b$. It should be noted that there can be but one integer c such that $ac = b$. If a divides b, then a is called a divisor of b, and b is called a multiple of a. We write $a \nmid b$ if a does not divide b. If

a divides b with $1 \leqslant a < b$, then we say that a is a proper divisor of b. The basic properties of division are listed below, where a, b and c represent integers.

Properties of division

(1) If $a \neq 0$, then $a|a$ and $a|0$.
(2) For any a, $1|a$.
(3) If $a|b$ and $a|c$ then for any integers x and y, $a|(bx + cy)$.
(4) If $a|b$ and $b|c$, then $a|c$.
(5) If $a > 0$, $b > 0$, $a|b$ and $b|a$, then $a = b$.
(6) If $a > 0$, $b > 0$, and $a|b$, then $a \leqslant b$.

The first two properties follow from the fact that $a \cdot 1 = a$ and $a \cdot 0 = 0$. In order to establish the third property, suppose that a divides b and c. There exist integers r and s such that $ar = b$ and $as = c$. Hence, $bx + cy = arx + asy = a(rx + sy)$. Since $bx + cy$ is a mulitple of a, a divides $bx + cy$. Proofs of the other properties are as straightforward and are left as exercises for the reader. From the third property, it follows that if $a|b$ and $a|c$, then $a|(b + c)$, $a|(b - c)$, and $a|(c - b)$. From the definition of division and the fact that divisions pair up, it follows that, for any positive integer n, there is a one-to-one correspondence between the divisors of n that are less than $\sqrt{n}$ and those which are greater than $\sqrt{n}$.

Example 2.1 Using induction, we show that 6 divides $7^n - 1$, for any positive integer n. Let $P(n)$ represent the variable proposition 6 divides $7^n - 1$. $P(1)$ is true since 6 divides $7 - 1$. Suppose for some positive integer k, $P(k)$ is true, that is, 6 divides $7^k - 1$ or, equivalently, there is an integer x such that $7^k - 1 = 6x$. We have $7^{k+1} - 1 = 7 \cdot 7^k - 1 = 7(6x + 1) - 1 = 6(7x + 1) = 6y$. Thus, $7^{k+1} - 1$ is a multiple of 6. Therefore, $P(k)$ implies $P(k + 1)$ and the result follows from the principle of mathematical induction.

Example 2.2 We determine three distinct positive integers a, b, c such that the sum of any two is divisible by the third. Without loss of generality, suppose that $a < b < c$. Since $c|(a + b)$ and $a + b < 2c$, $a + b$ must equal c. In addition, since $2a + b = a + c$ and $b|(a + c)$, $b|2a$. Since $2a < 2b$, b must equal $2a$. Hence, $c = a + b = 3a$. Therefore, n, $2n$ and $3n$, for any positive integer n, are three distinct integers with the property that the sum of any two is divisible by the third.

Many positive integers have interesting divisibility properties. For example, 24 is the largest integer divisible by all the positive integers less than its square root. It is also the only integer greater than unity such that the sum of the squares from 1 to itself is a square. One of the most basic tools for establishing divisibility properties is the division algorithm found in Book VII of Euclid's *Elements*. According to Euclid, given two line segments the shorter one can always be marked off a finite number of times on the longer length either evenly or until a length shorter than its own length remains and the process cannot continue. A more algebraic version of the division algorithm, one more appropriate for our use, is stated in the next theorem.

Theorem 2.2 (The division algorithm) *For any integer a and positive integer b there exist unique integers q and r with the property that* $a = bq + r$ with $0 \leqslant r < b$.

Proof Consider the set $S = \{a - sb\colon s \text{ is an integer and } a - sb \geqslant 0\}$. S consists of the nonnegative elements of the set $\{\ldots, a - 2b, a - b, a, a + b, a + 2b, \ldots\}$. If $a < 0$, then $a - ab = a(1 - b) \geqslant 0$, hence, $a - ab$ is in S. If $a \geqslant 0$, then $a - (0 \cdot b) = a \geqslant 0$, hence a is in S. In either case, S is a nonempty set of positive integers. By the well-ordering principle S contains a least element that we denote by $r = a - bq \geqslant 0$. In addition, $r - b = (a - bq) - b = a - (q + 1)b < 0$, hence, $0 \leqslant r < b$. In order to show that q and r are unique, suppose that there are two other integers u, v such that $a = bu + v$, with $0 \leqslant v < b$. If $u < q$, then since u and q are integers, we have $u + 1 \leqslant q$. Thus, $r = a - bq \leqslant a - b(u + 1) = (a - ub) - b = v - b < 0$, contradicting the fact that r is nonnegative. A similar contradiction arises if we assume $u > q$. Hence, from the law of trichotomy, $u = q$. Thus, $a = bq + r = bq + v$ implying that $v = r$, and the uniqueness of q and r is established. ■

Corollary *For an integer a and positive integer b, there exist unique integers q and r such that $a = bq + r$,* with $-|b|/2 < r \leqslant |b|/2$.

One of the most important consequences of the division algorithm is the fact that for any positive integer $n > 1$ every integer can be expressed in the form nk, $nk + 1$, $nk + 2$, ..., or $nk + (n - 1)$, for some integer k. Equivalently, every integer either is divisible by n or leaves a remainder 1, 2, ..., or $n - 1$ when divided by n. This fact is extremely useful in establishing results that hold for all integers.

Table 2.1.

n	n^2	n^3
$7k$	$7r$	$7s$
$7k+1$	$7r+1$	$7s+1$
$7k+2$	$7r+4$	$7s+1$
$7k+3$	$7r+2$	$7s+6$
$7k+4$	$7r+2$	$7s+1$
$7k+5$	$7r+4$	$7s+6$
$7k+6$	$7r+1$	$7s+6$

If we restrict our attention to division by the integer 2, the division algorithm implies that every integer is even or odd, that is, can be written in the form $2k$ or $2k+1$. Since $(2k)^2 = 4k^2$ and $(2k+1)^2 = 4k^2 + 4k + 1 = 4(k^2 + k) + 1$, we have established the following result.

Theorem 2.3 *Every square integer is of the form* $4k$ *or* $4k+1$, *where k is an integer.*

Since x^2 and y^2 must be of the form $4k$ or $4k+1$, $x^2 + y^2$, the sum of two squares, can only be of the form $4k$, $4k+1$, or $4k+2$ and we have established the next result.

Theorem 2.4 *No integer of the form* $4k+3$ *can be expressed as the sum of two squares.*

If we restrict ourselves to division by the integer 3, the division algorithm implies that every integer is of the form $3k$, $3k+1$, or $3k+2$. That is, division by 3 either goes evenly or leaves a remainder of 1 or 2. Using this fact, Theon of Smyrna claimed that every square is divisible by 3 or becomes so when 1 is subtracted from it.

Similarly, every integer is of the form $7k$, $7k+1$, $7k+2$, $7k+3$, $7k+4$, $7k+5$, or $7k+6$. That is, according to the division algorithm, the only remainders possible when dividing by 7 are 0, 1, 2, 3, 4, 5, and 6. From Table 2.1, it follows that any integer that is both a square and a cube must be of the form $7k$ or $7k+1$. For example, $(7k+2)^2 = 49k^2 + 28k + 4 = 7k(7k+4) + 4 = 7r + 4$, and $(7k+2)^3 = 343k^3 + 294k^2 + 8k + 8 = 7(49k^3 + 42k^2 + 12k + 1) + 1 = 7s + 1$. Therefore, any integer that is both a square and a cube cannot be of the form $7k+2$.

In *Theaetetus*, Plato remarks that his teacher, Theodorus of Cyrene, proved the irrationality of $\sqrt{3}$, $\sqrt{5}$, $\sqrt{7}$, $\sqrt{11}$, $\sqrt{13}$, and $\sqrt{17}$, but he gives no indication of Theodorus's method of proof. A number of proofs of the irrationality of $\sqrt{2}$ were known to ancient mathematicians. (Euclid included a generalization of the result in Book X of the *Elements*.) A proof that appears in Aristotle's *Prior Analytics* using the fact that integers are either even or odd, is demonstrated in the next example.

Example 2.3 We use the indirect method to show that $\sqrt{2}$ is irrational. Suppose that it is rational. Thus, there exist positive integers p and q, with no common factors, such that $\sqrt{2} = p/q$. Since $p^2 = 2q^2$, p^2 is even and, hence, p is even. Let $p = 2m$; then $p^2 = 4m^2$, hence, $q^2 = 2m^2$. Since q^2 is even, q must be even, contradicting the assumption that p and q have no common factors. Therefore, $\sqrt{2}$ is irrational.

In 1737, the irrationality of e, the base of the natural logarithm, was established by Euler. The irrationality of π was established by Johann Lambert in 1767. Lambert was self-educated and made significant contributions to physics, mathematics, and cartography. He developed the transverse Mercator projection by projecting onto a cylinder tangent to a meridian. In physics, the lambert is a unit of brightness. In non-Euclidean geometry, a Lambert quadrilateral is a four-sided figure having three right angles. A short proof of the irrationality of e is demonstrated in the next example.

Example 2.4 By definition,

$$\mathrm{e} = 1 + \frac{1}{1!} + \frac{1}{2!} + \frac{1}{3!} + \cdots.$$

Suppose that e is rational, that is, $\mathrm{e} = p/q$, where p and q are integers with no common factors. Let $\mathrm{e} = a + b$, where

$$a = 1 + \frac{1}{1!} + \frac{1}{2!} + \cdots + \frac{1}{q!}$$

and

$$b = \frac{1}{(q+1)!} + \frac{1}{(q+2)!} + \cdots.$$

Multiplying both sides of the expression for e by $q!$, we obtain $q! \cdot \mathrm{e} = q! \cdot a + q! \cdot b$. Since $q! \cdot a$ is an integer and $q! \cdot \mathrm{e}$ is an integer, it follows that $q! \cdot b$, the difference of two integers, is an integer. However,

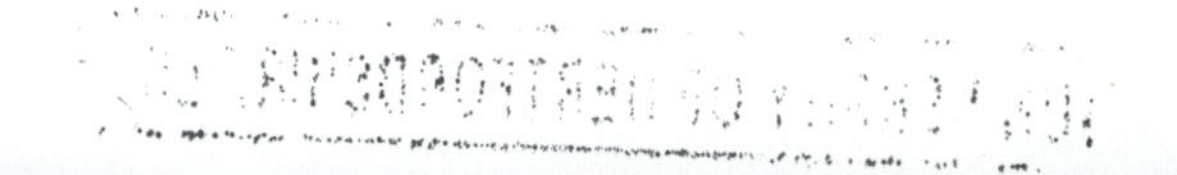

$$q! \cdot b = \frac{1}{(q+1)} + \frac{1}{(q+1)(q+2)} + \frac{1}{(q+1)(q+2)(q+3)} + \cdots < \tfrac{1}{2} + \tfrac{1}{4} + \tfrac{1}{8} + \cdots = 1,$$

implying that $0 < q! \cdot b < 1$, a contradiction. Therefore, e is irrational.

Most of our work in this book will be done in base 10. However, there are occasions when it is useful to consider other bases, in particular base 2. When $b \neq 10$, we use the notation n_b to denote the integer n written in base b. For example, $101\,101_2 = 45$, since $1 \cdot 2^5 + 0 \cdot 2^4 + 1 \cdot 2^3 + 1 \cdot 2^2 + 0 \cdot 2 + 1 = 45$. Representing integers in bases other than base 10 is useful if such representations are unique, which we establish with the next result.

Theorem 2.5 *If a and b are positive integers with $b > 1$, then a can be uniquely represented in the form $a = c_k b^k + c_{k-1} b^{k-1} + \cdots + c_1 b + c_0$, with integers c_i such that $0 \leqslant c_i < b$, for $i = 0, 1, 2, \ldots, k$ and $c_k \neq 0$.*

Proof From the division algorithm, we have that $a = bq_1 + c_0$, with $0 \leqslant c_0 < b$ and $q_1 < a$. If $q_1 \geqslant b$, we employ the division algorithm again to obtain $q_1 = bq_2 + c_1$, with $0 \leqslant c_1 < b$ and $q_2 < q_1$. If $q_2 \geqslant b$, we continue the process, obtaining a decreasing sequence of positive integers $q_1 > q_2 > \ldots$. Eventually, we obtain a positive number, say q_k, such that $q_k < b$. Set $q_k = c_k$. Eliminating $q_k, q_{k-1}, \ldots, q_1$ from the system

$$\begin{aligned} a &= bq_1 + c_0, \\ q_1 &= bq_2 + c_1, \\ &\cdots \\ q_{k-2} &= bq_{k-1} + c_{k-2}, \\ q_{k-1} &= bq_k + c_{k-1}, \\ q_k &= c_k, \end{aligned}$$

we obtain $a = c_k b^k + \cdots + c_1 b + c_0$, with $0 \leqslant c_i < b$, for $i = 0, 1, 2, \ldots, k-1$ and $c_k \neq 0$. The uniqueness of this expansion follows from the fact that if $a = d_k b^k + d_{k-1} b^{k-1} + \cdots + d_1 b + d_0$, then d_0 is the remainder when a is divided by b, hence, $d_0 = c_0$. Similarly, d_1 is the remainder when $q_1 = (a - d_0)/b$ is divided by b, hence, $d_1 = c_1$, and so forth. Therefore, it follows that $d_i = c_i$, for $i = 0, 1, 2, \ldots, k$, and the proof is complete. ■

Theorem 2.5 implies that every nonzero integer can be expressed uniquely in base 3, in the form $c_k 3^k + c_{k-1} 3^{k-1} + \cdots + c_1 3 + c_0$, when $c_i = 0, 1,$

or 2, for $i = 0, 1, 2, \ldots, k$, and $c_k \neq 0$, or equivalently, with $c_i = -1, 0$, or 1.

An elementary version of the game of nim consists of two players and a single pile of matches. Players move alternately, each player is allowed to take up to half the number of matches in the pile, and the player who takes the last match loses. A player can force a win by leaving $2^n - 1$ matches in the pile, where n is a positive integer. For example, if there were 73 matches in the pile a player attempting to force a win would remove 10 matches leaving $73 - 10 = 63 = (2^6 - 1)$ matches in the pile. In 1901, using properties of binary representations, Charles Bouton of Harvard developed several winning strategies for a more advanced version of nim where several piles of matches were involved and where players who moved alternately were allowed to remove matches from but a single pile on each move. His techniques were generalized by E. H. Moore in 1910.

Exercises 2.1

1. If $a = b + c$, and d divides both a and b, show that d divides c.
2. If $a|b$ and $b|c$, then show that $a|c$.
3. If $a > 0$, $b > 0$, $a|b$, and $b|a$, then show that $a = b$.
4. If $a > 0$, $b > 0$, and $a|b$, then show that $a \leqslant b$.
5. Use the definition of division to prove that if $a + b = c$ and $a|b$, then $a|c$.
6. Prove that if $a|b$ and $c|d$, then $ac|bd$.
7. True or false (if false give a counterexample):
   (a) if $a|bc$, then either $a|b$ or $a|c$,
   (b) if $a|(b + c)$, then either $a|b$ or $a|c$,
   (c) if $a^2|b^3$, then $a|b$,
   (d) if $a^2|c$ and $b^2|c$ and $a^2 \leqslant b^2$, then $a|b$,
   (e) if b is the largest square divisor of c and $a^2|c$, then $a|b$?
8. Prove that every rational number can be represented by a repeating decimal.
9. Determine the fractional representation for $0.\overline{123}$.
10. Use the fact that every integer is of the form $3k$, $3k + 1$, or $3k + 2$ to show that $\sqrt{3}$ is irrational. (Hint: Assume it is rational and get a contradiction.)

11. For any integer n, show that
    (a) 2 divides $n(n+1)$,
    (b) 3 divides $n(n+1)(n+2)$.
12. Prove that 6 divides $n(n+1)(2n+1)$ for any positive integer n.
13. Show that the sum of the squares of two odd integers cannot be a perfect square.
14. Prove that the difference of two consecutive cubes is never divisible by 2.
15. Show that if n is any odd integer then 8 divides $n^2 - 1$.
16. Show that if 3 does not divide the odd integer n then 24 divides $n^2 - 1$.
17. Use induction to prove that 3 divides $n(2n^2 + 7)$, for any positive integer n.
18. Show that 8 divides $5^{2n} + 7$, for any positive integer n.
19. Show that 7 divides $3^{2n+1} + 2^{n+2}$, for any positive integer n.
20. Show that 5 divides $3^{3n+1} + 2^{n+1}$, for any positive integer n.
21. Show that 4 does not divide $n^2 + 2$, for any integer n.
22. Show that the number of positive divisors of a positive integer is odd if and only if the integer is a square.
23. Show that any integer of the form $6k + 5$ is also of the form $3m + 2$, but not conversely.
24. Show that the square of any integer must be of the form $3k$ or $3k + 1$. [Theon of Smyrna]
25. Show that the cube of any integer is of the form $9k$, $9k + 1$, or $9k - 1$.
26. Show that the fourth power of any integer is of the form either $5k$ or $5k + 1$.
27. Prove that no integer of the form $8k + 7$ can be represented as the sum of three squares.
28. In an 1883 edition of *The Educational Times*, Emma Essennell of Coventry, England, showed for any integer n, $n^5 - n$ is divisible by 30, and by 240 if n is odd. Prove it.
29. Prove that $3n^2 - 1$ is never a square, for any integer n.
30. Show that no number in the sequence 11, 111, 1111, 11 111, ... is a square.
31. Prove that if a is a positive proper divisor of the positive integer b, then $a \leqslant b/2$.
32. If a and b are positive integers and $ab = n$, then show that either $a \leqslant \sqrt{n}$ or $b \leqslant \sqrt{n}$.
33. For any positive integer n, show that there is a one-to-one correspon-

dence between the divisors of n which are greater than or equal to $\sqrt{n}$ and the ways n may be expressed as the difference of two squares.

34. Determine the binary and ternary representations for 40, 173, and 5437.
35. Represent $101\,011_2$ and $201\,102_3$ in base 10.
36. Show that any integer of the form $111\,1\ldots1_9$ is triangular.
37. Given a scale with a single pan, determine the least number of weights and precisely the values of the weights necessary in order to weigh all integral weights in kilograms from 1 kilogram to 40 kilograms. [Bachet]
38. A number n is called a Niven number, named for Ivan Niven, a number theorist at the University of Oregon, if it is divisible by the sum of its digits. For example 24 is a Niven number since $2 + 4 = 6$ and 6 divides 24. In 1993, C. Cooper and R. E. Kennedy showed that it is not possible to have more than twenty consecutive Niven numbers. Niven numbers are also referred to as multidigital numbers or Harshard numbers. The latter name was given by D.R. Kaprekar and comes from the sanskrit word for 'great joy'. Determine the first twenty-five Niven numbers.
39. Let $s_d(n, b)$ denote the digital sum of the integer n expressed in base $b \geqslant 2$. That is, if $n = c_k b^k + c_{k-1} b^{k-1} + \cdots + c_1 b + c_0$, with integers c_i such that $0 \leqslant c_i < b$, for $i = 0, 1, 2, \ldots, k$, and $c_k \neq 0$, then $s_d(n, b) = \sum_{i=1}^{k} c_i$. For example, since $9 = 1001_2$, $s_d(9, 2) = 2$. For convenience, we denote $s_d(n, 10)$ by $s_d(n)$. Let $S_d(n, b)$, the extended digital sum of the integer n expressed in base $b \geqslant 2$, represent $s_d(n, b)$ summed over the digits of n. For example, since $3 = 11_2$, $6 = 110_2$ and $7 = 111_2$, $S_d(367, 2) = s_d(3, 2) + s_d(6, 2) + s_d(7, 2) = 2 + 2 + 3 = 7$. Determine $S_d(n, 2)$ for $n = 7$, 13, and 15.
40. For which values of n does $S_d(n, 2)$ divide n?
41. Find a positive integer n such that $n/2$ is square, $n/3$ is cube, and $n/5$ is a fifth power.

2.2 The greatest common divisor

If a and b are integers and d is a positive integer such that $d|a$ and $d|b$, then d is called a common divisor of a and b. If both a and b are zero then they have infinitely many common divisors. However, if one of them is nonzero, the number of common divisors of a and b is finite. Hence, there must be a largest common divisor. We denote the largest common divisor of a and b by $\gcd(a, b)$ and, following standard convention, call it the greatest common divisor of a and b. It follows straightforwardly from the definition that d is the greatest common divisor of a and b if and only if

(1) $d > 0$,
(2) $d|a$ and $d|b$,
(3) if $e|a$ and $e|b$ then $e|d$.

As pointed out in [Schroeder], physiological studies have shown that, with few exceptions, the brain, upon being presented with two harmonically related frequencies, will often perceive the greatest common divisor of the two frequencies as the pitch. For example, if presented with frequencies of 320 hertz and 560 hertz the brain will perceive a pitch of 80 Hz. One of the most important properties of the greatest common divisor of two numbers is that it is the smallest positive integer that can be expressed as a linear combination of the two numbers. We establish this result in the next theorem.

Theorem 2.6 *If a and b are not both zero and $d = \gcd(a, b)$, then d is the least element in the set of all positive linear combinations of a and b.*

Proof Let T represent the set of all linear combinations of a and b that are positive, that is, $T = \{ax + by\colon x \text{ and } y \text{ are integers and } ax + by > 0\}$. Without loss of generality, suppose that $a \neq 0$. If $a > 0$, then $a \cdot 1 + b \cdot 0 = a$ is in T. If $a < 0$, then $a(-1) + b \cdot 0 = -a$ is in T. Thus, in either case, T is a nonempty set of positive integers. By the well-ordering principle T contains a least element which we denote by $e = au + bv$. By the division algorithm, there exist integers q and r such that $a = eq + r$ with $0 \leqslant r < e$. Hence, $r = a - eq = a - (au + bv)q = a(1 - uq) + b(-vq)$. If $r \neq 0$ we have a contradiction since r is in T and $r < e$, the least element in T. Thus, $r = 0$ implying that e divides a. A similar argument shows that e divides b. Since e divides both a and b and d is the greatest common divisor of a and b, it follows that $e \leqslant d$. However, since $e = au + bv$ and d divides both a and b, it follows that d divides e, hence, $d \leqslant e$. Therefore, $e = d$ and the proof is complete. ■

Corollary *If d is the greatest common divisor of a and b, then there exist integers x and y such that $d = ax + by$.*

Example 2.5 Table 2.2 exhibits values for the linear combination $56x + 35y$, where $-4 \leqslant x \leqslant 4$ and $-4 \leqslant y \leqslant 4$. Note that all entries are

Table 2.2.

					x				
y	−4	−3	−2	−1	0	1	2	3	4
−4	−364	−308	−252	−196	−140	−84	−28	28	84
−3	−329	−273	−217	−161	−105	−49	7	63	119
−2	−294	−238	−182	−126	−70	−14	42	98	154
−1	−259	−203	−147	−91	−35	21	77	133	189
0	−224	−168	−112	−56	0	56	112	168	224
1	−189	−133	−77	−21	35	91	147	203	259
2	−154	−98	−42	14	70	126	182	238	294
3	−119	−63	−7	49	105	161	217	273	329
4	−84	−28	28	84	140	196	252	308	364

multiples of 7 and the least positive linear combination is 7. From Theorem 2.6, the greatest common divisor of 56 and 35 is 7.

Suppose d is the greatest common divisor and a and b, x and y are integers such that $d = ax + by$ and A and B are integers such that $a = Ad$ and $b = Bd$. It follows that $d = aX + bY$, where $X = x - Bt$ and $Y = y + At$, for any integer t. There are, therefore, an infinite number of ways to represent the greatest common divisor of two integers as a linear combination of the two given integers.

In Chapter 5, we show that the linear equation $ax + by = c$, where a, b and c are integers, has integer solutions if and only if the greatest common divisor of a and b divides c. Other properties of the greatest common divisor include the following, where a, b, c are positive integers.

(a) $\gcd(ca, cb) = c \cdot \gcd(a, b)$.

(b) If $d|a$ and $d|b$ then $\gcd\left(\frac{a}{d}, \frac{b}{d}\right) = \frac{\gcd(a, b)}{d}$.

(c) $\gcd\left(\frac{a}{\gcd(a, b)}, \frac{b}{\gcd(a, b)}\right) = 1$.

(d) If $\gcd(a, b) = 1$ then $\gcd(c, ab) = \gcd(c, a) \cdot \gcd(c, b)$.

(e) If $ax + by = m$, then $\gcd(a, b)$ divides m.

(f) If $\gcd(a, b) = 1$ and $a \cdot b = n^k$, then there exist integers r and s such that $a = r^k$ and $b = s^k$.

One of the most useful results in number theory is that if a linear combination of two integers is unity then the greatest common divisor of the two integers is unity. This result appears in Book VII of Euclid's *Elements*. We call two integers coprime (or relatively prime) if their greatest common divisor is unity.

Theorem 2.7 *Two integers a and b are coprime if and only if there exist integers x and y such that $ax + by = 1$.*

Proof This follows from Theorem 2.6. Sufficiency follows from the fact that no positive integer is less than 1. ■

For example, for any positive integer k, $6 \cdot (7k + 6) + (-7) \cdot (6k + 5) = 1$. Hence, from Theorem 2.7, $\gcd(7k + 6, 6k + 5) = 1$, for any positive integer k. In addition, suppose that $\gcd(n! + 1, (n + 1)! + 1) = d$, for some positive integer n. Since d divides $n! + 1$, d divides $(n + 1)! + 1$, and $n = (n + 1)[n! + 1] - [(n + 1)! + 1]$, d must divide n. However, if $d|n$ and $d|[n! + 1]$ then $d = 1$, since $1 = 1 \cdot (n! + 1) + (-n) \cdot (n - 1)!$. Therefore, $\gcd(n! + 1, (n + 1)! + 1) = 1$, for any positive integer n.

Theorem 2.8 *For integers a, b, and c, if $a|c$ and $b|c$ and a and b are coprime, then $ab|c$.*

Proof Since a and b divide c, there exist integers x and y such that $ax = by = c$. It follows from Theorem 2.7 that there exist integers u and v such that $au + bv = 1$. Multiplying both sides of the equation by c we obtain $c = auc + bvc = au(by) + bv(ax) = ab(uy) + ab(vx) = (ab)(uy + vx)$. Hence, $ab|c$. ■

Corollary *If $m_i|c$, for $1 \leqslant i < n$, $\gcd(m_i, m_j) = 1$, for $i \neq j$, and $m = \prod_{i=1}^{n} m_i$, then $m|c$.*

Example 2.6 Suppose gcd(a, b) = 1 and $d = \gcd(2a + b, a + 2b)$. Since d must divide any linear combination of $2a + b$ and $a + 2b$, d divides $[2(2a + b) + (-1)(a + 2b)]$ and d divides $[(-1)(2a + b) + 2(a + 2b)]$. Hence, $d|3a$ and $d|3b$. Since $\gcd(a, b) = 1$, d divides 3. Therefore, if $\gcd(a, b) = 1$, then $\gcd(2a + b, a + 2b) = 1$ or 3.

If a and b are integers such that both a and b divide m then m is called a common multiple of a and b. If a and b are nonzero then ab and $-ab$ are both common multiples of a and b and one of them must be positive. Hence, by the well-ordering principle, there exists a least positive common multiple of a and b. If m is the smallest positive common multiple of a and b, we call it the least common multiple of a and b, and denote it by $\text{lcm}(a, b)$. Thus, $m = \text{lcm}(a, b)$ if and only if

(1) $m > 0$,
(2) both a and b divide m,
(3) if both a and b divide n, then m divides n.

Theorem 2.9 *If either a or b is nonzero, then* $\text{lcm}(a, b) = |ab|/\gcd(a, b)|$, *where $|x|$ denotes the absolute value of x.*

Proof Let $d = \gcd(a, b)$, $a = Ad$, $b = Bd$, and $m = |ab|/d$. It follows that $m > 0$, $m = |Ab| = Ab$ and $m = |aB| = aB$. Hence, both a and b divide m. Suppose n is any other multiple of a and b. That is, there exist integers C and D such that $n = aC = bD$. We have $n = AdC = BdD$ so $AC = BD$. Hence, A divides BD. However, since $\gcd(A, B) = 1$, A must divide D. That is, there exists an integer E such that $AE = D$. Thus, $n = bD = bAE = mE$ implying that n is a multiple of m. Therefore, any multiple of both a and b is also a multiple of m. From the three-step criterion for least common multiple, we have that $m = \text{lcm}(a, b)$. ■

Note that $\gcd(56, 35) = 7$, $\text{lcm}(56, 35) = 280$, and $\gcd(56, 35) \cdot \text{lcm}(56, 35) = 7 \cdot 280 = 1960$. The greatest common divisor of more than two integers is defined as follows: $\gcd(a_1, a_2, \ldots, a_n) = d$ if and only if, for all $i = 1, 2, \ldots, n$, $d|a_i$ and if $e|a_i$, for all $i = 1, 2, \ldots, n$ then $e|d$. Similarly for the least common multiple, $\text{lcm}(a_1, a_2, \ldots, a_n) = m$ if and only if for $i = 1, 2, \ldots, n$, $a_i|m$ and if $a_i|e$ for all $i = 1, 2, \ldots, n$ then $m|e$. If $a_1, a_2, \ldots, a_n$ are coprime in pairs then $\gcd(a_1, a_2, \ldots, a_n) = 1$. For if $\gcd(a_1, a_2, \ldots, a_n) = d > 1$, then $d|a_1$ and $d|a_2$ contradicting the fact that $\gcd(a_1, a_2) = 1$. The converse is not true since $\gcd(6, 10, 15) = 1$ but neither 6 and 10, 6 and 15, nor 10 and 15 are coprime.

Given positive integers d and m then a necessary and sufficient condition for the existence of positive integers a and b such that

(a) $\gcd(a, b) = d$ and $\text{lcm}(a, b) = m$ is that $d|m$,
(b) $\gcd(a, b) = d$ and $a + b = m$ is that $d|m$, and
(c) $\gcd(a, b) = d$ and $a \cdot b = m$ is that $d^2|m$.

In order to establish (b), note that if $\gcd(a, b) = d$ and $a + b = m$, then there exist integers r and s such that $a = dr$ and $b = ds$. Hence, $m = a + b = dr + ds = d(r + s)$ and so $d|m$. Conversely, if $d|m$ then choose $a = d$ and $b = m - d$. Then, $a + b = m$. Since 1 and $m/d - 1$ are relatively prime, the greatest common divisor of $a = d \cdot 1$ and $b = d \cdot (m/d - 1)$ is d.

Exercises 2.2

1. Prove that if a divides bc and $\gcd(a, b) = 1$, then $a|c$.
2. Prove that for any positive integer n, $\gcd(n, n+1) = 1$.
3. Show that for any integer n, $\gcd(22n + 7, 33n + 10) = 1$.
4. Show that there cannot exist integers a and b such that $\gcd(a, b) = 3$ and $a + b = 65$.
5. Show that there are infinitely many pairs of integers a and b with $\gcd(a, b) = 5$ and $a + b = 65$.
6. If u_n represents the nth Fibonacci number then show that $\gcd(u_{n+1}, u_n) = 1$, for any positive integer n.
7. If $\gcd(a, b) = d$, and x and y are integers such that $a = xd$ and $b = yd$, show that $\gcd(x, y) = 1$.
8. Prove that if $\gcd(a, b) = 1$ and $\gcd(a, c) = 1$, then $\gcd(a, bc) = 1$. [Euclid]
9. Prove that if $\gcd(a, b) = 1$ then $\gcd(a^m, b^n) = 1$ for any positive integers m and n.
10. Prove that for integers a and b $\gcd(a, b)$ divides $\gcd(a + b, a - b)$.
11. Prove that if $\gcd(a, b) = 1$, then $\gcd(a + ab, b) = 1$.
12. Prove that if $\gcd(a, b) = 1$, then $\gcd(a + b, a - b) = 1$ or 2.
13. Suppose that $\gcd(a, b) = 1$. For what values of a and b is it true that $\gcd(a + b, a - b) = 1$?
14. If $c > 0$, then show that $\gcd(ca, cb) = c \cdot \gcd(a, b)$.
15. Show that for integers a and b, $\gcd(a, a + b)$ divides b.
16. Suppose that for integers a and b $\gcd(a, 4) = 2$ and $\gcd(b, 4) = 2$. Show that $\gcd(a + b, 4) = 2$.
17. If $c > 0$, then show that $\operatorname{lcm}(ac, bc) = c \cdot \operatorname{lcm}(a, b)$.
18. If a divides b determine $\gcd(a, b)$ and $\operatorname{lcm}(a, b)$.
19. Prove that $a|b$ if and only if $\operatorname{lcm}(a, b) = |b|$.
20. For any positive integer n, find $\operatorname{lcm}(n, n + 1)$.
21. For any positive integer n, show that $\operatorname{lcm}(9n + 8, 6n + 5) = 54n^2 + 93n + 40$.
22. Give an example to show that it is not necessarily the case that $\gcd(a, b, c) \cdot \operatorname{lcm}(a, b, c) = abc$.
23. Find all positive integers a and b such that $\gcd(a, b) = 10$, and $\operatorname{lcm}(a, b) = 100$, with $a \geqslant b$.

24. If a and b are positive integers such that $a+b=5432$ and $\text{lcm}(a, b)=223\,020$ then find a and b.
25. $\{30, 42, 70, 105\}$ is a set of four positive integers with the property that they are coprime when taken together, but are not coprime when taken in pairs. Find a set of five positive integers that are coprime when taken together, but are not coprime in pairs.

2.3 The Euclidean algorithm

A method to determine the greatest common divisor of two integers, known as the Euclidean algorithm, appears in Book VII of Euclid's *Elements*. It is one of the few numerical procedures contained in the *Elements*. It is the oldest algorithm that has survived to the present day. The method appears in India in the late fifth century Hindu astronomical work *Aryabhatiya* by Aryabhata. Aryabhata's work contains no equations. It includes 50 verses devoted to the study of eclipses, 33 to arithmetic, and 25 to time reckoning and planetary motion. Aryabhata called his technique the 'pulverizer' and used it to determine integer solutions x, y to the equation $ax - by = c$, where a, b and c are integers. We discuss Aryabhata's method in Chapter 5. In 1624, Bachet included the algorithm in the second edition of his *Problèmes plaisants et délectables*. It was the first numerical exposition of the method to appear in Europe.

The Euclidean algorithm, is based on repeated use of the division algorithm. Given two integers a and b where, say $a \geqslant b > 0$, determine the sequences $q_1, q_2, \ldots, q_{n+1}$ and $r_1, r_2 \ldots, r_{n+1}$ of quotients and remainders in the following manner.

$$\begin{aligned} a &= bq_1 + r_1, && \text{where } 0 \leqslant r_1 < b. \\ b &= r_1 q_2 + r_2, && \text{where } 0 \leqslant r_2 < r_1. \\ r_1 &= r_2 q_3 + r_3, && \text{where } 0 \leqslant r_3 < r_2. \\ & \cdots \\ r_{n-2} &= r_{n-1} q_n + r_n, && \text{where } 0 \leqslant r_n < r_{n-1}. \\ r_{n-1} &= r_n q_{n+1}. \end{aligned}$$

Suppose $r_n \neq 0$. Since $b > r_1 > r_2 \cdots \geqslant 0$, $r_1, r_2, \ldots, r_{n+1}$ is a decreasing sequence of nonnegative integers and must eventually terminate with a zero remainder, say $r_{n+1} = 0$. From the last equation in the Euclidean algorithm, we have that r_n divides r_{n-1} and from the penultimate equation it follows that r_n divides r_{n-2}. Continuing this process we find that r_n divides both a and b. Thus, r_n is a common divisor of a and b. Suppose that e is any positive integer which divides both a and b. From the first equation, it follows that e divides r_1. From the second equation, it follows that, since e divides b and e divides r_1, e divides r_2. Continuing this process, eventually, we find that e divides r_n. Thus, any common divisor of a and b is also a divisor of r_n. Therefore, r_n, the last nonzero remainder, is the greatest common divisor of a and b. We have established the following result.

Theorem 2.10 *Given two positive integers, the last nonzero remainder in the Euclidean algorithm applied to the two integers is the greatest common divisor of the two integers.*

According to the Euclidean algorithm the greatest common divisor of 819 and 165 is 3 since

$$\begin{aligned} 819 &= 165 \cdot 4 + 159, \\ 165 &= 159 \cdot 1 + 6, \\ 159 &= 6 \cdot 26 + 3, \\ 6 &= 3 \cdot 2. \end{aligned}$$

One of the most important and useful applications of the Euclidean algorithm is being able to express the greatest common divisor as a linear combination of the two given integers. In particular, to express the greatest common divisor of 819 and 165 as a linear combination of 819 and 165, we work backwards step by step from the Euclidean algorithm. Using brute force, we accomplish the feat in the following manner.

$$\begin{aligned} 3 &= 159 + (-26)6, \\ 3 &= (819 + 165(-4)) + (-26)(165 + 159(-1)), \\ 3 &= 819 + 165(-30) + 159(26), \\ 3 &= 819 + 165(-30) + (819 + 165(-4))(26), \\ 3 &= 819(27) + 165(-134). \end{aligned}$$

One of the earliest results in the field of computational complexity was established by Gabriel Lamé in 1845. Lamé, a graduate of the École Polytechnique in Paris, was a civil engineer who made several notable contributions to both pure and applied mathematics. He was considered by Gauss to be the foremost French mathematician of his generation. Lamé proved that the number of divisions in the Euclidean algorithm for two positive integers is less than five times the number of digits in the smaller of the two positive integers.

If we apply the Euclidean algorithm to integers a and b where

$a \geqslant b > 0$, then $q_i \geqslant 1$, for $1 \leqslant i \leqslant n$. Since $r_n < r_{n-1}$, $q_{n+1} > 1$. Let $a_1, a_2, \ldots$ denote the Fibonacci-type sequence with $a_1 = 1$ and $a_2 = 2$. We have

$$\begin{aligned}
r_n &\geqslant 1 &&= 1 = a_1,\\
r_{n-1} &= r_n q_{n+1} \geqslant 1 \cdot 2 &&= 2 = a_2,\\
r_{n-2} &= r_{n-1} q_n + r_n \geqslant 2 \cdot 1 + 1 &&= 3 = a_3,\\
r_{n-3} &= r_{n-2} q_{n-1} + r_{n-1} \geqslant 3 \cdot 1 + 2 &&= 5 = a_4,\\
r_{n-4} &= r_{n-3} q_{n-2} + r_{n-2} \geqslant 5 \cdot 1 + 3 &&= 8 = a_5,\\
&\cdots\\
b &= r_1 q_2 + r_2 \geqslant a_{n-1} \cdot 1 + a_{n-2} &&= a_n.
\end{aligned}$$

It follows that $b \geqslant a_n = u_{n+1} = (\tau^{n+1} - \sigma^{n+1})/(\tau - \sigma) > \tau^n$. Since $\log \tau > \frac{1}{5}$, $n < \log b / \log \tau < 5 \cdot \log b$. If m denotes the number of digits in b, then $b < 10^m$. Hence, $\log b < m$. Therefore, $n < 5m$ and we have established Lamé's result.

Theorem 2.11 (Lamé) *The number of divisions in the Euclidean algorithm for two positive integers is less than five times the number of digits in the smaller of the two positive integers.*

In 1970, John Dixon of Carleton University improved the bound by showing that the number of steps in the Euclidean algorithm is less than or equal to $(2.078)[\log a + 1]$, where a is the larger of the two positive integers. If there are a large number of steps in the Euclidean algorithm, expressing the greatest common divisor as a linear combination of the two integers by brute force can be quite tedious. In 1740, Nicholas Saunderson, the blind Lucasian Professor of Mathematics at Cambridge University, included an algorithm in his *Elements of Algebra* which greatly simplified the process. Saunderson attributed the origin of the method to Roger Cotes, the first Plumian Professor of Mathematics at Cambridge, who used the algorithm in the expansion of continued fractions. A similar technique can be traced back at least to the thirteenth century where it is found in Qin Jiushao's (CHIN JEW CHOW) *Mathematical Treatise in Nine Sections*.

Let a and b be integers, with $a \geqslant b > 0$. Utilizing the notation of the Euclidean algorithm let $d = \gcd(a, b) = r_n$ so $r_{n+1} = 0$ and $r_i = r_{i+1} q_{i+2} + r_{i+2}$, for $i = 1, 2, \ldots, n$. In addition, let $r_{-1} = a$, $r_0 = b$. Define $x_i = x_{i-2} + x_{i-1} q_i$, $y_i = y_{i-2} + y_{i-1} q_i$, for $i = 2, \ldots, n+1$. For completeness, let $x_0 = 0$, $x_1 = 1$, $y_0 = 1$, and $y_1 = q_1$. Using this notation, we establish the following result.

Theorem 2.12 (Saunderson's algorithm) *If d is the greatest common divisor of two integers a and b, with $a > b \geqslant 0$, then $d = a(-1)^{n-1}x_n + b(-1)^n y_n$.*

Proof Consider the variable proposition $P(n)$: $ax_n - by_n = (-1)^{n-1}r_n$. $P(0)$: $ax_0 - by_0 = 0 - b = (-1)^{-1}r_0$. $P(1)$: $ax_1 - by_1 = a \cdot 1 - bq_1 = r_1$. Hence, $P(1)$ is true. $P(2)$: $ax_2 - by_2 = a(x_0 + x_1q_2) - b(y_0 + y_1q_2) = ax_1q_2 - b(1 + q_1q_2) = (ax_1 - bq_1)q_2 - b = (-1)r_2$. Hence, $P(2)$ is true. Assume that $P(r)$ holds for all integers r between 1 and k for $k > 1$ and consider $P(k+1)$. We have $P(k+1)$:

$$\begin{aligned} ax_{k+1} - by_{k+1} &= a(x_{k-1} + x_kq_{k+1}) - b(y_{k-1} + y_kq_{k+1}) \\ &= (ax_{k-1} - by_{k-1}) + q_{k+1}(ax_k - by_k) \\ &= (-1)^k r_{k-1} + q_{k+1}(-1)^{k-1}r_k \\ &= (-1)^k(r_{k-1} - q_{k+1}r_k) \\ &= (-1)^k r_{k+1}. \end{aligned}$$

Hence, $P(k-1)$ and $P(k)$ imply $P(k+1)$ and, from the alternate principle of mathematical induction, $P(n)$ is true for all nonnegative integers. Therefore, $d = r_n = (-1)^{n-1}(ax_n - by_n) = a(-1)^{n-1}x_n + b(-1)^n y_n$. ■

Example 2.7 We use Saunderson's method to express the greatest common divisor of 555 and 155 as a linear combination of 555 and 155. From the Euclidean algorithm it follows that

$$\begin{aligned} 555 &= 155 \cdot 3 + 90, & a &= bq_1 + r_1, \\ 155 &= 90 \cdot 1 + 65, & b &= r_1q_2 + r_2, \\ 90 &= 65 \cdot 1 + 25, & r_1 &= r_2q_3 + r_3, \\ 65 &= 25 \cdot 2 + 15, & r_2 &= r_3q_4 + r_4, \\ 25 &= 15 \cdot 1 + 10, & r_3 &= r_4q_5 + r_5, \\ 15 &= 10 \cdot 1 + 5, & r_4 &= r_5q_6 + r_6, \\ 10 &= 5 \cdot 2 + 0, & r_5 &= r_6q_7. \end{aligned}$$

Hence, 5, the last nonzero remainder, is the greatest common divisor of 555 and 155. Table 2.3 contains the basic elements in applying Saunderson's algorithm, where $x_i = x_{i-2} + x_{i-1}q_i$, $y_i = y_{i-2} + y_{i-1}q_i$, for $i = 1, 2, \ldots, n+1$, $x_0 = 0$, $x_1 = 1$, $y_0 = 1$, and $y_1 = q_1$. A useful check when using Saunderson's algorithm arises from the fact that $r_{n+1} = 0$, hence, $ax_{n+1} = by_{n+1}$. For the case when $a = 55$ and $b = 155$, we fill in Table 2.3 with the appropriate terms to obtain Table 2.4. Therefore,

Table 2.3.

i	0	1	2	3	...	n	$n+1$
q_i		q_1	q_2	q_3	...	q_n	q_{n+1}
x_i	0	1	x_2	x_3	...	x_n	x_{n+1}
y_i	1	q_1	y_2	y_3	...	y_n	y_{n+1}

Table 2.4.

i	0	1	2	3	4	5	6	7
q_i		3	1	1	2	1	1	2
x_i	0	1	1	2	5	7	12	31
y_i	1	3	4	7	18	25	43	111

$5 = \gcd(155, 555) = (-12)555 + (43)155$. As a check, we have $a \cdot x_7 = 555 \cdot 31 = 17\,205 = 155 \cdot 111 = b \cdot y_7$.

In order to minimize the computations involved for his students, Saunderson devised an equivalent but more efficient algorithm illustrated in the next example. The simplified version determines the greatest common divisor to two natural numbers and expresses it as a linear combination of the two given integers in one fell swoop.

Example 2.8 Given $a = 555$ and $b = 155$ form the following sequence of equations.

1	$1 \cdot a - 0 \cdot b =$	555	
2	$0 \cdot a - 1 \cdot b =$	-155	3
3	$a - 3b =$	90	1
4	$a - 4b =$	-65	1
5	$2a - 7b =$	25	2
6	$5a - 18b =$	-15	1
7	$7a - 25b =$	10	1
8	$12a - 43b =$	-5	1
9	$-12a + 43b =$	5	

The first two equations are straightforward. Since 3 is the quotient when dividing 155 into 555, we multiply the second equation by 3 and add it to the first equation to obtain the third equation. We obtain the fourth equation by multiplying the third equation by unity, since 90 goes into 155 once, and adding it to the second equation, as so forth. After obtaining the eighth equation, $12a - 43b = -5$, we note that 5 divides into 10 evenly. Hence, $\gcd(555, 155) = 5$. Multiplying both sides of the eighth equation by -1 we obtain the desired result, $-12a + 43b = 5$.

Similarly if $a = 6237$ and $b = 2520$, we obtain

$$
\begin{array}{rrrl}
1 \cdot a - 0 \cdot b = & 6237 & \\
0 \cdot a - 1 \cdot b = & -2520 & 2 \\
a - \quad 2b = & 1197 & 2 \\
2a - \quad 5b = & -126 & 9 \\
19a - 47b = & 63 & 2 \\
40a - 99b = & 0 &
\end{array}
$$

Hence, $\gcd(6237, 2520) = 63 = 19(6237) + (-47)(2520)$. Furthermore, $\text{lcm}(6237, 2520) = 40a = 99b = 249\,480$.

At Cambridge Saunderson tutored algebra and lectured on calculus in the Newtonian style. Each year he gave a very acclaimed series of natural science lectures. Several copies of notes from students who attended his course are extant. However, it appears that their popularity may have rested on the fact that they were virtually devoid of mathematical content. Albeit he was an excellent teacher, he often wondered if his everlasting fate would include a stint in Hades teaching mathematics to uninterested students.

Saunderson was very diligent and forthright. He once told Horace Walpole, the author and third son of England's first Prime Minister Robert Walpole, that he would be cheating him to take his money, for he could never learn what he was trying to teach. Lord Chesterfield said of Saunderson that, 'He did not have the use of his eyes, but taught others to use theirs'.

Exercises 2.3

1. Find the greatest common divisors and the least common multiples for the following pairs of integers. Determine the Lamé and Dixon limits.
   (a) $a = 93$ and $b = 51$;
   (b) $a = 481$ and $b = 299$;
   (c) $a = 1826$ and $b = 1742$;
   (d) $a = 1963$ and $b = 1941$;
   (e) $a = 4928$ and $b = 1771$.

2. Express the greatest common divisor of each pair of integers as a linear combination of the two integers.
   (a) $a = 93$ and $b = 51$;
   (b) $a = 481$ and $b = 299$;
   (c) $a = 1826$ and $b = 1742$;
   (d) $a = 1963$ and $b = 1941$;
   (e) $a = 4928$ and $b = 1771$.

2.4 Pythagorean triples

One of the earliest known geometric applications of number theory was the construction of right triangles with integral sides by the Babylonians in the second millennium BC. In particular, if x, y and z are positive integers with the property that $x^2 + y^2 = z^2$ then the 3-tuple (x, y, z) is called a Pythagorean triple. In 1945 Otto Neugebauer and A. Sachs analyzed a nineteenth century BC Babylonian cuneiform tablet in the Plimpton Library archives at Columbia University. The tablet, designated Plimpton 322, lists 15 pairs (x, z) for which there is a y such that $x^2 + y^2 = z^2$ referring to Pythagorean triples ranging from (3, 4, 5) to (12 709, 13 500, 18 541). The Babylonians undoubtedly had an algorithm to generate such triples long before Pythagoras was born, but such are the whims of eponymy. The earliest appearance of Pythagorean triples in Europe was in the 1572 edition of Rafael Bombelli's *Algebra*. Twenty years later, they appear in François Viète's *Introduction to the Analytic Art*.

It will be convenient to restrict our attention to primitive Pythagorean triples, which are Pythagorean triples (x, y, z) with the additional property that x, y and z have no positive common divisor other than unity. For example, (3, 4, 5) is a primitive Pythagorean triple. In Theorem 3.3, we show the Pythagorean triple (x, y, z) is primitive if and only if $\gcd(x, y) = 1$, $\gcd(x, z) = 1$, and $\gcd(x, y) = 1$. We use this fact now to establish an algorithm, a version of which appears in Book X of Euclid's *Elements* that may have been used by the Babylonians to determine Pythagorean triples.

Theorem 2.13 *If (x, y, z) is a primitive Pythagorean triple, then there exist positive integers s and t, $s > t$, $\gcd(s, t) = 1$, one even and the other odd such that $x = 2st$, $y = s^2 - t^2$, and $z = s^2 + t^2$.*

Proof If (x, y, z) is a primitive Pythagorean triple, then x, y and z are coprime in pairs. If x and y are even then z is even. If x and y are odd, then

Table 2.5.

s	t	x	y	z
2	1	4	3	5
3	2	12	5	13
4	1	8	15	17
4	3	24	7	25
5	2	20	21	29

z^2 is not of the form $4k$ or $4k + 1$, a contradiction. Hence, x and y must be of different parity. Without loss of generality, let x be even and y be odd. Hence, z is odd. In addition, $x^2 = z^2 - y^2 = (z - y)(z + y)$. Since $z - y$ and $z + y$ must be even let $z - y = 2u$ and $z + y = 2v$. Now u and v must be coprime for if $\gcd(u, v) = d > 1$, then d divides both u and v implying that d divides both y and z, which contradicts the assumption that y and z are coprime. In addition, if u and v were both odd, then y and z would be even, a contradiction. So one of u and v is even and the other is odd. Since x is even, $x/2$ is an integer, and

$$\left(\frac{x}{2}\right)^2 = \left[\frac{(z-y)}{2}\right]\left[\frac{(z+y)}{2}\right] = uv.$$

Since $uv = (x/2)^2$ and $\gcd(u, v) = 1$, u and v must be perfect squares, say $u = s^2$ and $v = t^2$, where one of s and t is even and the other is odd. It follows that $x = 2st$, $y = s^2 - t^2$, and $z = s^2 + t^2$. ■

Example 2.9 Using Theorem 2.12, and several values of s and t, we obtain the primitive Pythagorean triplets shown in Table 2.5.

The next result implies that neither the equation $x^4 + y^4 = z^4$ nor the equation $x^{2n} + y^{2n} = z^{2n}$, with n a positive integer greater than 1, have integral solutions. We employ Fermat's method of descent to establish the result. In essence, Fermat's technique is a proof by contradiction. There are two paths we may take. Either we assume that a particular number is the least positive integer satisfying a certain property and proceed to find a smaller positive integer having the same property or we proceed to construct an infinitely decreasing sequence of positive integers. In either case, we arrive at a contradiction. The next result was arrived at independently by Fermat and his long-time correspondent Bernard Frenicle de Bessy. Frenicle, an official at the French Mint, discovered in 1634, that the frequency of a pendulum is inversely proportional to the square root of its length. Frenicle was a good friend of Galileo and offered to publish a French translation of his *Dialogue*.

Theorem 2.14 *The equation $x^4 + y^4 = z^2$ has no integral solutions.*

Proof Without loss of generality, we consider only primitive solutions to the equation. Let $a^4 + b^4 = c^4$ be the solution with $\gcd(a, b, c) = 1$ and least positive value for c. From Theorem 2.13, since $(a^2)^2 + (b^2)^2 = c^2$, there exist coprime integers s and t of opposite parity such that $s > t$, $a^2 = 2st$, $b^2 = s^2 - t^2$, and $c = s^2 + t^2$. Hence, $s < c$, $b^2 + t^2 = s^2$, with $\gcd(b, t) = \gcd(s, t) = 1$, with say s odd and t even. Applying Theorem 2.13 to $b^2 + t^2 = s^2$, we find that $t = 2uv$, $b = u^2 - v^2$, and $s = u^2 + v^2$, with u and v coprime, of opposite parity, and $u > v$. In addition, $(a/2)^2 = st/2$ and $\gcd(s, t/2) = \gcd(t, s) = 1$. Hence, $s = r^2$ and $t/2 = w^2$, with $(r, w) = 1$. Further, $w^2 = t/2 = uv$, so $u = m^2$ and $v = n^2$, with $\gcd(m, n) = 1$. Thus, $m^4 + n^4 = u^2 + v^2 = s = r^2$, with $r \leqslant s < c$ contradicting the minimality of c. Therefore $x^4 + y^4 = z^2$ has no integral solutions. ■

Problems concerning integral areas of rational right triangles go back to Diophantus. A right triangle whose sides form a primitive Pythagorean triple is called a Pythagorean triangle. The area of a Pythagorean triangle, *sans* the units of measurement, is called a Pythagorean number. It follows, from Theorem 2.13, that a Pythagorean number P can be represented as a product of the form $P = st(s + t)(s - t)$, where s and t are of different parity and $\gcd(s, t) = 1$. Among the properties of Pythagorean numbers are: every Pythagorean number is divisible by 6; for every integer $n > 12$ there is a Pythagorean number between n and $2n$; the units digit of a Pythagorian number is either 0, 4, or 6; there are infinitely many Pythagorean numbers of the form $10k$, $10k + 4$, and $10k + 6$; no Pythagorean number is square; no Pythagorean number is a Lucas number.

The Pythagorean triple (9999, 137 532, 137 895) is unusual since its associated Pythagorean triangle has area 687 591 234 which is almost pandigital. Note that the Pythagorean triangles (20, 21, 29) and (12, 35, 51) have different hypotenuses but the same area. In addition, for any positive integer k, triangles with sides $x = 20k^4 + 4k^2 + 1$, $y = 8k^6 - 4k^4 - 2k^2 + 1$, and $z = 8k^6 + 8k^4 + 10k^2$, have area $4k^2(2k^2 + 1)^2(2k^2 - 1)^2$; however, none is a right triangle. The incenter of a triangle is the center of the inscribed circle. The incenter is also the intersection of the angle bisectors.

Theorem 2.15 *The radius of the incircle of a Pythagorean triangle is an integer.*

Proof Denote the area, inradius, and incenter of the Pythagorean triangle ABC shown in Figure 2.1 by K, r, and I, respectively. Let $a = y + z$, $b = x + z$, and $c = x + y$. From Theorem 2.13, $K = \frac{1}{2}ab + \frac{1}{2}ra + \frac{1}{2}rb + \frac{1}{2}rc = \frac{1}{2}r(a + b + c) = \frac{1}{2}r(2st + s^2 - t^2 + s^2 + t^2) = rs(t + s)$. Similarly, $K = \frac{1}{2}xy = st(s^2 - t^2)$. Hence, $r = t(s - t)$, which is an integer. ■

It has been shown that no infinite set of noncollinear planar points exist whose pairwise distances are all integral. However, we can generate such finite sets with that property using primitive Pythagorean triples as shown in the next example.

Example 2.10 We use $n - 2$ primitive Pythagorean triples to determine n noncollinear points in the plane with the property that each is an integral distance from any other. Let (x, y) denote a point in the Cartesian plane with abscissa x and ordinate y. Suppose $n = 7$ and choose five different primitive Pythagorean triples, for example, those in Table 2.6. Let $p_1 = (0, 0)$, $p_2 = (0, 3 \cdot 5 \cdot 7 \cdot 15 \cdot 21) = (0, 33\,075)$, and $p_i = (x_i, 0)$, for $3 \leqslant i \leqslant 7$, where

$$\begin{aligned}
x_3 &= 4 \cdot 5 \cdot 7 \cdot 15 \cdot 21 &&= 44\,100,\\
x_4 &= 3 \cdot 12 \cdot 7 \cdot 15 \cdot 21 &&= 79\,380,\\
x_5 &= 3 \cdot 5 \cdot 24 \cdot 15 \cdot 21 &&= 113\,400,\\
x_6 &= 3 \cdot 5 \cdot 7 \cdot 8 \cdot 21 &&= 17\,640,\\
x_7 &= 3 \cdot 5 \cdot 7 \cdot 15 \cdot 20 &&= 31\,500.
\end{aligned}$$

The basic structure of x_i, the nonzero coordinate in p_i, for $i = 3, 4, 5, 6, 7$, derives from the product of the terms in the first column of Table 2.6. However, the $(n - 2)$nd term in the product is replaced by the correspond-

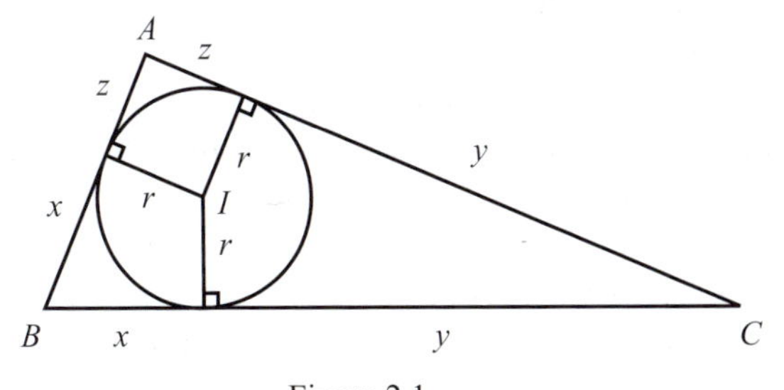

Figure 2.1.

Table 2.6.

3	4	5
5	12	13
7	24	25
15	8	17
21	20	29

ing term in the second column of Table 2.6. By construction $\{p_1, p_2, \ldots, p_7\}$ forms a set of seven noncollinear points in the plane with the property that any pair of points in the set are an integral distance apart. This follows by construction since the distance separating each pair is the length of a side of a Pythagorean triangle.

In 1900, D. H. Lehmer showed that the number of primitive Pythagorean triples with hypotenuse less than or equal to n is approximately $n/2\pi$. Pythagorean triangles can be generalized to Pythagorean boxes, rectangular parallelepipeds with length, width, height, and all side and main diagonals having integral values. It is an open question whether or not a Pythagorean box exists.

Exercises 2.4

1. For any positive integer n, show that $(2n^2 + 2n,\ 2n + 1,\ 2n^2 + 2n + 1)$ is a Pythagorean triple in which one side and the hypotenuse differ by one unit. Such triples were studied by Pythagoras, and rediscovered by Stifel when he was investigating properties of the mixed fractions $1\frac{1}{3}$, $2\frac{2}{5}$, $3\frac{3}{7}$, $4\frac{4}{9}$ $\ldots$, $n + n/(2n+1) = (2n^2 + 2n)/(2n+1)$.
2. For any positive integer $n > 1$, show that $(2n,\ n^2 - 1,\ n^2 + 1)$ is a Pythagorean triple in which one side differs from the hypotenuse by two units. Such triples were studied by Plato.
3. If (a, b, c) and (x, y, z) are Pythagorean triples, show that $(ax - by,\ ay + bx,\ cz)$ is a Pythagorean triple.
4. Prove that (3, 4, 5) is the only primitive Pythagorean triple whose terms are in arithmetic proportion, that is, they are of the form $(a,\ a + d,\ a + 2d)$.
5. Why is it not the case that the values $s = 3$ and $t = 5$ generate a primitive Pythagorean triple?
6. Show that if (x, y, z) is a primitive Pythagorean triple then the sum of the legs of the Pythagorean triangle generated is of the form $8m \pm 1$.

7. For any positive integer $n \geqslant 3$, show that there exists a Pythagorean triple (x, y, z) with n as one if its elements.
8. Define the Pell sequence, 0, 1, 2, 5, 12, 27, ..., a_n, ..., recursively such that $a_{n+2} = 2a_{n+1} + a_n$, with $a_0 = 0$ and $a_1 = 1$. Show that if $x_n = a_{n+1}^2 - a_n^2$, $y_n = 2a_{n+1}a_n$, and $z_n = a_{n+1}^2 + a_n^2$, with $n \geqslant 1$, then (x_n, y_n, z_n) is a Pythagorean triple with $x_n = y_n + (-1)^n$.
9. Show that if the pair (s, t), from Theorem 2.13, with $s > t$, generates a Pythagorean triple with $|x - y| = k \geqslant 0$, then $(2s + t, s)$ generates a Pythagorean triple with $|x - y| = k$.
10. Ignoring the dimensions of the units, find two Pythagorean triangles with the same area as perimeter.
11. Show that the Pythagorean triples (40, 30, 50), (45, 24, 51), and (48, 20, 52) have equal perimeters and their areas are in arithmetic proportion.
12. Prove that the product of three consecutive positive integers, with the first number odd, is a Pythagorean number.
13. Show that every Pythagorean number is divisible by 6.
14. Show that a Pythagorean number can never be a square.
15. What positive integers n are solutions to $x^2 - y^2 = n$?
16. Show that if (x, y, z) is a primitive Pythagorean triple then $12|xyz$.
17. Show that if (x, y, z) is a primitive Pythagorean triple then $60|xyz$. [P. Lenthéric 1830]
18. Find the coordinates of a set of eight noncollinear planar points each an integral distance from the others.
19. How many primitive Pythagorean triangles have hypotenuses less than 100? How accurate is Lehmer's prediction in this case?

2.5 Miscellaneous exercises

1. A raja wished to distribute his wealth among his three daughters Rana, Daya, and Cyndi such that Rana, the eldest, received half of his wealth, Daya received one-third, and Cyndi, the youngest, received one-ninth. Everything went well until the raja came to his seventeen elephants. He was in a quandary as to how to divide them amongst his daughters. To solve the problem he called in his lawyer who came riding her own elephant, which she, after surveying the situation, had coloured pink and placed among the seventeen elephants. The lawyer told Rana to take half or nine of the elephants, but not the pink one, which she did. The lawyer then told Daya to take a third or six of the elephants, but not the pink one, which Daya did. Then the lawyer told Cyndi to take

the two elephants remaining that were not pink. The raja and his daughters were happy and after collecting her fee the lawyer took her pink elephant and rode home. How was she able to accomplish this remarkable feat?

2. Elephantine triples are triples or 3-tuples of numbers of the form $(1/a, 1/b, 1/c)$ such that for the distinct positive integers a, b, c and some positive integer n, we have that $1/a + 1/b + 1/c = n/(n+1)$. For example, $(\frac{1}{2}, \frac{1}{3}, \frac{1}{9})$ is an elephantine triple. Find two more examples of elephantine triples.
3. A reciprocal Pythagorean triple (a, b, c) has the property that $(1/a)^2 + (1/b)^2 = (1/c)^2$. Show that (780, 65, 60) is a reciprocal Pythagorean triple.
4. Take three consecutive integers, with the largest a multiple of 3. Form their sum. Compute the sum of its digits, do the same for the result until a one-digit number is obtained. Iamblichus of Chalis claimed that the one-digit number obtained will always equal 6. For example, the sum of 9997, 9998 and 9999 is 29 994. The sum of the digits of 29 994 is 33 and the sum of the digits of 33 is 6. Prove Iamblichus's claim.
5. Given a scale with two pans, determine the least number of weights and the values of the weights in order to weigh all integral weights in kilograms from 1 kilogram to 40 kilograms. [Bachet]
6. Explain how the following multiplication rule works. To multiply two given numbers, form two columns, each headed by one of the numbers. Successive terms in the left column are halved, always rounding down, and successive terms in the right column are doubled. Now strike out all rows with even numbers in the left column and add up the numbers remaining in the right column to obtain the product of a and b. For example, to determine $83 \times 154 = 12\,782$, we have:

83	154
41	308
~~20~~	~~616~~
~~10~~	~~1 232~~
5	2 464
~~2~~	~~4 928~~
1	9 856
	12 782

7. In *The Educational Times* for 1882, Kate Gale of Girton College, Cambridge, proved that if $3n$ zeros are placed between the digits 3 and 7, then the number formed is divisible by 37. In addition, if $3n + 1$

zeros are placed between the digits 7 and 3, the number formed is divisible by 37. Prove these statements.

8. Let $f(n)$ be the smallest positive integer value of x_n such that $\sum_{k=1}^{n} x_k^{-1} = 1$, for some positive integers $x_1, x_2, \ldots, x_{n-1}$ such that $1 < x_1 < x_2 < \cdots < x_n$. Since $\frac{1}{2} + \frac{1}{3} + \frac{1}{6} = 1$ and 6 is the smallest positive integer with this property for $n = 3$ it follows that $f(3) = 6$. Determine $f(4)$.
9. If $a > 0$, $b > 0$, and $1/a + 1/b$ is an integer then show that $a = b$ and $a = 1$ or $a = 2$.
10. Show that in any set of $n + 1$ integers selected from the set $\{1, 2, \ldots, 2n\}$ there must exist a pair of coprime integers.
11. Show that the product of k consecutive natural numbers is always divisible by $k!$ [J.J. Sylvester]
12. Show that in any set of five consecutive positive integers there always exists at least one integer which is coprime to every other integer in the set.
13. A positive integer is called polite if it can be represented as a sum of two or more consecutive integers. For example, 7 is polite since $7 = 3 + 4$. Similarly, 2 is impolite since it cannot be written as a sum of two or more consecutive integers. Show that the only impolite positive integers are powers of 2.
14. Use Heron's formula for the area K of a triangle with sides a, b, c, namely, $K = \sqrt{s(s-a)(s-b)(s-c)}$, where $s = (a+b+c)/2$, to show that triangles with sides $x = 20k^4 + 4k^2 + 1$, $y = 8k^6 - 4k^4 - 2k^2 + 1$, and $z = 8k^6 + 8k^4 + 10k^2$, for k a positive integer, all have area $4k^2(2k^2+1)^2(2k^2-1)^2$.
15. Prove that every natural number belongs to one of three basic digital root sequences.
16. Use the principle of induction to show that if $c_1, c_2, \ldots, c_k$ are pairwise coprime integers and $c_i | n$, for $i = 1, 2, \ldots, k$, and $m = \prod_{i=1}^{k} c_i$, then $m | n$.
17. Prove, for all positive integers n, that $n^5/5 + n^3/3 + 7n/15$ is always an integer.
18. For n a positive integer, the Smarandache function $S(n)$ is the smallest positive integer m such that n divides $m!$. For example, $S(1) = 1$, $S(2) = 2$, and $S(3) = 3$. Determine $S(n)$ for $n = 4, 5, \ldots, 10$.
19. Let $h(p)$ denote the smallest positive integer such that $!h(p)$ is divisible by p, where $p > 3$ is prime and $!n = \sum_{k=0}^{n-1} k!$ for any positive integer n. For example, $!4 = 0! + 1! + 2! + 3! = 10$, hence, $h(5) = 4$. Determine $h(p)$ when $p = 7$ and $p = 11$.

20. Establish the following connection between Fibonacci-type sequences and Pythagorean triples discovered by A.F. Horadam in 1961. If $a_1, a_2, \ldots$ is a Fibonacci-type sequence then for $n \geqslant 3$, $(a_n a_{n+3}, 2a_{n+1}a_{n+2}, 2a_{n+1}a_{n+2} + a_n{}^2)$ is a Pythagorean triple.
21. If we were to extend the Fibonacci numbers, u_n, to include negative subscripts, that is $u_{n+2} = u_{n+1} + u_n$, where n is any integer, then determine a general rule for determining such an extended Fibonacci array.

2.6 Supplementary exercises

1. Show that if a divides c and $a + b = c$ then a divides b.
2. If $d \neq 0$, $c = ax + by$, and d divides b and c, must d divide a?
3. For every positive integer n show that 2 divides $n^2 - n$ and 6 divides $n^3 - n$.
4. For any positive integer n, show that 6 divides $7n^3 + 5n$.
5. For any positive integer n show that 15 divides $2^{4n} - 1$.
6. If n is an integer not divisible by 2 or 3, show that 24 divides $n^2 + 23$.
7. For any positive integer n show that 4 does not divide $n^2 + 2$.
8. Prove or disprove that for any positive integer n, 3 divides $2n^3 + 7$.
9. Show that if a and b are odd integers then 8 divides $a^2 - b^2$.
10. If n is an odd integer show that 12 divides $n^2 + (n+2)^2 + (n+4)^2 + 1$.
11. If n is an odd integer show that 32 divides $(n^2 + 3)(n^3 + 7)$.
13. Show that for any positive integer n, $1 + 2 + \cdots + n$ divides $3(1^2 + 2^2 + \cdots + n^2)$.
14. Find a six-digit number n such that when the digit on the left is removed and placed at the end the new number equals $3n$.
15. Show that the square of any odd integer must be of the form $8k + 1$.
16. A positive integer is called evil if it has an even number of ones in its base 2 representation. Determine the first fifteen evil numbers.
17. Show that if $y^2 = x^3 + 2$, then both x and y are odd.
18. Find $\gcd(2^3 \cdot 5^2 \cdot 7, 2 \cdot 3^4 \cdot 5^6)$.
19. Find the greatest common divisor of 1213 and 8658 and express it as a linear combination of 1213 and 8658.
20. Find the greatest common divisor of 198 and 243 and express it as a linear combination of 198 and 243.
21. Find the greatest common divisor of 527 and 765 and express it as a linear combination of 527 and 765.

22. Find the greatest common divisor of 6409 and 42823 and express it as a linear combination of 6409 and 42823.
23. Find the greatest common divisor of 2437 and 51329 and express it as a linear combination of 2437 and 51329.
24. Find the greatest common divisor of 1769 and 2378 and express it as a linear combination of 1769 and 2378.
25. Show that for any integer n, $\gcd(14n + 3, 21n + 4) = 1$.
26. Determine the greatest common divisor of $5n + 2$ and $7n + 3$, where n is any integer.
27. Determine the least positive integer in the set $\{341x + 527y\}$, where x and y are integers.
28. If $\gcd(a, c) = 1$ and b divides c, show that $\gcd(a, b) = 1$.
29. Show that if $\gcd(a, b) = 1$ and $a|bc$, than $a|c$.
30. If a divides b, show that $\gcd(a, c) = \gcd(a, b + c)$.
31. Show that $\gcd(u_n, u_{n+2}) = 1$ or 2, where u_n denotes the nth Fibonacci number.
32. If $\gcd(a, b) = 1$, show that $\gcd(2a + b, a + 2b) = 1$ or 3.
33. Fin a, b, c such that $\gcd(a, b, c) = 1$, but $\gcd(a, b) > 1$, $\gcd(a, b) > 1$, and $\gcd(b, c) > 1$.
34. Find two primitive Pythagorean triples that contain 17.
35. Find two primitive Pythagorean triples that contain 27.
36. If (x, y, z) and $(z, xy, xy + 1)$ are primitive Pythagorean triples show that x and y must be consecutive positive integers.
37. If $(x, x + 1, z)$ is a primitive Pythagorean triple show that $(3x + 2z + 1, 3x + 2z + 2, 4x + 3z + 2)$ is a primitive Pythagorean triple. Hence, there are an infinite number of primitive Pythagorean triples whose legs are consecutive positive integers.
38. Show that $(z, x(x + 1), x(x + 1) + 1)$ is a primitive Pythagorean triple if and only if $(x, x + 1, z)$ is a primitive Pythagorean triple.
39. If the sum of two consecutive integers is a square, show that the two integers form the side and a diagonal of a Pythagorean triple.
40. If $x^2 + y^2 = z^2$ show that 3 divides xy.
41. A positive integer n has the cycling digits property if every fraction m/n, with $1 \leqslant m < n$, has a block of repeated digits (the repetend of m/n). That is a cyclic permutation of the repetend of $1/n$. Show that 7, 17, and 19 have the cycling digits property.
42. Show that if an integer has the cycling digits property then it must be prime.
43. A positive integer n is called a Curzon number if $2n + 1$ divides $2^n + 1$. Determine the first ten Curzon numbers.

44. A positive integer n is called a Zuckerman number if it is divisible by the product of its digits. Show that 3111, 1311, 1131, and 1113 are Zuckerman numbers.
45. Given that 1, 2, ..., 9 are Zuckerman numbers. Determine the next fifteen Zuckerman numbers.
46. A positive integer n is called a riven (or reduced Niven) number if it is divisible by its digital root. Determine the first ten riven numbers.
47. A positive integer is called b-Niven (b-riven) if it is divisible by the sum of its digits (digital root) in base b. Show that every positive integer is 2-riven and 3-riven.
48. In the fourth, fifth and sixth row of Pascal's Triangle the hexagon

$$\begin{array}{cccc} & 4 & 6 & \\ 5 & & & 10 \\ & 15 & 20 & \end{array}$$

appears. Note that $5 \cdot 6 \cdot 20 = 4 \cdot 10 \cdot 15$ and $\gcd(5, 6, 20) = 1 = \gcd(4, 10, 15)$. Is this true in general? Namely, for positive integers n and k, with $n > k$, does

$$\binom{n}{k+1}\binom{n+1}{k+2}\binom{n+2}{k+3} = \binom{n}{k}\binom{n+1}{k+3}\binom{n+2}{k+2} \text{ and}$$

$$\gcd\left(\binom{n}{k+1}, \binom{n+1}{k+2}, \binom{n+2}{k+3}\right) = \gcd\left(\binom{n}{k}, \binom{n+1}{k+3}, \binom{n+2}{k+2}\right)?$$

49. If $S(n)$ denotes the Smarandache function show that the infinite series $\sum_{k=1}^{\infty} 1/S(n^k)$ diverges for any positive integer n.
50. For n a positive integer the pseudo-Smarandache function $S^*(n)$ is the smallest positive integer m such that n divides the mth triangular number. Determine $S^*(n)$ for $n = 2$ to 15.

3
Prime numbers

I was taught that the way of progress is neither swift nor easy.
Marie Curie

3.1 Euclid on primes

In this section, we investigate the fundamental structure of the integers. Playing the role of indivisible quantities are those integers designated as being prime. A positive integer, other than unity, is said to be prime if its only positive divisors are unity and itself. That is, a prime number is an integer greater than 1 with the minimal number of positive integral divisors. A positive integer which is neither unity nor prime is called composite. By considering unity as being neither prime or composite, we follow the custom of the Pythagoreans, the first group to distinguish between primes and composites. Unfortunately, there is no efficient method to determine whether or not a given number is prime. Eratosthenes of Cyrene (now in Libya) devised a technique, referred to as the sieve of Eratosthenes, to find prime numbers. Eratosthenes was a Greek mathematician–astronomer who served as director of the Library at Alexandria under Ptolemy III and was the first to calculate accurately the size of the earth and the obliquity of the earth's axis. He was also an athlete, a poet, a philosopher and an historian. He was called Pentathlus by his friends for his success in five Olympic sports. His enemies called him Beta for they considered him to be second in most fields of learning and first in none. Eratosthenes called himself Philologus, one who loves learning. According to legend, Eratosthenes, after all his accomplishments, ended his life at age 80 by starvation.

In order to determine all the primes less than or equal to the positive integer n using Eratosthenes's sieve, list all the integers from 2 to n. The smallest number, 2, must be prime making it the only even prime and perhaps the oddest prime of all. Every alternate number after 2 must be composite so cross them out. The smallest integer greater than 2 not crossed out, 3, must be prime. Every third number after 3 must be composite and if they have not been crossed out already, cross them out. The next smallest number greater than 3 not crossed out, 5, must be prime. Every fifth number after 5 is composite, if they have not already been crossed out, cross them out. Eratosthenes knew that one of the prime factors of a composite number must be less than or equal to the square root of the number. Thus, we continue the process until the largest prime less than $\sqrt{n}$ is reached. At this point, all composites up to n have been crossed out, only the primes from 2 to n remain. Nicomachus mentions Eratosthenes's method in his *Introduction*, but considers only odd numbers beginning the sieve process with 3.

Example 3.1 Figure 3.1 displays the results from applying the sieve of Eratosthenes to the set of positive integers between 2 and 99. All numbers not crossed out are prime.

In Proposition 32 of Book VII of the *Elements*, Euclid states that every integer greater than unity is divisible by at least one prime. Therefore, every number is prime or has a prime factor. The next result does not explicitly appear in the *Elements*, but it was undoubtedly known to Euclid. The result clearly indicates the importance of prime numbers and is instrumental in illustrating how they form the basic structure of the integers.

Theorem 3.1 *Every integer $n \geqslant 2$ is either prime or a product of primes.*

Proof Using induction, we begin with the case $n = 2$. Since 2 is a prime, the theorem is satisfied. Suppose that the hypothesis is true for all integers between 2 and k. Consider the integer $k + 1$. If $k + 1$ is prime then we are done, if it is not prime then it must factor into a product of two integers r

		2	3	~~4~~	5	~~6~~	7	~~8~~	~~9~~
~~10~~	11	~~12~~	13	~~14~~	~~15~~	~~16~~	17	~~18~~	19
~~20~~	~~21~~	~~22~~	23	~~24~~	~~25~~	~~26~~	~~27~~	~~28~~	29
~~30~~	31	~~32~~	~~33~~	~~34~~	~~35~~	~~36~~	37	~~38~~	~~39~~
~~40~~	41	~~42~~	43	~~44~~	~~45~~	~~46~~	47	~~48~~	~~49~~
~~50~~	~~51~~	~~52~~	53	~~54~~	~~55~~	~~56~~	~~57~~	~~58~~	59
~~60~~	61	~~62~~	~~63~~	~~64~~	~~65~~	~~66~~	67	~~68~~	~~69~~
~~70~~	71	~~72~~	73	~~74~~	~~75~~	~~76~~	~~77~~	~~78~~	79
~~80~~	~~81~~	~~82~~	83	~~84~~	~~85~~	~~86~~	~~87~~	~~88~~	89
~~90~~	~~91~~	~~92~~	~~93~~	~~94~~	~~95~~	~~96~~	97	~~98~~	~~99~~

Figure 3.1.

and s where both r and s are less than k. By the induction hypothesis both r and s must be primes or products of primes. Thus, $k + 1 = r \cdot s$ is a product of primes. Since $k + 1$ is either a prime or a product of primes, the result follows from induction. ■

In a caveat to his readers, Euclid notes that given three positive integers, a, b, c it is not always the case that if a divides the product of b and c, then either a divides b or a divides c. For example, 6 divides the product of 3 and 4, but 6 divides neither 3 nor 4. However, in Proposition 30 of Book VII of the *Elements*, Euclid proves that if a prime divides the product of two integers then it must divide at least one of them.

Theorem 3.2 (Euclid's Lemma) *If p is a prime and p divides ab, then either p divides a or p divides b.*

Proof Suppose that p is prime, p divides ab and p does not divide a. Since p divides ab there exists an integer c such that $pc = ab$. Since p does not divide a, p and a are coprime, so it follows from Theorem 2.7 that there exist integers x and y such that $1 = px + ay$. Hence, $b = b(px) + b(ay) = p(bx) + p(cy) = p(bx + cy)$. Thus, p divides b. ■

With a straightforward inductive argument it can be shown that if a prime p divides the product $m_1 m_2 \cdots m_n$, where each m_i is an integer, then p divides m_i for some i, $1 \leqslant i \leqslant n$. The importance of considering prime divisors becomes more evident with the proof of the next result, concerning a property of primitive Pythagorean triples. Lacking Euclid's Lemma at the time, we assumed it in the proof of Theorem 2.13.

Theorem 3.3 *The Pythagorean triple (x, y, z) is primitive if and only if x, y and z are coprime in pairs.*

Proof If $\gcd(x, y) = 1$, $\gcd(x, z) = 1$, and $\gcd(y, z) = 1$, then x, y and z have no common factor other than 1. Conversely, suppose that (x, y, z) is a primitive Pythagorean triple, $\gcd(x, y) = d > 1$, p is any prime which divides d. Since p divides x, p divides y, and $x^2 + y^2 = z^2$, p divides z^2. Hence, according to Euclid's Lemma, p divides z, contradicting the fact that x, y and z have no common factor other than 1. Similarly, it follows that $\gcd(x, z) = 1$, $\gcd(y, z) = 1$, and the result is established. ■

It is an open question whether or not there are infinitely many primitive Pythagorean triples with the property that the hypotenuse and one of the sides are prime. It is quite possible that the following result was known to Euclid. However, since he had no notation for exponents and could not express a number with an arbitrary number of factors, it was not included in the *Elements*. Nevertheless, it is very similar to Proposition 14 in Book IX. The result was first stated explicitly by Gauss, who included a proof of the result in his doctoral thesis.

Theorem 3.4 (The Fundamental Theorem of Arithmetic) *Except for the arrangement of the factors every positive integer $n > 1$ can be expressed uniquely as a product of primes.*

Proof Let n be the smallest positive integer for which the theorem is false, say $n = p_1 p_2 \cdots p_r = q_1 q_2 \cdots q_s$, where both r and s are greater than 1. If $p_i = q_j$, for some $1 \leqslant i \leqslant r$ and $1 \leqslant j \leqslant s$, then we could divide both sides of the equality by p_i to get two distinct factorizations of n/p_i, contradicting the minimality of n. Hence, the p_i and q_j are distinct. Without loss of generality, let $p_1 < q_1$. Then $m = (q_1 - p_1)q_2 \cdots q_s = (q_1 q_2 \cdots q_s) - (p_1 q_2 q_3 \cdots q_s) = (p_1 p_2 \cdots p_r) - (p_1 q_1 q_2 \cdots q_s) = p_1[(p_2 \cdots p_r) - (q_2 q_3 \cdots q_s)]$. Since p_1 does not divide $(q_1 - p_1)$, we have two distinct factorizations for m, one with p_1 as a factor and one without. Since $m < n$, this contradicts the minimality of n. Therefore, there is no smallest positive integer having two distinct prime factorizations and the theorem is proved. ■

Theorem 3.4 is fundamental in the sense that apart from a rearrangement of factors, it shows that a positive integer can be expressed as a product of primes in just one way. It would not be true if unity were considered to be prime. In addition, the Fundamental Theorem of Arithmetic does not hold if we restrict ourselves, say, to E, the set of even integers, albeit E, like the integers, is closed under the operations of addition and multiplication. The irreducible elements of E consist of all positive integers of the form $2 \cdot (2n + 1)$, where $n \geqslant 1$, hence, 6, 10, and 30 are irreducible in E. Thus, $2 \cdot 30$ and $6 \cdot 10$ are two distinct prime factorizations of 60 in E.

If n is a positive integer which is greater than 1, the canonical representation or prime power decomposition of n is given by $n = \prod_{i=1}^{r} p_i^{\alpha_i} = p_1^{\alpha_1} p_2^{\alpha_2} \cdots p_r^{\alpha_r}$, where $p_1 < p_2 < \cdots < p_r$ are prime and $\alpha_i > 0$, for $i = 1, \ldots, r$. We refer to $p_i^{\alpha_i}$ as the p_i component of n and employ the notation $p^\alpha \| n$ to signify that $p^\alpha | n$, and $p^{\alpha+1} \nmid n$. For example, $2^3 \| 2^3 3^5 7^4$, $3^5 \| 2^3 3^5 7^4$, and $7^4 \| 2^3 3^5 7^4$. If $p^\alpha \| m$, and $p^\beta \| n$, where p is prime and α, β, k, m, and n are positive integers, then $p^{\alpha+\beta} \| mn$.

The canonical notation for a positive integer is very useful in establishing number theoretic results and easing computation. For example, if we relax the conditions on the canonical representation and allow zero exponents with $m = \prod_{i=1}^{r} p_i^{\alpha_i}$ and $n = \prod_{i=1}^{r} p_i^{\beta_i}$ then, since $\max\{x, y\} + \min\{x, y\} = x + y$, the greatest common divisor and least common multiple of m and n are given respectively by $\gcd(m, n) = \prod_{i=1}^{r} p_i^{\gamma_i}$ and $\text{lcm}(m, n) = \prod_{i=1}^{r} p_i^{\delta_i}$, where $\gamma_i = \min\{\alpha_i, \beta_i\}$ and $\delta_i = \max\{\alpha_i, \beta_i\}$, for $i = 1, 2, \ldots, r$. For example, the canonical representation for 749 112 is given by $2^3 \cdot 3 \cdot 7^4 \cdot 13$ and that of 135 828 by $2^2 \cdot 3^2 \cdot 7^3 \cdot 11$. We alter the canonical notation slightly to represent 749 112 by $2^3 \cdot 3 \cdot 7^4 \cdot 11^0 \cdot 13^1$ and 135 828 by $2^2 \cdot 3^2 \cdot 7^3 \cdot 11^1 \cdot 13^0$. Accordingly, $\gcd(749\,112, 135\,828) = 2^2 \cdot 3 \cdot 7^3$ and $\text{lcm}(749\,112, 135\,828) = 2^3 \cdot 3^2 \cdot 7^4 \cdot 11 \cdot 13$.

In 1676, Wallis showed that the length of the period of the decimal expansion of $1/mn$ is the least common multiple of the length of the periods of $1/m$ and $1/n$. Primes play an important role in the decimal expansion of fractions. In particular, for prime denominators p, other than 2 or 5, all decimal expansions of fractions of the form m/p, for $1 \leqslant m < p$, repeat with cycles of the same length. In addition, the product of the number of distinct cycles with this length is $p - 1$. For example, there are five distinct cycles when $p = 11$, namely, $\overline{0.09}$, $\overline{0.18}$, $\overline{0.27}$, $\overline{0.36}$ and $\overline{0.45}$, each of length 2, and $2 \cdot 5 = 10 = 11 - 1$. Another problem that we will return to in Chapter 8 is determining which primes p have the property that $1/p$ has a decimal expansion of period $p - 1$. For example, 7 is such a prime since $\frac{1}{7} = \overline{0.142857}$.

Exercises 3.1

1. Use the sieve of Eratosthenes to determine all the primes from 100 to 250.
2. Charles de Bovilles, Latinized Carolus Bouvellus, a French philosopher and sometime mathematician, published *On Wisdom* in 1511, one of the first geometry texts written in French. Bouvellus claimed that for $n \geqslant 1$ one or both of $6n - 1$ and $6n + 1$ were prime. Show that his conjecture is false.
3. Bouvellus must have realized something was amiss for he soon revised his claim to read that every prime, except 2 and 3, can be expressed in the form $6n \pm 1$, for some natural number n. Show that this conjecture is true.
4. Show that every prime of the form $3k + 1$ can be represented in the form $6m + 1$.

5. In 1556, Tartaglia claimed that the sums $1+2+4$, $1+2+4+8$, $1+2+4+8+16$, etc. are alternately prime and composite. Show that his conjecture is false.
6. Determine the next three numbers and the general pattern in the sequence 4, 6, 9, 10, 14, 15, 21, 22, 25, 26, 33, 34, 35, 38, ...
7. Find the greatest common divisor and least common multiple of m and n if
   (a) $m = 540$ and $n = 3750$,
   (b) $m = 2^3 \cdot 3^2 \cdot 5 \cdot 7 \cdot 11^2$ and $n = 2 \cdot 5^2 \cdot 11^3$.
8. A positive integer is called squarefree if it is not divisible by the square of any prime. What can you deduce about the canonical representation of squarefree numbers?
9. Show that every positive integer greater than 1 is squarefree, square or a product of a squarefree integer and a square.
10. Determine the length of the longest sequence of consecutive squarefree integers.
11. A positive integer n is said to be powerful if $p^2|n$ for every prime divisor p of n. For example, $2^5 \cdot 3^6 \cdot 5^2$ is powerful, but $2^5 \cdot 3 \cdot 5^2$ is not. If $Q(x)$ denotes the number of powerful numbers less than x, determine $Q(100)$.
12. If p is irreducible in E, the set of even integers, and $p|ab$, does it follow that either $p|a$ or $p|b$? Justify your claim.
13. Consider the set $H = \{4n+1: n = 0, 1, 2, 3, \ldots\} = \{1, 5, 9, \ldots\}$. A number in H, other than 1, is called a Hilbert prime if it has no divisors *in* H other than 1 and itself, otherwise it is called a Hilbert composite. H is closed under multiplication. However, factorization in H is not unique since 9, 21, 33, 77 are Hilbert primes and $21 \cdot 33$ and $9 \cdot 77$ are two distinct irreducible factorizations of 693. Find the first 25 Hilbert primes. David Hilbert lectured at Göttingen University from 1892 to 1930. At the 1900 International Congress of Mathematicians in Paris, he challenged mathematicians with 23 problems, several of which remain unsolved.
14. Smith numbers, first defined by Albert Wilanski of Lehigh University, are composite numbers the sum of whose digits are equal to the sum of the digits in an extended prime factorization. For example, 27 is a Smith number since $27 = 3 \cdot 3 \cdot 3$ and $2+7 = 3+3+3$. In addition, $319 = 11 \cdot 29$ is a Smith number since $3+1+9 = 1+1+2+9$. The pair 728 and 729 are consecutive Smith numbers. It is an open question whether there are an infinite number of Smith numbers. Wilanski noted, in 1982, that the largest Smith number he knew of

belonged to his brother-in-law, Dr. Harold Smith, whose phone number was 4 937 775. Show that 4 937 775 is a Smith number.

15. Let $s_p(n, b)$ denote the prime digital sum of the composite integer n expressed in base $b \geqslant 2$. That is, if $n = p_1 p_2 \cdots p_r$, then $s_p(n, b) = \sum_{k=1}^{r} s_p(p_k, b)$, where $s_p(n, 10) = s_p(n)$. For example, $36 = 3 \cdot 3 \cdot 2 \cdot 2$, $3 = 11_2$, and $2 = 10_2$. Hence, $s_p(36, 2) = 2 + 2 + 1 + 1 = 6$. Determine $s_p(n, 2)$, for $1 \leqslant n \leqslant 16$.
16. A positive integer n is called a k-Smith number if $s_p(n) = k \cdot s_d(n)$, where k is also a positive integer. In 1987, Wayne McDaniel used the concept of k-Smith numbers to prove that there exist an infinite number of Smith numbers. Show that 104 is a 2-Smith number.
17. For n a positive integer, the nth Monica set M_n consists of all composite positive integers r for which n divides $s_d(r) - s_p(r)$. Show that if r is a Smith number that r belongs to M_n for all positive integers n.
18. Prove that if m and n are positive integers such that $m|n$, then M_n is a subset of M_m.
19. If $k > 1$ is a positive integer show that the set of k-Smith numbers is a subset of the $(k - 1)$st Monica set.
20. For a positive integer n, the nth Suzanne set S_n consists of all composite positive integers r for which n divides $s_d(r)$ and $s_p(r)$. In 1996, Michael Smith, who named Monica and Suzanne sets after his two cousins Monica and Suzanne Hammer, showed that there are an infinite number of elements in each Monica and Suzanne set. Clearly S_n is a subset of M_n. Show that it is not necessarily the case, however, that M_n is a subset of S_n.
21. Find all primes p such that $17p + 1$ is square.
22. Prove that every number of the form $4m + 3$ must have one prime factor of the form $4k + 3$.
23. Can a number of the form $4m + 1$ have a factor not of the form $4k + 1$? Justify your answer.
24. Prove that $n^4 - 1$ is composite for any positive integer $n > 1$.
25. Prove that if $n > 4$ is composite, then n divides $(n - 1)!$
26. Determine the number of distinct cycles and the length of each cycle, for decimal expansions of numbers of the form $m/13$, with $1 \leqslant m < 13$.
27. In 1968, T.S. Motzkin and E.G. Straus investigated the existence of pairs $\{m, n\}$ such that m and $n + 1$ have the same distinct prime factors and n and $m + 1$ have the same distinct prime factors. Show that $m = 5 \cdot 7$ and $n = 2 \cdot 3^7$ are such numbers.

28. Prove that if p is prime and α, β, m, and n are integers with α and β positive, $p^\alpha \| m$, and $p^\beta \| n$ then $p^{\alpha+\beta} \| mn$.
29. Give a counterexample to show that, in general, if $p^\alpha \| m$ and $p^\alpha \| n$ then $p^\alpha \nparallel (m+n)$, where p is prime, and m, n and α are integers with α positive.
30. Prove that $\sqrt[m]{n}$ is irrational unless n is the mth power of an integer.

3.2 Number theoretic functions

A function whose domain is the set of positive integers is called number theoretic or arithmetic. In many cases, the canonical representation of positive integers can be used to evaluate number theoretic functions. Two very important number theoretic functions are $\tau(n)$, the number of divisors of n, and $\sigma(n)$, the sum of the divisors of n. For convenience, we use the convention that $\sum_{d|n}$ and $\prod_{d|n}$ denote, respectively, the sum and product taken over all the divisors of n. For example, for $n = 12$, $\sum_{d|12} d = 1 + 2 + 3 + 4 + 6 + 12 = 28$ and $\prod_{d|12} d = 1 \cdot 2 \cdot 3 \cdot 4 \cdot 6 \cdot 12 = 1728$. It follows from the definitions of τ and σ that $\sum_{d|n} 1 = \tau(n)$ and $\sum_{d|n} d = \sigma(n)$. For completeness, we define $\tau(1)$ and $\sigma(1)$ to be 1. Unless a positive integer is square, its divisors pair up, hence, $\tau(n)$ is odd if and only if n is square. With the next result, we see how canonical representations can be used to compute number theoretic values.

Theorem 3.5 *If* $n = \prod_{i=1}^{r} p_i^{\alpha_i}$, then $\tau(n) = \prod_{i=1}^{r} (\alpha_i + 1)$.

Proof If $m = \prod_{i=1}^{r} p_i^{\beta_i}$ and $n = \prod_{i=1}^{r} p_i^{\alpha_i}$ then $m|n$ if and only if $0 \leqslant \beta_i \leqslant \alpha_i$, for $i = 1, 2, \ldots, r$. That is, if every p_i component of m is less than or equal to every p_i component of n. Thus, if $m = \prod_{i=1}^{r} p_i^{\beta_i}$ represents any divisor of $n = \prod_{i=1}^{r} p_i^{\alpha_i}$ then there are $\alpha_1 + 1$ choices for β_1, $\alpha_2 + 1$ choices for $\beta_2, \ldots,$ and $\alpha_r + 1$ choices for β_r. From the multiplication principle it follows that there are $(\alpha_1 + 1)(\alpha_2 + 1) \cdots (\alpha_r + 1)$ different choices for the $\beta_1, \beta_2, \ldots, \beta_r$, thus, that many divisors of n. Therefore, $\tau(n) = \prod_{i=1}^{r} (\alpha_i + 1)$. ■

For example, $\tau(13\,608) = \tau(2^3 3^5 7) = (3+1)(5+1)(1+1) = 4 \cdot 6 \cdot 2 = 48$. The history of the tau-function can be traced back to Girolamo Cardano, an Italian mathematician–physician, who noted in 1537 that the product of any k distinct primes has 2^k divisors. Cardano played a major role in popularizing the solution to cubic equations and wrote the first text devoted to the study of probability. Cardano's result was reestablished in 1544 by Michael Stifel and again in 1657 by the Dutch mathematician Frans van Schooten. In 1659 Van Schooten published an influential Latin translation of Descartes' *La Géométrie* that was highly regarded by Isaac Newton. The canonical formula for $\tau(n)$, in Theorem 3.5, is found in the 1719 edition of John Kersey's *Elements of that Mathematical Art Commonly Called Algebra*. Kersey was a London surveyor and highly respected teacher of mathematics. His book was very popular, went through several editions, and was recommended to students at Cambridge.

An equivalent representation for $\tau(n)$, based on the canonical representation for n, appeared in the 1732 edition of Newton's *Universal Arithmetic*. The canonical formula for $\tau(n)$, in Theorem 3.5, also appeared in the 1770 edition of Edward Waring's *Meditationes algebraicae* without justification, as was Waring's nature. Waring, Lucasian Professor of Mathematics at Cambridge University, succeeded Isaac Barrow, Isaac Newton, William Whiston, Nicholas Saunderson, and John Colson in that position. In 1919, Leonard Eugene Dickson, a number theorist at the University of Chicago, introduced the notation $\tau(n)$ to represent the number of divisors of the positive integer n and the notation $\sigma(n)$ to represent the sum of divisors of n.

Given a positive integer $n > 1$, there are infinitely many positive integers m such that $\tau(m) = n$. For example, if p is any prime then $\tau(p^{n-1}) = n$. It is possible for consecutive numbers to have the same number of divisors. For example, $\tau(14) = \tau(15) = 4$, $\tau(44) = \tau(45) = 6$, and $\tau(805) = \tau(806) = 8$. Richard K. Guy, of the University of Calgary, conjectured that $\tau(n) = \tau(n+1)$ for infinitely many positive integers. Three consecutive positive integers may also have the same number of divisors. For example, $\tau(33) = \tau(34) = \tau(35) = 4$ and $\tau(85) = \tau(86) = \tau(87) = 4$. An upper bound for $\tau(n)$ is given by $2\sqrt{n}$.

In 1838, P.G. Dirichlet proved that the average value of $\tau(k)$, $(1/n)\sum_{k=1}^{n}\tau(k)$, is approximately equal to $\ln(n) + 2\cdot\gamma - 1$, where $\ln(n)$ denotes the natural logarithm of n and γ denotes the Euler–Mascheroni constant, $\lim_{n\to\infty}(1 + \frac{1}{2} + \cdots + 1/n - \ln(n)) \approx 0.577\,215\,6 \ldots$. It is an open question as to whether γ is rational or irrational.

The nth harmonic number, denoted by H_n, is defined to be $1 + \frac{1}{2} + \frac{1}{3} + \cdots 1/n$. The Euler–Maclaurin Theorem states that for large values of n, H_n is approximately equal to $\ln(n) + \gamma + 1/2n$. An inductive argument can be used to show that $H_1 + H_2 + \cdots + H_n = (n+1)(H_{n+1} - 1)$. With this result and the Euler–Maclaurin Theorem, it follows that

$$
\begin{aligned}
\ln(n!) &= \ln(1) + \ln(2) + \cdots + \ln(n) \\
&\approx H_1 + H_2 + \cdots + H_n - n \cdot \gamma - \frac{1}{2} H_n \\
&= \left(n + \frac{1}{2}\right) H_n - n - n \cdot \gamma \\
&\approx \left(n + \frac{1}{2}\right)(\ln(n) + \gamma) - n - n \cdot \gamma \\
&\approx \left(n + \frac{1}{2}\right) \ln(n) - n.
\end{aligned}
$$

Hence, $n! \approx \sqrt{n} \cdot n^n \mathrm{e}^{-n}$. In 1730, the Scottish mathematican, James Stirling showed that $\sqrt{2\pi n}(n/\mathrm{e})^n$ gives a much better estimate of $n!$ even for small values of n. For example, 12! is 479 001 600. Stirling's formula yields 475 687 486.476.

A number n with the property that $\tau(n) > \tau(k)$, for all $k < n$, is called highly composite. For example, 2, 4, 6, 12, 24, 36, 48, 60, and 120 are highly composite. Highly composite numbers were studied extensively by Srinivasa Ramanujan and formed the basis of his dissertation at Cambridge. Ramanujan, a phenomenal self-taught Indian number theorist, was working as a clerk in an accounts department in Madras when his genius came to the attention of Gilbert Walker, head of the Indian Meteorological Department, and Mr E.H. Neville, Fellow of Trinity College, Cambridge. Walker was Senior Wrangler at Cambridge in 1889 and Neville was Second Wrangler in 1909. The examination for an honors degree at Cambridge is called the Mathematical Tripos. Up until 1910, the person who ranked first on the Tripos was called the Senior Wrangler. He was followed by the Second Wrangler, and so forth. The person who ranked last was referred to as the Wooden Spoon.

In his teens, Ramanujan independently discovered that if $S(x)$ denotes the number of squarefree positive integers less than or equal to x, then for large values of x, $S(x)$ is approximately equal to $6x/\pi^2$. A correspondence ensued between Ramanujan and the Cambridge mathematician G.H. Hardy. As a consequence, Ramanujan left India and went to England. He spent the period from 1914 to 1919 at Cambridge. Under the guidance of Hardy, Ramanujan published a number of remarkable mathematical results. Between December 1917 and October 1918, he was elected a Fellow of Trinity College, Cambridge, of the Cambridge Philosophical Society, of the Royal Society of London, and a member of the London Mathematical Society. His health deteriorated during his stay in England. He returned to India in 1920 and died later that year at the age of 32.

Let $D(k)$ denote the least positive integer having exactly k divisors. For example, $D(1) = 1$, $D(2) = 2$, $D(3) = 4$, $D(4) = 6$, and $D(5) = 16$. We say that n is minimal if $D(\tau(n)) = n$. All the highly composite numbers studied by Ramanujan are minimal. Normally, if $n = q_1 q_2 \cdots q_k$, where q_i is prime and $q_1 \leqslant q_2 \leqslant \cdots \leqslant q_k$, then $D(n) = 2^{q_1-1}3^{q_2-1} \cdots p_k^{q_k-1}$, where p_k denotes the kth prime. However, exceptions include the cases when $n = 8$, 16, 24, and 32.

In 1829, the German mathematician Carl Gustav Jacob Jacobi [yah KOH bee] investigated properties of the number theoretic function $E(n)$, the excess of the number of divisors of n of the form $4k + 1$ over the number of divisors of n of the form $4k + 3$. For example, the divisors of 105 of the form $4k + 1$ are, 5, 21, and 105, and the divisors of the form $4k + 3$ are 3, 7, 15, and 35. Hence, $E(105) = -1$. Since 2^α has no divisors of the form $4k + 3$ and only one of the form $4k + 1$, $E(2^\alpha) = 1$. If p is prime of the form $4k + 1$, $E(p^\alpha) = \alpha + 1$, and if p is a prime of the form $4k + 3$, $E(p^\alpha) = ((-1)^\alpha + 1)/2$. Jacobi claimed that if $n = 2^\alpha uv$, where each prime factor of u has the form $4k + 1$ and each prime factor of v the form $4k + 3$, then $E(n) = 0$ unless v is square and in that case $E(n) = \tau(u)$. Jacobi made important contributions to the theory of elliptic integrals before dying at age 47, a victim of smallpox.

In 1883, J.W.L. Glaisher [GLAY sure] established Jacobi's conjecture and showed that $E(n) - E(n-1) - E(n-3) + E(n-6) + E(n-10) - \cdots = 0$ or $(-1)^n[((-1)^k(2k+1) - 1)/4]$ depending, respectively, on whether n is not a triangular number or the triangular number $k(k+1)/2$. Glaisher, a Cambridge mathematican, was Senior Wrangler in 1871. He served as president of the London Mathematical Society and the Royal Astronomical Society. In 1901, Leopold Kronecker, the German mathematician who established an analogue to the Fundamental Theorm of Arithmetic for finite Abelian groups in 1858, showed that the mean value for $E(n)$ is approximately $\pi/4$.

In 1638, René Descartes remarked that the sum of the divisors of a prime to a power, say $\sigma(p^r)$, can be expressed as $(p^{r+1} - 1)/(p - 1)$. In 1658, Descartes, John Wallis, and Frenicle investigated properties of the sum of the divisors of a number assuming that if m and n are coprime then $\sigma(m \cdot n) = \sigma(m) \cdot \sigma(n)$. We establish this property in the next section.

Theorem 3.6 *If* $n = \prod_{i=1}^r p_i^{\alpha_i}$, *then the canonical formula for the sum of the divisors of a positive integer is given by*

$$\sigma(n) = \prod_{i=1}^{r}\left(\frac{p_i^{\alpha_i+1} - 1}{p_i - 1}\right).$$

Proof The sum of the divisors of the positive integer $n = p_1^{\alpha_1} p_2^{\alpha_2} \cdots p_r^{\alpha_r}$ can be expressed by the product

$$(1 + p_1 + p_1^2 + \cdots + p_1^{\alpha_1})(1 + p_2 + p_2^2 + \cdots + p_2^{\alpha_2}) \cdots$$
$$(1 + p_r + p_r^2 + \cdots + p_r^{\alpha_r}).$$

Using the formula for the sum of a finite geometric series,

$$1 + x + x^2 + \cdots + x^n = \frac{x^{n+1} - 1}{x - 1},$$

we simplify each of the r sums in the above product to find that the sum of the divisors of n can be expressed canonically as

$$\sigma(n) = \left(\frac{p_1^{\alpha_1+1} - 1}{p_1 - 1}\right)\left(\frac{p_2^{\alpha_2+1} - 1}{p_2 - 1}\right) \cdots \left(\frac{p_r^{\alpha_r+1} - 1}{p_r - 1}\right) = \prod_{i=1}^{r}\frac{p_i^{\alpha_i+1} - 1}{p_i - 1},$$

and the result is established. ■

For example,

$$\sigma(136\,608) = \sigma(2^3 3^5 7) = \left(\frac{2^4 - 1}{2 - 1}\right)\left(\frac{3^6 - 1}{3 - 1}\right)\left(\frac{7^2 - 1}{7 - 1}\right) = 43\,608.$$

The canonical formula for $\sigma(n)$ was first derived by Euler in 1750, who used $\int n$ to denote the sum of the divisors of n. Three years earlier, developing the theory of partitions, Euler derived an intriguing formula to evaluate $\sigma(n)$ involving pentagonal-type numbers, namely,

$$\sigma(n) = \sigma(n-1) + \sigma(n-2) - \sigma(n-5) - \sigma(n-7) + \sigma(n-12)$$
$$+ \sigma(n-15) + \cdots$$
$$+ (-1)^{k+1}\left[\sigma\left(n - \frac{3k^2 - k}{2}\right) + \sigma\left(n - \frac{3k^2 + k}{2}\right)\right] + \cdots,$$

where $\sigma(r) = 0$ if $r < 0$ and $\sigma(0) = n$. The result is elegant, but not very practical. For example, according to the formula, $\sigma(10) = \sigma(9) + \sigma(8) - \sigma(5) - \sigma(3) = 13 + 15 - 6 - 4 = 18$, and $\sigma(15) = \sigma(14) + \sigma(13) - \sigma(10) - \sigma(8) + \sigma(3) + \sigma(0) = 24 + 14 - 18 - 15 + 4 + 15 = 24$.

The function $\sigma_k(n)$ representing the sum of the kth powers of the divisors of n generalizes the number theoretic functions σ and τ since $\sigma_0(n) = \tau(n)$ and $\sigma_1(n) = \sigma(n)$. By definition, $\sigma_k(n) = \sum_{d|n} d^k$, with $\sigma_k(1) = 1$. Hence, $\sigma_2(15) = \sigma_2(3 \cdot 5) = 1^2 + 3^2 + 5^2 + 15^2 = 260$.

Clearly, a positive integer n is prime if and only if $\sigma(n) = n + 1$. In addition, for any positive integer n, $\sigma(n) - 2n$ is never an odd square and since $\sigma(n) > n$, there are only finitely many integers m such that $\sigma(m) = n$. There are 113 solutions to $\sigma(n) = \sigma(n+1)$ when $n < 10^7$, for example, $\sigma(14) = \sigma(15) = 24$ and $\sigma(206) = \sigma(207) = 312$. The Polish mathematician, Wastawa Sierpiński, conjectured that the equation $\sigma(n) = \sigma(n+1)$ is valid for infinitely many positive integers n.

Using our knowledge of harmonic numbers, we can determine an upper bound for $\sigma(n)$. For any positive integer n,

$$\sigma(n) = n \cdot \sum_{d|n} \frac{1}{d} \leqslant n \cdot \sum_{1 \leqslant k \leqslant n} \frac{1}{k} < n(1 + \ln(n)) = n + n \cdot \ln(n) < 2n \cdot \ln(n).$$

In 1972, U. Annapurna showed if $n > 12$ then $\sigma(n) < 6n^{3/2}/\pi^2$. Five years earlier R.L. Duncan had shown that $\sigma(n) < \frac{1}{6}(7n \cdot \omega(n) + 10n)$, where $\omega(n)$ denotes the number of distinct prime factors of n. That is, if $n = \prod_{i=1}^{r} p_i^{\alpha_i}$, $\omega(n) = \sum_{p|n} 1 = r$, with $\omega(1) = 0$. For example, $\omega(164\,640) = \omega(2^5 \cdot 3 \cdot 5 \cdot 7^3) = 1 + 1 + 1 + 1 = 4$.

The number theoretic function $\omega(n)$ has a number of interesting properties. For example, for any positive integer n, $2^{\omega(n)} \leqslant \tau(n) \leqslant n$. In addition, if n is a positive integer then there are $2^{\omega(n)}$ ordered pairs (r, s) such that $\gcd(r, s) = 1$ and $r \cdot s = n$. In general the number of ordered pairs of positive integers (r, s) such that $\operatorname{lcm}(r, s) = n$ is given by $\tau(n^2)$. For example, if $n = 12$, then $\tau(144) = 15$ and the ordered pairs are (1, 12), (2, 12), (3, 12), (3, 4), (4, 3), (4, 6), (4, 12), (6, 4), (6, 12), (12, 1), (12, 2), (12, 3), (12, 4), (12, 6), (12, 12). If r and s are positive integers such that r divides s, then the number of distinct pairs of positive integers x and y such that $\gcd(x, y) = r$ and $\operatorname{lcm}(x, y) = s$ is equal to 2^{k-1} where $k = \omega(s/r)$. For example, if $r = 2$ and $s = 60 = 2^2 \cdot 3 \cdot 5$, then $k = \omega(60/2) = \omega(2 \cdot 3 \cdot 5) = 3$. The four ordered pairs of solutions are (2, 60), (4, 30), (6, 20), and (10, 12). In 1838, P.G. Dirichlet showed that the average value of $\sigma(n)$, $(1/n)\sum_{k=1}^{n}\sigma(k)$, is approximately $\pi^2 n/6$ for large values of n.

Another number theoretic function of interest is the sum of aliquot parts of n, all the divisors of n except n itself, denoted by $s(n)$. Thus, $s(n) = \sigma(n) - n$. If p is prime then $s(p) = 1$. If $n > 1$; then the aliquot sequence generated by $n, a_1, a_2, \ldots,$ is defined recursively such that $a_1 = n$, and $a_{k+1} = s(a_k)$ for $k \geqslant 1$.

A sociable chain or aliquot cycle of length k, for k a positive integer, is an aliquot sequence with $s(a_{k+1}) = a_1$. A number is called sociable if it belongs to a sociable chain of length greater than 2. In 1918, Paul Poulet discovered that 12 496 generates a sociable chain of length 5 and 14 316 generates a sociable chain of length 28. In 1969, Henri Cohen discovered 7 new sociable chains of length 4. Currently, 45 sociable chains, having lengths 4, 5, 6, 8, 9, and 28, are known. In 1975, R.K. Guy and John Selfridge conjectured that infinitely many aliquot sequences never cycle but go off to infinity.

Sarvadaman Chowla studied properties of what is now called Chowla's function. The function, denoted by $s^*(n)$, represents the sum of all the divisors of n except 1 and the number itself. That is, $s^*(n) = \sigma(n) - n - 1 = s(n) - 1$. For example, since

$$\sigma(32\,928) = \sigma(2^5 \cdot 3 \cdot 7^3) = \left(\frac{2^6 - 1}{2 - 1}\right)\left(\frac{3^2 - 1}{3 - 1}\right)\left(\frac{7^4 - 1}{7 - 1}\right)$$
$$= 63 \cdot 4 \cdot 400 = 100\,800,$$

it follows that $s(32\,928) = \sigma(32\,928) - 32\,928 = 67\,872$ and $s^*(32\,928) = \sigma(32\,928) - 32\,928 - 1 = 67\,871$. If p is prime then $s^*(p) = 0$. Several pairs of integers m and n, including 48 and 75, 140 and 195, 1050 and 1925, 1575 and 1648, have the property that $s^*(m) = n$ and $s^*(n) = m$.

The functional value $\Omega(n)$, called the degree of n, represents the number of prime divisors of n counted with multiplicity. That is, if $n = \prod_{i=1}^{r} p_i^{\alpha_i}$, $\Omega(n) = \sum_{i=1}^{r} \alpha_i$, with the convention that $\Omega(1) = 0$. For example, $\Omega(164\,640) = \Omega(2^5 \cdot 3 \cdot 5 \cdot 7^3) = 5 + 1 + 1 + 3 = 10$. The average value of $\Omega(n)$, $(1/n)\sum_{k=1}^{n} \Omega(k)$, is approximately $\ln(\ln(n)) + 1.0346$ for large values of n.

Denote by E_n or O_n the number of positive integers k, $1 \leqslant k \leqslant n$, for which $\Omega(k)$ is even or odd, respectively. In 1919, George Polya conjectured that $O_n \geqslant E_n$, for $n \geqslant 2$. However, in 1958, C.B. Haselgrove showed that there were infinitely many positive integers n for which $O_n < E_n$. In 1966, R.S. Lehman showed that $n = 906\,180\,359$ is the smallest positive integer for which $O_n = E_n - 1$.

In 1657, Fermat challenged Frenicle and Sir Kenelm Digby to find, other than unity, a cube whose sum of divisors is square and a square whose sum of divisors is a cube. Before the existence of high-speed electronic computers, these were formidable problems. Digby was an author, naval commander, diplomat, and bon vivant, who dabbled in mathematics, natural science, and alchemy. His father was executed for his role in the Gunpowder Plot. Digby's elixir, 'powder of sympathy', was purported to heal minor wounds and cure toothaches. Digby passed the problem on to John Wallis who found five solutions to Fermat's first problem, namely the cubes $3^9 \cdot 5^3 \cdot 11^3 \cdot 13^3 \cdot 41^3 \cdot 47^3$, $2^3 \cdot 3^3 \cdot 5^3 \cdot 13^3 \cdot 41^3 \cdot 47^3$, $17^3 \cdot 31^3 \cdot 47^3 \cdot 191^3$, $2^3 \cdot 3^3 \cdot 5^3 \cdot 13^3 \cdot 17^3 \cdot 31^3 \cdot 41^3 \cdot 191^3$, and $3^9 \cdot 5^3 \cdot 11^3 \cdot 13^3 \cdot 17^3 \cdot 31^3 \cdot 41^3 \cdot 191^3$. Wallis countered with the problem of finding two squares other than 16 and 25 whose sums of divisors are equal. Wallis knew of four solutions to the problem, namely 788 544 and 1 214 404, 3 775 249 and 1 232 100, 8 611 097 616 and 11 839 180 864, and 11 839 180 864 and 13 454 840 025. Frenicle found two of the solutions given by Wallis to Fermat's first problem and two solutions to Fermat's second problem, namely the squares $2^4 \cdot 5^2 \cdot 7^2 \cdot 11^2 \cdot 37^2 \cdot 67^2 \cdot 163^2 \cdot 191^2 \cdot 263^2 \cdot 439^2 \cdot 499^2$ and $3^4 \cdot 7^6 \cdot 13^2 \cdot 19^2 \cdot 31^4 \cdot 67^2 \cdot 109^2$. Frenicle submitted no less than 48 solutions to the problem posed by Wallis including the pairs 106 276 and 165 649, 393 129 and 561 001 and 2 280 100 and 3 272 481. Wallis constructed tables of values for $\sigma(n)$ for n a square of a positive integer less than 500 or a cube of a positive integer less than 100. During the period from 1915 to 1917, A. Géradin found 11 new solutions to Fermat's first problem.

Exercises 3.2

1. Show that for any positive integer n
$$\sum_{d|n} d = \sum_{d|n} \frac{n}{d}.$$
2. Determine the number of divisors and sum of the divisors of (a) 122, (b) 1424, (c) 736, (d) 31, (e) $2^3 \cdot 3^5 \cdot 7^2 \cdot 11$.
3. Show that $\tau(242) = \tau(243) = \tau(244) = \tau(245)$.
4. Show that $\tau(n) = \tau(n+1) = \tau(n+2) = \tau(n+3) = \tau(n+4)$ if $n =$ 40 311.
5. In 1537, Girolamo Cardano claimed that if $n = \prod_{i=1}^{r} p_i = p_1 p_2 \cdots p_r$, where $p_1, p_2, \ldots, p_r$ are distinct primes, then $\tau(n) - 1 = 1 + 2 + 2^2 + 2^3 + \cdots + 2^{r-1}$. Prove his conjecture true.
6. Prove that $2^{\omega(n)} \leqslant \tau(n) < 2\sqrt{n}$, where $n > 1$ is any positive integer.
7. Show that $\prod_{d|n} d = n^{\tau(n)/2}$, for any positive integer n.
8. Determine the canonical structure of all positive integers having the property $\prod_{d|n} d = n^2$.
9. Determine the canonical structure of all positive integers having the property
$$\prod_{\substack{d|n \\ d \neq n}} d = n^2.$$
10. Use the Israilov–Allikov and Annapurna formulas to determine upper bounds for $\tau(n)$ and $\sigma(n)$ when $n = 1\,000\,000$.
11. Use Duncan's formula to obtain an upper bound for $\sigma(10^6)$.

12. Determine the average value of $\tau(n)$ for $1 \leqslant n \leqslant m = 25$, 50 and 100. Compare your results with Dirichlet estimates.
13. Determine the first five harmonic numbers.
14. For any positive integer n, prove that

$$H_1 + H_2 + \cdots + H_n = (n+1)(H_{n+1} - 1)$$

where H_n denotes the nth harmonic number.
15. Use Stirling's formula to estimate 16!
16. Show that 12 and 24 are highly composite.
17. In 1644, Mersenne asked his fellow correspondents to find a number with 60 divisors. Find $D(60)$, the smallest positive integer with 60 divisors.
18. Determine $D(n)$ when $n = 8$, 16, 24, and 32.
19. Determine $E(512)$, $E(24\,137\,569)$, $E(750)$, $E(2401)$. Use Glaisher's formula to determine $E(19)$.
20. Determine the average value of $E(n)$ for $1 \leqslant n \leqslant 25$. Compare it with $\pi/4$.
21. According to Liouville's formula, $(\sum_{d|n}\tau(d))^2 = \sum_{d|n}\tau^3(d)$. Check the validity of the formula for $n = 7$, 12, and 24.
22. Plato noted that 24 was the smallest positive integer equal to the sum of the divisors of three distinct natural numbers. That is, $n = 24$ is the smallest positive number such that the equation $\sigma(x) = n$ has exactly three solutions for x. What are the three solutions?
23. Use Euler's recursive formula for $\sigma(n)$ to show that $\sigma(36) = 91$.
24. Show that $\sigma(n)$ is odd if and only if n is a square or twice a square.
25. Determine the average value of $\sigma(n)$ for $1 \leqslant n \leqslant m$, for $m = 25$, 50 and 100. Compare your results with Dirichlet's estimate.
26. Determine the first 25 terms of the aliquot sequence generated by 276.
27. Determine all ordered pairs (r, s) such that $\text{lcm}(r, s) = 36$.
28. Determine the sociable chain of length 5 beginning with $n = 12\,496$.
29. Determine the terms in the social chain that begins with $n = 2\,115\,324$.
30. Show $\sigma^*(48) = 75$ and $\sigma^*(75) = 48$.
31. The Chowla sequence generated by n, denoted by $b_1, b_2, \ldots,$ is defined recursively as follows: $b_1 = n$, and $b_{k+1} = s^*(b_k)$ for $k \geqslant 1$. Determine the Chowla sequence generated by 36.
32. Calculate the average value of $\Omega(n)$ for $1 \leqslant n \leqslant 50$. Compare it with $\ln(\ln(50)) + 1.0346$.
33. Show that the sum of divisors of the cube $3^9 \cdot 5^3 \cdot 11^3 \cdot 13^3 \cdot 41^3 \cdot 47^3$ is a square. [Wallis]

34. Show that the sum of divisors of the square $2^4 \cdot 5^2 \cdot 7^2 \cdot 11^2 \cdot 37^2 \cdot 67^2 \cdot 163^2 \cdot 191^2 \cdot 263^2 \cdot 439^2 \cdot 499^2$ is a cube. [Frenicle]
35. Show that the sums of the divisors of 326^2 and 407^2 are equal. [Frenicle]
36. Show that 17, 18, 26, and 27 have the property that they equal the sum of the digits of their cubes.
37. Show that 22, 25, 28, and 36 have the property that they equal the sum of the digits of their fourth powers.
38. Show that 2 divides $[\sigma(n) - \tau(m)]$ for all positive integers n where m is the largest odd divisor of n.
39. If $n = \prod_{i=1}^{r} p_i^{\alpha_i}$ then prove that

$$\sigma_k(n) = \prod_{i=1}^{r} \left(\frac{p_i^{k(\alpha_i+1)}}{p_i - 1} \right).$$

40. Show that $\sum_{d|n}(1/d^2) = \sigma_2(n)/n^2$.
41. A number theoretic functions f is called additive if $f(m \cdot n) = f(m) + f(n)$ whenever $\gcd(m, n) = 1$, Show that ω, the number of distinct prime factors of n, is additive.
42. A number theoretic function f is called completely additive if $f(m \cdot n) = f(m) + f(n)$ for all positive integers m and n. Show that Ω, the degree function of n, is completely additive.
43. A number theoretic function f is called strongly additive if for all primes p, $f(p^\alpha) = f(p)$, where $\alpha \geqslant 1$. prove that ω is strongly additive.
44. Determine all positive integers that are divisible by 12 and have 14 divisors.
45. If $n = \prod_{i=1}^{r} p_i^{\alpha_i}$ is the canonical representation for n, let $\psi(n) = \alpha_1 p_1 + \alpha_2 p_2 + \cdots + \alpha_r p_r + 1$, with $\psi(1) = 1$. Define the psi-sequence, $a_1, a_2, \ldots,$ for n as follows: $a_1 = n$ and $a_k = \psi(a_{k-1})$ for $k > 1$. It is an open question whether for any positive integer greater than 6, the psi-sequence for that integer eventually contains the repeating pattern 7, 8, 7, 8, 7, 8, ... Prove that if $n > 6$ then $\psi(n) > 6$, and if $n > 8$ is composite then $\psi(n) \leqslant n - 2$.

3.3 Multiplicative functions

A number theoretic function f is said to be multiplicative if $f(m \cdot n) = f(m)f(n)$, where m and n are coprime. A number theoretic function f is said to be completely multiplicative if $f(m \cdot n) = f(m)f(n)$ for all positive integers m and n in the domain of f. By definition, every completely multiplicative number theoretic function is multiplicative.

Theorem 3.7 *If f is completely multiplicative and not the zero function, then $f(1) = 1$.*

Proof If f is not the zero function, then there exists a positive integer k such that $f(k) \neq 0$. Hence, $f(k) = f(k \cdot 1) = f(k)f(1)$. Dividing both sides by $f(k)$, we obtain $f(1) = 1$. ■

The next result illustrates the importance of multiplicative functions and shows that they are completely determined by their values on primes raised to powers.

Theorem 3.8 *Let $n = \prod_{i=1}^{r} p_i^{\alpha_i}$ be the canonical representation for n and let f be a multiplicative function; then $f(n) = \prod_{i=1}^{r} f(p_i^{\alpha_i})$.*

Proof Suppose that f is a multiplicative function and $\prod_{i=1}^{r} p_i^{\alpha_i}$ is the canonical representation of n. If $r = 1$, we have the identity, $f(p_i^{\alpha_1}) = f(p_i^{\alpha_1})$. Assume that the representation is valid whenever n has k or fewer distinct prime factors, and consider $n = \prod_{i=1}^{k+1} p_i^{\alpha_i}$. Since $\prod_{i=1}^{k} p_i^{\alpha_i}$ and $p_{k+1}^{\alpha_{k+1}}$ are relatively prime and f is multiplicative, we have

$$f(n) = f\left(\prod_{i=1}^{k+1} p_i^{\alpha_i}\right) = f\left(\prod_{i=1}^{k} p_i^{\alpha_i} \cdot p_{k+1}^{\alpha_{k+1}}\right) = f\left(\prod_{i=1}^{k} p_i^{\alpha_i}\right) \cdot f\left(p_i^{\alpha_{k+1}}\right)$$

$$= \prod_{i=1}^{k} (f(p_i^{\alpha_i})) \cdot f(p_{k+1}^{\alpha_{k+1}}) = \prod_{i=1}^{k+1} f(p_i^{\alpha_i}). \quad ■$$

It follows immediately from Theorem 3.8 that if f is a completely multiplicative function and $\prod_{i=1}^{r} p_i^{\alpha_i}$ is the canonical representation for n, then $f(n) = \prod_{i=1}^{r} [f(p_i)]^{\alpha_i}$. Thus, completely multiplicative functions are strictly determined when their values are known for primes. For example, if f is a completely multiplicative function, $f(2) = a$, $f(3) = b$, and $f(5) = c$, then $f(360) = f(2^3 \cdot 3^2 \cdot 5) = a^3 b^2 c$. There are several basic operations on functions in which the multiplicativity of the functions is preserved as shown in the next two results.

Theorem 3.9 *If f and g are multiplicative then so are $F = f \cdot g$ and $G = f/g$, the latter being true whenever g is not zero.*

Proof If m and n are coprime, then $F(mn) = f(mn) \cdot g(mn) = [f(m) \cdot$

$f(n)][g(m) \cdot g(n)] = [f(m) \cdot g(m)][f(n) \cdot g(n)] = F(m) \cdot F(n)$. A similar argument establishes the multiplicativity of $G = f/g$. ■

The Dirichlet product of two number theoretic functions f and g, denoted by $f^* g$, is defined as $\sum_{d|n} f(d)g(n/d)$. That is, $(f^* g)(n) = \sum_{rs=n} f(r)g(s)$. Hence, $(f^* g)(n) = (g^* f)(n)$, for positive integers n. The next result shows that if two functions are multiplicative then so is their Dirichlet product.

Theorem 3.10 *If f and g are multiplicative then so is $F(n) = \sum_{d|n} f(d)g(n/d)$.*

Proof If m and n are coprime, then $d|mn$ if and only if $d = d_1 d_2$, where $d_1|m$ and $d_2|m$, $\gcd(d_1, d_2) = 1$, and $\gcd(m/d_1, n/d_2) = 1$. Therefore,

$$\begin{aligned} F(mn) &= \sum_{d|mn} f(d) g\left(\frac{mn}{d}\right) = \sum_{d_1|m} \sum_{d_2|n} f(d_1 d_2) g\left(\frac{mn}{d_1 d_2}\right) \\ &= \sum_{d_1|m} \sum_{d_2|n} f(d_1) f(d_2) g\left(\frac{m}{d_1}\right) g\left(\frac{n}{d_2}\right) \\ &= \left[\sum_{d_1|m} f(d_1) g\left(\frac{m}{d_1}\right)\right] \left[\sum_{d_2|n} f(d_2) g\left(\frac{n}{d_2}\right)\right] = F(m)F(n). \quad ■ \end{aligned}$$

For any number theoretic function f, $\sum_{d|n} f(d) = \sum_{d|n} f(n/d)$. If we let g be the multiplicative function $g(n) = 1$ for any positive integer n, in Theorem 3.10, it follows that if f is multiplicative so is $F(n) = \sum_{d|n} f(d)$. In particular, the constant function $f(n) = 1$ and the identity function $f(n) = n$ are multiplicative. Hence, since $\tau(n) = \sum_{d|n} 1$ and $\sigma(n) = \sum_{d|n} n$, we have established the following result.

Theorem 3.11 *The number theoretic functions τ and σ are multiplicative.*

Example 3.2 Consider the multiplicative function $f(n) = n^k$, where k is a fixed positive integer. It follows from Theorem 3.10, with $f(n) = n^k$ and $g(n) = 1$, that the sum of the kth powers of the divisors of n, $\sigma_k(n) = \sum_{d|n} d^k$, is multiplicative. In addition, $\sigma_k(p^\alpha) = 1^k + p^k + p^{2k} + \cdots + p^{\alpha k} = (p^{k(\alpha+1)} - 1)/(p^k - 1)$. Therefore, if $n = \prod_{i=1}^r p_i^{\alpha_i}$,

$$\sigma_k(n) = \prod_{i=1}^r \left(\frac{p_i^{k(\alpha_i+1)} - 1}{p_i^k - 1} \right).$$

For any positive integer n, define the Möbius function, $\mu(n)$, as follows:

$$\mu(n) = \begin{cases} 1 & \text{if } n = 1, \\ (-1)^r & \text{if } n = p_1 p_2 \cdots p_r, \text{ is the} \\ & \text{product of } r \text{ distinct primes,} \\ 0 & \text{otherwise.} \end{cases}$$

For example, $\mu(42) = \mu(2 \cdot 3 \cdot 7) = (-1)^3 = -1$, $\mu(2805) = \mu(3 \cdot 5 \cdot 11 \cdot 17) = (-1)^4 = 1$, and $\mu(126) = \mu(2 \cdot 3^2 \cdot 7) = 0$. It is straightforward and left as an exercise to show that the Möbius function is multiplicative. Its properties were first investigated implicitly by Euler in 1748 and in 1832 by August Ferdinand Möbius, a professor of astronomy at the University of Leipzig, albeit neither used μ to denote the Möbius function. The symbol μ to denote the function was introduced by Frantz Mertens in 1874. In 1897, Mertens conjectured that, for all positive integers, $|\sum_{k=1}^{n} \mu(k)| < \sqrt{n}$. The conjecture has been verified for all $n < 10^9$. In 1984, Andrew Odlyzko and Herman te Riele proved that Merten's conjecture must be false for some value of $n \leqslant 3.21 \times 10^{64}$. The Möbius function has a number of useful properties. For instance, the average value of μ, $\sum_{n=1}^{\infty}(\mu(n)/n)$, is zero. In addition,

$$\sum_{n=1}^{\infty} \frac{\mu(n)}{n^2} = \frac{6}{\pi^2}.$$

Theorem 3.12 *For any positive integer n, if $\nu(n) = \sum_{d|n} \mu(d)$, then $\nu(1) = 1$, $\nu(n) = 0$ for other n.*

Proof If $n = 1$, then $\nu(1) = \sum_{d|n} \mu(n) = \mu(1) = 1$. If $n > 1$, since $\nu(n)$ is multiplicative, we need only evaluate ν on primes to powers. In addition, if p is prime, $\nu(p^\alpha) = \sum_{d|p^\alpha} \mu(d) = \mu(1) + \mu(p) + \mu(p^2) + \cdots + \mu(p^\alpha) = 1 + (-1) + 0 + \cdots + 0 = 0$. Thus, $\nu(n) = 0$ for any positive integer n greater than 1. ■

Theorem 3.13 (Möbius inversion formula) *If f is a number theoretic function and $F(n) = \sum_{d|n} f(d)$, then $f(n) = \sum_{d|n} \mu(d) F(n/d)$.*

Proof Suppose that f is a number theoretic function and $F(n) = \sum_{d|n} f(d)$. We have

$$\sum_{d|n}\mu(d)F\left(\frac{n}{d}\right) = \sum_{d|n}\mu(d)\sum_{a|n/d}f(a) = \sum_{d|n}\sum_{a|n/d}\mu(d)f(a)$$
$$= \sum_{a|n}\sum_{d|n/a}f(a)\mu(d) = \sum_{a|n}f(a)\sum_{d|n/a}\mu(d)$$
$$= f(n)\cdot 1 = f(n).$$

The switch of summands in the third equality is valid since d divides n and a divides n/d if and only if a divides n and d divides n/a. ■

From Theorems 3.10 and 3.13 and the fact that the Möbius function is multiplicative, we obtain the following result.

Corollary *If F is multiplicative and $F(n) = \sum_{d|n} f(d)$, then f is multiplicative.*

Exercises 3.3

1. If f is completely multiplicative and $n|m$, then show that
$$f\left(\frac{m}{n}\right) = \frac{f(m)}{f(n)}.$$
2. If k is a fixed positive integer, then show that $f(n) = n^k$ is completely multiplicative.
3. For any positive integer n, let $f(n) = c^{g(n)}$ with $c > 0$. Show that f is (completely) multiplicative if and only if g is (completely) additive.
4. Let $f(n) = k^{\omega(n)}$, where k is a fixed positive integer and $\omega(n)$ denotes the number of distinct prime divisors of n. Show that f is multiplicative but not completely multiplicative.
5. The Liouville lambda-function, λ, is defined as follows: $\lambda(1) = 1$ and $\lambda(n) = (-1)^{\Omega(n)}$ if $n > 1$, where Ω represents the degree function. Show that λ is multiplicative. Joseph Liouville [LYOU vill] published over 400 mathematical papers, edited the *Journal de mathématiques pures at appliquées* for 40 years. He also edited and published the works of short lived mathematical prodigy Evariste Galois.
6. For any positive integer n, let $F(n) = \sum_{d|n}\lambda(d)$, where λ represents the Liouville lambda-function. Determine the value of $F(n)$ when n is square and when n is not square.
7. If $n = \prod_{i=1}^{r} p_i^{\alpha_i}$ show that $F(n) = \sum_{d|n}\mu(d)\lambda(d) = 2^r$, for $n > 1$.
8. Let $\tau_e(n)$ denote the number of positive even divisors of the positive integer n, and let $\sigma_e = \sum_{d_e|n} d_e$, where d_e runs through the even

divisors of n. Let $\tau_o(n)$ denote the number of positive odd divisors of the positive integer n, and let $\sigma_o(n) = \sum_{d_o|n} d_o$ where d_o runs through the positive odd divisors of n. Evaluate $\tau_o(n)$, $\sigma_o(n)$, $\tau_e(n)$, and $\sigma_e(n)$, for $1 \leqslant n \leqslant 10$.

9. Show by counterexample that neither τ_e nor σ_e is multiplicative or completely multiplicative.
10. Show that τ_o and σ_o are multiplicative functions which are not completely multiplicative.
11. Prove that the Möbius function, $\mu(n)$, is multiplicative.
12. Prove that $\mu(n)\mu(n+1)\mu(n+2)\mu(n+3) = 0$, for any positive integer n.
13. Evaluate $\sum_{k=1}^{\infty} \mu(k!)$.
14. Find a positive integer n such that $\mu(n) + \mu(n+1) + \mu(n+2) = 3$.
15. Show that $\sum_{d|n} |\mu(d)| = 2^{\omega(n)}$, for all positive integers n.
16. Show that $\sum_{d|n} \mu(d)\tau(n/d) = 1$, for any positive integer n.
17. If n is an even integer, show that $\sum_{d|n} \mu(d)\sigma(d) = n$.
18. If $n = \prod_{i=1}^{r} p_i^{\alpha_i}$, with $\alpha_i \geqslant 1$, for $i = 1, \ldots, r$, show that $\sum_{d|n} \mu(d)\tau(d) = (-1)^r = (-1)^{\omega(n)}$.
19. Determine $f(n)$ if $\sum_{d|n} f(d) = 1/n$.
20. For any positive integer n, show that $n = \prod_{d|n} d^{\tau(d)\mu(n/d)/2}$.
21. Let $n = \prod_{i=1}^{r} p_i^{\alpha_i}$ and $\alpha(n) = \sum_{d|n} (\omega(d)/\tau(n))$, show that
$$\alpha(n) = \sum_{i=1}^{r} \frac{\alpha_i}{\alpha_i + 1}.$$
22. Von Mangolt's function, Λ, is defined on the positive integers as follows: $\Lambda(n) = \ln(n)$, if $n = p^\alpha$, and 0 otherwise, where p is prime and α a positive integer. Prove that $\sum_{d|n} \Lambda(d) = \ln(n)$.
23. For any positive integer n, prove that $\Lambda(n) = -\sum_{d|n} \mu(d)\ln(d)$.

3.4 Factoring

Devising an efficient technique to determine whether a large positive integer is prime or composite and if composite to find its prime factorization has been an ambitious goal of number crunchers for centuries. Primality tests are criteria used to determine whether or not a positive integer is prime. If a number passes a primality test then it may be prime. If it passes several primality tests it is more likely to be prime. However, if it fails any primality test then it is not prime. Brute force is reliable but not very efficient in determining whether or not a number is prime. The process of determining whether a number is divisible by any positive integers less than or equal to its square root is a very consuming process indeed. For example, if we wanted to determine if $2^{127} - 1$ is prime and estimated that only 10 percent, 1.3×10^{18}, of the numbers less than $\sqrt{2^{127} - 1}$ were prime, then at the rate of checking 10^9 prime factors a second it would take a high-speed computer 41 years to check all the prime factors of $2^{127} - 1$ that are less than $\sqrt{2^{127} - 1}$. We would find out that none of them divided $2^{127} - 1$.

Factoring a very large positive integer is a difficult problem. No practical factor algorithm currently exists. In Chapter 5, we discuss more elegant and sophisticated primality tests, including Fermat's Little Theorem and Wilson's Theorem. However, Monte Carlo methods, which employ statistical techniques to test for the primality of very large numbers, are beyond the scope of this book.

In 1202, Fibonacci's *Book of Calculations* contained a list of all the primes and composite natural numbers less than or equal to 100. Pietro Cataldi's *Treatise on Perfect Numbers* published in Bologna in 1603 contains factors of all positive integers less than 750. In 1657, Frans van Schooten listed all the primes up to 9929. In 1659, the first extensive factor tables were constructed and published by Johann Heinrich Rahn, Latinized Rohnius, in his *Algebra*. Rahn included all factors of the numbers from 1 to 24 000, omitting from the tables all multiples of 2 and 5. Rahn, who was a student of John Pell's at Zurich, introduced the symbol '÷' to denote division. In 1668, Thomas Brancker determined the least factor, greater than 1, of each integer less than 10^5. Johann Lambert, the first to show that π was irrational, published an extensive table of least factors of the integers up to 102 000 in 1770.

Others have not been so fortunate. In 1776, Antonio Felkel, a Viennese schoolteacher, constructed factor tables for the first 408 000 positive integers. The tables were published at the expense of the Austrian Imperial Treasury but, because of the disappointing number of subscribers, the Treasury confiscated all but a few copies and used the paper for cartridges in a war against the Turks, a dubious mathematical application to warfare at best. In 1856, A.L. Crelle, determined the first six million primes and, in 1861, Zacharias Dase extended Crelle's table to include the first nine million primes. Crelle founded and, for many years, edited and published the prestigious *Journal für die reine und angewandte Mathematik*.

In 1863, after 22 years of effort to complete the task, J.P. Kulik, a professor at the University of Prague, published factor tables that filled several volumes. His tables included the factors, except for 2, 3, and 5, of the first 100 million positive integers. He donated his work to the library at the University of Prague, but unfortunately, through someone's negligence, the second volume, the factorizations of the integers from 12 642 600 to 22 852 800, was lost. In 1910, D.H. Lehmer published factor tables for the integers up to 10 million. Lehmer worked on a long table equipped with rollers at each end. For small primes, he made paper stencils with holes through which he recorded multiples.

Fermat devised a number of ingenious methods to factor integers. We know of his work chiefly through his correspondence with Marin Mersenne, a Franciscan friar, number theory enthusiast and philosopher who corresponded with a number of mathematicians and scientists including Galileo and Torricelli. Mersenne was the leader of a group that met regularly in Paris in the 1630s to discuss scientific topics. He once asked Fermat whether he thought that 100 895 598 269 was prime. After a short period, Fermat replied that it was not and, in fact, it was the product of 898 423 and 112 303.

The basis for one of Fermat's factoring methods depends on the ability to write the integer to be factored as the difference of two integral squares. In this case, $100\,895\,598\,269 = 505\,363^2 - 393\,060^2 = (505\,363 + 393\,060)(505\,363 - 393\,060)$. Fermat assumed that the integer n to be factored was odd, hence its two factors u and v must also be odd. If $n = uv = x^2 - y^2 = (x + y)(x - y)$, $u = x + y$, and $v = x - y$, then $x = (u + v)/2$ and $y = (u - v)/2$. Fermat let k be the least integer for which $k^2 > n$ and formed the sequence

$$\begin{gathered} k^2 - n, \\ (k+1)^2 - n, \\ (k+2)^2 - n, \\ \ldots, \end{gathered}$$

until one of the terms, say $(k + m)^2 - n$, was a perfect square, which for many numbers may never be the case. He then let $(k + m)^2 - n = y^2$, so $x = k + m$ and $y = \sqrt{(k + m)^2 - n}$. Thus, a factorization of n is given by $n = (x - y)(x + y)$. For example, if $n = 931$, then $k = 31$ is the least integer such that $k^2 > 931$. We have

$$\begin{aligned} 31^2 - 931 &= 30, \\ 32^2 - 931 &= 93, \\ 33^2 - 931 &= 158, \\ 34^2 - 931 &= 225 = 15^2 = y^2. \end{aligned}$$

Hence, $y = 15$, $m = 3$, $x = k + m = 31 + 3 = 34$, and $931 = (34 -$

15)(34 + 15) = 19 · 49. Nevertheless, it is unlikely that Fermat used this method to factor 100 895 598 269 for he would have had to perform 75 000 iterations to arrive at his factorization.

In 1641, Frenicle asked Fermat if he could factor a number which can be written as the sum of two squares in two different ways. We do not have Fermat's answer but, in 1745, Euler showed that if $n = a^2 + b^2 = c^2 + d^2$, that is if n can be written as the sum of two squares in two distinct ways, then

$$n = \frac{[(a-c)^2 + (b-d)^2][(a+c)^2 + (b-d)^2]}{4(b-d)^2}.$$

For example, since $2501 = 50^2 + 1^2 = 49^2 + 10^2$, we have $a = 50$, $b = 1$, $c = 49$, and $d = 10$, hence

$$2501 = \frac{(1^2 + 9^2)(99^2 + 9^2)}{4 \cdot 9^2} = \frac{82 \cdot 9882}{4 \cdot 81} = \left(\frac{82}{2}\right)\left(\frac{9882}{2 \cdot 81}\right) = 41 \cdot 61.$$

In order to determine whether or not a very large number was prime, Euler used 65 numbers ranging from 1 to 1848 which he called *numeri idonei* (appropriate numbers) and are now sometimes referred to as convenient numbers. They had the property that if ab was one of the *numeri idonei*, $n = ax^2 + by^2$ uniquely, and $\gcd(ax, by) = 1$, then $n = p$, $2p$, or 2^k, where p is prime and k a positive integer. For example, using 57, one of the *numeri idonei*, Euler discovered the unique representation $1\,000\,003 = 19 \cdot 8^2 + 3 \cdot 577^2$, with $57 = 19 \cdot 3$ and $(19 \cdot 8, 3 \cdot 577) = 1$, hence, 1 000 003 is prime. In 1939, H.A. Heilbronn and S. Chowla showed that there were finitely many *numeri idonei*.

Exercises 3.4

1. Use Fermat's method to show that 12 971 is composite.
2. Use Euler's method to show that the following numbers are composite: (a) 493, and (b) $37\,673 = 187^2 + 52^2 = 173^2 + 88^2$.
3. Euler showed that if $N = a^2 + kb^2 = c^2 + kd^2$ then a factorization of N is given by $N = (km^2 + n^2)(kr^2 + s^2)/4$, where $a + c = kmr$, $a - c = ns$, $d + b = ms$, and $d - b = nr$. Show algebraically that the method is valid.
4. Use the factorization technique outlined in the previous exercise to factor 34 889 given that $34\,889 = 157^2 + (10 \cdot 32^2) = 143^2 + (10 \cdot 38^2)$.
5. Show that if the smallest prime factor p of n is greater than $n^{1/3}$, then the other factor of n must be prime.

6. Show that 2 027 651 281 is composite.

3.5 The greatest integer function

If x is any real number, then the greatest integer not greater than x, or integral part of x, denoted by $[\![x]\!]$, is the unique integer $[\![x]\!]$ such that $[\![x]\!] \leqslant x \leqslant [\![x]\!] + 1$. Equivalently, $[\![x]\!]$ is the integer such that $x - 1 < [\![x]\!] \leqslant x$. For example, $[\![2.5]\!] = 2$; $[\![10.1]\!] = 10$; $[\![0.4]\!] = 0$; $[\![-3.7]\!] = -4$.

Theorem 3.14 *If n is an integer and x any real number then $[\![x + n]\!] = [\![x]\!] + n$.*

Proof Since $x - 1 < [\![x]\!] \leqslant x$ it follows that $-x \leqslant -[\![x]\!] < -x + 1$. Combining this inequality with $x + n - 1 < [\![x + n]\!] \leqslant x + n$, we obtain $n - 1 < [\![x + n]\!] - [\![x]\!] < n + 1$. Hence, $[\![x + n]\!] - [\![x]\!] = n$. ∎

The greatest integer function has a number of useful properties. For instance, if a and b are integers with $0 < b \leqslant a$, then $[\![a/b]\!]$ is the number of positive integer multiples of b not exceeding a. That is, if $a = bq + r$, where $0 \leqslant r < q$, then $q = [\![a/b]\!]$. For example, there are $[\![3000/11]\!] = 272$ positive integers less than or equal to 3000 which are divisible by 11. In addition, if α and β are real numbers, with $\alpha > \beta$, then $[\![\alpha]\!] - [\![\beta]\!]$ represents the number of integers n such that $\beta < n \leqslant \alpha$. Furthermore, if $10^{k-1} \leqslant n < 10^k$, then the number of digits of n to the base b is given by $[\![\log_b(n)]\!] + 1$. For example, the number 3^{54} has 26 digits since $[\![\log(3^{54})]\!] + 1 = [\![54 \cdot \log(3)]\!] + 1 = [\![54 \cdot (0.477\,121\,3)]\!] + 1 = [\![25 \cdot 764\,55]\!] + 1 = 25 + 1 = 26$.

A point (x, y) in the Cartesian plane is called a lattice point if both coordinates x and y are integers. The greatest integer function can be used to determine the number of lattice points in a bounded region. In particular, if $y = f(x)$ is a nonnegative function whose domain is the closed interval $a \leqslant x \leqslant b$, where both a and b are integers and S denotes the region in the Cartesian plane consisting of all lattice points (x, y) for which $a \leqslant x \leqslant b$ and $0 < y \leqslant f(x)$, then the number of lattice points in the region S is given by $\sum_{n=a}^{b} [\![f(n)]\!]$.

Adrien Marie Legendre's *Théorie des nombres*, published in 1808, contains a wealth of number theoretic results. The book includes discussions of a number of topics that we will soon encounter including the Prime Number Theorem, the quadratic reciprocity law, and quadratic forms. It includes a nearly complete proof of Fermat's Last Theorem for the case when $n = 5$. In addition, Legendre used the greatest integer function to devise a method for determining the power of prime exponents in the canonical representation of factorials.

Theorem 3.15 (Legendre's Theorem) *If n is a positive integer and p is a prime such that p divides n then p appears in the canonical representation of $n!$ with exponent e_p, where $e_p = \sum_{k=1}^{\infty} [\![n/p^k]\!]$.*

Proof For a given integer k, the multiples of p^k that do not exceed n are $p^k, 2p^k, \ldots, qp^k$, where q is the largest integer such that $qp^k \leqslant n$. That is, q, the largest integer not exceeding n/p^k, equals $[\![n/p^k]\!]$. Thus, $[\![n/p^k]\!]$ is the number of positive multiples of p^k that do not exceed n. If $1 \leqslant m \leqslant n$ and $m = qp^k$, with $\gcd(p, q) = 1$ and $0 \leqslant k \leqslant r$, then m contributes exactly k to the total exponent e_p with which p appears in the canonical representation of $n!$ Moreover, m is counted precisely k times in the sum $[\![n/p]\!] + [\![n/p^2]\!] + [\![n/p^3]\!] + \cdots$, once as a multiple of p, once as a multiple of $p^2, \ldots,$ once as a multiple of p^k, and no more. If $k = 0$, then m is not counted in the sum. Therefore, $\sum_{k=1}^{\infty} [\![n/p^k]\!]$ equals the exponent of p in the canonical representation of $n!$ ■

Corollary *If $n = \prod_{i=1}^{r} p_i^{\alpha_i}$, then $n! = \prod_{i=1}^{r} p_i^{e_{p_i}}$.*

For example,

$$
\begin{array}{rcccccccccccccccccc}
16! & = & 1\cdot & 2\cdot & 3\cdot & 4\cdot & 5\cdot & 6\cdot & 7\cdot & 8\cdot & 9\cdot & 10\cdot & 11\cdot & 12\cdot & 13\cdot & 14\cdot & 15\cdot & 16 \\
 & = & 1\cdot & 2\cdot & 3\cdot & 2\cdot & 5\cdot & 2\cdot & 7\cdot & 2\cdot & 3\cdot & 2\cdot & 11\cdot & 2\cdot & 13\cdot & 2\cdot & 3\cdot & 2 \\
 & = & & & & 2 & & 3 & & 2 & 3 & 5 & & 2 & & 7 & 5 & 2 \\
 & = & & & & & & & & 2 & & & & 3 & & & & 2 \\
 & = & & & & & & & & & & & & & & & & 2
\end{array}
$$

There are $[\![16/2]\!]$ twos in the first row, $[\![16/4]\!]$ twos in the second row, $[\![16/8]\!]$ twos in the third row, and $[\![16/16]\!]$ twos in the fourth row. Hence, the exponent of 2 in the canonical representation of 16! is given by $[\![16/2]\!] + [\![16/4]\!] + [\![16/8]\!] + [\![16/16]\!] = 8 + 4 + 2 + 1 = 15$. In addition, from Legendre's Theorem, we have that $[\![452/3]\!] + [\![452/9]\!] + [\![452/27]\!] + [\![452/81]\!] + [\![452/243]\!] = 150 + 50 + 16 + 5 + 1 = 222$. Hence, 222 is the exponent of 3 in the canonical representation of 452!

Theorem 3.16 *If r is the exponent of 2 in the canonical representation of $n!$ and s is the number of ones in the binary representation of n, then*

$r + s = n$.

Proof Suppose $n = a_0 + a_1 \cdot 2 + a_2 \cdot 2 + \cdots + a_k \cdot 2^k$, where $0 \leqslant a_i \leqslant 1$, for $i = 1, 2, \ldots, k$, and $a_k \neq 0$.

$$\left[\!\left[\frac{n}{2}\right]\!\right] = \left[\!\left[\frac{a_0}{2} + a_1 + 2a_2 + \cdots + 2^{k-1}a_k\right]\!\right] = a_1 + 2a_2 + \cdots + 2^{k-1}a_k,$$

$$\left[\!\left[\frac{n}{4}\right]\!\right] = \left[\!\left[\frac{a_0}{4} + \frac{a_1}{2} + a_2 + \cdots + 2^{k-2}a_k\right]\!\right] = a_2 + 2a_3 + \cdots + 2^{k-2}a_k,$$

$$\cdots$$

$$\left[\!\left[\frac{n}{2^i}\right]\!\right] = a_i + 2a_{i+1} + \cdots + 2^{k-i}a_k.$$

Hence,

$$\begin{aligned} r &= \sum_{i=1}^{k} \left[\!\left[\frac{n}{2^i}\right]\!\right] \\ &= a_1 + a_2(1+2) + a_3(1+2+2^2) + \cdots + a_k(1+2+\cdots+2^k) \\ &= (a_0 + 2a_1 + 2^2a_2 + \cdots + 2^ka_k) - (a_0 + a_1 + \cdots + a_k) \\ &= n - (a_0 + a_1 + \cdots + a_k) = n - s. \end{aligned}$$

Therefore, $n = r + s$. ■

In general, if the representation of n to the base p, where p is prime, is given by $b_r p^k + b_{r-1}p^{k-1} + \cdots + b_1 p + b_0$, where $1 \leqslant b_i \leqslant p$, for $i = 1, 2, \ldots, k$, $b_p \neq 0$, and α is the exponent of p in the canonical representation of $n!$, then $\alpha(p-1) + n = b_0 + b_1 + \cdots + b_k$.

Exercises 3.5

1. Prove that for any real number x, $x - 1 < [\![x]\!] \leqslant x$.
2. Prove that $[\![x]\!] + [\![-x]\!] = 0$ if x is an integer and $[\![x]\!] + [\![-x]\!] = -1$ otherwise.
3. Prove that for any two real numbers x and y $[\![x+y]\!] \geqslant [\![x]\!] + [\![y]\!]$.
4. Find the most general sets of numbers for which the following equations in x hold:
   (a) $[\![x]\!] + [\![x]\!] = [\![2x]\!]$,
   (b) $[\![x+3]\!] = [\![x]\!] + 3$,
   (c) $[\![x+3]\!] = x + 3$,
   (d) $[\![9x]\!] = 9$.
5. Determine the exponents of 2, 3, and 5 in the canonical representation of 533!

6. Determine the smallest positive integer n such that 5^7 divides $n!$
7. Determine the number of terminal zeros in 1000!
8. Find the least positive integer n such that $n!$ terminates in 37 zeros.
9. How many integers strictly between 1000 and 10 000 are divisible by 7?
10. How many integers less than 1000 are divisible by 3 but not by 4?
11. Determine the number of integers less than or equal to 10 000 which are not divisible by 3, 5, or 7.
12. The largest number in decimal notation represented with just three digits and no additional symbols is 9^{9^9}. How many digits does 9^{9^9} have?
13. For any positive integer n, prove that $\sum_{k=1}^{n}\tau(k) = \sum_{k=1}^{n}[\![n/k]\!]$ and $\sum_{k=1}^{n}\sigma(k) = \sum_{k=1}^{n} k[\![n/k]\!]$. [Dirichet 1849]
14. If p is prime and $p|n$, determine the power that p appears to in the canonical representation of $(2n)!/(n!)^2$.
15. Show that

$$\sum_{k=1}^{n}\mu(k)\left[\!\!\left[\frac{n}{k}\right]\!\!\right] = 1.$$

3.6 Primes revisited

In Proposition 20 in Book IX of the *Elements*, Euclid proved that there is no largest prime. Specifically, he established the following result.

Theorem 3.17 (Euclid's Theorem) *The number of primes is infinite.*

Proof Suppose that the number of primes is finite and p is the largest prime. Consider $N = p! + 1$. N cannot be composite because division of N by any prime 2, 3, ..., p leaves a remainder 1, hence, it has no prime factors. However, N cannot be prime since, $N > p$. Since N cannot be either prime or composite, we have a contradiction. Hence, our assumption is incorrect and the number of primes must be infinite. ■

The largest prime known having only 0 and 1 for digits is $\frac{1}{9}(10^{640} - 1)\cdot 10^{640} + 1$. Even with an infinitude of primes the six-millionth prime has only nine digits. Nevertheless, large prime gaps exist. In particular, if n is any positive integer then $(n+1)! + 2$, $(n+1)! + 3$, ..., $(n+1)! + (n+1)$ is a sequence of n consecutive composite integers.

In 1748, Euler devised a proof of the infinitude of primes using the fact that if $m > 1$ and $n > 1$ are natural numbers with $\gcd(m, n) = 1$, then

$$\left(\frac{1}{1-\frac{1}{m}}\right)\cdot\left(\frac{1}{1-\frac{1}{n}}\right) = \left(\sum_{k=0}^{\infty}\left(\frac{1}{m}\right)^k\right)\cdot\left(\sum_{k=0}^{\infty}\left(\frac{1}{n}\right)^k\right)$$
$$= 1 + \frac{1}{m} + \frac{1}{n} + \frac{1}{m^2} + \frac{1}{mn} + \frac{1}{n^2} + \cdots.$$

Because of the unique factorization of positive integers into products of primes, this series is precisely the sum of the reciprocals of all the positive integers of the form $1/m^{\alpha}n^{\beta}$ with α and β nonnegative, each counted only once. He reasoned that if $p_1 < p_2 < \cdots < p_r$ constituted all the primes then for each i, $1 \leqslant i \leqslant r$,

$$\sum_{k=0}^{\infty}\left(\frac{1}{p_i}\right)^k = \left(\frac{1}{1-\frac{1}{p_i}}\right).$$

Therefore,

$$\sum_{n=1}^{\infty}\frac{1}{n} = \prod_{i=1}^{r}\left(\sum_{k=0}^{\infty}\left(\frac{1}{p_i}\right)^k\right) = \prod_{i=1}^{r}\left(\frac{1}{1-\frac{1}{p_i}}\right) < \infty,$$

which is impossible since $\sum_{n=1}^{\infty} n^{-1}$ is the divergent harmonic series. Hence, the number of primes must be infinite.

In 1775, Euler claimed that for a fixed positive integer a, the sequence $a + 1, 2a + 1, 3a + 1, \ldots$ contains infinitely many primes. In 1785, Legendre conjectured that for coprime positive integers a and b there are infinitely many primes which leave a remainder of a when divided by b. Hence, if a and b are coprime the arithmetic progression $a, a + b, a + 2b, a + 3b, \ldots$ contains infinitely many primes. The validity of Legendre's conjecture was established in 1837 by Peter Gustav Lejeune Dirichlet, Gauss's successor at Göttingen and father of analytic number theory. Not only does Legendre's result give yet another proof of the infinitude of primes, but it indicates that there are an infinite number of primes of the form $4k + 1$, of the form $4k + 3$, of the form $6k + 5$, and so forth. In 1770, Edward Waring conjectured that if $a, a + b, a + 2b$ are three primes in arithmetic progression and $a \neq 3$ then 6 must divide b. The result was established in 1771 by J.L. Lagrange.

In 1845, Joseph Louis François Bertrand, a French mathematician and educator, conjectured that for any positive integer $n \geqslant 2$, there is a prime p for which $n \leqslant p \leqslant 2n$. Bertrand's postulate was first proven by P.L.

Table 3.1.

x	$x/\ln(x)$	$\mathrm{Li}(x)$	$R(x)$	$\pi(x)$
10	4.3	5.12	4.42	4
100	21.7	29.1	25.6	25
500	80.4	100.8	94.4	95
1 000	144.7	176.6	165.6	168
5 000	587.0	683.2	669.1	669
10 000	1 087.0	1 245.11	1 226.4	1 230
15 000	1 559.9	1 775.6	1 755.57	1 754
10^6	72 381.9	78 632	78 555.9	78 498
10^9	48 254 630	50 849 240	50 847 465	50 847 478

Chebyshev in 1852. Bertrand had verified his conjecture for all positive integers less than 3×10^6. Bertrand's postulate acquired its name because Bertrand had assumed it to prove that the number of primes is infinite. If p were the largest prime, then by Bertrand's postulate there would be a larger prime between $p + 1$ and $2(p + 1)$, contradicting the hypothesis that p was largest.

Two functions $f(x)$ and $g(x)$ are said to be asymptotically equivalent if $\lim_{x\to\infty}(f(x)/g(x)) = 1$. For example, if $A(x)$ denotes the average distance between the first x primes, for example, $A(20) = 3$, $A(150) = 5$, and $A(10^{50}) = 155$, it can be shown that $A(x)$ and $\ln(x)$ are asymptotically equivalent.

One of the more intriguing functions in number theory is the prime counting function, denotes by $\pi(x)$. It represents the number of primes less than or equal to x, where x is any real number. It is related to the Möbius function and the distinct prime factor function by the equation $\pi(x) \leqslant \sum_{mn\leqslant x}\mu(m)\omega(n)$. In 1798, Legendre conjectured that $\pi(x)$ is asymptotically equivalent to $x/(\ln(x) - 1.083\,66)$. When he was 15 years old, Gauss attempted to prove what we now call the Prime Number Theorem, namely, that $\pi(x)$ is asymptotically equivalent to $x/\ln(x)$. Using L'Hôpital's rule, Gauss showed that the logarithmic integeral $\int_2^x \mathrm{d}t/(\ln(t))^{-1}$, denoted by $\mathrm{Li}(x)$, is asymptotically equivalent to $x/\ln(x)$, and, if the Prime Number Theorem is true, to $\pi(x)$. Gauss felt that $\mathrm{Li}(x)$ gave better approximations to $\pi(x)$ than $x/\ln(x)$ for large values of x. Bernhard Riemann, a nineteenth century German mathematician who studied under Dirichlet and Jacobi, believed that $R(x) = \sum_{k=1}^{\infty}(\mu(k)/k)\mathrm{Li}(k^{1/k})$ gave better approximations to $\pi(x)$ than either $x/\ln(x)$ or $\mathrm{Li}(x)$. The data in Table 3.1 seem to indicate that he was correct. Riemann made significant contributions to non-Euclidean geometry and analysis before dying at age 39 of tuberculosis.

The first proofs of the Prime Number Theorem were given independently in 1896 by the French mathematician Jacques Hadamard and the Belgian mathematician C.J. de la Vallée Poussin. Hadamard was a firm believer that the sole purpose of mathematical rigor was to legitimitize 'conquests of intuition'. Both proofs entail the use of complex number theory applied to the Riemann zeta-function, $\zeta(s) = \sum_{n=1}^{\infty} n^{-s}$, where s is a complex number. A great deal of theory regarding functions of a complex variable was developed in attempts to prove the Prime Number Theorem. The first proof using only elementary properties of numbers was given by Paul Erdös and Atle Selberg in 1948.

The real valued Riemann zeta-function, where s is real, has a number of interesting properties. For example, using the integral test from calculus, we find that the infinite series $\sum_{n=1}^{\infty} n^{-s}$ converges when $s > 1$. Hence, the real Riemann zeta-function is well-defined. In 1736, Euler showed that

$$\zeta(2k) = \sum_{k=1}^{\infty} \frac{1}{n^{sk}} = \frac{2^{2k-2}\pi^{2k}|B_{2k}|}{2k!},$$

where B_m denotes the mth Bernoulli number. In particular, $\zeta(2) = \pi^2/6$, $\zeta(4) = \pi^4/90$, and $\zeta(6) = \pi^6/945$. In 1885, Ernesto Cesàro proved that the probability that n has no mth power divisors larger than 1 is $1/\zeta(r)$. There are a number of identities between the real Riemann zeta-function and number theoretic functions we encountered earlier. The first three identities shown below were established by Cesàro in 1883.

(a) $(\zeta(s))^2 = \sum_{n=1}^{\infty} \frac{\tau(n)}{n^s}$, $s > 1$,

(b) $\zeta(s) \cdot \zeta(s-1) = \sum_{n=1}^{\infty} \frac{\sigma(n)}{n^s}$, $s > 2$,

(c) $\zeta(s) \cdot \zeta(s-k) = \sum_{n=1}^{\infty} \frac{\sigma_k(n)}{n^s}$, $s > k+1$,

(d) $\frac{1}{\zeta(s)} = \sum_{n=1}^{\infty} \frac{\mu(n)}{n^s}$, $s > 1$,

(e) $-\frac{\frac{\mathrm{d}(\zeta(s))}{\mathrm{d}s}}{\zeta(s)} = \sum_{n=1}^{\infty} \frac{\Lambda(n)}{n^s}$, $s > 1$, where $\mathrm{d}/\mathrm{d}s$ denotes the derivative with respect to s,

(f) $\zeta(s) = \prod_p \left(1 - \frac{1}{p^s}\right)^{-1}$, where $s > 1$ and p runs through all primes.

For example, to establish (f)

$$\begin{aligned}\zeta(s) &= 1 + \frac{1}{2^s} + \frac{1}{3^s} + \frac{1}{4^s} + \frac{1}{5^s} + \cdots \\ &= 1 + \frac{1}{3^s} + \frac{1}{5^s} + \cdots + \frac{1}{2^s}\left(1 + \frac{1}{2^s} + \frac{1}{3^s} + \cdots\right) \\ &= 1 + \frac{1}{3^s} + \frac{1}{5^s} + \cdots + \frac{1}{2^s}\zeta(s).\end{aligned}$$

Thus,

$$\zeta(s)\left(1 - \frac{1}{2^s}\right) = 1 + \frac{1}{3^s} + \frac{1}{5^s} + \cdots = 1 + \frac{1}{5^s} + \frac{1}{7^s} + \cdots + \frac{1}{3^s}\zeta(s)\left(1 - \frac{1}{2^s}\right)$$

and transposing we have

$$\zeta(s)\left(1 - \frac{1}{2^s}\right)\left(1 - \frac{1}{3^s}\right) = 1 + \frac{1}{5^s} + \frac{1}{7^s} + \cdots.$$

Continuing this process, we obtain

$$\zeta(s)\prod_p \left(1 - \frac{1}{p^s}\right) = 1,$$

where p runs through the primes. Therefore,

$$\zeta(s) = \prod_p \left(1 - \frac{1}{p^s}\right)^{-1}.$$

The expression in the right side of (f) is called the Euler product. As we noted earlier in this section, Euler used it to prove the infinitude of primes. When $s = x + yi$ is complex, the identity implies that the Riemann zeta-function has no zeros for $x > 1$. If $x < 0$, $\zeta(s)$ has only trivial zeros at $s = -2, -4, -6, \ldots$. All other zeros of the zeta-function must therefore occur when $0 \leqslant x \leqslant 1$. In 1860, Riemann conjectured that all zeros occur on the line $x = \frac{1}{2}$. This conjecture, known as the Riemann hypothesis, is one of 23 outstanding unsolved problems posed by Hilbert in 1900. Nearly three million zeros of the zeta-function have been found on the line $x = \frac{1}{2}$ and none off it. In 1951, G.H. Hardy showed that an infinite number of zeros of the zeta-function lie on the critical line $x = \frac{1}{2}$.

The pairs, 3 and 5, 5 and 7, 11 and 13, 17 and 19, and 1 000 000 000 061 and 1 000 000 000 063, are examples of consecutive odd primes, called twin primes. By the spring of 2005, the largest known pair of twin primes was given by $33\,218\,925 \times 2^{169\,690} \pm 1$. In 1737, Euler proved that the infinite series of reciprocals of primes,

$$\sum \frac{1}{p} = \frac{1}{2} + \frac{1}{3} + \frac{1}{5} + \frac{1}{7} + \frac{1}{11} + \cdots,$$

diverges. In 1919, Viggo Brun showed that the infinite series of reciprocals of twin primes,

$$\sum \frac{1}{q} = \left(\frac{1}{3} + \frac{1}{5}\right) + \left(\frac{1}{5} + \frac{1}{7}\right) + \left(\frac{1}{11} + \frac{1}{13}\right) \cdots,$$

converges to 1.902 160 577 832 78..., called Brun's constant. Brun also proved that for every positive integer n these exist n consecutive primes none of which are twin primes. In 1949, P. Clement showed that $(n, n+2)$ forms a pair of twim primes if and only if $n(n+2)$ divides $[4((n-1)!+1)+n]$. In 1849, A. Prince de Polignac conjectured that for a fixed positive even integer n, there are infinitely many prime pairs p and $p+n$. Polignac's conjecture when $n = 2$ is the twin prime conjecture.

If we let p denote an odd prime then there is only one triple of consecutive odd primes $(p, p+2, p+4)$, namely (3, 5, 7). Hence, we define prime triplets to be 3-tuples of the form $(p, p+2, p+6)$ or $(p, p+4, p+6)$, where p and $p+6$ are odd primes and one of $p+2$ and $p+4$ is an odd prime. That is, a sequence of four consecutive odd integers forms a prime triplet if the first and last are prime and one of the two other numbers is prime. For example, (5, 7, 11) and (7, 11, 13) are prime triplets. It is an open question whether or not there are an infinite number of prime triplets. Of course, if there were then there would be an infinite number of twin primes.

The smallest prime quartet, that is a 4-tuple of the form $(p, p+2, p+6, p+8)$, where p, $p+2$, $p+6$, and $p+8$ are odd primes, is (5, 7, 11, 13). The next smallest is (11, 13, 17, 19). It is not known whether the number of prime quartets is infinite but the 8-tuple (11, 13, 17, 19, 23, 29, 31, 37) is the only example known of a prime octet, a set of eight primes beginning with p and ending with $p+26$ both of which are odd primes.

Many primes have interesting and surprising properties. For example, 43 and 1987 are primes in cyclic descending order, that is, in the cyclic order –9–8–7–6–5–4–3–2–1–9–8–7–. The largest prime in cyclic descending order is 76 543. There are 19 primes with their digits in cyclic ascending order. The smallest being 23. The largest known prime in cyclic ascending order, 1 234 567 891 234 567 891 234 567 891, was discovered in 1972 by Ralph Steiner and Judy Leybourn of Bowling Green State University. Some primes, called right-truncatable primes, remain prime when they are right truncated. The largest known left-truncatable prime is 357 686 312 646 216 567 629 137. See Table 3.2.

Table 3.2.

Right-truncatable primes	Left-truncatable primes
73939133	46232647
7393913	6232647
739391	232647
73939	32647
7393	2647
739	647
73	47
7	7

Table 3.3.

Number of digits	Number of primes	Number of palindromic primes
1	4	4
2	21	1
3	143	15
4	1 061	0
5	8 363	93
6	68 906	0
7	586 081	668
8	5 096 876	0

The numbers 313, 383 and 757 are examples of three-digit palindromic primes while 12 421 is an example of a five-digit palindromic prime. The number of primes versus the number of palindromic primes is illustrated in Table 3.3.

When the digits of a prime are reversed, sometimes a square results as in the case of 163, since $361 = 19^2$. In several cases when the digits of a prime are reversed the result is another prime as is the case with 13, 17, 37 and 1193. Primes whose reverse is also prime are called reversible primes. Some primes, such as 113 and 79, have the property that any permutation of their digits is prime. The prime 113 also has the property that the sum and product of its digits are primes. A prime is called a permutation prime if at least one nontrivial permutation of its digits yields another prime. Since 3391 is prime, 1933 is a permutation prime. In 1951, H.-E. Richert showed that, except for numbers whose digits are all ones no prime number exists with more than 3 and less than $6 \cdot 10^{175}$ digits such that every permutation of its digits is prime.

The primes 13 331, 15 551, 16 661, 19 991, 72 227 are examples of primes of the form *ab bba*. The primes 1 333 331, 1 777 771, 3 222 223 and 3 444 443 are all of the form *a bbb bba*. For a number to be prime and of the form $aaa \cdots a$, it must be the case that $a = 1$. An integer written in decimal notation using only ones is called a repunit, short for repeated unit. The nth repunit, R_n, is given by $(10^n - 1)/9$. There is a scarcity of primes among the repunits. The only known repunit primes for n less than 10^4 are R_2, R_{19}, R_{23}, R_{317}, and R_{1031}. A necessary condition for R_n to be prime is that n be prime. Properties of repunits were first discussed by William Shanks in 1874.

A prime p is called a Sophie Germain prime if $2p + 1$ is also prime. It is an open question whether there are an infinite number of Sophie Germain primes. In 1995, the largest known Sophie Germain prime, $2\,687\,145 \cdot 3003 \cdot 10^{5072} - 1$, was discovered by Harvey Dubner. Sophie Germain managed to obtain the mathematical lecture notes from the Ecole Polytechnique and taught herself calculus. She corresponded with Gauss, Legendre and Cauchy, under the pseudonym Monsieur Le Blanc. She won numerous prizes for her work in mathematical physics and number theory. In 1823, she established Fermat's Last Theorem for a class of prime exponents. In particular, she showed that if p is a Sophie Germain prime then $x^p + y^p = z^p$ has no nontrivial integer solutions.

Let $p_n^\#$, the nth primorial number, denote the product of the first n primes. For example, $p_1^\# = 2$, $p_2^\# = 2 \cdot 3 = 6$, $p_3^\# = 2 \cdot 3 \cdot 5 = 30$, and so forth. Reverend Reo F. Fortune, an anthropologist at Cambridge University once married to Margaret Mead, the sociologist, devised an algorithm to generate what we now call fortunate numbers. In order to generate the fortunate numbers, determine the smallest prime p greater than $p_n^\# + 1$, then f_n, the nth fortunate number, is given by $p - p_n^\#$. The first three fortunate primes, 3, 5, and 7, are derived in Table 3.4, where p denotes the smallest prime greater than $p_n^\# + 1$. It is an open question whether every fortunate number is prime. By the end of the twentieth century $p_{24\,029}^\# + 1$ was the largest known prime of the form $p_n^\# + 1$, and $p_{15\,877}^\# - 1$ was the largest known prime of the form $p_n^\# - 1$.

Finding any type of pattern that will enable one to determine prime numbers is a much more difficult task. Euler and the Russian mathematician Christian Goldbach proved that no polynomial $f(x) = a_0 + a_1x + a_2x^2 + \cdots + a_nx^n$ can ever yield primes for all positive integer values of x. For if b is a positive integer such that $f(b) = p$, a prime, then p divides

Table 3.4.

n	$p_n^{\#}$	$p_n^{\#}+1$	p	$f_n = p - p_n^{\#}$
1	2	3	5	3
2	6	7	11	5
3	30	31	37	7
4	210	211	223	13
5	2310	2311	2333	23

$f(b+mp)$, for $m = 1, 2, \ldots$, so there are infinitely many values of n for which $f(n)$ is composite. Nevertheless, there have been some notable attempts to devise such polynomials. For example, in 1772 Euler noted that $f(x) = x^2 + x + 17$ yields primes for $x = 1, \ldots, 15$, but not for $x = 16$. That same year, he and Legendre showed that $f(x) = x^2 + x + 41$ yields primes for $-41 \leqslant x < 40$, but not for $x = 40$. Euler claimed that $f(x) = 2x^2 + p$, for $p = 3, 5, 11$, or 29, assumes prime values for $x = 0, 1, \ldots, p-1$. In 1899, E.B. Escott showed that $f(x) = x^2 = 79x + 1601$ yields primes for $x = 0, \ldots, 79$, but not for $x = 80$.

Going up a dimension, for any natural numbers x and y let $f(x, y) = \frac{1}{2}(y-1)[|A^2 - 1| - (A^2 - 1)] + 2$, where $A = x(y+1) - (y! + 1)$. Hence, when $n = 2k+1$, $f(n, n) = k + 2$. The image of $f(x, y)$ includes all prime numbers as x and y run through the positive integers. The function generates the prime 2 an infinite number of times but each odd prime only once. Dirichlet conjectured that if $\gcd(a, b, c) = 1$, then as x, y, and z range over the positive integers, $ax^2 + bxy + cy^2$ generates infinitely many primes.

In 1958 Norman Galbreath conjectured that in the table of absolute values of the rth difference of the primes, shown in Figure 3.2, the leading

```
2   3   5   7  11  13  17  19  23 ...
  1   2   2   4   2   4   2   4 ...
    1   0   2   2   2   2   2 ...
      1   2   0   0   0   0 ...
        1   2   0   0   0 ...
          1   2   0   0 ...
            1   2   0 ...
              1   2 ...
                1 ...
```

Figure 3.2.

diagonal consists of only ones. Galbreath showed that the conjecture was valid for the first 60 thousand primes.

In 1956, a sieve process, similar to that of Eratosthenes, was devised by Verna Gardiner and Stanislaw Ulam. The process is as follows: from a list of positive integers strike out all even numbers, leaving the odd numbers. Apart from 1, the smallest remaining number is 3. Beginning the count with the number 1, pass through the list of remaining numbers striking out every third number. The next smallest number not crossed out is 7. Beginning the process again with the number 1, pass through the list of remaining numbers striking out every seventh number. The smallest number not crossed out greater than 7 is 9. Strike out every ninth number from what is left, and so on. The numbers that are not struck out are called lucky numbers. The lucky numbers between 1 and 99 are shown in Figure 3.3.

Lucky numbers have many properties similar to those of primes. For example, for large values of n, the number of lucky numbers between 1 and n compares favorably with the number of primes between 1 and n. There are 715 numbers between 1 and 48 000 that are both prime and lucky. Every even integer less than or equal to 10^5 can be expressed as the sum of two lucky numbers. Ulam noted that there appear to be just as many lucky numbers of the form $4n + 1$ as of the form $4n + 3$.

In 1775, Lagrange conjectured that every odd positive integer can be expressed as $p + 2q$ where p and q are prime. In 1848, Polignac conjectured that every positive odd integer is expressible as $p + 2^k$, where p is prime and k a positive integer. However, neither 509 nor 877 can be expressed in such a manner.

Every even positive integer is of the form $10k$, $10k + 2$, $10k + 4$, $10k + 6$, or $10k + 8$. Hence, since $10k = 15 + (10k - 15)$, $10k + 2 = 10 + (10k - 8)$, every even integer greater than 38 can be written as the sum of two composite numbers. In 1724, Christian Goldbach showed that

	1	~~2~~	3	~~4~~	~~5~~	~~6~~	7	~~8~~	9
~~10~~	~~11~~	~~12~~	13	~~14~~	15	~~16~~	~~17~~	~~18~~	~~19~~
~~20~~	21	~~22~~	~~23~~	~~24~~	25	~~26~~	~~27~~	~~28~~	~~29~~
~~30~~	31	~~32~~	33	~~34~~	~~35~~	~~36~~	37	~~38~~	~~39~~
~~40~~	~~41~~	~~42~~	43	~~44~~	~~45~~	~~46~~	~~47~~	~~48~~	49
~~50~~	51	~~52~~	~~53~~	~~54~~	~~55~~	~~56~~	~~57~~	~~58~~	~~59~~
~~60~~	~~61~~	~~62~~	63	~~64~~	~~65~~	~~66~~	67	~~68~~	69
~~70~~	~~71~~	~~72~~	73	~~74~~	75	~~76~~	~~77~~	~~78~~	79
~~80~~	~~81~~	~~82~~	~~83~~	~~84~~	85	~~86~~	~~87~~	~~88~~	~~89~~
~~90~~	~~91~~	~~92~~	93	~~94~~	~~95~~	~~96~~	~~97~~	~~98~~	99

Figure 3.3.

the product of three consecutive integers can never be a square. In 1742, Goldbach wrote to Euler in St Petersburg, asking whether or not every positive integer greater than 1 was the sum of three or fewer primes. The query, known as Goldbach's conjecture, is another of Hilbert's problems that remains unsolved. Goldbach taught in St Petersburg and tutored Peter II in Moscow before accepting a post in the Russian Ministry of Foreign Affairs. Euler responded to Goldbach saying that the problem was difficult and equivalent to that of representing every even positive integer, greater than 2, as the sum of two primes. Goldbach's letter to Euler was not published until 1843. Oddly enough, the conjecture first appeared in print in 1770 in Edward Waring's *Meditationes algebraicae*, an abstruse algebraic work. G.H. Hardy said that Goldbach's conjecture was one of the most difficult problems in mathematics. H.S. Vandiver jested that if he came back to life after death and was told that the problem had been solved he would immediately drop dead again. In 1930, the Russian mathematician L. Schnirelmann proved that there is a positive integer S such that every positive integer is the sum of at most S primes. Seven years later, I.M. Vinogradov proved that from some point on every odd number is the sum of three odd primes. Hardy and J.E. Littlewood devised a formula to determine the number of such representations given that one such representation exists. In 1966, Chen Jing-Run proved that every sufficiently large even integer can be expressed as the sum of a prime and an integer having at most 2 prime factors. In 2005, Tomás Oliveira e Silva showed the Goldbach conjecture to be true for all positive integers less than 2×10^{17}.

The only solution to $p^m - q^n = 1$, where m and n are positive integers and p and q are prime, is given by $3^2 - 2^3 = 1$. It is an open question whether $n! + 1$ is prime for infinitely many integral values of n, likewise whether there always exists a prime between two consecutive squares, and whether there is a prime of the form $a^2 + b$ for each positive integer b. In 1993, the largest prime of the form $n! + 1$ known, $1477! + 1$, was found by Dubner, and the largest known prime of the form $n! - 1$, $3601! - 1$ was found by Caldwell. In 1922, Hardy and Littlewood conjectured that there are infinitely many prime numbers of the form $n^2 + 1$. In 1978, Hendrik Iwaniec showed that there are infinitely many numbers of the form $n^2 + 1$ which are either prime or the product of two primes. It remains an open question whether the sequence, 2, 5, 17, 37, 101, 197, 257, ..., with general term $n^2 + 1$, contains an infinite number of primes.

Suppose that we have a large urn containing all the positive integers, from which we select two integers a and b and ask the question, 'What is the probability that a and b are coprime?' The answer relies on a result established by Euler concerning the Riemann zeta-function, namely that $\zeta(2) = \sum_{n=1}^{\infty} n^{-2} = \pi^2/6$. The question was answered first by Cesàro in 1881 and independently by J.J. Sylvester two years later. Given two positive integers a, b, and a prime p, since p divides every pth integer, the probability that p divides a is given by $1/p$. Similarly, the probability that p divides b is $1/p$. Since the two events are independent, the probability that p divides both a and b is the product of the probabilities. That is, $(1/p)(1/p) = 1/p^2$. Therefore, the probability of the complementary event, that either $p \nmid a$ or $p \nmid b$, is given by $1 - 1/p^2$. Now a and b are coprime if and only if $p \nmid a$ or $p \nmid b$ for every prime p. So the probability that $\gcd(a, b) = 1$ is given by the infinite product $(1 - (\frac{1}{2})^2)(1 - (\frac{1}{3})^2)(1 - (\frac{1}{5})^2)(1 - (\frac{1}{7})^2) \cdots$, where the product is taken over all the primes. However, from a property of the Riemann zeta-function, we have

$$\left[1 + \left(\frac{1}{2}\right)^2 + \left(\frac{1}{3}\right)^2 + \cdots\right]\left[\left(1 - \left(\frac{1}{2}\right)^2\right)\left(1 - \left(\frac{1}{3}\right)^2\right)\left(1 - \left(\frac{1}{5}\right)^2\right)\cdots\right] = 1.$$

Dividing both sides by $1 + (\frac{1}{2})^2 + (\frac{1}{3})^2 + \cdots$ and using Euler's result we obtain

$$\left(1 - \left(\frac{1}{2}\right)^2\right)\left(1 - \left(\frac{1}{3}\right)^2\right)\left(1 - \left(\frac{1}{5}\right)^2\right)\left(1 - \left(\frac{1}{7}\right)^2\right)\cdots = \frac{6}{\pi^2}.$$

Thus, the probability of randomly selecting two coprime numbers is just over 61%.

We end this section with a remarkable result established by Euler in 1738, namely that

$$\frac{3}{4} \cdot \frac{5}{4} \cdot \frac{7}{8} \cdot \frac{11}{12} \cdot \frac{13}{12} \cdot \frac{17}{16} \cdot \frac{19 \cdots}{20 \cdots} = \frac{\pi}{4}.$$

Since the infinite geometric series $\sum_{k=1}^{\infty} x^k$ converges to $1/(1 - x)$, when $|x| < 1$, we have

$$\frac{3}{4} = \left(\frac{1}{1 + \frac{1}{3}}\right) = 1 - \frac{1}{3} + \left(\frac{1}{3}\right)^2 - \left(\frac{1}{3}\right)^3 + \cdots,$$

$$\frac{5}{4} = \left(\frac{1}{1 - \frac{1}{5}}\right) = 1 + \frac{1}{5} + \left(\frac{1}{5}\right)^2 + \left(\frac{1}{5}\right)^3 + \cdots,$$

$$\frac{7}{8} = \left(\frac{1}{1+\frac{1}{7}}\right) = 1 - \frac{1}{7} + \left(\frac{1}{7}\right)^2 - \left(\frac{1}{7}\right)^3 + \cdots,$$

$$\ldots.$$

Hence,

$$\frac{3}{4}\cdot\frac{5}{4}\cdot\frac{7}{8}\cdot\frac{11}{12}\cdot\frac{13}{12}\cdot\frac{17}{16}\cdot\frac{19\cdots}{20\cdots} = \sum_{n=0}^{\infty}\frac{(-1)^n}{2n+1} = \arctan(1) = \frac{\pi}{4}.$$

Exercises 3.6

1. Prove that $\{3, 5, 7\}$ is the only set of three consecutive odd numbers that are all prime.
2. Are there an infinite number of primes of the form $n^2 - 1$, where $n > 2$?
3. Prove that the number of primes of the form $4k + 3$ is infinite.
4. Prove that the number of primes of the form $4k + 1$ is infinite. (Hint: Suppose that there are only finitely many primes of the form $4k + 1$, say $q_1, \ldots, q_r$, and consider $N = 4(q_1 \cdots q_r)^2 + 1$.
5. Does the sequence 31, 331, 3331, 33 331, ... always yield a prime?
6. If $\mathscr{P}_n$ denotes the nth prime and $A(n) = (\mathscr{P}_n - 2)/(n - 1)$ denotes the average distance between the first n primes, determine $A(50)$ and compare it with $\ln(50)$.
7. Use L'Hôpital's rule to prove that $\mathrm{Li}(x)$ and $x/\ln(x)$ are asymptotically equivalent.
8. Show that $\zeta(6) = \pi^2/945$.
9. If p and $p + 2$ are twin primes, show that $\sigma(p + 2) = \sigma(p) + 2$.
10. Show that $n(n + 2)$ divides $[4((n - 1)! + 1) + n]$, when $n = 17$. Hence, the twin primes 17 and 19 satisfy Clement's formula.
11. If p and $p + 2$ are twin primes, with $p > 3$, prove that 12 divides $2(p + 1)$.
12. Does the product of twin primes always differ from a square by 1?
13. Odd primes which are not in a set of twin primes are called isolated primes. Find the first ten isolated primes.
14. Determine a prime triple with all terms greater than 13.
15. Determine a prime quartet with all terms greater than 100.
16. Show that 76 883 is a left-truncatable prime.
17. Show that 59 393 339 is a right-truncatable prime.
18. Find three primes such that the reverse of their digits yields a square or a cube.

19. Find all two-digit reversible primes.
20. Find all 15 three-digit palindromic primes.
21. Show that the palindromic numbers 1441 and 3443 factor into palindromic primes.
22. Show that 113 is a panpermutation prime, that is, all the permutations of its digits yield primes.
23. Show that 1423 and 1847 belong to permutation sets.
24. Find ten four-digit reversible primes.
25. Let $R_n = (10^n - 1)/9$ for n a positive integer denote the nth repunit. Show that $3304 \cdot R_4$ is a Smith number.
26. Find all Sophie Germain primes between 11 and 200.
27. A Cunningham chain of length k is a finite sequence of primes $p_1, p_2, \ldots, p_k$ such that either $p_{i+1} = 2p_i + 1$ or $p_{i+1} = 2p_i - 1$, for $i = 1, 2, \ldots k$. Determine a Cunningham chain that begins with 5.
28. Determine the next fortunate number f_6.
29. Show that $f(x) = x^2 + x + 17$ yields primes for $x = 1, \ldots, 15$, but not for $x = 16$.
30. Show that $f(x) = 2x^2 + p$ generates primes for $p = 11$ and $x = 1, 2, \ldots, 10$, but not for $x = 11$.
31. Show that $f(x) = x^2 - 79x + 1601$ generates primes for $x = 25, 30, 40, 60$, but not for $x = 80$.
32. Show that $f(x, y) = \frac{1}{2}(y - 1)[|A^2 - 1| - (A^2 - 1)] + 2$, where $A = x(y + 1) - (y! + 1)$, yields a prime when $x = [(p - 1)! + 1]/p$ and $y = p - 1$, where p is prime.
33. With $f(x, y)$ as defined in the previous exercise, evaluate $f(n, n)$ for any positive integer n.
34. Determine the first 50 lucky numbers.
35. Show that every even integer greater than 4 and less than or equal to 50 is the sum of two lucky numbers.
36. Show that Goldbach's conjecture and Euler's restatement of it are equivalent.
37. Verify Goldbach's conjecture for all even integers between 4 and 50.
38. A copperbach number is a positive integer which can be expressed as the sum of two primes in exactly two different ways. For example, $14 = 7 + 7 = 11 + 3$. Find three other copperbach numbers.
39. A silverbach number is a positive integer which can be expressed as the sum of two primes in at least three different ways. For example, $26 = 3 + 23 = 7 + 19 = 13 + 13$. Find three other silverbach numbers.

40. Paul Levy conjectured that every odd number greater than 5 can be expressed in the form $2p + q$, where p and q are prime. Show that the conjecture is true for all odd numbers between 7 and 49.
41. If the first 10^9 positive integers were put into a very large urn, estimate the probability that a number drawn from the urn is prime.
42. Prove that

$$(\zeta(s))^2 = \sum_{n=1}^{\infty} \frac{\tau(n)}{n^s},$$

where $s > 1$ and n is a positive integer.

43. Prove that

$$\zeta(s) \cdot \zeta(s-1) = \sum_{n=1}^{\infty} \frac{\sigma(n)}{n^s},$$

where $s > 2$ and n is a positive integer.

44. Prove that

$$\zeta(s) \cdot \zeta(s-k) = \sum_{n=1}^{\infty} \frac{\sigma_k(n)}{n^s},$$

where $s > k + 1$ and n is a positive integer.

45. Prove that

$$\frac{1}{\zeta(s)} = \sum_{n=1}^{\infty} \frac{\mu(n)}{n^s},$$

where $s > 1$ and n is a positive integer.

3.7 Miscellaneous exercises

1. Given that $\gcd(a, b) = p$, where p is prime, determine $\gcd(a^m, b^n)$, where m and n are positive integers.
2. If p is a prime and a and b are positive integers such that $\gcd(a, p^2) = p$ and $\gcd(b, p^3) = p^2$, determine $\gcd(a + b, p^4)$ and $\gcd(ab, p^4)$.
3. In 1951, Alfred Moessner devised a sieve process that generates integral powers. According to Moessner's algorithm, in order to obtain the nth powers of the natural numbers, begin with the sequence of natural numbers and strike out every nth natural number. Form the sequence of partial sums of the remaining terms and from it strike out each $(n - 1)$st term. Form the sequence of partial sums of the remaining terms and from it strike out each $(n - 2)$nd term. Repeat the process $n - 1$ times to obtain the sequence of nth powers of natural

numbers. The validity of Moessner's process was established by Oskar Perrone in 1951. For example, in order to generate third powers, we have

1	2	~~3~~	4	5	~~6~~	7	8	~~9~~	10	11	~~12~~	13	14	~~15~~
1	~~3~~		7	~~12~~		19	~~27~~		37	~~48~~		61	~~75~~	
1			8			27			64			125		

Use Moessner's algorithm to generate the first five fourth powers of the natural numbers.

4. If we take the sequence of nth powers of the positive integers, the nth differences, Δ_n, will all be equal to $n!$ For example if $n = 3$, we have

1		8		27		64		125		216
	7		19		37		61		91	
		12		18		24		30		
			6		6		6			

Show that the fourth difference of the fourth powers of the positive integers are 4!.

5. V. Ramaswami Aiyer founded the Indian Mathematical Society in 1907 and the *Journal of the Indian Mathematical Society* in 1909. In 1934, he discovered that, if a positive integer n appears in the array shown in Table 3.5, then $2n + 1$ is composite, and if n does not appear in the array then $2n + 1$ is prime, and all odd primes can be obtained in this manner. Show that this is the case for the Aiyer array.
6. Determine a necessary and sufficient condition for the product of the first n positive integers to be divisible by the sum of the first n positive integers.
7. Let $d_k(n)$ represent the number of distinct solutions to the equation $x_1 \cdot x_2 \cdots x_k = n$, where $x_1, x_2, \ldots, x_k$ run independently through the set of positive integers. Show that $d_2(n) = \tau(n)$. Determine $d_1(n)$.
8. Let $t(n - k, k)$ represent the number of divisors of $n - k$ greater than k where $n > k \geqslant 0$. In 1887, M. Lerch showed that $\tau(n) = n - \sum_{k=1}^{n-1} t(n - k, k)$. According to Lerch's formula with $n = 10$, we have that $\tau(10) = 10 - [t(9, 1) + t(8, 2) + t(7, 3) + t(6, 4) + t(5, 5) + t(4, 6) + t(3, 7) + t(2, 8) + t(1, 9)] = 10 - [2 + 2 + 1 + 1 + 0 + 0 + 0 + 0 + 0] = 4$. Use Lerch's formula to show that $\tau(24) = 8$.
9. In 1878, Cesàro showed that the mean difference between the number of odd and even divisors of any integer is ln(2). In 1883, J.W.L Glaisher showed that if $\theta(n)$ represents the excess of the sum of the odd divisors of n over the even divisors of n, then $\theta(n) + \theta(n - 1) + \theta(n - 3) + \theta(n - 6) + \theta(n - 10) + \cdots = 0$, where 1, 3, 6, ... are

Table 3.5.

4	7	10	13	16	19	...
7	12	17	22	27	32	...
10	17	24	31	38	45	...
13	22	31	40	49	58	...
16	27	38	49	60	71	...
19	32	45	58	71	84	...
...	...	...	...	...	...	...

the triangular numbers, and $\theta(n - n) = 0$. For example, $\theta(6) + \theta(5) + \theta(3) + \theta(0) = \theta(6) + 6 + 4 + (-6) = 0$. Thus, $\theta(6) = -4$. Use Glaisher's formula to determine $\theta(10)$ and $\theta(24)$.

10. Recall that a positive integer is called polite if it can be written as a sum of two or more consecutive positive integers. Prove that the number of ways of writing the polite positive integer n as a sum of two or more consecutive positive integers is $\tau(m) - 1$, where m is the largest odd divisor of n. For example, if $n = 30$ then its largest odd divisor is 15, $\tau(15) = 4$. We obtain $9 + 10 + 11$, $6 + 7 + 8 + 9$, $4 + 5 + 6 + 7 + 8$ as the three ways to represent 30 as a sum of two or more consecutive positive integers.
11. Show that for any positive integer $n > 1$ the sum $1 + \frac{1}{2} + \frac{1}{3} + \frac{1}{4} + \cdots + 1/n$ is never an integer.
12. Let P be a polygon whose vertices are lattice points. Let I denote the number of lattice points inside the polygon and B denote the number of lattice points on the boundary of P. Determine a formula for the area of the region enclosed by P as a function of I and B. [G. Pick 1899]
13. Generalize Pick's formula to the case where the region contains a polygonal hole whose vertices are lattice points.
14. Ulam's spiral if formed as shown in Table 3.6. Continue the pattern for several more revolutions of the spiral and color the primes red. Can you detect any patterns? Ulam's spiral appeared on the cover of the March 1964 issue of *Scientific American*.
15. Given $[\![\sqrt{2}]\!] = 1$, $[\![2\sqrt{2}]\!] = 2$, and $[\![3\sqrt{2}]\!] = 4$, if n is a positive integer, find the first 16 terms of the sequence generated by $[\![n\sqrt{2}]\!]$.
16. If $a = 2 + \sqrt{2}$ then $[\![a]\!] = 3$, $[\![2a]\!] = 6$, and $[\![3a]\!] = 10$. If n is a positive integer, find the first 20 terms of the sequence generated by $[\![na]\!]$.
17. Show that if $f(n) = (1 + \sqrt{8n - 7})/2$ then the nth term of the

Table 3.6.

...	...	...	...	...	...	...	...
...	36	35	34	33	32	31	...
...	17	16	15	14	13	30	...
...	18	5	4	3	12	29	...
...	19	6	1	2	11	28	...
...	20	7	8	9	10	27	...
...	21	22	23	24	25	26	...
...	...	...	...	...	...	...	...

sequence 1, 2, 2, 3, 3, 3, 4, 4, 4, 4, 5, 5, 5, 5, 5, ... is given by $[\![f(n)]\!]$.

18. The sequence 1, 2, 4, 5, 7, 9, 10, 12, 14, 16, ... is formed by taking the first odd number, the next two even numbers, the next three odd numbers, the next four even numbers, and so forth. Show that the general term of the sequence is given by $a_n = 2n - [\![(1 + \sqrt{8n-7})/2]\!]$.
19. In *The Educational Times* for 1881, Belle Easton of Buffalo, New York, showed the highest power of p dividing the product $p^n!$ is given by $(p^n - 1)/(p - 1)$. Prove it.
20. In *The Educational Times* for 1883, Belle Easton determined the greatest value of x for which $2^n!/2^x$ is an integer. What value did she find for x?
21. In *The Educational Times* for 1892, Emily Perrin of Girton College, Cambridge, showed that if n is a positive integer, A is the sum of the divisors of n whose quotient is odd (the divisors d such that d times an odd number is n), B is the sum of the divisors of n having even quotient, and C is the sum of the odd divisors of n, then $A = B + C$. Prove it.
22. In 1898, C.J. de la Vallée Poussin showed that if a large number, say n, is divided by all the primes up to n, then the average fraction by which the quotient falls short of the next whole number is given approximately by γ, the Euler–Mascheroni constant. For example, if $n = 43$, then $21\frac{1}{2}$, $14\frac{1}{3}$, $8\frac{3}{5}$, $6\frac{1}{7}$, $3\frac{10}{11}$, $3\frac{4}{13}$, $2\frac{9}{17}$, $2\frac{5}{19}$, $1\frac{20}{23}$, $1\frac{14}{29}$, $1\frac{12}{31}$, $1\frac{6}{37}$, $1\frac{2}{41}$, will fall short of 22, 15, 9, 7, 4, 4, 3, 3, 2, 2, 2, 2, 2, respectively by $\frac{1}{2}$, $\frac{2}{3}$, $\frac{2}{5}$, $\frac{6}{7}$, $\frac{1}{11}$, $\frac{9}{13}$, $\frac{8}{17}$, $\frac{14}{19}$, $\frac{3}{23}$, $\frac{15}{29}$, $\frac{19}{31}$, $\frac{31}{37}$, $\frac{39}{41}$. The average value of these 13 numbers is approximately γ. Use de la Vallée Poussin's technique with $n = 67$ to obtain an estimate for the Euler–Mascheroni number.
23. For $n > 0$ and $k \geqslant 2$, let $\tau_k(n) = \sum_{d|n} \tau_{k-1}(d)$, where $\tau_1(n) = \tau(n)$. Show that if $n = \prod_{i=1}^{r} p_i^{\alpha_i}$,

$$\tau_2(n) = \prod_{i=1}^{r} \binom{\alpha_i + 2}{2}.$$

In general,

$$\tau_k(n) = \prod_{i=1}^{r} \binom{\alpha_i + k}{k}.$$

24. Let $S = \{(x, y): 0 \leqslant x \leqslant 1, 0 \leqslant y \leqslant 1\}$ and $T = \{(u, v): u + v \leqslant \pi/2\}$. Use the transformation $x = \sin u/\cos v$, $y = \sin v/\cos u$ to show that $\int\int_T \mathrm{d}u\mathrm{d}v = \int\int_S (1 - x^2y^2)^{-1}\mathrm{d}x\mathrm{d}y$. Use the latter equality to show that $\zeta(2) = \pi^2/6$.

3.8 Supplementary exercises

1. If p and $p + 2$ are twin primes, with $p > 3$, show that 6 divides $p + 1$.
2. A positive integer is called a biprime, semiprime, or 2-almost prime if it is the product of two primes. Determine the first fifteen biprimes.
3. A positive integer is called an n-almost prime if it is the product of n primes. Determine the first ten 3-almost primes, the first ten 4-almost primes, and the first ten 5-almost primes.
4. A positive integer is called a superbiprime if it is the product of two distinct primes. Determine the first fifteen superbiprimes.
5. A positive integer is called bicomposite if it is composite and has at least four prime factors which need not be distinct. Determine the first fifteen bicomposite numbers.
6. For what positive integral values of n is $n^{34} - 9$ prime.
7. Prove that no positive integer of the form $n^2 + 1$ is divisible by a prime of the form $4k + 3$.
8. A positive integer is called squarefull, or nonsquarefree, if it contains at least one square in its prime factorization. Determine the first ten squarefull numbers.
9. Show that every squarefull number can be written as the product of a square and a cube, each greater than unity.
10. Prove that every powerful number is of the form m^2n^3, where m and n are greater than unity.
11. Which of the binomial coefficients $\binom{n}{k}$, for $1 \leqslant n \leqslant 20$ and $3 \leqslant k \leqslant n/2$, are powerful?
12. Show that $\binom{50}{3}$ is powerful.
13. Find necessary conditions that $\sum_{k=1}^{n} k$ divides $\prod_{k=1}^{n} k$.
14. Given a positive integer n expressed as a product of individual primes,

remove the multiplication sign and consider the resulting integer. The primeness of n, denoted by $\eta(n)$ is the number of iterations it takes to reach a prime number. For example, $\eta(2) = \eta(3) = 1$, and $\eta(4) = 2$, since $4 = 2 \cdot 2 \rightarrow 22 = 2 \cdot 11 \rightarrow 211$, which is prime. Determine $\eta(n)$, for $n = 5, 6, \ldots, 12$.

15. Show that 4 divides the sum of two twin primes.
16. A positive integer is called an emirp number if it is a nonpalindromic prime whose reversal is also prime. Determine the first fifteen emirp numbers.
17. Show that a positive integer is square if and only if it has an odd number of divisors.
18. How many divisors does 4825 have and what is their sum?
19. Show that $\tau(n) = \tau(n+1) = \tau(n+2) = \tau(n+3)$ if $n = 3655$.
20. Show that $\sigma(n+2) = \sigma(n) + 2$ if $n = 8575$.
21. Show that $\sigma(523116) = 3 \cdot 523116$.
22. Evaluate $\tau(p^2q^3)$ and $\sigma(p^2q^3)$ when p and q are prime.
23. Show that $\sigma(n)$ is odd if and only if $n = m^2$ or $n = 2m^2$.
24. Show that n is composite if and only if $\sigma(n) > n + \sqrt{n}$.
25. Determine the smallest positive integer with thirty divisors, with forty-two divisors.
26. Determine the first fifteen minimal numbers. Hint: The seventh number in the sequence will be the smallest positive integer with seven divisors.
27. A positive integer n with the property that it and $n + 1$ have the same sum of prime divisors (taken with multiplicity) is called a Ruth-Aaron number. For example, 714 is a Ruth-Aaron number since $714 = 2 \cdot 3 \cdot 7 \cdot 17$, $715 = 5 \cdot 11 \cdot 13$, and $2 + 3 + 7 + 17 = 5 + 11 + 13 = 29$. Determine the first fifteen Ruth-Aaron numbers.
28. A positive integer n is called refactorable, or a tau number, if $\tau(n)$ divides n. Determine the first twenty refactorable numbers.
29. Prove that there are an infinite number of refactorable numbers.
30. Prove that any odd refactorable number is square.
31. Prove that n is odd and refactorable if and only if $2n$ is refactorable.
32. Determine $\mu(136) + \mu(212)$.
33. Evaluate $\sum_{d|36}\mu(d)$.
34. Evaluate $\sum_{d|28}(1/d)$.
35. Let $n = 2^r p_1^{\alpha_1} p_2^{\alpha_2} \cdots p_k^{\alpha_k}$, where p_i for $1 \leqslant i \leqslant k$ is an odd prime and $f(n) = 2^r$. Is f completely multiplicative?
36. Let $n = 2^r p_1^{\alpha_1} p_2^{\alpha_2} \cdots p_k^{\alpha_k}$, where p_i for $1 \leqslant i \leqslant k$ is an odd prime and $f(n) = 2^r$. Is f multiplicative?

37. Use Fermat's method to factor 2184.
38. Use Fermat's method to factor 171366.
39. Use Euler's method to factor 306. (Hint: $306 = 16^2 + 2 \cdot 5^2 = 12^2 + 2 \cdot 9^2$.)
40. Determine the smallest positive integer n such that $n!$ has 500 terminal zeros.
41. Determine the largest power of 15 that divides 60!
42. Determine the number of terminal zeros of $500!/200!$.
43. Determine the exponent of 7 in the canonical representation of 1079!.
44. Determine the exponent of 3 in the canonical representation of 91!.
45. The ceiling function $\lceil x \rceil$ is defined to be the least integer greater than x. Let $f(x) = x\lceil x \rceil$. Show that a finite number of iterations of f applied to 8/7 yields an integer. It is an open question as to whether iterations of f on any rational number greater than unity will eventually result in an integer.
46. A positive integer n is called a vampire number if it has a factorization using its own digits, e.g., 1395 is a vampire number since $1395 = 31 \cdot 9 \cdot 5$. Determine five vampire numbers less than 5000.
47. A number is called a Ramanujan number if it contains an odd number of distinct prime divisors. Ramanujan showed that the sum of the reciprocals of the squares of all such numbers is $9/2\pi^2$ and the sum of the reciprocals of their fourth powers is $15/2\pi^4$. Find the first fifteen Ramanujan numbers.
48. Find the first five primes p for which p divides $1 + (p - 1)!$. Are all your answers squarefree?
49. Given a positive integer n greater than unity and a prime p greater than 3, show that $6n$ divides $p^n + (6n - 1)$.
50. A integer n is called a Cullen number if $n = 2^m + 1$ where m is a positive integer. Except for when $m = 141$, all Cullen numbers are composite for $0 \leqslant m \leqslant 1000$. An integer n is called a Woodall number if $n = 2^m - 1$. Find the first three Woodall numbers that are prime.

4

Perfect and amicable numbers

It is always better to ask some of the questions than to try to know all the answers.

James Thurber

4.1 Perfect numbers

History is replete with numbers thought to have mystical or anodynical powers. One set of such is that of the perfect numbers. A positive integer n is said to be perfect if the sum of its divisors is twice the number itself, that is, if $\sigma(n) = 2n$. The concept of perfect numbers goes back to Archytas of Tarentum, a colleague of Plato, who claimed that if $2^n - 1$ is prime then the sum of the first $2^n - 1$ positive integers is a perfect number. An equivalent statement, Theorem 4.1, appears as the final proposition in Book IX of Euclid's *Elements*, the culmination of the three books in the *Elements* Euclid devotes to number theory.

Theorem 4.1 *If* $2^n - 1$ *is a prime number then* $2^{n-1}(2^n - 1)$ *is perfect.*

Proof The only divisors of 2^{n-1} are $1, 2, 2^2, \ldots, 2^{n-1}$. If $2^n - 1$ is prime its only divisors are itself and 1. Since 2^{n-1} and $2^n - 1$ are coprime, the sum of the divisors of $2^{n-1}(2^n - 1)$ can be represented as the product of the sums of the divisors of 2^{n-1} and $2^n - 1$. Hence,

$$\begin{aligned}(1 + 2 + 2^2 + \cdots + 2^{n-1})[(2^n - 1) + 1] &= \left(\frac{2^n - 1}{2 - 1}\right) \cdot 2^n \\ &= (2^n - 1)(2^n) \\ &= 2(2^{n-1})(2^n - 1).\end{aligned}$$

Therefore, $2^{n-1}(2^n - 1)$ is perfect as claimed. ■

We call numbers of the form $2^{n-1}(2^n - 1)$, where $2^n - 1$ is prime, Euclidean perfect numbers. It is important to note, however, that Euclid did

Table 4.1.

n	$2^{(n-1)}$	$2^n - 1$	$2^{(n-1)}(2^n - 1)$
2	2	3	6
3	4	7	28
5	16	31	496
7	64	127	8128

not claim that all perfect numbers are of the form $2^{p-1}(2^p - 1)$, where p is prime, or that all even perfect numbers are of that form.

The first four even perfect numbers were known to the ancients and can be found in the second century works of Nicomachus and Theon of Smyrna. They appear in the last column of Table 4.1.

Perfect numbers have generated a wealth of conjectures in number theory. In *Introduction to Arithmetic*, Nicomachus partitioned the positive integers into perfect, abundant, and deficient numbers. He defined a positive integer n to be abundant if $\sigma(n) > 2n$ and to be deficient if $\sigma(n) < 2n$. He claimed that abundant and deficient numbers were numerous, but knew of no way to generate them.

Abundant numbers, like lucky numbers, have some Goldbach-type properties. For example, every number greater than 46 can be expressed as the sum of two abundant numbers. In the early seventeenth century, Bachet showed that 945 was the only odd abundant number less than 1000 and claimed that biprimes, except $2 \cdot 3$, are deficient numbers.

With respect to perfect numbers, Nicomachus conjectured that there is only one perfect number between 1 and 10, only one between 10 and 100, and only one between 1000 and 10 000. That is, the nth perfect number has exactly n digits. He also conjectured that Euclidean perfect numbers end alternately in 6 and 8.

Iamblichus, two centuries later, reiterated Nicomachus's claim that there is exactly one perfect number in the interval $10^k \leqslant n \leqslant 10^{k+1}$ for any nonnegative integer k. Boethius noted that perfect numbers were rare, but thought that they could be easily generated in a regular manner. In the late seventh century, Alcuin [AL kwin] of York, a theologian and advisor to Charlemagne, explained the occurrence of the number 6 in the creation of the universe on the grounds that 6 was a perfect number. He added that the second origin of the human race arose from the deficient number 8 since there were eight souls on Noah's ark from which the entire human race sprang. Alcuin concluded that the second origin of humanity was more imperfect than the first.

In 950, Hrotsvita [ros VEE tah], a Benedictine nun in Saxony, mentioned the first four perfect numbers in a treatise on arithmetic. She was the author of the earliest known Faustian-type legend where the protagonist sells his soul to the devil for worldly gain. In 1202, Fibonacci listed the first three perfect numbers in *Liber abaci*. In the early thirteenth century Jordanus de Nemore claimed, in *Elements of Arithmetic*, that every multiple of a perfect or abundant number is abundant and every divisor of a perfect number is deficient. Nemore, Latinized Nemorarius, was the head of a Teutonic monastic order. He perished in a shipwreck in 1236. About 1460, the fifth perfect number, $2^{12}(2^{13} - 1)$, appeared in a Latin codex. In the late fifteenth century, Regiomontanus listed the first six perfect numbers as 6, 28, 496, 8128, 33 550 336, and 8 589 869 056.

In 1510, Bouvellus, in *On Perfect Numbers*, discovered the odd abundant number, 45 045. He showed that every even perfect number is triangular and conjectured, as did Tartaglia 50 years later, that the sum of the digits of every Euclidean perfect number larger than 6 leaves a remainder 1 when divided by 9. The conjecture was proven by Cataldi in 1588 and independently, in 1844, by Pierre Laurent Wantzel when he showed the digital root of a Euclidean perfect number is unity. Seven years earlier Wantzel had given the first rigorous proof of the impossibility of trisecting a given angle with only a straight edge and a collapsing compass. The trisection of a general angle, the duplication of a cube and the squaring of a circle, three great problems bequeathed to us by the Greeks of antiquity, have all been shown to be impossible.

In 1536, in *Arithmetic*, Hudalrichus Regius showed that $2^{11} - 1 = 23 \cdot 89$ and, in doing so, established that it is not always the case that $2^p - 1$ is prime when p is prime. In 1544, in *Complete Arithmetic*, Michael Stifel stated that all Euclidean perfect numbers greater than 6 are triangular and multiples of 4, which did little to enhance his mathematical reputation. In 1575, Francesco Maurolico, Latinized Franciscus Maurolycus, a Benedictine and professor of mathematics at Messina, showed that Euclidean perfect numbers are hexagonal. In 1599, Pierre de la Ramée, Latinized Petrus Ramus, author of a system of logic opposed in many respects to the Aristotelian system, claimed that there is at most one k-digit perfect number, resurrecting Nicomachus's conjecture. In 1638, in *On Perfect Numbers*, Jan Brozek, Latinized Broscius, a professor of theology, astronomy, and rhetoric at Krakow, showed that $2^{23} - 1$ is composite and claimed that there are no perfect numbers between 10^4 and 10^5.

In 1588, in *Treatise on Perfect Numbers*, Cataldi showed that Euclidean perfect numbers end in either 6 or 8, but not alternately as Nicomachus had claimed. In 1891, Lucas proved that every even perfect number, except for 6 and 496, ends in 16, 28, 36, 56, or 76 and all but 28 can be expressed as $7k \pm 1$. Cataldi showed that $2^{17} - 1$ was prime and discovered the sixth perfect number, $2^{16}(2^{17} - 1)$. Fifteen years later, he discovered the seventh perfect number, $2^{18}(2^{19} - 1)$, and conjectured that $2^n - 1$ was prime for $n = 23$, 29, 31 and 37. However, in 1640, Fermat factored $2^{23} - 1$ and $2^{37} - 1$. A century later Euler showed that $2^{29} - 1$ was composite. Cataldi, professor of mathematics and astronomy at Florence, Perugia, and Bologna, founded the first modern mathematics academy, in Bologna. He wrote his mathematical works in Italian and, in an effort to create interest in the subject, distributed them free of charge.

In 1638, René Descartes wrote to Marin Mersenne, the French cleric who kept up a prodigious mathematical correspondence in the seventeenth century, to the effect that he thought all even perfect numbers were of the form $2^{n-1}(2^n - 1)$, with $2^n - 1$ prime. He added, however, that he could see no reason why an odd perfect number could not exist. In correspondnce between Frenicle and Fermat in 1640 several major results concerning perfect numbers were established. Using Mersenne as a conduit, Frenicle asked Fermat to produce a perfect number of 20 or 21 digits or more. Two months later, Fermat replied that there were none.

Fermat began his research on perfect numbers by determining all the primes of the form $a^n - 1$, where a and n are positive integers. His conclusion is stated as Theorem 4.2.

Theorem 4.2 *If $a^n - 1$ is prime for integers $n > 1$ and $a > 1$, then $a = 2$ and n is prime.*

Proof Since $a^n - 1 = (a - 1)(a^{n-1} + a^{n-2} + \cdots + a + 1)$ is prime, $a - 1 = 1$, hence $a = 2$. Moreover, if n is a composite number, say $n = rs$, with $r > 1$ and $s > 1$, then $2^n - 1 = 2^{rs} - 1 = (2^r - 1)(2^{r(s-1)} + 2^{r(s-2)} + \cdots + 1)$. However, each factor on the right exceeds 1 contradicting the fact that $2^n - 1$ is prime. Hence, n is prime and the result is established. ∎

Frenicle wrote that $2^{37} - 1$ was composite but he could not find its factors. Fermat replied that its factors were 223 and 616 318 177. Fermat discovered that if p is prime and $2^p - 1$ is composite then all the prime factors of $2^p - 1$ must be of the form $np + 1$, where n is a positive integer and

$p > 2$. Hence, any prime divisor of $2^{37} - 1$ is of the form $37n + 1$. In order to verify $2^{37} - 1$ is prime, Fermat had only to check to see if $149 = 37 \cdot 4 + 1$ and $223 = 37 \cdot 6 + 1$ were factors. In 1732, Euler extended Fermat's work and claimed that if $n = 4k - 1$ and $8k - 1$ are prime then $2^n - 1$ has the factor $8k - 1$. Euler used the result to show $2^n - 1$ is composite for $n = 11$, 23, 83, 131, 179, 191, 239 and found factors of $2^n - 1$ when $n = 29$, 37, 43, 47, and 73. Lagrange gave a formal proof of Euler's claim in 1775 as did Lucas in 1878. In 1772, Euler showed that $2^{31} - 1$ was prime and generated the eighth perfect number, $2^{30}(2^{31} - 1)$.

Euler, in a posthumous work entitled *On Amicable Numbers*, established the converse of Euclid's theorem on perfect numbers by showing that all even perfect numbers are Euclidean.

Theorem 4.3 *Every even perfect number is of the form* $2^{n-1}(2^n - 1)$, *where* $2^n - 1$ *is prime.*

Proof Suppose that r is an even perfect number, say $r = 2^{n-1}s$, where $n \geqslant 2$ and s is odd. Since r is perfect $\sigma(r) = 2r$. We have $\sigma(r) = \sigma(2^{n-1}s) = 2(2^{n-1}s) = 2^n s$. Since 2^{n-1} and s have no common factors, the sum of the divisors of $2^{n-1}s$ is given by $(2^n - 1)/(2 - 1)$ times the sum of the divisors of s, that is, $\sigma(r) = (2^n - 1)\sigma(s)$. Hence, $2^n s = (2^n - 1)\sigma(s)$. Let $\sigma(s) = s + t$ where t denotes the sum of the divisors of s which are strictly less than s. Thus, $2^n s = (2^n - 1)(s + t)$ and we have that $s = (2^n - 1)t$. Thus, t divides s and thus must be one of the divisors of s, which could only be the case if $t = 1$. Therefore, $s = 2^n - 1$, and the result is established. ■

According to Theorem 4.3, in order to find even perfect numbers, we need only find primes of the form $2^p - 1$, where p is also a prime. Such primes, denoted by M_p, are called Mersenne primes. In 1644, in the preface of his *Cogitata physico-mathematica*, Mersenne claimed M_p is prime for $p = 2$, 3, 5, 7, 13, 17, 19, 31, 67, 127, and 257. The number of combinations of M_p things taken two at a time is given by $2^{p-1}(2^p - 1)$. Hence, all even perfect numbers are triangular and, as such, lie on the third diagonal of Pascal's triangle.

In 1869, F. Landry showed that $2^n - 1$ was composite if $n = 53$ or 59. In 1876 Lucas discovered a technique that was improved by D. H. Lehmer in 1930, called the Lucas–Lehmer test. Let p be prime, $a_1 = 4$, and a_{n+1} be the remainder when $(a_n)^2 - 2$ is divided by M_p. According to the test, if M_p divides a_{p-1}, that is if $a_{p-1} = 0$, then M_p is prime. For example, the

Lucas–Lehmer sequence for $31 = 2^5 - 1$ is given by 4, 14, 8, 0. Hence, M_5 is prime. In 1877, Lucas discovered the 9th perfect number $2^{126}(2^{127} - 1)$ when he verified that M_{127} was prime. In 1883, I. Pervushin discovered the 10th perfect number when he established that M_{61} was prime.

At a special session on number theory at a meeting of the American Mathematical Society in October 1903, Frank Nelson Cole of Columbia University presented a paper entitled 'On the factorization of large numbers'. When his turn came to speak, he went to the blackboard, multiplied 761 838 257 287 by 193 707 721 and obtained 147 573 952 589 676 412 927, which is $2^{67} - 1$. Cole put down the chalk and, amid vigorous applause, returned to his seat without ever uttering a word. There were no questions. He later said that it took him several years, working Sunday afternoons, to find the factors of $2^{67} - 1$. Cole served as Secretary to the AMS from 1896 to 1920 and as editor of the *AMS Bulletin* for 21 years.

In 1911, R.E. Powers verified that M_{89} was prime and, in 1914, showed M_{107} was prime. Hence, up to the First World War, only 12 perfect numbers were known corresponding to the Mersenne primes M_p, for $p = 2$, 3, 5, 7, 13, 17, 19, 31, 61, 89, 107, and 127. When the age of electronic computers dawned in the early 1950s, mathematicians applied the new technology to the search for Mersenne primes. In 1952, Raphael M. Robinson, using the SWAC computer at the National Bureau of Standards, now the National Institute of Standards and Technology, showed that M_{521}, M_{607}, M_{1279}, M_{2203}, and M_{2281} were prime. It took 66 minutes of computer time to confirm that M_{2281} is prime. In 1957, Hans Riesel, with the help of a BESK computer, discovered that M_{3217} was prime. In 1961, Alexander Hurwitz of UCLA showed that M_{4253} and M_{4423} were prime using an IBM 7090. In 1963, Don Gillies, using the ILLIAC computer at the University of Illinois, generated the Mersenne primes M_{9689}, M_{9941}, and $M_{11\,213}$. The last generates the 23rd perfect number and for a time $2^{11\,213} - 1$ appeared in the University of Illinois's metered stamp cancellation. In 1971, Bryant Tuckerman took 39.44 minutes of computer time using an IBM 360/91 at the Watson Research Center to discover the 24th Mersenne prime, $M_{19\,937}$.

In 1978, after three years of hard work using a Control Data CYBER 174, Laura Nickel and Curt Noll, 18-year-old undergraduates at California State University at Hayward, discovered that $M_{21\,701}$ is prime. In 1979, Noll showed that $M_{23\,209}$ was prime. Later that year, Harry Nelson and David Slowinski of Cray Research discovered the 27th Mersenne prime

$M_{44\,497}$. In the early 1980s, using a Cray X-MP, Slowinski determined that $M_{86\,243}$ and $M_{132\,049}$ were Mersenne primes. It took three hours of computer time to establish that $M_{132\,049}$ was indeed prime. In 1985, Slowinski, using a Cray X-MP 24 at Cheveron Geoscience in Houston, discovered that $M_{216\,091}$ was prime. In 1988, Walter N. Colquitt and Luther Welsh, Jr, with the help of a NEC SX 2 supercomputer at the Houston Area Research Center, discovered that $M_{110\,503}$ was prime. In 1992, Slowinski and Paul Gage of Cray Research established that $M_{756\,839}$ was prime. In 1994, using the Lucas–Lehmer test and 7.2 hours on a Cray Y-MP M90 series computer Slowinski and Gage showed that $M_{859\,433}$ and $M_{1\,257\,787}$ were prime. Given the present data, it appears that roughly every three-thousandth prime is a Mersenne prime. In 1996, George Woltman established the Great Internet Mersenne Prime Search (GIMPS). Volunteers using their own personal computers aid in the search for large prime numbers. In November 1996, Joel Armengaud, a 29-year-old programmer from Paris, France, using a Lucas–Lehmer program written by Woltman and the help of 750 programmers scattered across the internet, established that $M_{1\,398\,269}$ is prime. In 1997, Gordon Spencer using Woltman's GIMPS program showed that $M_{2\,976\,221}$ is prime. In 1998, Roland Clarkson, a student at California State University, Dominguez Hill, using Woltman's GIMPS program and a networking software written by Scott Kuratowsi, a software development manager and entrepreneur from San Jose, California, showed that the 909 526 digit number $M_{3\,021\,377}$ is prime.

In June 1999, Nayan Hajratwala, a director of an Internet consulting firm from Saline, Michigan, discovered the first million-digit prime, $M_{6972593}$. In doing so he was awarded $50,000 from the Electronic Frontier Foundation. In December of 2001 Michael Cameron, a 20-year-old college student at Georgian College in Ontario, Canada discovered the thirty-ninth Mersenne prime, $m_{13466917}$. In November 2003, Michael Shafer, a 26-year-old chemical engineering graduate student at Michigan State University, discovered the fortieth Mersenne prime, $M_{20996011}$. In May 2004, Josh Findley, a consultant to the National Oceanic and Atmospheric Administration from Orlando, Florida, discovered the forty-first Mersenne prime, $M_{24036583}$. In February 2005, Martin Nowak, a German eye surgeon and member of the GIMPS Project, discovered the forty-second Mersenne prime $M_{25964951}$. The last six Mersenne primes were discovered through the very successful GIMPS program, however, it remains an open question as to whether there are an infinite number of Mersenne primes.

Let $V(x)$ represent the number of perfect numbers n such that $n \leqslant x$. In 1954, H.-J. Kanold showed that the natural density of perfect numbers,

Table 4.2. *Known Mersenne primes M_p*

Number	Value of p	Discoverer	Year
1	2	anonymous	4th cent. BC
2	3	anonymous	4th cent. BC
3	5	anonymous	4th cent. BC
4	7	anonymous	4th cent. BC
5	13	anonymous	1456
6	17	Cataldi	1588
7	19	Cataldi	1603
8	31	Euler	1772
9	61	Pervushin	1883
10	89	Powers	1911
11	107	Powers	1914
12	127	Lucas	1876
13	521	Robinson	1952
14	607	Robinson	1952
15	1 279	Robinson	1952
16	2 203	Robinson	1952
17	2 281	Robinson	1952
18	3 217	Riesel	1957
19	4 253	Hurwitz	1961
20	4 423	Hurwitz	1961
21	9 689	Gillies	1963
22	9 941	Gillies	1963
23	11 213	Gillies	1963
24	19 937	Tuckerman	1971
25	21 701	Noll, Nickel	1978
26	23 209	Noll	1979
27	44 497	Nelson, Slowinski	1979
28	86 243	Slowinski	1982
29	110 503	Colquitt, Welsh	1988
30	132 049	Slowinski	1983
31	216 091	Slowinski	1985
32	756 839	Slowinski, Gage	1992
33	859 433	Slowinski, Gage	1994
34	1 257 787	Slowinski, Gage	1996
35	1 398 269	Armengaud, Woltman, Kuratowski	1996
36	2 976 221	Spencer, Woltman, Kuratowski	1997
37	3 021 377	Clarkson, Woltman, Kuratowski	1998
38	6 972 693	Hajratwala, Woltman, Kuratowski	1999
39	13 466 917	Cameron, Woltman, Kuratowski	2001
40	20 996 011	Schafer, Woltman, Kuratowski	2003
41	24 036 583	Findley, Woltman, Kuratowski	2004
42	25 964 951	Nowak, Woltman, Kuratowski	2005

$\lim_{x\to\infty}(V(x)/x)$, equals zero, implying that $V(x)$ goes to infinity slower than x does. In a posthumous work, *Tractatus de numerorum ductrina*, Euler proved that there are no odd perfect numbers of the form $4k + 3$, and if an odd perfect number exists it must be of the form $p^{4a+1}N^2$, where p is a prime of the form $4k + 1$, $a \geqslant 0$, N is odd and p does not divide N. In 1888, J.J. Sylvester showed that no odd perfect number exists with less than six distinct prime factors and no odd perfect number exists, not divisible by 3, with less than eight distinct prime factors. In 1991, R.P. Brent, G.L. Cohen, and H.J.J. te Riele showed that if n is odd and perfect then $n \geqslant 10^{300}$.

Exercises 4.1

1. In 1700, Charles de Neuvéglise claimed the product of two consecutive integers $n(n + 1)$ with $n \geqslant 3$ is abundant. Prove or disprove his claim.
2. In 1621, Bachet claimed that every multiple of a perfect number or an abundant number is abundant. Prove that his claim is true thereby establishing Neuvéglise's conjecture that there are an infinite number of abundant numbers.
3. Prove that there are an infinite number of odd deficient numbers and an infinite number of even deficient numbers.
4. Show that every proper divisor of a perfect number is deficient.
5. Determine the binary representations for the first four perfect numbers. Generalize your answers.
6. Show that the digital root of the seventh perfect number is 1.
7. Show that every Euclidean perfect number is triangular.
8. Show that every Euclidean perfect number is hexagonal.
9. Prove that $\Sigma_{d|n} d^{-1} = 2$ if and only if n is perfect. [Carlo Bourlet 1896]
10. Show that the product of the divisors of the even perfect number $n = 2^{p-1}(2^p - 1)$ is given by n^p.
11. Show that $M_{1\,398\,269}$ has 420 921 digits.
12. Show that no perfect number greater than 6 can be either a product of two primes or a power of a prime.
13. Show that 6 is the only squarefree perfect number.
14. Show that every Euclidean perfect number greater than 6 can be expressed as the sum of consecutive odd cubes beginning with unity cubed. For example, $28 = 1^3 + 3^3$, $496 = 1^3 + 3^3 + 5^3 + 7^3$, and $8128 = 1^3 + 3^3 + \cdots + 15^3$.
15. Prove that the units digit of any Euclidean perfect number is either 6 or 8. [Cataldi 1588]

16. Prove that the sum of the digits of every Euclidean perfect number larger than 6 always leaves a remainder of 1 when divided by 9. (Hint: it suffices to show that every Euclidean perfect number is of the form $9k + 1$.) [Cataldi 1588]
17. Show that 6 is the only positive integer n with the property that n and $\sigma(\sigma(n))$ are perfect.
18. Use the Lucas–Lehmer test to show that M_7 is prime.
19. Show that a number of the form $2 \cdot 3^{\alpha}$ cannot be perfect, unless $\alpha = 1$.
20. A positive integer n is called multiplicatively perfect or product perfect if the product of its divisors is equal to n^2. For example, 6 and 15 are product perfect. Find the first 15 product perfect numbers.
21. Use the number theoretic function τ to succinctly classify all product perfect numbers.
22. What is the length of the aliquot cycle generated by a perfect number?

4.2 Fermat numbers

Fermat, after discovering the conditions on the integers a and n for $a^n - 1$ to be prime, determined under what conditions $a^n + 1$ is prime.

Theorem 4.4 *If $a^n + 1$, with $a > 1$ and $n > 0$, is prime then a is even and $n = 2^r$ for some positive integer r.*

Proof Suppose that $a^n + 1$ is prime. If a were odd, then $a^n + 1$ would be even, greater than 3, and hence not prime. Therefore, a is even. Suppose that n has an odd factor which is greater than 1, say $n = rs$, with s odd and greater than 1. Hence, $a^n + 1 = a^{rs} + 1 = (a^r + 1)(a^{r(s-1)} - a^{r(s-2)} + \cdots - a^r + 1)$. Since $s \geqslant 3$, both factors of $a^n + 1$ are greater than 1, contradicting the fact that $a^n + 1$ is prime. Hence, n has no odd factors and must be a power of 2. ■

If n is a nonnegative integer, $2^{2^n} + 1$, denoted by F_n, is called a Fermat number. The first five Fermat numbers, corresponding to $n = 0, 1, 2, 3, 4$, are, respectively, 3, 5, 17, 257, and 65 537 and are prime. Fermat conjectured that F_n was prime for every nonnegative integer n. However, one of Euler's first number theoretic discoveries, was that F_5 is composite. Specifically, he showed the $4\,294\,967\,297 = 641 \cdot 6\,700\,417$. Later, he proved that every prime divisor of F_n for $n \geqslant 2$ must be of the form $k \cdot 2^{n+2} + 1$. He used this discovery to show that $19 \cdot 2^{9450} + 1$ divides F_{9448} and $5 \cdot 2^{23\,473} + 1$ divides $F_{23\,471}$. Currently, the only Fermat numbers

$$\begin{array}{ccccccccccccc} &&&&&1&&&&& &=1\\ &&&&1&&1&&&& &=3\\ &&&1&&0&&1&&& &=5\\ &&1&&1&&1&&1&& &=15\\ &1&&0&&0&&0&&1& &=17\\ 1&&1&&0&&0&&1&&1 &=51\\ \end{array}$$

....

Figure 4.1.

known to be prime are F_0, F_1, F_2, F_3, and F_4. In addition, 274 177 divides F_6, 596 495 891 274 977 217 divides F_7, $1575 \cdot 2^{19} + 1$ divides F_{16}, and $F_5 = (2^9 + 2^7 + 1)(2^{27} - 2^{21} + 2^{19} - 2^{17} + 2^{14} - 2^9 - 2^7 + 1)$. The only F_n whose prime status remains undecided are those with $n \geqslant 24$. Even F_{3310}, which has 10^{990} digits, has been shown to be composite. In 1877, J.T.F. Pepin proved that F_n is prime if and only if it does not divide $3^{2^{2^n-1}} + 1$. In 1905 J.C. Morehead and A.E. Western, using Pepin's test (Theorem 6.14), showed that F_7 was composite. Four years later, they proved that F_8 was composite. In 1977, Syed Asadulla established that the digital root of F_n is 5 or 8 according as $n > 1$ is odd or even. It is an open question whether every Fermat number is squarefree.

In 1796, Gauss renewed interest in Fermat numbers when, as the capstone of his *Disquisitiones arithmeticae*, he proved that a regular polygon of $n = 2^k p_1 p_2 \cdots p_r$ sides can be constructed using a straightedge and compasses if and only if the primes p_i, for $1 \leqslant i \leqslant r$, are distinct and each is a Fermat prime. The only such polygons known with an odd number of sides are those for which n equals 3, 5, 17, 257, 65 537 or a product of these numbers. William Watkins of California State University, Northridge, discovered that the binary number represented by the rows of the Pascal triangle, where even numbers are represented by 0 and odd numbers by 1, generates these odd numbers for which constructable regular polygons exist. (See Figure 4.1.)

Gauss requested that a 17-sided regular polygon should be inscribed on his tombstone, but his request was thought by local stonemasons to be too difficult to construct even without being restricted to using only straightedge and collapsing compasses.

Exercises 4.2

1. Find the digital roots of the first six Fermat numbers.
2. Using Gauss's result concerning regular polygons, for which numbers

Table 4.3.

5	11	23	47	95	191	...	$3 \cdot 2^n - 1$
2	4	8	16	32	64	...	2^n
6	12	24	48	96	192	...	$3 \cdot 2^n$
	71	287	1151	4607	18431		

n less than 26 can regular polygons of n sides be constructed using only Euclidean tools?

3. Show that $\prod_{i=0}^{n-1} F_i = F_n - 2$.
4. Prove that the last digit of any Fermat number, F_n, for $n \geqslant 3$ is always 7.
5. Prove that if $m \neq n$ then $\gcd(F_m, F_n) = 1$.
6. Prove that if $m < n$ then F_m divides $F_n - 2$.
7. Show that F_n for $n > 0$ is of the form $12k + 5$.
8. Prove that no Fermat number is square.
9. Prove that no Fermat number is a cube.
10. Prove that no Fermat number greater than 3 is a triangular number.

4.3 Amicable numbers

Distinct positive integers m and n are called amicable if each is the sum of the proper divisors of the other, that is, if $\sigma(m) = m + n = \sigma(n)$. Perfect numbers are those numbers which are amicable with themselves. Iamblichus ascribed the discovery of the first pair of amicable numbers, 220 and 284, to Pythagoras, who when asked what a friend was, answered, 'another I', which in a numerical sense, is just what these numbers are to each other. Reference to the number 220 can be found in the Book of Genesis.

Amicable numbers appear repeatedly in Islamic works where they play a role in magic, astrology, the casting of horoscopes, sorcery, talismans, and the concoction of love potions. Ibn Khaldun, a fourteenth century Islamic historian, stated in the *Muqaddimah* (*Introduction to History*) that 'persons who have concerned themselves with talismans affirm that the amicable numbers 220 and 284 have an influence to establish a union or close friendship between two individuals.' Khaldun, who developed the earliest nonreligious philosophy of history, was rescued by and served for a time in the court of the Turkish conqueror Tamerlane.

Thabit ibn Qurra, a ninth century mathematician, devised the first method to construct amicable pairs. He formed the sequence $a_0 = 2$, $a_1 = 5$, $a_2 = 11$, $a_3 = 23, \ldots,$ in which each term is obtained by doubling

Table 4.4.

a	b
	c
	d

the preceding term and adding 1 to it. If any two successive odd terms a and p of the sequence are primes, and if $r = pq + p + q$ is also prime, then Thabit concluded that $2^n pq$ and $2^n r$ are amicable. According to Thabit's method $2^n pq$ and $2^n r$ are amicable if $p = (3 \cdot 2^n) - 1$, $q = (3 \cdot 2^{n-1}) - 1$, and $r = (9 \cdot 2^{2n-1}) - 1$, with $n > 1$, are all odd primes. Thabit's method generates the three amicable pairs 220 and 284, 17 296 and 18 416, and 9 363 584 and 9 437 056 and no others of that type less than 2×10^{10} have been discovered.

Thabit's rule was rediscovered on a number of occasions. In 1646, Fermat constructed a table in which the second row consisted of the powers of 2, the third row three times the number on the second row, the first row the number on the third row less 1, and the fourth row the product of two successive numbers on the third row less 1, as shown in Table 4.3. Fermat claimed that if the number d on the fourth row is prime, the number b directly above it on the first row and the number a directly preceding b on the first row are both prime, and if c is the number on the second row above d, then $c \cdot d$ and $a \cdot b \cdot c$ are amicable, as shown in Table 4.4. For example, in Table 4.3, '71' on the fourth row is prime. The number on the first row directly above '71' is '11'; it and the number immediately preceding it on the first row '5' are both prime. The number '4' on the second row is above '71'. Hence, $4 \cdot 71$ and $4 \cdot 5 \cdot 11$ form an amicable pair.

In 1742, Euler devised a method for generating amicable pairs. At that time, only three pairs of amicable numbers were known. He listed 30 new pairs of amicable numbers in *On Amicable Numbers* and eight years later found 59 more pairs. In 1866, 16-year-old Nicoló Paganini discovered an amicable pair, 1184 and 1210, which Euler had overlooked. Unfortunately, Paganini gave no indication whatsoever of how he found the pair. In 1884, P. Seelhoff used Euler's method to discover two new pairs of amicable numbers, $(3^2 \cdot 7^2 \cdot 13 \cdot 19 \cdot 23 \cdot 83 \cdot 1931, 3^2 \cdot 7^2 \cdot 13 \cdot 19 \cdot 23 \cdot 162\,287)$ and $(2^6 \cdot 139 \cdot 863, 2^6 \cdot 167 \cdot 719)$. In 1911, L.E. Dickson discovered two new pairs of amicable numbers, $(2^4 \cdot 12\,959 \cdot 50\,231, 2^4 \cdot 17 \cdot 137 \cdot 262\,079)$ and $(2^4 \cdot 10\,103 \cdot 735\,263, 2^4 \cdot 17 \cdot 137 \cdot 2\,990\,783)$. In 1946, E.B. Escott added 233 pairs to the list. In 1997, at age 79, Mariano Garcia discovered an

Table 4.5. *Some amicable pairs*

$2^2 \cdot 5 \cdot 11$	$2^2 \cdot 71$	Pythagoreans
$2^4 \cdot 23 \cdot 47$	$2^4 \cdot 1151$	Fermat (1636)
$2^7 \cdot 191 \cdot 383$	$2^7 \cdot 73\,727$	Descartes (1636)
$2^2 \cdot 5 \cdot 23 \cdot 137$	$2^2 \cdot 23 \cdot 827$	Euler (1747)
$3^2 \cdot 5 \cdot 7 \cdot 13 \cdot 17$	$3^2 \cdot 7 \cdot 13 \cdot 107$	Euler (1747)
$3^2 \cdot 5 \cdot 7 \cdot 1317$	$3^2 \cdot 7 \cdot 13 \cdot 107$	Euler (1747)
$3^2 \cdot 5 \cdot 11 \cdot 13 \cdot 19$	$3^2 \cdot 5 \cdot 13 \cdot 239$	Euler (1747)
$3^2 \cdot 5 \cdot 7^2 \cdot 13 \cdot 41$	$3^2 \cdot 7^2 \cdot 13 \cdot 251$	Euler (1747)
$3^2 \cdot 5 \cdot 7 \cdot 53 \cdot 1889$	$3^2 \cdot 5 \cdot 7 \cdot 102\,059$	Euler (1747)
$2^2 \cdot 13 \cdot 17 \cdot 389\,509$	$2^2 \cdot 13 \cdot 17 \cdot 198\,899$	Euler (1747)
$3^2 \cdot 5 \cdot 7 \cdot 19 \cdot 37 \cdot 887$	$3^2 \cdot 5 \cdot 19 \cdot 37 \cdot 7103$	Euler (1747)
$3^4 \cdot 5 \cdot 11 \cdot 29 \cdot 89$	$3^4 \cdot 5 \cdot 11 \cdot 2699$	Euler (1747)
$3^2 \cdot 7^2 \cdot 11 \cdot 13 \cdot 41 \cdot 461$	$3^2 \cdot 7^2 \cdot 11 \cdot 13 \cdot 19\,403$	Euler (1747)
$3^2 \cdot 5 \cdot 13 \cdot 19 \cdot 29 \cdot 569$	$3^2 \cdot 5 \cdot 13 \cdot 19 \cdot 17\,099$	Euler (1747)
$3^2 \cdot 5 \cdot 7^2 \cdot 13 \cdot 97 \cdot 193$	$3^2 \cdot 7^2 \cdot 13 \cdot 97 \cdot 1163$	Euler (1747)
$3^2 \cdot 5 \cdot 7 \cdot 13 \cdot 41 \cdot 163 \cdot 977$	$3^2 \cdot 7 \cdot 13 \cdot 41 \cdot 163 \cdot 5867$	Euler (1747)
$2^3 \cdot 17 \cdot 79$	$2^3 \cdot 23 \cdot 59$	Euler (1747)
$2^4 \cdot 23 \cdot 1367$	$2^4 \cdot 53 \cdot 607$	Euler (1747)
$2^4 \cdot 47 \cdot 89$	$2^4 \cdot 53 \cdot 79$	Euler (1747)
$2^5 \cdot 37$	$2 \cdot 5 \cdot 11^2$	Paganini (1866)

amicable pair each of whose members has 4829 digits. Currently, about 1200 amicable pairs are known. (See Table 4.5 for some.)

There are a number of unanswered questions concerning amicable pairs; for example, whether there are an infinite number of amicable pairs, or whether there exists a pair of amicable numbers of opposite parity. It appears plausible that the sum of the digits of amicable pairs taken together is divisible by 9 and that every pair of amicable numbers has unequal remainders when each component is divided by 4. Charles Wall has conjectured that if there exists and odd pair of amicable numbers with equal remainders when divided by four then no odd perfect numbers exist. In 1988, the amicable pair $(A \cdot 140\,453 \cdot 85\,857\,199,\ A \cdot 56\,099 \cdot 214\,955\,207)$, with $A = 5^4 \cdot 7^3 \cdot 11^3 \cdot 13^2 \cdot 17^2 \cdot 19 \cdot 61^2 \cdot 97 \cdot 307$, was discovered proving that it is not the case that all odd amicable pairs are divisible by 3.

Exercises 4.3

1. Show that (220, 284), (1184, 1210), (17 296, 18 416) and $(2^4 \cdot 23 \cdot 479, 2^4 \cdot 89 \cdot 127)$ are amicable pairs.

2. Prove that if (m, n) is an amicable pair then
$$\frac{1}{\sum_{d|m}\frac{1}{d}}+\frac{1}{\sum_{d|n}\frac{1}{d}}=1.$$
3. Show that for the amicable pairs $(2^2\cdot 5\cdot 23\cdot 137,\ 2^2\cdot 23\cdot 827)$ and $(2^3\cdot 17\cdot 4799,\ 2^3\cdot 29\cdot 47\cdot 59)$ the sum of the digits taken together is divisible by 9.
4. A pair of numbers (m, n), with $m<n$, is called betrothed if $\sigma(m)=m+n+1=\sigma(n)$. In 1979, 11 betrothed pairs were known. Show that (48, 75), (140, 195), and (1575, 1648) are betrothed pairs.
5. A triple (a, b, c) is called an amicable triple if $\sigma(a)=\sigma(b)=\sigma(c)=a+b+c$. Show that $(2^5\cdot 3^3\cdot 47\cdot 109,\ 2^5\cdot 3^2\cdot 7\cdot 659,\ 2^5\cdot 3^2\cdot 5279)$ is an amicable triple.
6. Show that $(2^2\cdot 3^2\cdot 5\cdot 11,\ 2^5\cdot 3^2\cdot 7,\ 2^2\cdot 3^2\cdot 71)$ is an amicable triple.
7. Show that (123 228 768, 103 340 640, 124 015 008) is an amicable triple.
8. Determine the length of the aliquot cycle generated by an amicable number.

4.4 Perfect-type numbers

A positive integer n is called multiperfect or, more precisely, k-perfect if $\sigma(n)=kn$, where $k\geqslant 2$ is a positive integer. Thus, a perfect number is a 2-perfect number. The term multiperfect was coined by D.H. Lehmer in 1941. The first multiperfect number, with $k>2$, was discovered by the Cambridge mathematician Robert Recorde in 1557, when he noted in his *Whetstone of Witte* that 120 is 3-perfect. In *Whetstone*, Recorde introduced the modern symbol of two horizontal line seqments for equals, '=' adding that 'no 2 things can be more equal'. In 1556, Recorde's *The Castle of Knowledge* introduced English readers to the Copernican theory.

Multiperfect numbers were studied extensively by French mathematicians in the seventeenth century. In 1631, Mersenne challenged Descartes to find a 3-perfect number other than 120. Six years later, Fermat discovered that 672 is 3-perfect. Fermat constructed an array similar to that found in Table 4.6, where the second row consists of the powers of 2, the top row numbers one less than the numbers on the second row, and the third row one more. Fermat claimed that if the quotient of a number in the top row of the $(n+3)$rd column and the bottom row of the nth column is prime, for $n>1$, then three times the product of the quotient and the

Table 4.6.

	1	2	3	4	5	6	7	8	
$2^n - 1$	1	3	7	15	31	63	127	255	...
2^n	2	4	8	16	32	64	128	256	...
$2^n + 1$	3	5	9	17	33	65	129	257	...

number in the $(n + 2)$nd column is a 3-perfect number. In essence, Fermat claimed that if $q = (2^{n+3} - 1)/(2^n + 1)$ is prime then $3 \cdot q \cdot 2^{n+2}$ is 3-perfect. For example, from Table 4.6, with $n = 3$, $q = \frac{63}{9} = 7$, hence, $3 \cdot 7 \cdot 2^{3+2} = 672$ is 3-perfect.

In 1638, André Jumeau, prior of Sainte Croix, Oloron-Ste-Marie, showed that 523 776 was a 3-perfect number, and issued a second challenge to Descartes to find another 3-perfect number. Descartes responded that 1 476 304 896 is 3-perfect and listed six 4-perfect numbers, and two 5-perfect numbers. Descartes claimed that if n was 3-perfect and not divisible by 3 then $3n$ is 4-perfect; if 3 divides n and both 5 and 9 do not divide n then $45n$ is 4-perfect; if 3 divides n and 57, 9, and 13 do not divide n then $3 \cdot 7 \cdot 13 \cdot n$ is 4-perfect. He added that Fermat's method only yields the 3-perfect numbers 120 and 672.

In 1639, Mersenne discovered the fifth 3-perfect number 459 818 240. Eight years later, Fermat found the 3-perfect number 51 001 180 160, 2 4-perfect numbers, 2 5-perfect numbers, and the first 2 6-perfect numbers. In 1647, Mersenne claimed that if n were 5-perfect and 5 did not divide n then $5n$ would be 6-perfect. In 1929, Poulet listed 36 4-perfect numbers, 55 5-perfect numbers, 166 6-perfect numbers, 69 7-perfect numbers and 2 8-perfect numbers, one of them being $2^{62} \cdot 3^{22} \cdot 5^{10} \cdot 7^4 \cdot 11^3 \cdot 13^7 \cdot 17^2 \cdot 19 \cdot 23 \cdot 29^2 \cdot 31 \cdot 37^2 \cdot 43 \cdot 47 \cdot 53 \cdot 61^2 \cdot 67^2 \cdot 73 \cdot 89 \cdot 97^2 \cdot 127^2 \cdot 139 \cdot 167 \cdot 181 \cdot 193 \cdot 271 \cdot 307 \cdot 317 \cdot 337 \cdot 487 \cdot 521 \cdot 1523 \cdot 3169 \cdot 3613 \cdot 5419 \cdot 9137 \cdot 14281 \cdot 92737 \cdot 649657 \cdot 2384579 \cdot 12207031 \cdot 1001523179$. In the 1950s, Benito Franqui and Mariano García at the University of Puerto Rico and Alan Brown independently generated about 100 multiperfect numbers, albeit there were a few numbers common to both lists and some overlap with the multiperfect numbers generated by Poulet 25 years earlier. No multiperfect numbers have been discovered with $k > 10$. (See Tables 4.7 and 4.8.) Two open questions concerning multiperfect numbers are whether there are infinitely many multiperfect numbers and whether an odd multiperfect number exists.

Table 4.7.

Multiperfect type	Number known
2-perfect	37
3-perfect	6
4-perfect	36
5-perfect	65
6-perfect	245
7-perfect	516
8-perfect	1097
9-perfect	1086
10-perfect	25

Table 4.8. *Some multiperfect numbers (in order of discovery)*
(a) 3-perfect numbers

1	$2^3 \cdot 3 \cdot 5$	Recorde (1557)
2	$2^5 \cdot 3 \cdot 7$	Fermat (1637)
3	$2^9 \cdot 3 \cdot 11 \cdot 31$	Jumeau (1638)
4	$2^{13} \cdot 3 \cdot 11 \cdot 43 \cdot 127$	Descartes (1638)
5	$2^8 \cdot 5 \cdot 7 \cdot 19 \cdot 37 \cdot 73$	Mersenne (1639)
6	$2^{14} \cdot 5 \cdot 7 \cdot 19 \cdot 31 \cdot 151$	Fermat (1643)

Table 4.8. *(b) 4-perfect numbers*

1	$2^5 \cdot 3^3 \cdot 5 \cdot 7$	Descartes (1638)
2	$2^3 \cdot 3^2 \cdot 5 \cdot 7 \cdot 13$	Descartes (1638)
3	$2^9 \cdot 3^3 \cdot 5 \cdot 11 \cdot 31$	Descartes (1638)
4	$2^9 \cdot 3^2 \cdot 7 \cdot 11 \cdot 13 \cdot 31$	Descartes (1638)
5	$2^{13} \cdot 3^3 \cdot 5 \cdot 11 \cdot 43 \cdot 127$	Descartes (1638)
6	$2^{13} \cdot 3^2 \cdot 7 \cdot 11 \cdot 13 \cdot 43 \cdot 127$	Descartes (1638)
7	$2^8 \cdot 3 \cdot 5 \cdot 7 \cdot 19 \cdot 37 \cdot 73$	Mersenne (1639)
8	$2^7 \cdot 3^3 \cdot 5^2 \cdot 17 \cdot 31$	Mersenne (1639)
9	$2^{10} \cdot 3^3 \cdot 5^2 \cdot 23 \cdot 31 \cdot 89$	Mersenne (1639)
10	$2^{14} \cdot 3 \cdot 5 \cdot 7 \cdot 19 \cdot 31 \cdot 151$	Fermat (1643)
11	$2^7 \cdot 3^6 \cdot 5 \cdot 17 \cdot 23 \cdot 137 \cdot 547 \cdot 1093$	Fermat (1643)
12	$2^2 \cdot 3^2 \cdot 5 \cdot 7^2 \cdot 13 \cdot 19$	Lehmer (1900)
13	$2^8 \cdot 3^2 \cdot 7^2 \cdot 13 \cdot 19^2 \cdot 37 \cdot 73 \cdot 127$	Lehmer (1900)
14	$2^{14} \cdot 3^2 \cdot 7^2 \cdot 13 \cdot 19^2 \cdot 31 \cdot 127 \cdot 151$	Carmichael (1910)
15	$2^{25} \cdot 3^3 \cdot 5^2 \cdot 19 \cdot 31 \cdot 683 \cdot 2731 \cdot 8191$	Carmichael (1910)
16	$2^{25} \cdot 3^6 \cdot 5 \cdot 19 \cdot 23 \cdot 137 \cdot 547 \cdot 683 \cdot 1093 \cdot 2731 \cdot 8191$	Carmichael (1910)
17	$2^5 \cdot 3^4 \cdot 7^2 \cdot 11^2 \cdot 19^2 \cdot 127$	Poulet (1929)
18	$2^5 \cdot 3^4 \cdot 7^2 \cdot 11^2 \cdot 19^4 \cdot 151 \cdot 911$	Poulet (1929)
19	$2^7 \cdot 3^{10} \cdot 5 \cdot 7 \cdot 23 \cdot 107 \cdot 3851$	Poulet (1929)
20	$2^8 \cdot 3^2 \cdot 7^2 \cdot 13 \cdot 19^4 \cdot 37 \cdot 73 \cdot 151 \cdot 911$	Poulet (1929)

Table 4.8. *(c) 5-perfect numbers*

1	$2^7 \cdot 3^4 \cdot 5 \cdot 7 \cdot 11^2 \cdot 17 \cdot 19$	Descartes (1638)
2	$2^{10} \cdot 3^5 \cdot 5 \cdot 7^2 \cdot 13 \cdot 19 \cdot 23 \cdot 89$	Frenicle (1638)
3	$2^7 \cdot 3^5 \cdot 5 \cdot 7^2 \cdot 13 \cdot 17 \cdot 19$	Descartes (1638)
4	$2^{11} \cdot 3^3 \cdot 5^2 \cdot 7^2 \cdot 13 \cdot 19 \cdot 31$	Mersenne (1639)
5	$2^{20} \cdot 3^3 \cdot 5 \cdot 7^2 \cdot 13^2 \cdot 19 \cdot 31 \cdot 61 \cdot 127 \cdot 337$	Fermat (1643)
6	$2^{17} \cdot 3^5 \cdot 5 \cdot 7^3 \cdot 13 \cdot 19^2 \cdot 37 \cdot 73 \cdot 127$	Fermat (1643)
7	$2^{10} \cdot 3^4 \cdot 5 \cdot 7 \cdot 11^2 \cdot 19 \cdot 23 \cdot 89$	Fermat (1643)
8	$2^{21} \cdot 3^6 \cdot 5^2 \cdot 7 \cdot 19 \cdot 23^2 \cdot 31 \cdot 79 \cdot 89 \cdot 137 \cdot 547 \cdot 683 \cdot 1093$	Lehmer (1900)
9	$2^{11} \cdot 3^5 \cdot 5 \cdot 7^2 \cdot 13^2 \cdot 19 \cdot 31 \cdot 61$	Poulet (1929)
10	$2^{11} \cdot 3^5 \cdot 5^2 \cdot 7^3 \cdot 13^2 \cdot 31^2 \cdot 61 \cdot 83 \cdot 331$	Poulet (1929)
11	$2^{11} \cdot 3^5 \cdot 5^3 \cdot 7^3 \cdot 13^3 \cdot 17$	Poulet (1929)
12	$2^{11} \cdot 3^6 \cdot 5 \cdot 7^2 \cdot 13 \cdot 19 \cdot 23 \cdot 137 \cdot 547 \cdot 1093$	Poulet (1929)
13	$2^{11} \cdot 3^{10} \cdot 5 \cdot 7^2 \cdot 13 \cdot 19 \cdot 23 \cdot 107 \cdot 3851$	Poulet (1929)
14	$2^{14} \cdot 3^2 \cdot 5^2 \cdot 7^3 \cdot 13 \cdot 19 \cdot 31^2 \cdot 83 \cdot 151 \cdot 331$	Poulet (1929)
15	$2^{15} \cdot 3^7 \cdot 5 \cdot 7 \cdot 11 \cdot 17 \cdot 41 \cdot 43 \cdot 257$	Poulet (1929)
16	$2^{17} \cdot 3^5 \cdot 5 \cdot 7^3 \cdot 13 \cdot 19^2 \cdot 37 \cdot 73 \cdot 127$	Poulet (1929)
17	$2^{17} \cdot 3^5 \cdot 5 \cdot 7^3 \cdot 13 \cdot 19^4 \cdot 37 \cdot 73 \cdot 151 \cdot 911$	Poulet (1929)
18	$2^{19} \cdot 3^6 \cdot 5 \cdot 7 \cdot 11 \cdot 23 \cdot 31 \cdot 41 \cdot 137 \cdot 547 \cdot 1093$	Poulet (1929)
19	$2^{19} \cdot 3^7 \cdot 5^2 \cdot 7 \cdot 11 \cdot 31^2 \cdot 41^2 \cdot 83 \cdot 331 \cdot 431 \cdot 1723$	Poulet (1929)
20	$2^{19} \cdot 3^{10} \cdot 5 \cdot 7 \cdot 11 \cdot 23 \cdot 31 \cdot 41 \cdot 107 \cdot 3851$	Poulet (1929)

Table 4.8. *(d) 6-perfect numbers*

1	$2^{23} \cdot 3^7 \cdot 5^3 \cdot 7^4 \cdot 11^3 \cdot 13^3 \cdot 17^2 \cdot 31 \cdot 41 \cdot 61 \cdot 241 \cdot 307 \cdot 467 \cdot 2801$	Fermat (1643)
2	$2^{27} \cdot 3^5 \cdot 5^3 \cdot 7 \cdot 11 \cdot 13^2 \cdot 19 \cdot 29 \cdot 31 \cdot 43 \cdot 61 \cdot 113 \cdot 127$	Fermat (1643)
3	$2^{23} \cdot 3^7 \cdot 5^5 \cdot 11 \cdot 13^2 \cdot 19 \cdot 31^2 \cdot 43 \cdot 61 \cdot 83 \cdot 223 \cdot 331 \cdot 379 \cdot 601 \cdot 757 \cdot 1201 \cdot 7019 \cdot 823\,543 \cdot 616\,318\,177 \cdot 100\,895\,598\,169$	Fermat (1643)
4	$2^{19} \cdot 3^6 \cdot 5^3 \cdot 7^2 \cdot 11 \cdot 13 \cdot 19 \cdot 23 \cdot 31 \cdot 41 \cdot 137 \cdot 547 \cdot 1093$	Lehmer (1900)
5	$2^{24} \cdot 3^8 \cdot 5 \cdot 7^2 \cdot 11 \cdot 13 \cdot 17 \cdot 19^2 \cdot 31 \cdot 43 \cdot 53 \cdot 127 \cdot 379 \cdot 601 \cdot 757 \cdot 1801$	Lehmer (1900)
6	$2^{62} \cdot 3^8 \cdot 5^4 \cdot 7^2 \cdot 11 \cdot 13 \cdot 19^2 \cdot 23 \cdot 59 \cdot 71 \cdot 79 \cdot 127 \cdot 157 \cdot 379 \cdot 757 \cdot 43\,331 \cdot 3\,033\,169 \cdot 715\,827\,883 \cdot 2\,147\,483\,647$	Cunningham (1902)
7	$2^{15} \cdot 3^5 \cdot 5^2 \cdot 7^2 \cdot 11 \cdot 13 \cdot 17 \cdot 19 \cdot 31 \cdot 43 \cdot 257$	Carmichael (1906)
8	$2^{36} \cdot 3^8 \cdot 5^5 \cdot 7^7 \cdot 11 \cdot 13^2 \cdot 19 \cdot 31^2 \cdot 43 \cdot 61 \cdot 83 \cdot 223 \cdot 331 \cdot 379 \cdot 601 \cdot 757 \cdot 1201 \cdot 7019 \cdot 112\,303 \cdot 898\,423 \cdot 616\,318\,177$	Gérardin (1908)
9	$2^{15} \cdot 3^5 \cdot 5^4 \cdot 7^3 \cdot 11^2 \cdot 13 \cdot 17 \cdot 19 \cdot 43 \cdot 71 \cdot 257$	Poulet (1929)
10	$2^{15} \cdot 3^7 \cdot 5^3 \cdot 7^2 \cdot 11 \cdot 13 \cdot 17 \cdot 19 \cdot 41 \cdot 43 \cdot 257$	Poulet (1929)

Table 4.8. *(e) 7-perfect numbers*

1	$2^{46} \cdot 3^{15} \cdot 5^3 \cdot 7^5 \cdot 11 \cdot 13 \cdot 17 \cdot 19^2 \cdot 23 \cdot 31 \cdot 37 \cdot 41 \cdot 43 \cdot 61 \cdot 89 \cdot 97 \cdot 127 \cdot 193 \cdot 2351 \cdot 4513 \cdot 442\,151 \cdot 13\,264\,529$	Cunningham (1902)
2	$2^{46} \cdot 3^{15} \cdot 5^3 \cdot 7^5 \cdot 11 \cdot 13 \cdot 17 \cdot 19^4 \cdot 23 \cdot 31 \cdot 37 \cdot 41 \cdot 43 \cdot 61 \cdot 89 \cdot 97 \cdot 151 \cdot 193 \cdot 911 \cdot 2351 \cdot 4513 \cdot 442\,151 \cdot 13\,264\,529$	Cunningham (1902)
3	$2^{32} \cdot 3^{11} \cdot 5^4 \cdot 7^5 \cdot 11^2 \cdot 13^2 \cdot 17 \cdot 19^3 \cdot 23 \cdot 31 \cdot 37 \cdot 43 \cdot 61 \cdot 71 \cdot 73 \cdot 89 \cdot 181 \cdot 2141 \cdot 599\,479$	Poulet (1929)
4	$2^{32} \cdot 3^{11} \cdot 5^4 \cdot 7^8 \cdot 11^2 \cdot 13^2 \cdot 17^2 \cdot 19^3 \cdot 23 \cdot 31 \cdot 37^2 \cdot 61 \cdot 67 \cdot 71 \cdot 73 \cdot 89 \cdot 181 \cdot 307 \cdot 1063 \cdot 2141 \cdot 599\,479$	Poulet (1929)
5	$2^{35} \cdot 3^{13} \cdot 5^2 \cdot 7^5 \cdot 11^3 \cdot 13 \cdot 17 \cdot 19^2 \cdot 31^2 \cdot 37^2 \cdot 41 \cdot 43 \cdot 61 \cdot 67 \cdot 73 \cdot 83 \cdot 109 \cdot 127 \cdot 163 \cdot 307 \cdot 331 \cdot 547^2 \cdot 613 \cdot 1093$	Poulet (1929)
6	$2^{35} \cdot 3^{13} \cdot 5^2 \cdot 7^5 \cdot 11^3 \cdot 13 \cdot 17 \cdot 19^4 \cdot 31^2 \cdot 37^2 \cdot 41 \cdot 43 \cdot 61 \cdot 67 \cdot 73 \cdot 83 \cdot 109 \cdot 151 \cdot 163 \cdot 307 \cdot 331 \cdot 547^2 \cdot 613 \cdot 911 \cdot 1093$	Poulet (1929)
7	$2^{35} \cdot 3^{13} \cdot 5^2 \cdot 7^5 \cdot 11^3 \cdot 17 \cdot 19^2 \cdot 31 \cdot 37^2 \cdot 41 \cdot 47 \cdot 61^2 \cdot 67 \cdot 73 \cdot 97 \cdot 109 \cdot 127 \cdot 163 \cdot 307 \cdot 547^2 \cdot 613 \cdot 1093$	Poulet (1929)
8	$2^{35} \cdot 3^{13} \cdot 5^2 \cdot 7^5 \cdot 11^3 \cdot 17 \cdot 19^4 \cdot 31 \cdot 37^2 \cdot 41 \cdot 47 \cdot 61^2 \cdot 67 \cdot 73 \cdot 97 \cdot 109 \cdot 151 \cdot 163 \cdot 307 \cdot 547^2 \cdot 613 \cdot 911 \cdot 1093$	Poulet (1929)
9	$2^{35} \cdot 3^{13} \cdot 5^3 \cdot 7^4 \cdot 11^2 \cdot 13^3 \cdot 17^2 \cdot 19^2 \cdot 23 \cdot 37^2 \cdot 41 \cdot 43 \cdot 67 \cdot 73 \cdot 109 \cdot 127 \cdot 163 \cdot 307^2 \cdot 367 \cdot 467 \cdot 547^2 \cdot 613 \cdot 733 \cdot 1093 \cdot 2801$	Poulet (1929)
10	$2^{35} \cdot 3^{13} \cdot 5^3 \cdot 7^4 \cdot 11^2 \cdot 13^3 \cdot 17^2 \cdot 19^4 \cdot 23 \cdot 37^2 \cdot 41 \cdot 43 \cdot 67 \cdot 73 \cdot 109 \cdot 151 \cdot 163 \cdot 307^2 \cdot 367 \cdot 467 \cdot 547^2 \cdot 613 \cdot 733 \cdot 911 \cdot 1093 \cdot 2801$	Poulet (1929)

A positive integer n is called k-hyperperfect if $k \cdot \sigma(n) = (k+1)n + k - 1$. For example, 21, 2133, and 19 521 are 2-hyperperfect and 325 is 3-hyperperfect. In 1974, Daniel Minoli and Robert Bear described a number of properties of hyperperfect numbers. For example, if $3^n - 1$ is prime then $3^{n-1}(3^n - 2)$ is 2-hyperperfect. They conjectured that for each positive integer k there exists a k-hyperperfect number.

A positive integer n is called semiperfect or pseudoperfect if there exists a collection of distinct proper divisors of n such that their sum is n. For example, 20 is semiperfect since its divisors include 1, 4, 5, 10, and $20 = 10 + 5 + 4 + 1$. Every multiple of a semiperfect number is semiperfect, hence, there are infinitely many semiperfect numbers. It is an open question whether every odd abundant number is semiperfect. A positive integer is called primitive semiperfect if it is semiperfect and is not divisible by any other semiperfect number. All numbers of the form $2^m p$, where $m \geqslant 1$, p is prime, and $2^m < p < 2^{m+1}$, are primitive semiperfect as are 770 and 945. The smallest odd primitive semiperfect number is 945. An abundant number which is not semiperfect is called a weird number. There are 24 weird numbers known, all even and less than 10^6.

In 1680, Leibniz conjectured that if n was not prime then n did not divide $2^n - 2$. In 1736, Euler proved that if p was prime then it divided $2^p - 2$. It was thought for a while that if a positive integer n divided $2^n - 2$, then it was prime. However, in 1819, F. Sarrus showed 341 divides $2^{341} - 2$, yet $341 = 11 \cdot 31$. Hence, there exist composite numbers n, called pseudoprimes, which divide $2^n - 2$. Even though all composite Fermat numbers are pseudoprime, pseudoprimes are much rarer than primes. In 1877, Lucas showed that 2701 is a pseudoprime. Thc smallest even pseudoprime, 161 038, was discovered in 1950 by D.H. Lehmer. In 1903, E. Malo showed that if $n > 1$ was an odd pseudoprime then so was $2^n - 1$. In 1972, A. Rotkiewicz showed that if p and q were distinct primes then $p \cdot q$ is pseudoprime if and only if $(2^p - 1)(2^q - 1)$ is pseudoprime. Hence, there are an infinite number of pseudoprimes. For example, $2^{341} - 2$ is a pseudoprime since $2^{341} - 2 = 2(2^{340} - 1) = 2[(2^{10})^{34} - 1^{34}] = 2[(2^{10} - 1)(\ldots)] = 2[(1023)(\ldots)] = 2[(3)(341)(\ldots)]$. Thus, the composite 341, divides $2^{341} - 2$.

A composite integer m is called a k-pseudoprime if m divides $k^m - k$. For example, 341 is a 2-pseudoprime. a 2-pseudoprime is often referred to simply as a pseudoprime. A composite integer m is called a Carmichael number if m divides $k^m - k$ whenever $1 < k < m$ and $\gcd(k, m) = 1$. Hence, a Carmichael number m is a number that is a k-pseudoprime for all values of k, where $\gcd(k, m) = 1$. In Example 5.9, we show that $561 = 3 \cdot 11 \cdot 17$ is a Carmichael number.

All Carmichael numbers are odd and the product of at least three prime factors. In 1939, J. Chernick showed that if $m \geqslant 1$ and $n = (6m + 1)(12m + 1)(18m + 1)$, and $6m + 1$, $12m + 1$, and $18m + 1$ are prime, then n is a Carmichael number. For example, $1729 = 7 \cdot 13 \cdot 19$ is a Carmichael number. A. Korselt devised a criterion in 1899 for such numbers showing that a positive integer n is Carmichael if and only if n is squarefree and $p - 1$ divides $n - 1$ for all primes p which divide n. In 1993, W.R. Alford, A. Granville, and C. Pomerance showed that there are no more than $n^{2/7}$ Carmichael numbers less than or equal to n. Richard Pinch of Cambridge University calculated all 105 212 Carmichael numbers less than 10^{15}. In 1994, Alford, Granville, and Pomerance proved that there are an infinite number of Carmichael numbers. See Table 4.9.

In 1948, A.K. Srinivasan defined a positive integer n to be practical if every positive integer less than n can be expressed as a sum of distinct

Table 4.9. *The 20 smallest Carmichael numbers*

$561 = 3 \cdot 11 \cdot 17$
$1\,105 = 5 \cdot 13 \cdot 17$
$1\,729 = 7 \cdot 13 \cdot 19$
$2\,465 = 5 \cdot 17 \cdot 29$
$2\,821 = 7 \cdot 13 \cdot 31$
$6\,601 = 7 \cdot 23 \cdot 41$
$8\,911 = 7 \cdot 19 \cdot 67$
$10\,585 = 5 \cdot 29 \cdot 73$
$15\,841 = 7 \cdot 31 \cdot 73$
$29\,341 = 13 \cdot 37 \cdot 61$
$41\,041 = 7 \cdot 11 \cdot 13 \cdot 41$
$46\,657 = 13 \cdot 37 \cdot 97$
$52\,633 = 7 \cdot 73 \cdot 103$
$62\,745 = 3 \cdot 5 \cdot 47 \cdot 89$
$63\,973 = 7 \cdot 13 \cdot 19 \cdot 37$
$75\,361 = 11 \cdot 13 \cdot 17 \cdot 31$
$101\,101 = 7 \cdot 11 \cdot 13 \cdot 101$
$115\,921 = 13 \cdot 37 \cdot 241$
$126\,217 = 7 \cdot 13 \cdot 19 \cdot 73$
$162\,401 = 17 \cdot 41 \cdot 233$

divisors of n. If n is a positive integer, then $2^{n-1}(2^n - 1)$ is practical. There are 49 practical numbers less than 200. The integer 10 is not practical since 4 cannot be expressed as a sum of distinct divisors of 10. However, 8 is practical since $1 = 1$, $2 = 2$, $3 = 2 + 1$, $4 = 4$, $5 = 4 + 1$, $6 = 4 + 2$, and $7 = 4 + 2 + 1$.

A positive integer n is called unitary nonrepetitive, if, excluding the divisors 1 and n, it is possible to express $n - 1$ as a sum of some or all of the remaining divisors of n using each divisor once and only once. For example, 6 and 20 are unitary nonrepetitive since $5 = 2 + 3$ and $19 = 10 + 5 + 4$. In fact, every perfect number is unitary nonrepetitive.

A positive integer is called an Ore number if the harmonic mean of its divisors is an integer. That is, n is Ore if $H(n) = n \cdot \tau(n)/\sigma(n)$ is an integer. Every Euclidean perfect number is Ore. The smallest Ore number which is not perfect is 140.

Thabit ibn Qurra introduced two terms that describe the deviation of a number from being perfect. He defined the abundancy of an abundant number, denoted by $\alpha(n)$, as $\sigma(n) - 2n$ and the deficiency of a deficient number, denoted by $\delta(n)$, as $2n - \sigma(n)$. A positive integer n is called quasiperfect if it has an abundancy of 1 and almost perfect if it has a deficiency of 1.

A more modern definition of abundancy, denoted by $a(n)$, is given by $a(n) = \sigma(n)/n$. Two positive integers are called friendly if they have the same abundancy, for example 12 and 234 are friendly since $a(12) = a(234) = 7/3$. More precisely, two positive integers are called k-friendly if they have the same abundancy k, for example 120 and 672 are 3-friendly. A clique is a set of three or more friendly numbers. The perfect numbers form a clique as do the k-multiperfect numbers for $k > 2$. A positive integer with no friends is called solitary. A primitive friendly pair of positive integers consists of a pair of friendly numbers with no common factor of the same multiplicity, for example 6 and 28 are primitive friendly while 30 and 140 are friendly but not primitive friendly.

Every quasiperfect number n is the square of an odd integer, is greater than 10^{20}, and $\omega(n) \geqslant 5$, but so far none has been found. The only examples of almost perfect numbers are powers of 2. A positive integer n is called superperfect if $\sigma(\sigma(n)) = 2n$. In 1969, D. Suryanarayana showed that all even superperfect numbers are of the form 2^{p-1}, where $2^p - 1$ is a Mersenne prime. That same year, H.-J. Kanold showed that odd superperfect numbers must be square numbers. In 1975, Carl Pomerance showed that there are no odd superperfect numbers less than $7 \cdot 10^{24}$. In 1944, Paul Erdös and Leon Alaoglu defined a positive integer n to be superabundant if $\sigma(n)/n > \sigma(k)/k$, for all positive integers $k < n$. For example, 2 and 4 are superabundant, but 3 and 5 are not. There exist an infinite number of superabundant numbers.

A positive integer n is called m-superperfect if $\sigma^m(n) = 2n$. For $m \geqslant 3$, no even m-superperfect number exists. Paul Erdös defined a positive integer n to be untouchable if there does not exist a positive integer x such that $\sigma(x) = n$. For example, 2, 52, 88, 96, and 120 are untouchable. A divisor d of a natural number n is said to be unitary if $\gcd(d, n/d) = 1$. The sum of the unitary divisors of n is denoted by $\sigma^*(n)$. A natural number is said to be unitary perfect if $\sigma^*(n) = 2n$. Since $\sigma^*(60) = 1 + 3 + 4 + 5 + 12 + 15 + 20 + 60 = 120$, 60 is unitary perfect. In 1975, Charles Wall showed that there are no odd unitary perfect numbers. The only unitary perfect numbers known are 6, 60, 90, 87 360, and 146 361 946 186 458 562 560 000 $(2^{18} \cdot 3 \cdot 5^4 \cdot 7 \cdot 11 \cdot 13 \cdot 19 \cdot 37 \cdot 79 \cdot 109 \cdot 157 \cdot 313)$.

In 1971, Peter Haggis defined a pair of positive integers (m, n) to be unitary amicable if $\sigma^*(m) = \sigma^*(n) = m + n$. Nineteen unitary amicable pairs have been discovered including (114, 126), (1140, 1260), and

(18 018, 22 302). No coprime pair of unitary amicable numbers has been discovered. It is an open question whether there are infinitely many pairs of unitary amicable numbers.

Exercises 4.4

1. Show that 120, 672, and $523\,776 = 2^9 \cdot 3 \cdot 11 \cdot 31$ are 3-perfect.
2. Prove that there are no squarefree 3-perfect numbers.
3. Show that $30\,240 = 2^5 \cdot 3^3 \cdot 5 \cdot 7$ is 4-perfect. [Descartes]
4. Show that $14\,182\,439\,040 = 2^7 \cdot 3^4 \cdot 5 \cdot 7 \cdot 11^2 \cdot 17 \cdot 19$ is 5-perfect. [Descartes]
5. Let $(\sigma(n) - n)/n = h$. If h is an integer we call n an h-fold perfect number. Show that n is an h-fold perfect number if and only if n is $(h-1)$-perfect.
6. Show that 21, 2133, and 19 521 are 2-hyperperfect.
7. Show that 325 is 3-hyperperfect.
8. Show that 36, 40, 770, and 945 are pseudoperfect.
9. Show that 770 and 945 are primitive semiperfect.
10. Show that 70 is weird.
11. Show that $161\,038 = 2 \cdot 73 \cdot 1103$ is a pseudoprime.
12. Show that 24 is a practical number.
13. Show that Euclidean perfect numbers are practical.
14. Show that 24 is unitary nonrepetitive.
15. Show that all perfect numbers are unitary nonrepetitive.
16. Show that 140 is an Ore number.
17. Prove that every perfect number is Ore.
18. Determine the abundancy $\alpha(n)$ of 60 and the deficiency $\beta(n)$ of 26.
19. The arithmetic mean of the divisors of a positive integer is denoted by $A(n)$ and given by $A(n) = \sigma(n)/\tau(n)$. Determine the arithmetic mean of the divisors of p^α, where p is prime and α is a positive integer.
20. A positive integer n is called arithmetic if the arithmetic mean of its divisors is an integer. Determine the first 10 arithmetic numbers.
21. Determine the harmonic mean, $H(n) = n \cdot \tau(n)/\sigma(n)$, of the divisors of p^α, where p is prime and α is a positive integer.
22. Oystein Ore of Yale conjectured that $H(n)$ is never an integer when n is odd, if $n > 1$, then $H(n) > 1$ and except for $n = 1$, 4, 6, or a prime, $H(n) > 2$. Determine $H(1)$, $H(4)$, $H(6)$, and $H(p)$, where p is prime.
23. Determine $H(2^{n-1}(2^n - 1))$ where $2^{n-1}(2^n - 1)$ is a Euclidean perfect number.
24. Determine the geometric mean, $G(n) = (\prod_{d|n} d)^{1/\tau(n)}$, of the divisors

of p^α, where p is prime and α is a positive integer.

25. Show that $A(n)$ and $H(n)$ are multiplicative. Is $G(n)$ multiplicative?
26. Show that 2^n, for n a positive integer, is almost perfect.
27. Show that 16 is a superperfect number.
28. Show that 90 and 87 360 are unitary perfect.
29. Show that if $n = \prod_{i=1}^r p_i^{\alpha_i}$, then $\sigma^*(n) = \prod_{i=1}^r (p_i^{\alpha_i} + 1)$.
30. A positive number is called primitive abundant if it is abundant, but all of its proper divisors are deficient. Find a primitive abundant positive integer.
31. Show that 114 and 126 are a unitary amicable pair.
32. Find five solitary numbers that are not prime or a power of a prime.
33. Find a number that is primitive friendly to 24.
34. Excluding multiperfect numbers, find a clique consisting of three numbers.
35. Show that if a and b are friendly and c is coprime to a and b then ac and bc are friendly.

4.5 Supplementary exercises

1. Determine the number of digits in the first ten Euclidean perfect numbers.
2. Determine the abundancy of 132, 160, and 186.
3. Determine the deficiency of 38, 46, and 68.
4. Show that all biprimes, except $2 \cdot 3$, are deficient.
5. Determine all biprimes that are perfect.
6. Classify the first 30 positive integers as being abundant, deficient, or perfect.
7. Show that 945 is abundant.
8. How many digits does $M_{20996011}$ have?
9. Show that every Euclidean perfect number is hexagonal.
10. Use the Lucas–Lehmer test to show that M_{13} is prime.
11. Use the Lucas–Lehmer test to show that M_{17} is prime.
12. Use the Lucas–Lehmer test to show that M_{19} is prime.
13. If n is an odd perfect number show that $n = p \cdot m^2$, where p is prime.
14. Determine $G_n = 3^{3^n}$, where $n = 0, 1, 2, 3, 4$.
15. Does the units digit of G_n end in 3 and 7, alternately?
16. Is the digital root of G_n always equal to 3 or 9?
17, Show that $2^2 \cdot 5 \cdot 251$ and $2^2 \cdot 13 \cdot 107$ are amicable.
18. Show that 1184 and 1210 are amicable.
19. Show that 17296 and 18416 are amicable.

20. Show that 1050 and 1295 are betrothed.
21. Show that 2024 and 2295 are betrothed.
22. Are 503056 and 514736 amicable, betrothed, or neither?
23. Show that the sum of the reciprocals of the divisors of a k-perfect number is k.
24. For what value of k is 32760 k-perfect?
25. For what value of k is 459818240 k-perfect?
26. For what value of k is 14290848 k-perfect?
27. For what value of k is 523776 k-perfect?
28. For what value of k is 1379454720 k-perfect?
29. Show that 84 is a pseudoperfect number.
30. Show that 18 is practical, pseudoperfect, and unitary nonrepetitive.
31. Which of 20, 26, and 38 are Ore numbers?
32. Determine the first fifteen practical numbers.
33. Show that 30, 56, 556, and 96 are practical numbers.
34. An even abundant number is called impractical if it is not practical. Show that 70, 102, and 114 are impractical.
35. For what value of k are 301, 325, and 756 k-hyperperfect?
36. Show that 108, 126, and 160 are pseudoperfect.
37. Show that 1105, 2465, and 6601 are Carmichael numbers.
38. Show that 270, 496, and 672 are Ore numbers.
39. For any positive integer n, let $P^*(n)$ denote the product of the unitary divisors of n. We say that n is multiplicatively unitary perfect if $P^*(n) = n^2$ and multiplicatively unitary superperfect if $P^*(P^*(n)) = n^2$. show that the product of two distinct primes is multiplicative unitary perfect.
40. Find a multiplicatively unitary perfect number greater than unity that is not a product of two primes.

5

Modular arithmetic

Even if you are on the right track, you'll get run over if you just sit there.

Will Rogers

5.1 Congruence

In this section, we introduce a concept of fundamental importance that will revolutionize the way we regard problems concerning divisibility. Albeit the underlying ideas have Indian and Chinese origins and Euler investigated some basic properties of remainders, it was Gauss who, in 1801, introduced the modern concepts of congruence and the arithmetic of residue classes to European audiences in *Disquisitiones arithmeticae* (*Arithmetical Investigations*) when he was 24. Gauss considered number theory to be the queen of mathematics. To him, its magical charm and inexhaustible wealth of intriguing problems placed it on a level way above other branches of mathematics. We owe a debt of gratitude to mathematicians such as Euler, Lagrange, Legendre, and Gauss for treating number theory as a branch of mathematics and not just a collection of interesting problems.

Given three integers a, b, and m, with $m \geqslant 2$, we say that a is congruent to b modulo m, denoted by $a \equiv b \pmod{m}$, if a and b yield the same remainder or residue when divided by m. Equivalently, $a \equiv b \pmod{m}$, if there is an integer k such that $a - b = km$, that is, their difference is divisible by m. If a is not congruent to b modulo m we write $a \not\equiv b \pmod{m}$. For example, $52 \equiv 38 \pmod{7}$ since $52 - 38 = 14 = 2 \cdot 7$. If $a = mq + r$, with $0 \leqslant r < m$, then r is called the least residue of a modulo m. The least residue of 58 modulo 4 is 2 since $58 = 4 \cdot 14 + 2$ and $0 \leqslant 2 < 4$. If the columns for the residue classes modulo 4 in Table 5.1 below were extended, 58 would appear in the penultimate column. The ability to effectively replace congruences with equalities and vice versa will be of crucial importance in solving problems. For example, $5x \equiv 6 \pmod{11}$ if and only if there is an integer k such that $5x = 6 + 11k$. Similarly, if $3x + 5y = 7$, then $3x \equiv 7 \pmod{5}$ and $5y \equiv 7 \pmod{3}$.

By a partition of a set S, we mean a collection of disjoint subsets of S

Table 5.1.

(a) Residue classes modulo 3

$(0)_3$	$(1)_3$	$(2)_3$
...	...	...
−12	−11	−10
−9	−8	−7
−6	−5	−4
−3	−2	−1
0	1	2
3	4	5
6	7	8
9	10	11
12	13	14
...	...	...

(b) Residue classes modulo 4

$(0)_4$	$(1)_4$	$(2)_4$	$(3)_4$
...	...	...	...
−16	−15	−14	−13
−12	−11	−10	−9
−8	−7	−6	−5
−4	−3	−2	−1
0	1	2	3
4	5	6	7
8	9	10	11
12	13	14	15
16	17	18	19
...	...	...	...

whose union is S. Given a set S, a relation R on S is a subset of $S \times S = \{(a, b)\colon a \in S \text{ and } b \in S\}$. We say that a is related to b, denoted by aRb, if (a, b) is in R. For example, 'divides' is a relation on $\mathbb{Z} \times \mathbb{Z}$. A relation R is reflexive on S if, for all a in S, aRa; symmetric, if aRb implies bRa; and transitive, if aRb and bRc imply aRc. An equivalence relation R on S is a subset of $S \times S$ which is reflexive, symmetric, and transitive. Given an equivalence relation R on a set S, the subsets $R_a = \{x\colon xRa\}$ form a partition of S. Conversely, given a partition of S, the relation R such that aRb if a and b are in the same subset of the partition is an equivalence relation on S. In Theorem 5.1, we show that congruence is an equivalence relation on the set of integers and, hence, splits the integers into disjoint residue classes. The disjoint residue classes modulo 3 and 4 are represented by the columns in Table 5.1.

Theorem 5.1 *Congruence is an equivalence relation on the set of integers.*

Proof Let R correspond to the relation 'is congruent to modulo m', where $m \geqslant 2$ is a positive integer. That is, aRb signifies that $a \equiv b \pmod{m}$. For any integer a, $a = a + 0 \cdot m$, hence, $a \equiv a \pmod{m}$ implying that aRa. Therefore, congruence is a reflexive relation. If a and b are integers such that aRb, then $a \equiv b \pmod{m}$. Hence, for some integer k, $a = b + km$. Thus, $b = a + (-k)m$ implying that $b \equiv a \pmod{m}$. Hence, bRa. Therefore, congruence is symmetric. If a, b, and c are integers such that aRb and bRc, then $a \equiv b \pmod{m}$ and $b \equiv c \pmod{m}$. Hence, there exist integers s

and t such that $a = b + sm$ and $b = c + tm$. Thus, $a = c + (s + t)m$ implying that $a \equiv c \pmod{m}$, hence, aRc and congruence is transitive. Therefore, we have established that congruence is an equivalence relation. ■

Each residue class modulo m is infinite and consists of all the integers having the same remainder when divided by m. Let $(a)_m = \{a + km\colon k \in Z\}$, for example $(3)_5 = \{\ldots, -2, -7, -2, 3, 8, 13, \ldots\}$. In Table 5.1(a), the three disjoint residue classes modulo 3, $(0)_3$, $(1)_3$, $(2)_3$, constitute the three columns. In Table 5.1(b), the four disjoint residue classes modulo 4, $(0)_4$, $(1)_4$, $(2)_4$, $(3)_4$, constitute the four columns. Every integer appears in one of the three columns in Table 5.1(a) and in one of the four columns in Table 5.1(b).

A complete residue system modulo m consists of any set of m integers, no two of which are congruent modulo m. For example, $\{-12, -2, 8\}$ and $\{7, 15, 23\}$ form complete residue systems modulo 3. The set $\{1, 2, 3, \ldots, m\}$ forms a complete residue system modulo m as does the set $\{0, \pm 1, \pm 2, \ldots, \pm (m - 1)/2\}$ when m is odd. Usually, the most convenient complete residue system modulo m to work with is the least residue system $\{0, 1, 2, 3, \ldots, m - 1\}$.

The next result illustrates the property that two integers are congruent modulo m, that is, belong to the same residue class modulo m, if and only if they have the same remainder when each is divided by m.

Theorem 5.2 *The integers a and b have the same least residue modulo m if and only if $a \equiv b \pmod{m}$.*

Proof Let r and s be the least residues of a and b modulo m, respectively. From the division algorithm there exist integers t and u such that $a = mt + r$ and $b = mu + s$, with $0 \leqslant r < m$ and $0 \leqslant s < m$. Thus $a - b = m(t - u) + (r - s)$. Hence m divides $a - b$ if and only if m divides $r - s$. Since both r and s are less than m, m divides $r - s$ if and only if $r - s = 0$. Therefore, $a \equiv b \pmod{m}$ if and only if $r = s$. ■

If $a \equiv b \pmod{m}$ and $c \equiv d \pmod{m}$, there exist integers r and s such that $a = b + rm$ and $c = d + sm$, hence $a + c = b + d + (r + s)m$ and $ac = (b + rm)(d + sm) = bd + (rd + bs + rsm)m$. Hence, $a + c \equiv b + d \pmod{m}$ and $ac \equiv bd \pmod{m}$. We generalize these two results in the next two theorems. The proofs are straightforward and are left as exercises.

Theorem 5.3 *If $a_i \equiv b_i \pmod{m}$, for $i = 1, 2, \ldots, n$, then*

$$\text{(a)} \quad \sum_{i=1}^{n} a_i \equiv \sum_{i=1}^{n} b_i \pmod{m}, \qquad \textit{and } \text{(b)} \quad \prod_{i=1}^{n} a_i \equiv \prod_{i=1}^{n} b_i \pmod{m}.$$

Theorem 5.4 *If $a \equiv b \pmod{m}$, for any integer c and nonnegative integer n,*

(a) $a \pm c \equiv b \pm c \pmod{m}$,
(b) $ac \equiv bc \pmod{m}$,
(c) $a^n \equiv b^n \pmod{m}$.

For example, since $(27)(98) + (13)(15)^{77} \equiv 6 \cdot 0 + (-1)(1)^{77} \equiv 6 \pmod{7}$, it follows from Theorem 5.4 that the least positive residue of $(27)(98) + (13)(15)^{77}$ modulo 7 is 6. Equivalently, the remainder when $(27)(98) + (13)(15)^{77}$ is divided by 7 is 6.

Halley's comet appears in our skies approximately every 76 years. It visited us in 1835, 1910, and most recently in 1986. It will return in 2061. From Theorem 5.4, $1835^{1910} + 1986^{2061} \equiv 1^{1910} + 5^{2061} \equiv 1 + (5^6)^{343} \cdot 5^3 \equiv 1 + (1)^{343} \cdot 6 \equiv 1 + 6 \equiv 0 \pmod{7}$. Hence, 7 divides $1835^{1910} + 1986^{2061}$.

In the seventeenth century, English spelling was not as uniform as it is now. Halley spelt his name differently on a number of occasions. In 1985, Ian Ridpath, a British astronomer, used the London telephone directory to conduct an informal survey to determine how people with the surname Halley pronounced their name. The majority of those surveyed preferred [HAL ee]. However, some used [HALL ee], some [HAIL ee], and some preferred not to be disturbed. How Edmond Halley pronounced his name remains an open question.

Example 5.1 If p is a prime greater than 3, then $p \equiv \pm 1 \pmod{3}$. Hence, $p^2 \equiv 1 \pmod{3}$ and $p^2 + 2 \equiv 0 \pmod{3}$. Since $2 \equiv -1 \pmod{3}$, for any positive integer n, $2^{2^n} \equiv 1 \pmod{3}$. Hence, $2^{2^n} + 5 \equiv 6 \equiv 0 \pmod{3}$. Thus 3 divides $2^{2^n} + 5$. Therefore, if p is a prime greater than 3, $p^2 + 2$ is composite and, for any positive integer n, $2^{2^n} + 5$ is composite.

The following result follows from Theorem 2.8 using a straightforward inductive argument that we omit.

Theorem 5.5 *If $a \equiv b \pmod{m_i}$, for $i = 1, 2, \ldots, k$, where $m_1, m_2, \ldots, m_k$ are pairwise coprime, then $a \equiv b \pmod{m}$, where $m = \prod_{i=1}^{k} m_i$.*

If $\gcd(m, n) = 1$, the system of congruences $x \equiv a \pmod{m}$ and $x \equiv b \pmod{n}$ can be written as a single congruence of the form $x \equiv c \pmod{mn}$.

For example, if $x \equiv 1 \pmod{5}$ and $x \equiv 3 \pmod{4}$ then there is an integer k such that $x = 1 + 5k$. Since $1 + 5k \equiv 3 \pmod{4}$, $k \equiv 2 \pmod{4}$ or $k = 2 + 4t$. Substituting, we obtain $x = 1 + 5(2 + 4t) = 11 + 20t$. Therefore, $x \equiv 11 \pmod{20}$.

In modular arithmetic the cancellation law, if $ac \equiv bc \pmod{m}$ then $a \equiv b \pmod{m}$, does not necessarily hold. For example, $4 \cdot 5 \equiv 4 \cdot 8 \pmod{6}$ but $5 \not\equiv 8 \pmod{6}$. However, we can establish the following result.

Theorem 5.6 *If* $ac \equiv bc \pmod{m}$ *then* $a \equiv b \pmod{m/d}$, *where d is the greatest common divisor of c and m.*

Proof If $ac \equiv bc \pmod{m}$, there exists an integer k such that $ac - bc = km$. Let $d = \gcd(c, m)$; then $(a - b)(c/d) = k(m/d)$, with $\gcd(c/d, m/d) = 1$. Hence, m/d divides $a - b$ or, equivalently, $a \equiv b \pmod{m/d}$. ■

Corollary *If* $ac \equiv bc \pmod{m}$ *and* $\gcd(c, m) = 1$, *then* $a \equiv b \pmod{m}$.

Example 5.2 Raising both sides of the congruence $5 \cdot 2^7 = -1 \pmod{641}$ to the fourth power yields $5^4 \cdot 2^{28} \equiv 1 \pmod{641}$. Since $641 = 625 + 16$, $5^4 \equiv -2^4 \pmod{641}$ and, hence, $2^{32} \equiv -1 \pmod{641}$. The latter congruence implies that there is an integer k such that $2^{32} + 1 = 641 \cdot k$. Hence, 641 divides $2^{32} + 1$. Therefore, the Fermat number F_5 is composite.

Example 5.3 (The binary-square technique) Consider the composite number $161\,038 = 2 \cdot 73 \cdot 1103$. Since 161 037 can be represented in binary notation as $100\,111\,010\,100\,001\,101_2$, $161\,037 = 2^{17} + 2^{14} + 2^{13} + 2^{12} + 2^{10} + 2^8 + 2^3 + 2^2 + 2^0$ and, hence, $2^{161\,037} = 2^{131\,072} \cdot 2^{16\,384} \cdot 2^{8192} \cdot 2^{4096} \cdot 2^{1024} \cdot 2^{256} \cdot 2^8 \cdot 2^4 \cdot 2^1$. Beginning with $2^1 \equiv 2 \pmod{73}$ and $2^1 \equiv 2 \pmod{1103}$ and squaring both sides of the congruence in each succeeding step, we obtain the following array.

$2^1 \equiv 2 \pmod{73}$	$2^1 \equiv 2 \pmod{1103}$
$2^2 \equiv 4 \pmod{73}$	$2^2 \equiv 4 \pmod{1103}$
$2^4 \equiv 16 \pmod{73}$	$2^4 \equiv 16 \pmod{1103}$
$2^8 \equiv 37 \pmod{73}$	$2^8 \equiv 256 \pmod{1103}$
$2^{16} \equiv 55 \pmod{73}$	$2^{16} \equiv 459 \pmod{1103}$
$2^{32} \equiv 32 \pmod{73}$	$2^{32} \equiv 8 \pmod{1103}$
$2^{64} \equiv 2 \pmod{73}$	$2^{64} \equiv 64 \pmod{1103}$
$2^{128} \equiv 4 \pmod{73}$	$2^{128} \equiv 787 \pmod{1103}$

$$
\begin{array}{ll}
2^{256} \equiv 16 \pmod{73} & 2^{256} \equiv 586 \pmod{1103} \\
2^{512} \equiv 37 \pmod{73} & 2^{512} \equiv 363 \pmod{1103} \\
2^{1024} \equiv 55 \pmod{73} & 2^{1024} \equiv 512 \pmod{1103} \\
2^{2048} \equiv 32 \pmod{73} & 2^{2048} \equiv 733 \pmod{1103} \\
2^{4096} \equiv 2 \pmod{73} & 2^{4096} \equiv 128 \pmod{1103} \\
2^{8192} \equiv 4 \pmod{73} & 2^{8192} \equiv 942 \pmod{1103} \\
2^{16\,384} \equiv 16 \pmod{73} & 2^{16\,384} \equiv 552 \pmod{1103} \\
2^{32\,768} \equiv 37 \pmod{73} & 2^{32\,768} \equiv 276 \pmod{1103} \\
2^{65\,536} \equiv 55 \pmod{73} & 2^{65\,536} \equiv 69 \pmod{1103} \\
2^{131\,072} \equiv 32 \pmod{73} & 2^{131\,072} \equiv 349 \pmod{1103}
\end{array}
$$

Therefore,

$$
\begin{aligned}
2^{161\,037} &\equiv 2^{131\,072} \cdot 2^{16\,384} \cdot 2^{8192} \cdot 2^{4096} \cdot 2^{1024} \cdot 2^{256} \cdot 2^{8} \cdot 2^{4} \cdot 2 \\
&\equiv 32 \cdot 16 \cdot 4 \cdot 2 \cdot 55 \cdot 16 \cdot 37 \cdot 16 \cdot 2 \equiv 4\,267\,704\,320 \equiv 1 \pmod{73},
\end{aligned}
$$

and

$$
\begin{aligned}
2^{161\,037} &\equiv 2^{131\,072} \cdot 2^{16\,384} \cdot 2^{8192} \cdot 2^{4096} \cdot 2^{1024} \cdot 2^{256} \cdot 2^{8} \cdot 2^{4} \cdot 2 \\
&\equiv (349 \cdot 552 \cdot 942 \cdot 128) \cdot (512 \cdot 586 \cdot 256 \cdot 16 \cdot 2) \\
&\equiv 23\,228\,725\,248 \cdot 2\,457\,862\,144 \equiv 787 \cdot 918 \equiv 1 \pmod{1103}.
\end{aligned}
$$

Thus, $2^{161\,038} \equiv 2 \pmod{2}$, $2^{161\,038} \equiv 2 \pmod{73}$, and $2^{161\,038} \equiv 2 \pmod{1103}$. Thus, 2, 73, and 1103 each divide $2^{161\,038} - 2$. Therefore, 161 038 divides $2^{161\,038} - 2$ and, hence, 161 038 is a pseudoprime.

Harold Davenport of Cambridge University investigated properties of systems of congruences, called Davenport coverings, such that each integer satisfies at least one of the congruences. Davenport coverings having the property that each integer satisfies exactly one congruence are called exact Davenport coverings. For example, $x \equiv 0 \pmod{2}$ and $x \equiv 1 \pmod{2}$ is an exact Davenport covering of the integers. A necessary condition that a system of congruences be an exact Davenport covering is that the sum of the reciprocals of the moduli is unity and the greatest common divisor of the moduli is greater than one. Paul Erdös, the peripatetic Hungarian mathematician, proposed the following open question: for any positive integer n, does there exist a Davenport covering with distinct moduli all greater than n? Three such examples of Davenport coverings with $n = 2$ are given in the columns of Table 5.2.

Easter, named for Ostura, a pagan goddess of spring, was celebrated by the early Christian Church. However, there was no uniform method for determining Easter. The Council of Nicaea convened by Constantine the Great on June 1, 325, to solve the problem caused by Arianism, formulated the doctrine of the Trinity, ordered bishops to establish hospitals in every

Table 5.2.

$x \equiv 0 \pmod 2$	$x \equiv 0 \pmod 2$	$x \equiv 0 \pmod 2$
$x \equiv 0 \pmod 3$	$x \equiv 0 \pmod 3$	$x \equiv 0 \pmod 3$
$x \equiv 1 \pmod 4$	$x \equiv 1 \pmod 4$	$x \equiv 1 \pmod 4$
$x \equiv 1 \pmod 6$	$x \equiv 5 \pmod 6$	$x \equiv 3 \pmod 8$
$x \equiv 11 \pmod{12}$	$x \equiv 7 \pmod{12}$	$x \equiv 7 \pmod{12}$
		$x \equiv 23 \pmod{24}$

cathedral city, and fixed the date of Easter. They decreed that Easter would henceforth occur on the first Sunday after the full moon that occurs on or after March 21, the date of the vernal equinox. As a consequence, each year Easter falls between March 22 (in 2285) and April 25 (in 2038), the least common occurrence being March 22 and the most common being April 19. Gauss's method for determining the date of Easter is illustrated below.

In Table 5.3, m and n are given by $m \equiv 15 + C - [\![C/4]\!] - [\![(8C + 13)/25]\!] \pmod{30}$ and $n \equiv 4 + C - [\![C/4]\!] \pmod 7$, where C denotes the century year. For example, for 1941, $C = 19$. Gauss let

$$\begin{aligned}
a &= \text{YEAR} \pmod 4,\\
b &= \text{YEAR} \pmod 7,\\
c &= \text{YEAR} \pmod{19},\\
d &= 19c + m \pmod{30},\\
e &= 2a + 4b + 6d + n \pmod 7,
\end{aligned}$$

According to Gauss's algorithm, Easter is either March $(22 + d + e)$ or April $(d + e - 9)$. Gauss noted two exceptions to his rule: if $d = 29$ and $e = 6$, Easter falls one week earlier, on April 19; if $d = 28$, $e = 6$, and $m = 2, 5, 10, 13, 16, 21, 24$, or 39, Easter falls one week earlier, on April

Table 5.3.

Period	m	n
1583–1699	22	2
1700–1799	23	3
1800–1899	23	4
1900–1999	24	5
2000–2099	24	5
2100–2199	24	6

18. For example, for the year 2020, $a = 0$, $b = 4$, $c = 6$, $d = 18$, and $e = 3$. Hence, in 2020, Easter will fall on April 12.

Exercises 5.1

1. If $a \equiv b \pmod{m}$, prove for any integer c and nonnegative integer n that
   (a) $a \pm c \equiv b \pm c \pmod{m}$,
   (b) $ac \equiv bc \pmod{m}$, and
   (c) $a^n \equiv b^n \pmod{m}$.
2. If $a_i \equiv b_i \pmod{m}$, for $i = 1, 2, \ldots, n$, prove that
   (a) $\sum_{i=1}^{n} a_i = \sum_{i=1}^{n} b_i \pmod{m}$ and (b) $\prod_{i=1}^{n} a_i = \prod_{i=1}^{n} b_i \pmod{m}$.
3. If $a \equiv b \pmod{m_1}$ and $a \equiv b \pmod{m_2}$ where $\gcd(m_1, m_2) = 1$, prove that $a \equiv b \pmod{m_1 m_2}$.
4. Show that if $a \equiv b \pmod{m}$ and d divides m, where $d > 0$, then $a \equiv b \pmod{d}$.
5. If $a \equiv b \pmod{m}$ and $a \equiv b \pmod{n}$ then show that $a \equiv b \pmod{\operatorname{lcm}(m, n)}$.
6. Show that if $a \equiv b \pmod{m}$ and $c \equiv d \pmod{m}$ then for any integers x and y, $(ax + cy) \equiv (bx + dy) \pmod{m}$.
7. Prove that if $a \equiv b \pmod{m}$, then $\gcd(a, m) = \gcd(b, m)$.
8. Show that if $a^2 \equiv b^2 \pmod{p}$, where p is prime, then either p divides $a + b$ or p divides $a - b$.
9. Show that $\{47, 86, 22, -14, 32, 20, 143\}$ is a complete residue system modulo 7.
10. Find all integers x such that $-100 \leqslant x \leqslant 100$, and $x \equiv 7 \pmod{19}$.
11. Find a complete residue system modulo 11 composed of multiples of 7.
12. Show that $\{2, 4, 6, \ldots, 2m\}$ is a complete residue system modulo m if and only if m is odd.
13. Show that $\{1^2, 2^2, 3^2, \ldots, m^2\}$ is never a complete residue system modulo m if $m > 2$.
14. Show that 7 divides $1941^{1963} + 1963^{1991}$.
15. Determine the last two digits of 9^{9^9}.
16. Show that 39 divides $53^{103} + 103^{53}$.
17. Show that 7 divides $111^{333} + 333^{111}$.
18. What is the least positive remainder when 19^{385} is divided by 31?
19. Find the units digit of 3^{97}.
20. What are the last two digits of 3^{1000}?

21. Find the remainder when $1! + 2! + \cdots + 100!$ is divided by 15.
22. Find the remainder when $1^5 + 2^5 + \cdots + 100^5$ is divided by 4.
23. Show that $61! + 1 \equiv 63! + 1 \pmod{71}$.
24. Show that 7 divides $5^{2n} + 3 \cdot 2^{5n-2}$ for any positive integer n.
25. Show that 13 divides $3^{n+2} + 4^{2n+1}$ for any positive integer n.
26. If n is odd then show that $n^2 \equiv 1 \pmod 8$.
27. What was the date of Easter in 1916?
28. What day does Easter fall in the current year?
29. Show that $x \equiv 0 \pmod 2$, $x \equiv 0 \pmod 3$, $x \equiv 1 \pmod 4$, $x \equiv 1 \pmod 6$, and $x \equiv 11 \pmod{12}$ form a Davenport covering for the integers.
30. Show that the cube of any positive integer leaves a remainder 0, 1, or 8 when divided by 9.
31. Show that the sum of three consecutive cubes is a multiple of 9.
32. If $n = c_k b^k + \cdots + c_1 b + c_0$, where $0 < c_k < b$, $0 \leqslant c_i < b$, for $i = 1, 2, \ldots, k-1$, and $b > 1$ is a positive integer, show that $b-1$ divides n if and only if $b-1$ divides $c_0 + \cdots + c_k$.
33. If the positive integer n has the remainders r and s when divided by the positive integers m and $m+1$, respectively, show that n has the remainder $(m+1)r + m^2 s$ when divided by $m(m+1)$. [Stifel 1544]

5.2 Divisibility criteria

Before the age of calculators and computers a number of very practical criteria were used to test for divisibility. For example, in the *Talmud* it is written, if a and b are positive integers and 7 divides $2a + b$, then 7 divides $100a + b$. Other rules can be found in the works of al-Khwarizmi and Fibonacci, who included divisibility criteria for 7, 9, and 11 in *Liber abaci*. Some are very straightforward, for example, for any integer n, 2 divides n if and only if the last digit of n is even, and 5 divides n if and only if the last digit of n is either 0 or 5. The next result is helpful in establishing divisibility criteria for other positive integers.

Theorem 5.7 *Let $f(x) \equiv \sum_{i=0}^{n} c_i x^i \pmod m$, where the c_i are integers, for $i = 1, 2, \ldots, n$. If $a \equiv b \pmod m$, then $f(a) \equiv f(b) \pmod m$.*

Proof It follows from Theorem 5.4 that, since $a \equiv b \pmod m$, $a^i \equiv b^i \pmod m$, and $c_i a^i \equiv c_i b^i \pmod m$, for $i = 1, 2, \ldots, n$. Hence, $\sum_{i=0}^{n} c_i a^i = \sum_{i=0}^{n} c_i b^i$, and the result is established. ■

Before assuming the chair of mathematics at Montpellier, Joseph Diez Gergonne was an artillery officer and taught at the Lyceum in Nîmes. The Gergonne point of a triangle, the intersection of the Cevians joining the vertices of the triangle with the points of contact of the incircle, is named for him. He founded the mathematics journal *Annales de Mathématiques* and in 1814, devised the following divisibility criteria.

Theorem 5.8 *Let $\sum_{i=0}^{n} a_i(10)^i$ be the decimal representation of an integer a, $s = \sum_{i=1}^{n} a_i$, the sum of the digits of a, and $t = \sum_{i=0}^{n}(-1)^i a_i$, the alternating sum of the digits of a; then*

(a) $9|a$ *if and only if* $9|s$,
(b) $3|a$ *if and only if* $3|s$, *and*
(c) $11|a$ *if and only if* $11|t$.

Proof If $f(x) = \sum_{i=0}^{n} a_i x^i$, then $a = f(10)$, $s = f(1)$, and $t = f(-1)$. Since $10 \equiv 1 \pmod 9$, $a \equiv s \pmod 9$ or $a - s = 9k$. Hence, $9|a$ if and only if $9|s$. Similarly, $3|a$ if and only if $3|s$. Since $10 \equiv (-1) \pmod{11}$, $a \equiv t \pmod{11}$, so $11|a$ if and only if $11|t$. ■

Example 5.4 Suppose we wish to determine x, y, z, given that 5, 9, and 11 divide $2x1642y032z$. Since 5 divides the number $z = 0$ or 5. From Theorem 5.8, $x + y + z \equiv 7 \pmod 9$ and $-x + y + z \equiv 0 \pmod{11}$. If $z = 0$, $x + y \equiv 7 \pmod 9$ and $-x + y \equiv 0 \pmod{11}$, with $x = y = 8$ as a solution. If $z = 5$, $x + y \equiv 2 \pmod 9$ and $-x + y \equiv 6 \pmod{11}$, then $x = 8$ and $y = 3$ is a solution. Therefore, solutions are given by 28 164 280 320 ($2^{12} \cdot 3^2 \cdot 5 \cdot 11 \cdot 29 \cdot 479$) and 28 164 230 325 ($5 \cdot 9 \cdot 11 \cdot 56\,897\,435$).

Example 5.5 Using divisibility criteria, we show that each term of the sequence 49, 4489, 444 889, 44 448 889, 4 444 488 889, ... is a square. The general term of the sequence is given by

$$
\begin{aligned}
&9 + 8 \cdot 10 + 8 \cdot 10^2 + \cdots + 8 \cdot 10^n + 4 \cdot 10^{n+1} + \cdots + 4 \cdot 10^{2n+1} \\
&\quad = 1 + 4(1 + 10 + 10^2 + \cdots + 10^n) + 4(1 + 10 + \cdots + 10^{2n+1}) \\
&\quad = 1 + 4 \cdot \frac{10^{n+1} - 1}{9} + 4 \cdot \frac{10^{2n+2} - 1}{9} = \frac{4 \cdot 10^{2n+2} + 4 \cdot 10^{n+1} + 1}{9} \\
&\quad = \left(\frac{2 \cdot 10^{n+1} + 1}{3}\right)^2.
\end{aligned}
$$

From Theorem 5.8, 3 divides $2 \cdot 10^{n+1} + 1$. Therefore, $(2 \cdot 10^{n+1} + 1)/3$ is an integer and the result is established.

Example 5.6 (A divisibility rule for 7) Given a positive integer n, truncate n by deleting the tens and units digits, then double the number that remains and add to it the two-digit number that was truncated. The result is divisible by 7 if and only if n is divisible by 7. Repeat the process until divisibility or nondivisibility by 7 is obvious. Consider $n = 13\,295\,\mathbf{476}$. We have

$$\begin{aligned} 2(132\,954) + 76 &= 265\,9\mathbf{84}, \\ 2(2659) + 84 &= 54\mathbf{02}, \\ 2(54) + 02 &= 110. \end{aligned}$$

Since 110 is not divisible by 7, 13 295 476 is not divisible by 7.

Example 5.7 (A divisibility rule for 13) Given a positive integer n, truncate n by deleting the units digit. Four times the units digit added to the remaining number is divisible by 13 if and only if n is divisible by 13. Repeat the process until divisibility or nondivisibility by 13 is obvious. Consider $n = 53\,699\,13\mathbf{9}$; we have

$$\begin{aligned} 5\,369\,913 + 4(9) &= 5\,369\,94\mathbf{9}, \\ 536\,994 + 4(9) &= 537\,03\mathbf{0}, \\ 53\,703 + 4(0) &= 53\,70\mathbf{3}, \\ 5370 + 4(3) &= 538\mathbf{2}, \\ 538 + 4(2) &= 54\mathbf{6}, \\ 54 + 4(6) &= 7\mathbf{8}, \\ 7 + 4(8) &= 3\mathbf{9}. \end{aligned}$$

Since 39 is divisible by 13, 53 699 139 is divisible by 13.

The process of casting out nines can be traced to the tenth century Islamic physician and philosopher, Avicenna [AVE eh SEN ah]. It was popular in medieval schools as a check of arithmetical calculations and is based on properties of digital roots. For instance, the digital root of 9785 is 2 and the digital root of 4593 is 3. Hence, the digital root of their sum must be 5. That is, $\rho(9785) + \rho(4593) = 2 + 3 = 5 = \rho(14\,378)$. Analogously, using congruence notation, we have $9785 \equiv 2 \pmod 9$ and $4593 \equiv 3 \pmod 9$, and $9785 + 4593 = 14\,378 \equiv 5 \pmod 9$. The technique of casting out nines is most beneficial in finding errors in addition and multiplication.

In a number of medieval schools a method called the cross bones check, based on digital roots, was employed. For example, suppose we wish to multiply 3253, whose digital root is 4, by 4912, whose digital root is 7. We begin by making a cross and placing a 4 in the west position and a 7 in the east position. Since $4 \cdot 7 = 28$ has digital root 1, we put a 1 in the north position. If after calculating, we found the product to be 15 978 836, we put its digital root 2 in the south location, as shown in Figure 5.1. However, the 2 in the south position does not equal the 1 in the north position indicating we have made a mistake in our calculation. The process of casting out nines and the cross bones check are equivalent and both will pick up errors, but neither will guarantee calculations are error free.

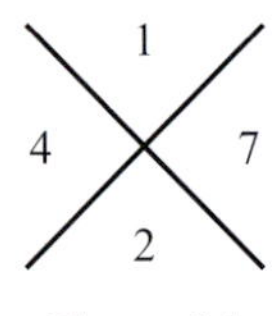

Figure 5.1

Exercises 5.2

1. Prove that if 7 divides $100a + b$, then 7 divides $2a + b$. Is the converse true?
2. Show that if the sum of the digits of a number is subtracted from the number, then the difference is always divisible by 9.
3. Without performing the indicated operations determine the digit x in each of the following calculations.
   (a) $(65\,248) \cdot (124\,589) = 8\,1x9\,183\,07x$.
   (b) $(x12) \cdot (1\,9x3\,12x) = 1\,000\,000\,000$.
   (c) $6\,x56\,681 = (3(843 + x))^2$.
4. Show that $9|R_n$ if and only if $9|n$, where $R_n = (10^n - 1)/9$.
5. Show that 11 divides R_n if and only if n is even.
6. Use the divisibility rule outlined in Example 5.6 to check if 691 504 249 989 is divisible by 7.
7. Use the divisibility rule outlined in Example 5.7 to check if 67 911 603 138 353 is divisible by 13.
8. Use the cross bones check to show that 125 696 times 458 does not equal 57 569 585.
9. We can check the divisibility by 7 of a positive integer having more

than two digits by deleting the units digit of the number and subtracting twice the units digit from what remains. The result is divisible by 7 if and only if the original number is divisible by 7. We can check the divisibility by 13 of a positive integer having more than three digits by deleting the units digit of the number and subtracting 9 times the units digit from what remains. The result is divisible by 13 if and only if the original number is divisible by 13. Devise a similar rule for divisibility by 17.

10. Show that when $7 \cdot 5^{41}$ is written out decimally at least one digit appears more than three times.

5.3 Euler's phi-function

We now introduce a very important and useful number theoretic function. For any positive integer n, the Euler phi-function represents the number of positive integers not exceeding n that are coprime to n, where by convention $\phi(1) = 1$. For example, $\phi(12) = 4$, since 1, 5, 7, and 11 are the only integers that are positive, less than 12, and coprime to 12. Properties of the function were first investigated by Euler in 1760, who at one time used $\pi(n)$ to denote the function. In *Disquisitiones* Gauss introduced the notation $\phi(n)$.

Euler's phi-function has many interesting properties. For example, except for $n = 1$ and 2, $\phi(n)$ is even. Except when $\gcd(n, 10) \neq 1$ the periods of the base 10 decimal expansions of the unit fractions $1/n$ are divisors of $\phi(n)$. In addition, $\sigma(n) + \phi(n) = n \cdot \tau(n)$ is a necessary and sufficient condition for n to be prime. In 1857, Liouville showed that

$$\frac{\zeta(s-1)}{\zeta(s)} = \sum_{n=1}^{\infty} \frac{\phi(n)}{n^s},$$

where $s > 1$ and ζ denotes the real Riemann zeta-function. Bounds for the phi-function are given by

$$\frac{\sqrt{n}}{2} < \phi(n) \leqslant \frac{n}{e^{\gamma} \cdot \ln\ln(n)},$$

where γ denotes the Euler–Mascheroni constant. The average value of the first n values for $\phi(n)$ can be approximated, for large values of n, by $6n/\pi^2$. De la Vallée-Poussin showed that if a and b are coprime positive integers and $\Pi_{a,b}(x)$ denotes the number of primes of the form $a \cdot k + b$ less than or equal to x, for k a positive integer, then

$$\lim_{x\to\infty}\frac{\Pi_{a,b}(x)}{x\cdot\ln(x)}=\frac{1}{\phi(a)}.$$

In 1950, H. Gupta showed that for all $k \geqslant 1$ there is a positive integer n such that $\phi(n) = k!$ There are several open questions concerning the phi-function. For example, in 1922 R.D. Carmichael asked, if given a natural number n does there exist another natural number m such that $\phi(m) = \phi(n)$? In 1994, A. Shalafly and Stan Wagon showed that given a positive integer n, if $\phi(m) \neq \phi(n)$ for all $m \neq n$, then $n > 10^{10\,000\,000}$. In 1932, D.H. Lehmer asked, if $\phi(n)$ divides $n - 1$ does that always imply that n is prime? Lehmer showed that if such a composite positive integer n existed it would be odd, squarefree, and $\omega(n) \geqslant 7$.

Let $\xi(n)$ denote the number of positive integers k, $1 \leqslant k \leqslant n$, such that k is not a divisor of n and $\gcd(k, n) \neq 1$. For example, $\xi(n) = 0$, for $n = 1$, and $\xi(n) = 1$, for $n = 6$ and 9, and $\xi(p) = 0$ whenever p is prime. By construction, $n = \tau(n) + \phi(n) + \xi(n) - 1$. From the Möbius inversion formula, if $\chi(n) = \sum_{d|n}\xi(n)$, then $\xi(n) = \sum_{d|n}\mu(d)\chi(n/d)$. Several number theoretic functions are related by the identity

$$\chi(n) = \sigma(n) + \tau(n) - \left(\frac{1}{2}\right)^{\omega(n)}\prod_{r=1}^{\omega(n)}\frac{(\alpha_i + 2)!}{\alpha_i!} - n,$$

where $n = \prod_{i=1}^{r} p_i^{\alpha_i}$.

The subset of the least residue system modulo n consisting of only those integers which are less than n and are coprime to n is called a reduced residue system modulo n. For example, the set $\{1, 5, 7, 11\}$ forms the reduced residue system modulo 12. For any positive integer n, the set $\{x\colon 1 \leqslant x \leqslant n,\ \gcd(x, n) = 1\}$ forms a multiplicative group with $\phi(n)$ elements.

Theorem 5.9 *If* $\{a_1, a_2, \ldots, a_{\phi(m)}\}$ *is a reduced residue system modulo* m, *and* $\gcd(c, m) = 1$, *then* $\{ca_1, ca_2, \ldots, ca_{\phi(m)}\}$ *is also a reduced residue system modulo* m.

Proof Let $\{a_1, a_2, \ldots, a_{\phi(m)}\}$ be a reduced residue system modulo m, and $\gcd(c, m) = 1$. Since $\gcd(c, m) = 1$ and $\gcd(a_i, m) = 1$, it follows from Theorem 2.7 that $\gcd(ca_i, m) = 1$, for $i = 1, 2, \ldots, \phi(m)$. If $ca_i \equiv ca_j \pmod{m}$, for some $1 \leqslant i < j \leqslant \phi(m)$, it follows from the corollary to Theorem 5.6 that $a_i = a_j$, contradicting the fact that $\{a_1, a_2, \ldots, a_{\phi(m)}\}$ is a set of $\phi(m)$ distinct elements. Therefore, $\{ca_1, ca_2, \ldots, ca_{\phi(m)}\}$ is a set of $\phi(m)$ incongruent integers coprime to m. ■

One of the most elegant results concerning the Euler phi-function, due to Gauss, is that $\varphi(d)$ summed over all the divisors d of a positive integer n equals n.

Theorem 5.10 (Gauss) *For any positive integer n,* $\sum_{d|n}\varphi(d) = n$.

Proof Let n^d denote the number of elements in $\{1, 2, \ldots, n\}$ having a greatest common divisor of d with n, then

$$n = \sum_{d|n} n_d = \sum_{d|n} \left(\frac{n}{d}\right) = \sum_{d|n} \varphi(d). \quad \blacksquare$$

One of the most important properties concerning the phi-function is its multiplicativity.

Theorem 5.11 *The Euler phi-function is multiplicative, that is, if* $\gcd(m, n) = 1$, *then* $\phi(mn) = \phi(m)\phi(n)$.

Proof Since $g(n) = n$ is multiplicative and $n = \sum_{d|n}\phi(n)$, it follows from the corollary to Theorem 3.13 that the phi-function is multiplicative. ■

Gauss based his proof of the multiplicativity of ϕ on the fact that if a is any one of the $\phi(m)$ positive integers less than m and coprime to m, and b is any one of the $\phi(n)$ positive integers less than n and coprime to n, then there is exactly one positive integer x less than mn, such that $x \equiv a \pmod{m}$ and $x \equiv b \pmod{n}$. Since x is coprime to m and to n, it is coprime to mn. Thus, there are $\phi(m)$ choices for a and $\phi(n)$ choices for b and each pair of choices uniquely determines a value for x that is coprime to mn. Therefore, Gauss reasoned, there are $\phi(m) \cdot \phi(n)$ choices for x.

We now use the multiplicative property of the Euler phi-function to develop a method to calculate $\phi(n)$ for any given positive integer n.

Theorem 5.12 *If p is a prime and α is a positive integer then* $\phi(p^\alpha) = p^\alpha(p-1)/p = p^{\alpha-1}(p-1)$.

Proof Among the p^α positive integers less than or equal to p^α, those $p^{\alpha-1}$ which are not coprime to p^α are exactly $p, 2p, \ldots, (p^{\alpha-1}-1)p, p^{\alpha-1}p$. That is, they are precisely the $p^{\alpha-1}$ multiples of p which are less than or equal to p^α. Hence, the number of positive integers less than p^α and coprime to p^α is given by

$$p^\alpha - p^{\alpha-1} = p^{\alpha-1}(p-1). \quad \blacksquare$$

Corollary *If* $n = \prod_{i=1}^{r} p_i^{\alpha_i}$, *then*

$$\phi(n) = n \cdot \prod_{i=1}^{r} \left(\frac{p_i - 1}{p_i} \right).$$

For example,

$$\begin{aligned}
&\phi(304\,920) \\
&\quad = \phi(2^3 \cdot 3^2 \cdot 5 \cdot 7 \cdot 11^2) \\
&\quad = 2^3 \cdot 3^2 \cdot 5 \cdot 7 \cdot 11^2 \cdot \left(\frac{2-1}{2}\right)\left(\frac{3-1}{3}\right)\left(\frac{5-1}{5}\right)\left(\frac{7-1}{7}\right)\left(\frac{11-1}{11}\right) \\
&\quad = 2^2 \cdot 3 \cdot 11 \cdot 2 \cdot 4 \cdot 6 \cdot 10 = 63\,360.
\end{aligned}$$

In June of 1640, Fermat wrote to Mersenne that if p is prime and divides $2^q - 1$, then q divides $p - 1$. In a letter to Frenicle, in October 1640, Fermat claimed that he could prove that if p is prime with $0 \leqslant a < p$, then p divides $a^p - a$; however, he added, the proof was too long to be included in the letter.

About 30 years later, in an unpublished manuscript discovered in 1863, Leibniz used the fact that if p is prime then p divides the binomial coefficient $\binom{p}{k}$ to show that if p is prime then p divides $(a_1 + a_2 + \cdots + a_n)^p - (a_1^p + a_2^p + \cdots + a_n^p)$. Letting $a_i = 1$, for $i = 1, 2, \ldots, n$, Leibniz showed that p divides $n^p - n$, for any positive integer n. The first published proof of the corollary to Theorem 5.13, Fermat's Little Theorem, was given by Euler in 1736. Euler proved the generalized result, the Euler–Fermat Theorem, in 1760.

Theorem 5.13 (Euler–Fermat Theorem) *If* $\gcd(a, m) = 1$, *then* $a^{\phi(m)} \equiv 1 \pmod{m}$.

Proof Let $a_1, a_2, \ldots, a_{\phi(m)}$ form a reduced residue system modulo m. Since $\gcd(a, m) = 1$, it follows from Theorem 5.9 that the products $a \cdot a_1, a \cdot a_2 \ldots, a \cdot a_{\phi(m)}$ also form a reduced residue system modulo m. Thus, for each i, $1 \leqslant i \leqslant \phi(m)$, there is an integer j, $1 \leqslant j \leqslant \phi(m)$, such that $a \cdot a_i \equiv a_j \pmod{m}$. Thus, $\prod_{i=1}^{\phi(m)} a \cdot a_i \equiv \prod_{j=1}^{\phi(m)} a_j \pmod{m}$, or $a^{\phi(m)} \prod_{i=1}^{\phi(m)} a_i \equiv \prod_{j=1}^{\phi(m)} a_j \pmod{m}$. Since $\gcd(a_i, m) = 1$, for $1 \leqslant i \leqslant \phi(m)$, we cancel $\prod_{k=1}^{\phi(m)} a_k$ from both sides of the equation to obtain $a^{\phi(m)} \equiv 1 \pmod{m}$. ■

Corollary (Fermat's Little Theorem) *If p is a prime, and* $\gcd(a, p) = 1$, *then* $a^{p-1} \equiv 1 \pmod{p}$.

Since 1, 5, 7, and 11 are coprime to 12 and $\phi(12) = 4$, the Euler–Fermat Theorem implies that 1^4, 5^4, 7^4, and 11^4 are all congruent to 1 modulo 12. In addition, since, $a^{p-1} - 1 = (a^{(p-1)/2} - 1)(a^{(p-1)/2} + 1)$, an immediate consequence of Fermat's Little Theorem is that if p is an odd prime and $\gcd(a, p) = 1$, then $a^{(p-1)/2} \equiv \pm 1 \pmod{p}$. The converse of Fermat's Little Theorem is false since $a^{560} \equiv 1 \pmod{561}$ for all a such that $\gcd(a, 561) = 1$, yet $561 = 3 \cdot 11 \cdot 17$ is not prime. The contrapositive of Fermat's Little Theorem may be used as a primality test. That is, if for some positive integer a less than n, we find that $a^{n-1} \not\equiv 1 \pmod{n}$ then n is not prime. For example, $2^{2146} \equiv 662 \pmod{2147}$, hence, 2147 is not prime. A primality test devised by Lucas, based on Fermat's Little Theorem, states that if m is a positive integer such that $a^{m-1} \equiv 1 \pmod{m}$ and $a^{(m-1)/p} \equiv 1 \pmod{m}$ for every prime divisor p of $m - 1$, then m is prime.

Example 5.8 Let us apply the Euler–Fermat Theorem to solve the linear equation $x^{341} \equiv 127 \pmod{893}$. We have $\phi(893) = \phi(19 \cdot 47) = 18 \cdot 46 = 828$. From either the Euclidean or the Saunderson algorithm, we find that $\gcd(828, 341) = 1$ and $(-7) \cdot 828 + 17 \cdot 341 = 1$. Hence, $(x^{341})^{17} = x^{1+828 \cdot 7} = x(x^{828})^7 = x \cdot (1)^7 = x$. Using the binary-square method and the fact that $17 = 16 + 1$ we obtain

$$
\begin{aligned}
127^2 &\equiv 55 \pmod{893}, \\
127^4 &\equiv 346 \pmod{893}, \\
127^8 &\equiv 54 \pmod{893}, \\
127^{16} &\equiv 237 \pmod{893}.
\end{aligned}
$$

Therefore, $x \equiv (x^{341})^{17} \equiv (127)^{17} \equiv 127^{16} \cdot 127^1 \equiv 237 \cdot 127 \equiv 630 \pmod{893}$.

Example 5.9 Recall that a composite positive integer n is called a Carmichael number if $a^n \equiv a \pmod{n}$, whenever a is less than and coprime to n. Suppose $1 < a < 561 = 3 \cdot 11 \cdot 17$ and $\gcd(a, 561) = 1$. We have $a^{561} - a = a(a^{560} - 1) = a[(a^{10})^{56} - 1^{56}] = a[(a^{10} - 1) \cdot f(a)]$, where $f(a)$ is a polynomial in a. Since 11 divides $a^{10} - 1$, 11 divides $a^{561} - a$. In addition, $a^{561} - a = a[(a^{16})^{35} - 1^{35}] = a(a^{16} - 1) \cdot g(a)$ and $a^{561} - a = a[(a^2)^{280} - 1^{280}] = a(a^2 - 1) \cdot h(a)$, where $g(a)$ and $h(a)$ are polynomials in a. Since 17 divides $a^{16} - 1$ and 3 divides $a^2 - 1$, it follows that 561 divides $a^{561} - a$ whenever $\gcd(a, 561) = 1$. Therefore, 561 is a Carmichael number.

The set of Farey fractions $\mathscr{F}_n$ of order n consist of the ascending sequence of irreducible fractions between 0 and 1 whose denominators do not exceed n. That is, k/m is in $\mathscr{F}_n$ if and only if $0 \leqslant k \leqslant m \leqslant n$, and k and m are coprime. For example $\mathscr{F}_1 = \{\frac{0}{1}, \frac{1}{1}\}, \mathscr{F}_2 = \{\frac{0}{1}, \frac{1}{2}, \frac{1}{1}\}$. The middle term of $\mathscr{F}_n$ is always $\frac{1}{2}$, since the number of irreducible fractions with denominator m is given by $\phi(m)$, the number of Farey fractions of order n, the number of irreducible fractions $0 \leqslant k/m \leqslant 1$, with $0 \leqslant m < n$, is $1 + \sum_{k=1}^{n} \phi(k)$. Two fractions a/b and c/d in $\mathscr{F}_n$ are called complementary if their sum is unity. The two fractions adjacent to $\frac{1}{2}$ are complementary. If a/b and c/d are complementary and $a/b < c/d$, the fractions preceding a/b and following c/d are complementary. In 1883, J.J. Sylvester proved that the sum of the Farey fractions of order n is $\frac{1}{2}[1 + \sum_{k=1}^{n} \phi(k)]$. For example, $\mathscr{F}_6 = \{\frac{0}{1}, \frac{1}{6}, \frac{1}{5}, \frac{1}{4}, \frac{1}{3}, \frac{2}{5}, \frac{1}{2}, \frac{3}{5}, \frac{2}{3}, \frac{4}{5}, \frac{5}{6}, \frac{1}{1}\}$ and $\frac{1}{2}(1 + \sum_{k=1}^{6} \phi(k)) = 6.5$. For larger values of n, the sum can be approximated by $3(n/\pi)^2$.

In 1802, C.H. Haros discovered several basic properties of Farey fractions. In 1816, those and other properties appeared in an article by John Farey. Farey, a geologist, wrote a letter to the *Philosophical Magazine* noting several properties of such fractions he observed in Henry Goodwyn's *Complete Decimal Quotients*, a privately circulated manuscript. That same year Cauchy offered proofs to most of the results mentioned by Haros and Farey. A pair of Farey fractions $(a/b, c/d)$ is said to be a Farey pair if $bc - ad = 1$. Adjacent Farey fractions are examples of Farey pairs. The

$$\frac{1}{1} \qquad \phi(1) = 1$$

$$\frac{1}{2} \quad \frac{2}{2} \qquad \phi(2) = 2$$

$$\frac{1}{3} \quad \frac{2}{3} \quad \frac{3}{3} \qquad \phi(3) = 2$$

$$\frac{1}{4} \quad \frac{2}{4} \quad \frac{3}{4} \quad \frac{4}{4} \qquad \phi(4) = 2$$

$$\frac{1}{5} \quad \frac{2}{5} \quad \frac{3}{5} \quad \frac{4}{5} \quad \frac{5}{5} \qquad \phi(5) = 4$$

$$\frac{1}{6} \quad \frac{2}{6} \quad \frac{3}{6} \quad \frac{4}{6} \quad \frac{5}{6} \quad \frac{6}{6} \qquad \phi(6) = 2$$

Figure 5.2

mediant of a Farey pair $(a/b, c/d)$ is given by $(a+c)/(b+d)$. The mediant of two Farey fractions of order n is a Farey fraction of order $n+1$.

Exercises 5.3

1. Find $\phi(n)$ for the following values of n.
   (a) 406; (b) 756; (c) 1228; (d) 7642.
2. Find the reduced residue system modulo 18.
3. Show that $\phi(25\,930) = \phi(25\,935) = \phi(25\,940) = \phi(25\,942)$.
4. If p and $p+2$ are twin primes, show that $\phi(p+2) = \phi(p) + 2$.
5. Show that $(\phi(n)\sigma(n) + 1)/n$ is an integer if n is prime.
6. If p is prime then show that $1 + \phi(p) + \phi(p^2) + \cdots + \phi(p^n) = p^n$.
7. Show that $f(n) = \phi(n)/n$ is strongly multiplicative. That is, show that $f(p^k) = f(p)$, where p is prime and k is a positive integer.
8. Give a characterization of n if
   (a) $\phi(n)$ is odd,
   (b) $\phi(n) = n - 1$,
   (c) $\phi(n)$ divides n,
   (d) 4 divides $\phi(n)$,
   (e) $\phi(n) = 2^k$, for some positive integer k,
   (f) $\phi(n) = n/2$,
   (g) $\phi(n) = n/4$,
   (h) 2^k divides $\phi(n)$ for some positive integer k.
9. Show that $\phi(n^2) = n\phi(n)$, for $n \geqslant 1$.
10. Show that if $n = 11^k \cdot p$, where $k \geqslant 1$ and p is prime, then $10|\phi(n)$. Hence, there are infinitely many positive integers for which 10 divides $\phi(n)$.
11. Show that if $n = 2^{2k+1}$, where $k \geqslant 1$, then $\phi(n)$ is square. Hence, there are infinitely many integers n for which $\phi(n)$ is square.
12. Determine the possible remainders when the hundredth power of an integer is divided by 125.
13. Estimate upper and lower bounds for $\phi(n)$ when $n = 100$ and $n = 1000$.
14. Find the average value of $\phi(n)$, for $1 \leqslant n \leqslant 100$. How does the average value compare with $6 \cdot 100/\pi^2$?
15. If $n \geqslant 2$ then show that

$$\sum_{\substack{\gcd(x,n)=1 \\ x<n}} x = \frac{n \cdot \phi(n)}{2}.$$

16. Show that $\phi(n) \leqslant n - \sqrt{n}$ if n is composite.
17. Show for any positive integer n that $\sum_{k=1}^{n}\phi(k) \cdot [\![n/k]\!] = n(n+1)/2$. [Dirichlet 1849]
18. Evaluate $\sum_{d|36}\phi(d)$.
19. Show that $\phi(p^\alpha) + \sigma(p^\alpha) \geqslant 2p^\alpha$ where p is prime and α a positive integer.
20. Find all positive integers n such that $\phi(n) + \sigma(n) = 2n$.
21. Show that $f(n) = \sigma(n) \cdot \phi(n)/n^2$ is multiplicative.
22. If p is prime then show that p divides $\binom{p}{k}$, where $1 \leqslant k \leqslant p-1$.
23. If p is an odd prime, then show that
    (a) $1^{p-1} + 2^{p-1} + \cdots + (p-1)^{p-1} \equiv (-1) \pmod{p}$, and
    (b) $1^p + 2^p + \cdots + (p-1)^p \equiv 0 \pmod{p}$.
24. If $\gcd(m, n) = 1$ show that $m^{\phi(n)} + n^{\phi(m)} \equiv 1 \pmod{mn}$.
25. Use the Euler–Fermat Theorem to solve for x if $41x \equiv 53 \pmod{62}$.
26. Show that 6601 is a Carmichael number.
27. Verify the Ramanujan sum

$$\sum_{d|\gcd(m,n)} d \cdot \mu\left(\frac{n}{d}\right) = \frac{\mu\left(\dfrac{n}{\gcd(m,\, n)}\right) \cdot \phi(n)}{\phi\left(\dfrac{n}{\gcd(m,\, n)}\right)},$$

    for the case when $m = 90$ and $n = 105$.
28. Show that $\sum_{d|n}\phi(d) \cdot \tau(n/d) = \sigma(n)$.
29. Show that $\sum_{d|n}\phi(d) \cdot \sigma(n/d) = n \cdot \tau(n)$.
30. Prove that n is prime if and only if $\sigma(n) + \phi(n) = n \cdot \tau(n)$.
31. For $n = 12$, show that $n = \tau(n) + \phi(n) + \xi(n) - 1$.
32. For $n = 12$, show that

$$\chi(n) = \sigma(n) + \tau(n) - \left(\frac{1}{2}\right)^{\omega(n)} \prod_{i=1}^{\omega(n)} \frac{(\alpha_i + 2)!}{\alpha_i!} - n,$$

    where $n = \prod_{i=1}^{r} p_i^{\alpha_i}$.
33. For $n = 12$, show that $\xi(n) = \sum_{d|n}\mu(d)\chi(n/d)$.
34. Prove that

$$\sum_{d|\,p^\alpha} \frac{\mu^2(d)}{\phi(d)} = \frac{p^\alpha}{\phi(p^\alpha)},$$

    where p is prime and α a positive integer.
35. Compare the values of $\frac{1}{2}\sum_{k=1}^{10}\phi(k)$ and $3(10/\pi)^2$.
36. Determine $\mathscr{F}_7$.
37. If a/b and c/d are two successive terms of $\mathscr{F}_n$, show that $bc - ad = 1$. [Haros]

38. If a/b and c/d are any two fractions such that $a/b < c/d$, show that
$$\frac{a}{b} < \frac{a+c}{b+d} < \frac{c}{d}.$$
39. If $(a/b,\ c/d)$ is a Farey pair, the closed interval $[a/b,\ c/d]$ is called a Farey interval. Show that the length of a Farey interval is $1/bd$.
40. If $x/y = (a+c)/(b+d)$, then $a/b < x/y < c/d$, with $bx - ay = cy - dx = 1$. Find x/y such that $a/b < x/y < c/d$, $bx - ay = m$ and $cy - dx = n$.
41. If $n+1$ is a cube show that 504 divides $n(n+1)(n+2)$.

5.4 Conditional linear congruences

The object of this section and the next chapter will be to develop techniques to enable us to solve integral polynomial congruences in one variable. More precisely, if $f(x)$ is a polynomial whose coefficients are integers, we say that a is a root of the conditional congruence $f(x) \equiv 0 \pmod{m}$ if $f(a) \equiv 0 \pmod{m}$. Since $f(a) \equiv f(b) \pmod{m}$ if $a \equiv b \pmod{m}$, all solutions of the conditional congruence $f(x) \equiv 0 \pmod{m}$ will be known provided we find all the solutions in any complete residue system modulo m. Therefore, we restrict ourselves to finding solutions to $f(x) \equiv 0 \pmod{m}$ in the least residue system modulo m, $\{0, 1, \ldots, m-1\}$. We say that $f(x) \equiv 0 \pmod{m}$ has r incongruent solutions modulo m, when exactly r elements in the set $\{0, 1, \ldots, m-1\}$ are solutions to $f(x) \equiv 0 \pmod{m}$.

Diophantus's name is immortalized in the designation of indeterminate integral equations, even though he considered only positive rational solutions to equations and, long before his time, the Pythagoreans and Babylonians found positive integral solutions to $x^2 + y^2 = z^2$. Nevertheless, we call an integral equation from which we require only integer solutions a Diophantine equation.

In 1900, at the International Congress of Mathematicians in Paris, one of the 23 problems posed by David Hilbert of Göttingen to challenge mathematicians in the twentieth century asked if there were any uniform method for solving all Diophantine equations. In 1970, Yuri Matiasevich, of the Steklov Institute of Mathematics, using earlier results of Martin Davis, Hillary Putnam, and Julia Robinson, answered Hilbert's query in the negative. Robinson of the University of California at Berkeley was the first woman to serve as president of the American Mathematical Society.

Let us consider solutions to the simplest polynomial congruences, namely, linear congruences of the form $ax \equiv b \pmod{m}$. The following result was established by Bachet in 1612.

Theorem 5.14 (Bachet's Theorem) *If a and b are integers, m is a positive integer, and $\gcd(a, m) = 1$, then a unique solution to $ax \equiv b \pmod{m}$ exists. If $\gcd(a, m) = d$ and $d|b$, then d incongruent solutions exist. If $d \nmid b$, then no solution exists.*

Proof Suppose $ax \equiv b \pmod{m}$ and $d|b$, then there exists an integer t such that $td = b$. Since $\gcd(a, m) = d$, there exist integers r and s such that $d = ar + ms$. Thus, $b = td = tar + tms$, so $a(tr) \equiv b \pmod{m}$ and tr is a solution to the congruence $ax \equiv b \pmod{m}$. Suppose that x_0 is such that $ax_0 \equiv b \pmod{m}$, hence, $ax_0 - b = km$ for some integer k. Since $d|a$ and $d|m$ it follows that $d|b$. By contraposition, if $d \nmid b$, then no solution exists to $ax \equiv b \pmod{m}$. Thus if x_0 is a solution to $ax \equiv b \pmod{m}$, so is $x_0 + k(m/d)$, since $d|a$ and $a(x_0 + k(m/d)) \equiv ax_0 + km(a/d) \equiv ax_0 \equiv b \pmod{m}$, for $k = 1, 2, \ldots, d - 1$. ■

The proof of Theorem 5.14 is constructive and implies that if x_0 is a solution to $ax \equiv b \pmod{m}$, then so is $x_0 + k(m/d)$, for $k = 1, 2, \ldots, d - 1$. In order to obtain solutions to linear equations a combination of brute force and cleverness must often be applied. The three possible cases for a first order linear congruence are illustrated in the following example.

Example 5.10 Solve for x if

(a) $22x \equiv 4 \pmod{29}$,
(b) $51x \equiv 21 \pmod{36}$, and
(c) $35x \equiv 15 \pmod{182}$.

Solutions:

(a) $22x \equiv 4 \pmod{29}$, divide both sides by 2 to obtain $11x \equiv 2 \pmod{29}$, multiply both sides by 8 to obtain $88x \equiv 16 \pmod{29}$, reduce modulo 29 to obtain $x \equiv 16 \pmod{29}$.
(b) $51x \equiv 21 \pmod{36}$, reduce modulo 36 to obtain $15x \equiv 21 \pmod{36}$, divide by 3 to obtain $5x \equiv 7 \pmod{12}$, multiply both sides by 5 to obtain $25x \equiv 35 \pmod{12}$, reduce modulo 12 to obtain $x \equiv 11 \pmod{12}$, hence $x = 11 + 12t$, which implies that the answers to the original congruence are $x \equiv 11 \pmod{36}$, $x \equiv 23 \pmod{36}$ and $x \equiv 35 \pmod{36}$.

(c) $35x \equiv 15 \pmod{182}$, since $\gcd(35, 182) = 7$ and $7 \nmid 15$ the congruence has no solutions.

Around 1900, the Russian mathematician Georgi Voroni devised a formula to solve a special case of first order linear congruences, namely if $\gcd(a, m) = 1$, the solution to $ax \equiv 1 \pmod{m}$ is given by $x \equiv (3 - 2a + 6\sum_{k=1}^{a-1}[\![mk/a]\!]^2) \pmod{m}$. Voroni's formula works best when a is small and m is large. For example, the solution to $4x \equiv 1$ (mod 37) is given by $x \equiv 3 - 8 + 6([\![\frac{37}{4}]\!]^2 + [\![\frac{74}{4}]\!]^2 + [\![\frac{111}{4}]\!]^2) \pmod{37} = 6799 \pmod{37} \equiv 28 \pmod{37}$. According to the next result, our knowledge of first order linear congruences may be applied to solve linear Diophantine equations of the form $ax + by = n$.

Theorem 5.15 *The Diophantine equation $ax + by = n$ is solvable if and only if d divides n, where $d = \gcd(a, b)$, and if (x_0, y_0) is any solution, then every solution is given by*

$$\left(x_0 + k\left(\frac{b}{d}\right),\ y_0 - k\left(\frac{a}{d}\right)\right), \quad \textit{where } k = 0, \pm 1, \pm 2, \ldots.$$

Proof Solving $ax + by = n$ is equivalent to solving either $ax \equiv n \pmod{b}$ or $by \equiv n \pmod{a}$. A solution to either of these congruences is possible if and only if $d|n$, where $d = \gcd(a, b)$. If x_0 is any solution to $ax \equiv n$ (mod b), every solution to $ax \equiv n \pmod{b}$ is given by $x_0 + k(b/d)$. Hence, if $y_0 = (n - ax_0)/b$, $y = y_0 - k(a/d)$ and $x = x_0 + k(b/d)$. Therefore,

$$\begin{aligned} n - ax = n - a\left(x_0 + k\left(\frac{b}{d}\right)\right) &= b\left[\frac{n - ax_0}{b} - k\left(\frac{a}{d}\right)\right] \\ &= b\left[y_0 - k\left(\frac{a}{d}\right)\right] = by. \ \blacksquare \end{aligned}$$

For example, in order to find integral solutions to the linear equation $15x + 7y = 110$, we solve either $7y \equiv 110 \pmod{15}$ or $15x \equiv 110$ (mod 7). Without loss of generality, consider $15x \equiv 110 \pmod{7}$. Reducing modulo 7, we obtain $x \equiv 5 \pmod{7}$. Hence, $x = 5 + 7k$, for $k = 0, \pm 1, \pm 2, \ldots$. Thus, $15(5 + 7k) + 7y = 110$ or $75 + 15(7k) + 7y = 110$. Thus, $7y = 35 - 15(7k)$. It follows that $y = 5 - 15k$, for $k = 0, \pm 1, \pm 2, \ldots$. We could just as well have used the Euclidean algorithm to obtain integers a and b such that $15a + 7b = 1$ and then multiplied both sides of the equation by 110. In fact, an alternate technique to solve $ax + by = n$, with $d = \gcd(a, b)$, noted by P. Barlow in 1811, follows from the fact that, since

$d|n$ and $\gcd(a/d, b/d) = 1$, there exist integers x and y such that $(a/d)x + (b/d)y = 1$. Therefore, $a(nx/d) + b(ny/d) = n$ and all solutions are given by $x = nx/d + k(b/d)$, and $y = ny/d + k(a/d)$, for k an integer.

In 1826, generalizing Theorem 5.15 to higher order linear equations of the form $ax + by + cz = d$, Cauchy showed that if the greatest common divisor of a, b, c is unity, every integral solution to $ax + by + cz = 0$ is of the form $x = bt - cs$, $y = cr - at$, and $z = as - br$. In 1859, V.A. Lebesgue showed that if the greatest common divisor of a, b, c is unity then every integral solution to $ax + by + cz = d$ is given by $x = deg + ces + bt/D$, $y = dfg + cfu - at/D$, and $z = dh - Ds$, where s and t are arbitrary, $D = \gcd(a, b)$, $ae + bf = D$, and $Dg + ch = 1$. Thus, in order to find integral solutions to the equation $ax + by + cz = n$, let $ax + by = n - cz$, solve for z where $cz \equiv n \pmod{d}$ and $d = \gcd(a, b)$, and plug the solution back into the original equation. In 1774, T. Moss listed 412 solutions to $17x + 21y + 27z + 36w = 1000$ in the *Ladies' Diary*. In 1801, Gauss noted that if the greatest common divisor of the coefficients of $ax + by + cz + dw = e$ divides e then an integral solution exists.

Example 5.11 Let us determine a solution to the linear equation $6x + 8y + 5z = 101$. Since $\gcd(6, 8) = 2$, $5z \equiv 101 \pmod{2}$, implying that $z \equiv 1 \pmod{2}$ $z = 1 + 2t$. Substituting, we obtain $6x + 8y + 5 + 10t = 101$ or $6x + 8y + 10t = 96$. Hence, $3x + 4y + 5t = 48$. Considering the equation modulo 3, we obtain $4y + 5t \equiv 48 \pmod{3}$, implying that $y \equiv -2t \pmod{3}$ or $y = -2t + 3s$. Thus, $6x - 16t + 24s + 5 + 10t = 101$ or $x = 16 + t - 4s$. Therefore, the complete solution is given by $x = 16 + t - 4s$, $y = -2t + 3s$, and $z = 1 + 2t$.

Astronomical problems dealing with periodic motions of celestial bodies have been prevalent throughout history. One method for solving such problems originated in China. *Master Sun's Mathematical Manual* written in the late third century repeats many of the results found in the earlier *Nine Chapters on the Mathematical Art*, but presents in verse a new rule called 'the great generalization' for determining, in particular, a number having the remainders 2, 3, 2 when divided by 3, 5, 7 respectively. The method was clearly outlined and disseminated in the Sichuan mathematician–astronomer Qin Jiushao's *Mathematical Treatise in Nine Sections* in 1247. Quite remarkably, as an indication of the transmission of knowledge in the ancient world, Nicomachus included the same example in his *Introduction to Arithmetic*. The rule, known as the Chinese Remainder

Theorem, offers a practical method for determining the solution of a set of first order linear congruences. We credit the first modern statement of the theorem to Euler. Gauss discovered the result independently around 1801. In 1852, the method was popularized in a treatise, *Jottings on the Science of Chinese Arithmetic*, by Alexander Wylie.

Theorem 5.16 (Chinese Remainder Theorem) *If $m_1, m_2, \ldots, m_k$ are given moduli, coprime in pairs, then the system of linear congruences $x \equiv a_i \pmod{m_i}$, for $1 \leqslant i \leqslant k$, has a unique solution modulo $m = \prod_{i=1}^{k} m_i$.*

Proof In order to solve the system $x \equiv a_i \pmod{m_i}$, for $i = 1, 2, \ldots, k$, let $M_i = m/m_i$, where $m = \prod_{i=1}^{k} m_i$, and b_i be such that $M_i b_i \equiv 1 \pmod{m_i}$. We have $m_j | M_i$ and $\gcd(m_i, m_j) = 1$ for $i \neq j$. Since $\gcd(m_i, M_i) = 1$, the congruence $M_i y \equiv 1 \pmod{m_i}$ has a unique solution b_i, for $1 \leqslant i \leqslant k$. Hence, for each i there exists an integer b_i such that $M_i b_i \equiv 1 \pmod{m_i}$. Let $x_0 \equiv \sum_{i=1}^{k} M_i b_i a_i \pmod{m}$. Since $a_i M_i b_i \equiv a_i \pmod{m_i}$ and $M_i \equiv 0 \pmod{m_j}$, for $i \neq j$, it follows that $x_0 \equiv a_i \pmod{m_i}$, for $i = 1, 2, \ldots, k$. Hence, x_0 is a solution of the system of linear congruences. Suppose that x_1 is any other solution of the system. We have $x_0 \equiv x_1 \equiv a_i \pmod{m_i}$, for $i = 1, 2, \ldots, k$. Hence, $m_i | (x_1 - x_0)$, for $i = 1, 2, \ldots, k$. Since $\gcd(m_i, m_j) = 1$, for $i \neq j$, it follows that $m | (x_1 - x_0)$, and thus $x_1 \equiv x_0 \pmod{m}$. Therefore, if a solution exists, it is unique modulo m. ■

Example 5.12 Let us use the Chinese Remainder Theorem to solve the system

$$x \equiv 2 \pmod{3},$$
$$x \equiv 3 \pmod{5},$$
$$x \equiv 2 \pmod{7}.$$

Let $m = 3 \cdot 5 \cdot 7 = 105$, then

$$\begin{array}{lll} m_1 = 3, & m_2 = 5, & m_3 = 7, \\ a_1 = 2, & a_2 = 3, & a_3 = 2, \\ M_1 = 35, & M_2 = 21, & M_3 = 15. \end{array}$$

Solve the following congruences for b_i, for $i = 1, 2$, and 3.

$$\begin{array}{lll} 35b_1 \equiv 1 \pmod{3}, & 21b_2 \equiv 1 \pmod{5}, & 15b_3 \equiv 1 \pmod{7}, \\ 2b_1 \equiv 1 \pmod{3}, & b_2 \equiv 1 \pmod{5}, & b_3 \equiv 1 \pmod{7}, \\ b_1 \equiv 2 \pmod{3}, & b_2 \equiv 1 \pmod{5}. & \end{array}$$

Therefore, $x \equiv \sum_{i=1}^{3} M_i b_i a_i = 35 \cdot 2 \cdot 2 + 21 \cdot 1 \cdot 3 + 15 \cdot 1 \cdot 2 \equiv 233 \equiv 23 \pmod{105}$.

Simultaneous first order linear equations encountered in Chinese remainder-type problems may be solved directly (and often more efficiently) using brute force. For example, suppose we are given the following system of linear equations:

$$x \equiv 3 \pmod{2},$$
$$x \equiv 1 \pmod{5},$$
$$x \equiv 2 \pmod{7}.$$

From the first equation $x = 3 + 2k$, for some integer k. Substituting into the second equation for x yields $3 + 2k \equiv 1 \pmod{5}$ or $2k \equiv 3 \pmod{5}$ so $k \equiv 4 \pmod{5}$ or $k = 4 + 5r$, for some integer r. Substituting, we obtain $x = 3 + 2k = 3 + 2(4 + 5r) = 11 + 10r$. Substituting into the third equation for x yields $11 + 10r \equiv 2 \pmod{7}$, implying that $3r \equiv 5 \pmod{7}$ or $r \equiv 4 \pmod{7}$. Hence, $r = 4 + 7s$, for some integer s. Substituting, we obtain $x = 11 + 10r = 11 + 10(4 + 7s) = 51 + 70s$. Therefore, $x \equiv 51 \pmod{70}$.

The Chinese Remainder Theorem is a special case of a more general result, illustrated by the Buddhist monk Yi Xing (YEE SHING) around 700, which states that the system $x \equiv a_i \pmod{m_i}$, for $i = 1, 2, \ldots, k$, is solvable if and only if $\gcd(m_i, m_j) | (a_j - a_i)$, for $1 \leqslant i \leqslant j \leqslant k$, and, if a solution exists, it is unique modulo $m = \text{lcm}(m_1, m_2, \ldots, m_k)$. Qin Jiushao outlined a method for solving such problems by finding integers $c_1, c_2, \ldots, c_k$ which are coprime in pairs such that c_i divides m_i, for $i = 1, 2, \ldots, k$, and $\text{lcm}(c_1, c_2, \ldots, c_k) = \text{lcm}(m_1, m_2, \ldots, m_k)$. He let $M_i = m/c_i$ and b_i be such that $M_i b_i \equiv 1 \pmod{c_i}$; then a solution is given by $x \equiv \sum_{i=1}^{k} M_i b_i c_i \pmod{m}$.

Example 5.13 Using Qin Jiushao's method, let us solve the system

$$x \equiv 1 \pmod{4},$$
$$x \equiv 5 \pmod{6},$$
$$x \equiv 4 \pmod{7}.$$

We have $m = \text{lcm}(4, 6, 7) = 84$. Hence

$$\begin{array}{lll} a_1 = 1, & a_2 = 5, & a_3 = 4, \\ m_1 = 4, & m_2 = 6, & m_3 = 9, \\ c_1 = 4, & c_2 = 3, & c_3 = 7, \\ N_1 = 21, & N_2 = 28, & N_3 = 12. \end{array}$$

$21b_1 \equiv 1 \pmod{4}$, $28b_2 \equiv 1 \pmod{3}$, and $12b_3 \equiv 1 \pmod{7}$ imply that

$b_1 \equiv 1 \pmod 4$, $b_2 \equiv 1 \pmod 3$, and $b_3 \equiv 1 \pmod 7$. Hence, $x \equiv \sum_{i=1}^{k} N_i b_i a_i = 21 \cdot 1 \cdot 1 + 28 \cdot 1 \cdot 5 + 12 \cdot 4 \cdot 3 \equiv 305 \equiv 53 \pmod{84}$.

The following problem has a long history. It appears in the work of the sixth century Indian mathematician Bhaskara and the eleventh century Egyptian mathematician al-Hasan. In 1202, Fibonacci included it in his *Liber abaci*.

Example 5.14 A woman went to market and a horse stepped on her basket and crushed her eggs. The rider offered to pay her for the damage. He asked her how many she had brought. She did not know, but when she took them out two at a time there was one left. The same thing happened when she took them out 3, 4, 5, and 6 at a time, but when she took them out 7 at a time there were none left. What is the smallest number that she could have had? In essence we are being asked to solve the system of congruences

$$\begin{aligned} x &\equiv 1 \pmod 2,\\ x &\equiv 1 \pmod 3,\\ x &\equiv 1 \pmod 4,\\ x &\equiv 1 \pmod 5,\\ x &\equiv 1 \pmod 6,\\ x &\equiv 0 \pmod 7. \end{aligned}$$

The system is redundant and reduces to the equivalent system

$$\begin{aligned} x &\equiv 1 \pmod{12},\\ x &\equiv 1 \pmod 5,\\ x &\equiv 0 \pmod 7, \end{aligned}$$

with solution $x \equiv 301 \pmod{420}$.

Methods for solving systems of linear Diophantine equations for integral solutions date to at least the fifth century when the 'hundred fowl' problem appeared in Zhang Quijian's (JANG CHEE SHE ANN) *Mathematical Manual* which appeared around 475. Specifically, the problem asks how one can use exactly 100 coins to purchase 100 fowl, where roosters cost 5 coins, hens cost 3 coins, and one coin will fetch 3 chickens. The problem is equivalent to solving the equations $5x + 3y + \frac{1}{3}z = 100$ and $x + y + z = 100$. Multiplying the first equation by 3 and subtracting the second equation leads to the equation $7x + 4y = 100$. Among the solutions to the system are $x = 4$, $y = 18$, $z = 78$; $x = 8$, $y = 11$, $z = 81$; and $x = 12$, $y = 4$, $z = 84$.

In 800, Alcuin (Flaccus Albinus) authored a book of exercises and included the problem: if one distributes 100 bushels evenly among 100 people such that men get 3, women get 2, and children get half a bushel, how many people are there of each kind? Around 1211, Abu Kamil ibn Aslam found positive integral solutions to a set of equations that date back to the second century, namely, $x + y + z = 100$ and $5x + y/20 + z = 100$. He determined almost a hundred solutions to the system $x + y + z + w = 100$ and $4x + y/10 + z/2 + w = 100$.

In 1867, extending an 1843 result of De Morgan, A. Vachette showed that one of n^2, $n^2 - 1$, $n^2 - 4$, $n^2 + 3$ is divisible by 12 and the quotient is the number of positive solutions of $x + y + z = n$. In 1869 V. Schlegel proved that the number of positive integral solutions to $x + y + z = n$, where $x \leqslant y + z$, $y \leqslant x + z$, $z \leqslant x + y$, is $(n^2 - 1)/8$ or $(n + 2)(n + 4)/8$ according as n is odd or even.

A method, known to Islamic and Hindu mathematicians, called the rule of the virgins, can be employed to determine the number of nonnegative integral solutions to a system of linear equations. According to the rule, the number of such solutions to the equations $\sum_{i=1}^{k} a_i x_i = m$ and $\sum_{i=1}^{k} b_i x_i = n$ is given by the coefficient of $x^m y^n$ in the expansion of $\prod_{i=1}^{k}(1 - x^{a_i} y^{b_i})^{-1}$.

Exercises 5.4

1. Solve the following linear congruences:
   (a) $16x \equiv 27 \pmod{29}$,
   (b) $20x \equiv 16 \pmod{64}$,
   (c) $131x \equiv 21 \pmod{77}$,
   (d) $22x \equiv 5 \pmod{12}$,

(e) $17x \equiv 6 \pmod{29}$.

2. Find all solutions to $4x + 51y = 9$.
3. Find all solutions to $2x + 3y = 4$.
4. Someone wishes to purchase horses and cows spending exactly \$1770. A horse costs \$31 and a cow \$21. How many of each can the person buy? [Euler 1770]
5. A person pays \$1.43 for apples and pears. If pears cost 17¢ and apples 15¢, how many of each did the person buy?
6. Divide 100 into two parts, one divisible by 7 and the other divisible by 11.
7. Use Voroni's formula to solve for x if $5x \equiv 1 \pmod{61}$.
8. If one distributes 100 bushels evenly among 100 people such that men get 3, women get 2, and children get half a bushel, how many people are there of each kind? [Alcuin c. 800]
9. A duck costs 5 drachmas, a chicken costs 1 drachma, and 20 starlings cost 1 drachma. With 100 drachmas, how can one purchase 100 birds? [c. 120]
10. A group of 41 men, women, and children eat at an inn. The bill is for 40 sous. Each man pays 4 sous, women 3 sous, and three children eat for a sou. How many men, women, and children were there? [Bachet]
11. Show that the system $3x + 6y + z = 2$ and $4x + 10y + 2z = 3$ has no integral solutions.
12. Solve the following system:
$$x + y + z = 30, \frac{x}{3} + \frac{y}{2} + 2z = 30.$$
[Fibonacci 1228]
13. Find an integer having the remainders 1, 2, 5, 5 when divided by 2, 3, 6, 12 respectively. [Yi Xing c. 700]
14. Find an integer having the remainders 5, 4, 3, 2 when divided by 6, 5, 4, and 3 respectively. [Brahmagupta, Bhaskara, and Fibonacci]
15. Find a number with remainders of 3, 11, and 15, when divided by 10, 13, and 17, respectively. [Regiomontanus]
16. US Senator Riley was first elected in 1982. Her reelection is assured unless her campaign coincides with an attack of the seven-year itch such as hit her in 1978. When must she worry first? [For non-American readers: US Senators are elected for a fixed term of six years.]
17. A band of 17 pirates upon dividing their gold coins found that three coins remain after the coins have been apportioned evenly. In an ensuing brawl, one of the pirates was killed. The wealth was again redistributed equally, and this time ten coins remained. Again an

argument broke out and another pirate was killed. This time the fortune was distributed evenly among the survivors. What was the least number of gold coins the pirates had to distribute?

18. According to the biorhythm theory, a person has a physical cycle of 23 days, with a maximum after 5.75 (6) days; an emotional cycle of 28 days, with a maximum after 7 days; and an intellectual cycle of 33 days, with a maximum after 8.25 (8) days. When does a person first have all the maxima on the same day, and after how many days will that occur again?
19. Find a nonzero solution to $49x + 59y + 75z = 0$. [Euler 1785]
20. Find a solution to $5x + 8y + 7z = 50$. [Paoli 1794]
21. Find a solution to the system $x + y + z = 240$ and $97x + 56y + 3z = 16\,047$. [Regiomontanus]
22. Find a five-digit number n with the property that the last five digits of n^2 are exactly the same and in the same order as the last five digits of n.
23. According to the rule of the virgins, how many nonnegative integral solutions should the system $2x + y = 2$ and $x + 3y = 7$ have?

5.5 Miscellaneous exercises

1. According to the Dirichlet principle if n boxes contain $n + 1$ items, then one box must contain at least two items. Given any set S of n integers, use the Dirichlet principle to prove that for pairs of integers selected from S, n divides either the sum or the difference of two numbers. (Hint: Let the integers be $a_1, \ldots, a_n$ and consider $a_1 + a_2$, $a_1 + a_3, \ldots, a_1 + a_n$ modulo n.)
2. Given n integers $a_1, a_2, \ldots, a_n$, use the Dirichlet principle to prove that there exists a nonempty subset whose sum is a multiple of n. [Hint: Let the integers be $a_1, a_2, \ldots, a_n$ and consider $a_1 + a_2$, $a_1 + a_2 + a_3, \ldots, a_1 + a_2 + \cdots + a_n$.]
3. Show that if $a_1, a_2, \ldots, a_{\phi(m)}$ and $b_1, b_2, \ldots, b_{\phi(n)}$ are reduced residue systems modulo m and n respectively with $\gcd(m, n) = 1$, then $T = \{na_i + mb_j\colon 1 \leqslant i \leqslant \phi(m) \text{ and } 1 \leqslant j \leqslant \phi(n)\}$ is a set of $\phi(m)\phi(n)$ integers forming a reduced residue system modulo mn.
4. With T defined as in the previous exercise, show that no two elements in T can be congruent. Hence, every integer coprime to mn is counted exactly once, hence, $\phi(m)\phi(n) = \phi(mn)$.
5. Show that

$$\phi(2n) = \begin{cases} \phi(n) & \text{if } n \text{ is odd,} \\ 2 \cdot \phi(n) & \text{if } n \text{ is even.} \end{cases}$$

6. Show that

$$\phi(3n) = \begin{cases} 3 \cdot \phi(n) & \text{if } 3 | n, \\ 2 \cdot \phi(n) & \text{if } 3 \nmid n. \end{cases}$$

7. Carmichael's lambda function $\Lambda_c(n)$ is defined as follows:

$$\Lambda_c(1) = \Lambda_c(2) = 1,$$
$$\Lambda_c(4) = 2,$$
$$\Lambda_c(2^r) = 2^{r-2}, \text{ for } r \geqslant 3,$$
$$\Lambda_c(p^k) = \phi(p^k) \text{ if } p \text{ is an odd prime, and}$$
$$\Lambda_c(2^r p_1^{\alpha_1} \ldots p_r^{\alpha_r}) = \text{lcm}(\phi(2^k), \phi(p_1^{\alpha_1}), \ldots, \phi(p_r^{\alpha_r})).$$

A composite number n is called a Carmichael number if and only $\Lambda_c(n)$ divides $n - 1$. Find
(a) $\Lambda_c(24)$,
(b) $\Lambda_c(81)$,
(c) $\Lambda_c(341)$,
(d) $\Lambda_c(561)$,
(e) $\Lambda_c(2^6 \cdot 3^4 \cdot 5^2 \cdot 7 \cdot 19)$.

8. Find a solution to $7x + 5y + 15z + 12w = 149$.
9. Solve $27x + 33y + 45z + 77w = 707$.
10. Solve $10x + 11y + 12z = 200$. [*The Gentleman's Diary*, 1743]
11. A farmer buys 100 birds for \$100. If chickens cost \$0.50 each, ducks \$3 each, and turkeys \$10 each, and the farmer buys at least one bird of each type, how many of each type did he buy?
12. Show that 42 divides $n^7 - n$ for any integer n.
13. For any positive integer n, prove that

$$\sum_{d|n} d \cdot \phi(d) \cdot \sigma\left(\frac{n}{d}\right) = \sum_{d|n} d^2. \text{ [Liouville 1857]}$$

14. For any positive integer n, prove that

$$\sum_{d|n} \mu(d) \cdot \phi(d) = \prod_{p|n} (2 - p),$$

where p is prime.

15. Prove that if $2^{64} + 1$ is divisible by $1071 \cdot 2^8 + 1$, then $1071^2 + 16\,777\,216^2$, $1071^4 + 256^4$, and $1071^8 + 1^8$ are composite. [Hint: show that if $(-1071)^n + 2^{64-8n} \equiv 0 \pmod{1071 \cdot 2^8 + 1}$, then $(-1071)^{n+1} + 2^{64-8(n+1)} \equiv 0 \pmod{1071 \cdot 2^8 + 1}$.] This problem appeared in *The Educational Times*, in 1882, and was solved by Sarah

Marks (Hertha Ayrton), of Girton College, Cambridge. Ayrton, an English experimenter, was the first woman nominated to be a Fellow of the Royal Society. She was ruled ineligible since she was a married woman and, hence, had no rights of her own under English law. She was awarded the Hughes Medal from the Society for her work with electric arcs and determining the cause of sand ripples on the seashore. She remains the only woman to be awarded a medal from the Royal Society in her own right.

16. A nonempty set G on which there is defined a binary operation, denoted by juxtaposition, is called a group if G is closed, associative, there is an element e (the identity) such that for all a in G, $ea = ae = e$, and for each element a in G there is an element a^{-1} in G such that $aa^{-1} = a^{-1}a = e$. In addition, if G is commutative then it is called an Abelian group. The order of a group is the number of elements in the group. The least residue system modulo m, $\{0, 1, 2, \ldots, m-1\}$, under the operation of addition modulo m, denoted by Z_m, is an Abelian group of order m. Find the inverse for each element in Z_{10}.
17. If p is a prime, the least residue system modulo p, less zero, denoted by Z_p^*, is an Abelian group of order $p-1$ under multiplication modulo p. Find the inverse of each element in Z_{11}^*.
18. The reduced residue system modulo m, $\{a_1, a_2, \ldots, a_{\phi(m)}\}$, forms an Abelian group of order $\phi(m)$ under multiplication modulo m. Find the inverse of each element in Z_{12}^*.
19. A subgroup H of a group G is a nonempty subset of G that is a group under the same operation. Show that H is a subgroup of G if, for all a and b in H, ab^{-1} is in H.
20. Describe all the subgroups of Z_m.
21. A ring is a nonempty set with two binary operations, called addition and multiplication, that is an Abelian group under addition and is closed and associative under multiplication. If a ring is commutative under multiplication it is called a commutative ring. If there is a multiplicative identity it is called a ring with unity. The least residue system modulo m, $\{0, 1, 2, \ldots, m-1\}$, denoted also by Z_m, under addition and multiplication modulo m is a commutative ring with unity. Which elements in Z_6 fail to have multiplicative inverses?
22. A field is a nonempty set with two binary operations, say addition and multiplication, that is distributive, an Abelian group under addition, and whose nonzero elements form an Abelian group under multiplication. For p a prime, Z_p under the operations of addition and

multiplication modulo p is an example of a finite field. Find the multiplicative inverses for all nonzero elements in Z_1.

23. If $0 < a$, $b < m$, $\gcd(a, m) = 1$ and x runs through a complete residue system modulo m, then show that $ax + b$ runs through a complete residue system modulo m.

5.6 Supplementary exercises

1. Which of the following are true:
   (a) $57 \equiv 21 \pmod 6$
   (b) $11 \equiv -14 \pmod{17}$
   (c) $k^2 \equiv k \pmod k$, for k a positive integer.
2. What is the remainder when 3^{29} is divided by 23?
3. What is the remainder when $5^{128} \cdot 3^{173}$ is divided by 13?
4. What is the remainder when $3^{78} \cdot 5^{167}$ is divided by 17?
5. Use mathematical induction to prove that for all positive integers n, $4^n \equiv 1 + 3n \pmod 9$.
6. Determine the date of Easter in 2010 and 2025.
7. Does 7 divide $888^{999} + 999^{888}$?
8. Find the units digit of 3^{714}.
9. Determine the units digit of 666^{1984}.
10. Determine the last two digits of 98^{89}.
11. Use the '7' divisibility rule to show that 7 divides 38278621551023 but does not divide 168780379625.
12. Use the '13' divisibility rule to show that 13 divides 71088868594757 but does not divide 253560062125.
13. Use the '7' and '13' divisibility rules to show that 7 and 13 divide 2307396569853375.
14. A positive integer with repeated digit is called a repdigit, for example 222, 55555, and all repunits are repdigits. When does 7 divide a repdigit.
15. A positive integer n is called balanced if the number of integers less than or equal to n and coprime to n divides the sum of the digits of n. Determine the first twenty balanced numbers.
16. If p is prime show that $\sigma(p) + \varphi(p) = p \cdot \tau(p)$.
17. Determine the value of $\sigma(n) + \varphi(n)$, when $n = p \cdot q$ and both p and q are prime.
18. Determine $\varphi(n)$ for $n = 3780$, 4200, 29601, and 115830.
19. Evaluate $\sum_{d|72} \phi(d)$.
20. Evaluate $\sum_{d|126} \phi(d)$.

21. Exhibit the Farey fractions of order 11.
22. Solve for x if $36x \equiv 27 \pmod{51}$.
23. Solve for x if $19x \equiv 21 \pmod{77}$.
24. Solve for x if $8x \equiv 5 \pmod{12}$.
25. Solve for x if $14x \equiv 21 \pmod{77}$.
26. Solve for x if $36x \equiv 27 \pmod{51}$.
27. Find all solutions to $2x + 3y \equiv 1 \pmod{7}$.
28. Find all solutions to $60x + 18y = 97$.
29. Solve for x if $x \equiv 5 \pmod{2}$, $x \equiv 1 \pmod{3}$, and $x \equiv 2 \pmod{5}$.
30. Solve for x if $x \equiv 3 \pmod{5}$, $x \equiv 2 \pmod{7}$, and $x \equiv 1 \pmod{4}$.
31. Solve for x if $x \equiv 5 \pmod{2}$, $x \equiv 1 \pmod{3}$, $x \equiv 2 \pmod{5}$, and $x \equiv 5 \pmod{7}$.
32. Find the least positive integer that leaves a remainder 3 when divided by 7, a remainder 2 when divided by 11, and is divisible by 5.
33. At a clambake, the total cost of a lobster dinner is $31 and a chicken dinner is $13. What can we conclude if the total bill was $666?
34. A shopper spends a total of $8.39. If apples cost 25¢ each and oranges cost 18¢ each, how many of each type of fruit were purchased?
35. Is it possible to have 50 coins, all pennies, dimes, and quarters worth $3?
36. If eggs are removed from a basket two, three, and five at a time, there remain, respectively, one. But if the eggs are removed seven at a time no eggs remain. What is the least number of eggs that could have been in the basket? What is the next smallest number of eggs that could have been in the basket?
37. A class in number theory is divided up to study the Chinese Remainder Theorem. When divided up into groups of three, two students were left out; into groups of four, one student was left out; into groups of five, the students found out that if the professor was added to one of the groups no one was left out. What is the least number of students possible in the class?
38. Is $\{(0)_2, (0)_3, (1)_4, (5)_6, (7)_{12}\}$ a Davenport covering of the integers?
39. Find a Davenport covering that includes $(0)_2$ and $(0)_3$.
40. Show that the sum of the reciprocals of the moduli in a Davenport covering is at least unity.
41. Prove that if two residue classes in a Davenport covering have a common element, then they have a common arithmetic progression, and hence the sum of the reciprocals of the moduli is greater than unity.
42. For what values of m and k $(k > 1)$ is the set $\{0^k, 1^k, 2^k, \ldots,$

$(m-1)^k\}$ a reduced residue system modulo m?

43. If n is composite, determine c such that $(n-1)! + 1 \equiv c \pmod{n}$.
44. Show that the sequence of Fibonacci numbers modulo 7 is periodic.
45. Determine a formula for the length of the period of the Fibonacci sequence modulo m.

6

Congruences of higher degree

Never send to know for whom the bell tolls; it tolls for thee.

John Donne

6.1 Polynomial congruences

We now develop techniques, introduced by Gauss in *Disquisitiones*, for solving polynomial congruences of the form $f(x) \equiv 0 \pmod{m}$, where $f(x)$ is a polynomial with integer coefficients of degree greater than one whose solutions come from the least residue system $\{0, 1, \ldots, m-1\}$. In the late eighteenth century, Lagrange developed techniques to solve polynomial equations where m was prime. Polynomial equations with nonprime moduli can be solved using the Chinese Remainder Theorem.

Theorem 6.1 *If* $m = \prod_{i=1}^{k} m_i$ and $\gcd(m_i, m_j) = 1$, *for* $1 \leqslant i < j \leqslant k$, *then any solution of* $f(x) \equiv 0 \pmod{m}$ *is simultaneously a solution of the system* $f(x) \equiv 0 \pmod{m_i}$, for $i = 1, 2, \ldots, k$, *and conversely.*

Proof Suppose $f(x_0) \equiv 0 \pmod{m}$. Since $m_i | m$, $f(x) \equiv 0 \pmod{m_i}$, for $i = 1, 2, \ldots, k$. Hence, any solution of $f(x) \equiv 0 \pmod{m}$ is a solution to the system of equations $f(x) \equiv 0 \pmod{m_i}$, for $i = 1, 2, \ldots, k$. Conversely, suppose that $f(x_0) \equiv 0 \pmod{m_i}$, for $i = 1, 2, \ldots, k$. Then, $m_i | f(x_0)$ for $i = 1, 2, \ldots, k$. Since $\gcd(m_i, m_j) = 1$, for $i \neq j$, from the corollary to Theorem 2.8, $m | f(x_0)$. Therefore, $f(x_0) \equiv 0 \pmod{m}$. ■

If $f(x) \equiv 0 \pmod{p_i^{\alpha_i}}$ has n_i solutions, for $i = 1, \ldots, k$, from the multiplication principle, $f(x) \equiv 0 \pmod{n}$, where $n = \prod_{i=1}^{k} p_i^{\alpha_i}$ has at most $\prod_{i=1}^{k} n_i$ solutions. According to Theorem 6.1, in order to solve the polynomial equation $f(x) \equiv 0 \pmod{n}$, where $n = \prod_{i=1}^{k} p_i^{\alpha_i}$, where $\alpha_i \geqslant 1$, for $i = 1, 2, \ldots, k$, we first solve the equations $f(x) \equiv 0 \pmod{p_i^{\alpha_i}}$, for $i = 1, \ldots, k$. Then use the Chinese Remainder Theorem or brute force to obtain the solution modulo n. In either case, we need a technique to solve polynomial congruences of the form $f(x) \equiv 0 \pmod{p^\alpha}$, where p is prime and $\alpha \geqslant 2$ is a natural number. The next result shows that solutions to $f(x) \equiv 0 \pmod{p^\alpha}$ are generated from solutions to $f(x) \equiv 0 \pmod{p^{\alpha-1}}$.

Theorem 6.2 *Let $f(x)$ be a polynomial with integral coefficients, p a prime, and $\alpha \geqslant 1$ an integer. If $x_{\alpha+1} = x_\alpha + kp^\alpha$, where x_α is a solution to $f(x) \equiv 0 \pmod{p^\alpha}$, and k is a solution to $(f(x_\alpha)/p^\alpha) + k \cdot f'(x_\alpha) \equiv 0 \pmod{p}$ where $0 \leqslant x_\alpha < p^\alpha$, $0 \leqslant k < p$, and $f'(x)$ denotes the derivative of the function $f(x)$, then $x_{\alpha+1}$ is a solution to $f(x) \equiv 0 \pmod{p^{\alpha+1}}$.*

Proof For p a prime, if $p^{\alpha+1}|a$ then $p^\alpha|a$. Hence, each solution of $f(x) \equiv 0 \pmod{p^{\alpha+1}}$ is also a solution of $f(x) \equiv 0 \pmod{p^\alpha}$. More precisely, if $f(x_{\alpha+1}) \equiv 0 \pmod{p^{\alpha+1}}$, then there exists an x_α such that $f(x_\alpha) \equiv 0 \pmod{p^\alpha}$ with $x_{\alpha+1} \equiv x_\alpha \pmod{p^\alpha}$ or, equivalently, $x_{\alpha+1} = x_\alpha + kp^\alpha$. Using a Taylor expansion, $f(x_{\alpha+1}) = f(x_\alpha + kp^\alpha) = f(x_\alpha) + kp^\alpha f'(x_\alpha) + k^2N$, where N is an integer divisible by $p^{\alpha+1}$. Thus, $f(x_\alpha) + kp^\alpha f'(x_\alpha) \equiv 0 \pmod{p^{\alpha+1}}$. Since $f(x_\alpha) \equiv 0 \pmod{p^\alpha}$, $f(x_\alpha)/p^\alpha = M$ is an integer. Thus, $f(x_\alpha) = Mp^\alpha$, implying that $Mp^\alpha + kp^\alpha f'(x_\alpha) \equiv 0 \pmod{p^{\alpha+1}}$. Upon division by p^α, it follows that $M + kf'(x_\alpha) \equiv 0 \pmod{p}$. ■

Example 6.1 In order to solve $53x \equiv 282$ modulo 11^3, set $f(x) = 53x - 282$. Thus, $f'(x) = 53$. Any solution to $53x \equiv 282 \pmod{11^2}$ will be of the form $x_1 = x_0 + k \cdot 11$, where $53x_0 \equiv 282 \pmod{11}$ and $f(x_0)/11 + 53k \equiv 0 \pmod{11}$. The only solution to $53x_0 \equiv 282 \pmod{11}$ is given by $x_0 \equiv 2 \pmod{11}$. Since $f(2) = 53(2) - 282 = -176$, we obtain $-176/11 + 53k \equiv -16 + 53k \equiv 0 \pmod{11}$, implying that $k \equiv 3 \pmod{11}$. Therefore, a solution to $53x \equiv 282 \pmod{11^2}$ is given by $x_1 = x_0 + k \cdot 11 = 2 + 3 \cdot 11 = 35$. A solution to $53x \equiv 282 \pmod{11^3}$ is given by $x_2 = x_1 + r \cdot 11^2$, where $f(35)/11^2 + 53r \equiv 1573/11^2 + 53r \equiv 13 + 53r \equiv 0 \pmod{11}$, implying that $r \equiv 1 \pmod{11}$. Therefore, $x_2 = x_1 + r \cdot 11^2 = 35 + 1 \cdot 11^2 = 156$ is a solution to $53x \equiv 282 \pmod{11^3}$.

With Theorems 6.1 and 6.2 established, we now restrict ourselves to methods of solving polynomial congruences of the form $f(x) \equiv 0 \pmod{p}$, where p is prime. When Euler accepted Catherine the Great's offer and moved to St Petersburg, Joseph Louis Lagrange succeeded him in Berlin. Even though they probably never met, there was an extensive correspondence between the two mathematicians. Lagrange's most productive period with respect to number theory was the period from 1766 to 1777, the time he spent in Berlin. Lagrange's works are very readable and are noted for their well-organized presentation and the clarity of their style.

Lagrange in the late eighteenth century determined an upper limit on the number of solutions to polynomial equations as a function of the degree of the polynomial. In particular, he established that a polynomial equation can have at most p incongruent solutions modulo p. According to Fermat's Little Theorem, $x^p - x \equiv 0 \pmod{p}$ has exactly p solutions. Hence, Lagrange's Theorem is a best possible result.

Theorem 6.3 (Lagrange's Theorem) *The number of incongruent solutions of the polynomial equation $f(x) \equiv 0 \pmod{p}$ is never more than the degree of $f(x)$.*

Proof Given $f(x) \equiv 0 \pmod{p}$, where p is prime and n denotes the degree of $f(x)$, we reason inductively. If $n = 1$, consider congruences of the form $ax + b \equiv 0 \pmod{p}$, where $a \not\equiv 0 \pmod{p}$ so $ax \equiv -b \pmod{p}$. Since $\gcd(a, p) = 1$, Theorem 5.14 implies that the equation has exactly one solution. Suppose the theorem is true for all polynomials of degree less than or equal to n. Consider $f(x) \equiv 0 \pmod{p}$, with p prime and $\deg(f(x)) = n + 1$. Suppose further that $f(x)$ has $n + 2$ incongruent roots modulo p, and r is one of those roots. It follows that $f(x) = g(x)(x - r)$, where $\deg(g(x)) = n$. If s is any other root of $f(x) \equiv 0 \pmod{p}$, then $f(s) \equiv g(s)(s - r) \equiv 0 \pmod{p}$. Now $s - r \not\equiv 0 \pmod{p}$, since $\gcd(s - r, p) = 1$, and p is prime. Hence, $g(s) \equiv 0 \pmod{p}$, and s is a root of $g(x) \equiv 0 \pmod{p}$. Thus $g(x) \equiv 0 \pmod{p}$, a polynomial equation of degree n, has $n + 1$ roots, contradicting the induction assumption. ■

If $n > 4$ is composite then n divides $(n - 1)!$ or, equivalently, $(n - 1)! \equiv 0 \pmod{n}$. In 1770, in *Meditationes algebraicae*, Edward Waring stated that one of his students, John Wilson, had conjectured that if p is a prime then it divides $(p - 1)! + 1$, but the proof seemed difficult due to a lack of notation to express prime numbers. In 1761, Wilson, like Waring, before him, was Senior Wrangler at Cambridge. Wilson, however, left mathematics quite early to study law, became a judge, and was later knighted. Leibniz conjectured the result as early as 1683, but was also unable to prove it. Having been sent a copy of *Meditationes algebraicae* by Waring, Lagrange gave the first proof of the theorem and its converse in 1771. Gauss reportedly came up with the gist of a proof in five minutes while walking home one day. His classic riposte to Waring's comment was that proofs should be 'drawn from notions rather than from notations'.

Since

$$\sin\left(\frac{(n-1)!+1}{n}\right)\pi = 0$$

if and only if n is prime, Wilson's Theorem provides an interesting but not very practical criterion for determining whether or not a number is prime. The proof shown below is due to the Russian mathematician Pafnuti Chebyshev, propounder of the law of large numbers. We noted earlier that it is an open question whether $n! + 1$ is prime for infinitely many values of n. The next result shows that $n! + 1$ is composite for infinitely many values of n.

Theorem 6.4 (Wilson's Theorem) *The natural number n is prime if and only if $(n-1)! \equiv -1 \pmod{n}$.*

Proof Suppose that p is prime. By Fermat's Little Theorem solutions to $g(x) = x^{p-1} - 1 \equiv 0 \pmod{p}$ are precisely $1, 2, \ldots, p-1$. Consider $h(x) = (x-1)(x-2) \cdots (x-(p-1)) \equiv 0 \pmod{p}$, whose solutions by construction are the integers $1, 2, \ldots, p-1$. Since $g(x)$ and $h(x)$ both have degree $p-1$ and the same leading term, $f(x) = g(x) - h(x) \equiv 0 \pmod{p}$ is a congruence of degree at most $p-2$ having $p-1$ incongruent solutions, contradicting Lagrange's Theorem. Hence, every coefficient of $f(x)$ must be a multiple of p, and thus $\deg(f(x)) = 0$. However, since $f(x)$ has no constant term, $f(x) \equiv 0 \pmod{p}$ is also satisfied by $x \equiv 0 \pmod{p}$. Therefore, $0 \equiv f(0) = g(0) - h(0) = -1 - (-1)^{p-1}(p-1)! \pmod{p}$. If p is an odd prime, then $(-1)^{p-1} \equiv 1 \pmod{p}$, and if $p = 2$, then $(-1)^{p-1} \equiv -1 \equiv 1 \pmod{2}$. Hence, for any prime p, we have $(p-1)! \equiv -1 \pmod{p}$. Conversely, if n is composite, then there exists an integer d, $1 < d < n$, such that $d|n$. Hence, $d|(n-1)!$, and $(n-1)! \equiv 0 \pmod{d}$, implying that $(n-1)! \not\equiv -1 \pmod{n}$. ■

Let $f(x, y) = \frac{1}{2}(y-1)[|A^2 - 1| - (A^2 - 1)] + 2$, where $A = x(y+1) - (y! + 1)$, x and y are positive integers. If p is an odd prime, $x_0 = [(p-1)! + 1]/p$ and $y_0 = p - 1$, then,

$$A = \frac{1}{p}[(p-1)! + 1][p - 1 + 1] - [(p-1)! + 1] = 0.$$

Hence,

$$f(x_0, y_0) = \frac{(p-1)-1}{2}[|1| - |-1|] + 2 = p.$$

Hence, $f(x, y)$ is an example of a prime generating function.

Exercises 6.1

1. Solve for x
   (a) $2x^9 + 2x^6 - x^5 - 2x^2 - x \equiv 0 \pmod{5}$,
   (b) $x^4 + x + 2 \equiv 0 \pmod{7}$.
2. Solve for x
   (a) $x^3 + 3x^2 + 31x + 23 \equiv 0 \pmod{35}$,
   (b) $x^3 + 2x - 3 \equiv 0 \pmod{45}$,
   (c) $x^3 - 9x^2 + 23x - 15 \equiv 0 \pmod{77}$.
3. Use Theorem 6.2 to solve for x
   (a) $x^2 + 8 \equiv 0 \pmod{121}$,
   (b) $5x^3 - 2x + 1 \equiv 0 \pmod{343}$,
   (c) $x^2 + x + 7 \equiv 0 \pmod{81}$.
4. Solve the modular system
   $$\begin{cases} 5x^2 + 4x - 3 \equiv 0 \pmod{6}, \\ 3x^2 + 10 \equiv 0 \pmod{17}. \end{cases}$$
5. Use Wilson's Theorem to show that 17 is prime.
6. Find the remainder when 15! is divided by 17.
7. Show that $18! \equiv -1 \pmod{437}$.
8. For any odd prime p, show that $1^2 \cdot 3^2 \cdots (p-2)^2 \equiv 2^2 \cdot 4^2 \cdots (p-1)^2 \equiv (-1)^{(p+1)/2} \pmod{p}$.
9. If p is an odd prime, show that $x^2 \equiv 1 \pmod{p}$ has exactly two incongruent solutions modulo p.
10. Modulo 101, how many solutions are there, to the polynomial equation $x^{99} + x^{98} + x^{97} + \cdots + x + 1 = 0$? [Hint: multiply the polynomial by $x(x-1)$.]
11. Use the fact that Z_p^*, the nonzero residue classes modulo a prime p, is a group under multiplication to establish Wilson's Theorem. [Gauss]
12. Prove that if $p > 3$ is a prime then
    $$1 + \tfrac{1}{2} + \tfrac{1}{3} + \cdots + \frac{1}{p-1} \equiv 0 \pmod{p}.$$
    [J. Wolstenholme 1862]
13. If p is prime Wilson's Theorem implies that $(p-1)! + 1 = kp$ for some k. When does $k = 1$ and when does $k = p$?

6.2 Quadratic congruences

In the previous section, we showed that solutions to $ax^2 + bx + c \equiv 0 \pmod{m}$ depend on the solution to $ax^2 + bx + c \equiv 0 \pmod{p}$, where p is a prime and $p|m$. If p is an odd prime with $\gcd(a, p) = 1$, then

$\gcd(4a, p) = 1$. Multiplying both sides of $ax^2 + bx + c \equiv 0 \pmod p$ by $4a$ we obtain $4a^2x^2 + 4abx + 4ac \equiv 0 \pmod p$ or $(2ax + b)^2 \equiv (b^2 - 4ac) \pmod p$. Therefore, to solve the quadratic equation $ax^2 + bx + c \equiv 0$ modulo a prime p, we need only find solutions to

$$(2ax + b) \equiv y \pmod p,$$

where y is a solution to

$$y^2 \equiv (b^2 - 4ac) \pmod p.$$

Since $\gcd(2a, p) = 1$, the first of these equations always has a unique solution. Hence, as Gauss realized, a solution to the original problem depends solely on solving congruences of the form $x^2 \equiv k \pmod p$.

Example 6.2 In order to solve $3x^2 + 15x + 9 \equiv 0 \pmod{17}$ we first solve $y^2 \equiv b^2 - 4ac = 225 - 108 = 117 \equiv 15 \pmod{17}$. Since $7^2 \equiv 10^2 \equiv 15 \pmod{17}$, we obtain the solutions $y \equiv 7 \pmod{17}$ and $y \equiv 10 \pmod{17}$. If $y \equiv 7 \pmod{17}$ then $2ax + b = 6x + 15 \equiv 7 \pmod{17}$, implying that $x \equiv 10 \pmod{17}$. If $y \equiv 10 \pmod{17}$ then $2ax + b = 6x + 15 \equiv 10 \pmod{17}$, implying that $x \equiv 2 \pmod{17}$. Therefore, the solutions to $3x^2 + 15x + 9 \equiv 0 \pmod{17}$ are given by $x \equiv 2 \pmod{17}$ and $x \equiv 10 \pmod{17}$.

Our goal at this point is twofold. We aim to determine which equations of the form $x^2 \equiv a \pmod p$ have solutions, for p an odd prime, and to find a technique to obtain such solutions. If the equation $x^2 \equiv a \pmod p$ has a solution then a is called a quadratic residue (QR) of p, otherwise a is called a quadratic nonresidue (QNR) of p. The integer 0 is usually excluded from consideration since it is a trivial quadratic residue of p, for every prime p. Since $(p - b)^2 \equiv b^2 \pmod p$, if b is a QR of p, then $p - b$ is a QR of p.

For example, modulo 17, we find that

$$\begin{array}{llll} 1^2 \equiv 1, & 2^2 \equiv 4, & 3^2 \equiv 9, & 4^2 \equiv 16, \\ 5^2 \equiv 8, & 6^2 \equiv 2, & 7^2 \equiv 15, & 8^2 \equiv 13, \\ 9^2 \equiv 13, & 10^2 \equiv 15, & 11^2 \equiv 2, & 12^2 \equiv 8, \\ 13^2 \equiv 16, & 14^2 \equiv 9, & 15^2 \equiv 4, & 16^2 \equiv 1. \end{array}$$

Therefore, the quadratic residues of 17 are 1, 2, 4, 8, 9, 13, 15, and 16. The quadratic nonresidues of 17 are 3, 5, 6, 7, 10, 11, 12 and 14. Euler, Lagrange, Legendre, and Gauss developed the theory of quadratic residues in attempting to prove Fermat's Last Theorem.

For convenience, we introduce the Legendre symbol $(\frac{a}{p})$, which is defined as follows: for p an odd prime, and a an integer with $\gcd(a, p) = 1$,

$$\left(\frac{a}{p}\right) = \begin{cases} 1 & \text{if } a \text{ is a quadratic residue of } p, \\ -1 & \text{if } a \text{ is a quadratic nonresidue of } p. \end{cases}$$

Adrien Marie Legendre studied mathematics at Collège Mazarin in Paris. He taught for five years with Pierre-Simon Laplace at the Ecole Militaire in Paris. His treatise on ballistics was awarded a prize from the Berlin Academy. Legendre was financially independent but lost a fortune during the French Revolution. In 1798, Legendre introduced the symbol $(\frac{a}{p})$ in *Essai sur la théorie des nombres*. In *Essai*, the first modern work devoted to number theory, Legendre mentioned many of the number theoretic contributions of Euler and Lagrange. In the next theorem, we show that modulo an odd prime p half the integers between 1 and $p - 1$ are quadratic residues and half are quadratic nonresidues.

Theorem 6.5 *If p is an odd prime, then there are precisely $(p - 1)/2$ incongruent quadratic residues of p given by*

$$1^2, 2^2, \ldots, \left(\frac{p-1}{2}\right)^2$$

Proof Let p be an odd prime. We wish to determine the values for a, $1 \leqslant a \leqslant p - 1$, for which the equation $x^2 \equiv a \pmod{p}$ is solvable. Since $x^2 \equiv (p - x)^2 \pmod{p}$, squares of numbers in the sets $\{1, 2, \ldots, (p - 1)/2\}$ and $\{(p - 1)/2 + 1, \ldots, p - 1\}$ are congruent in pairs. Thus, we need only consider values of x for which $1 \leqslant x \leqslant (p - 1)/2$. But the squares $1^2, 2^2, \ldots, ((p - 1)/2)^2$ are all incongruent modulo p, otherwise $x^2 \equiv a \pmod{p}$ would have four incongruent solutions, contradicting Lagrange's Theorem. Thus, the $(p - 1)/2$ quadratic residues of p are precisely the integers

$$1^2, 2^2, \ldots, \left(\frac{p-1}{2}\right)^2. \blacksquare$$

According to Theorem 6.5, the quadratic residues of 19 are given by 1^2, 2^2, 3^2, 4^2, 5^2, 6^2, 7^2, 8^2, and 9^2. Modulo 19, they are respectively, 1, 4, 9, 16, 6, 17, 11, 7 and 5. Knowing that half the numbers are quadratic residues of a prime, we still need to find an efficient method to distinguish between QRs and QNRs for large primes. One of the first such methods was devised by Euler in 1755. Before establishing Euler's method to determine whether an integer is a quadratic residue of a prime, we establish the following result.

Lemma *If p is an odd prime and* $\gcd(a, p) = 1$, *then either* $a^{(p-1)/2} \equiv 1$ *or* $a^{(p-1)/2} \equiv -1$ *modulo p.*

Proof From Fermat's Little Theorem, if p is an odd prime and $\gcd(a, p) = 1$, then $a^{p-1} - 1 = (a^{(p-1)/2} - 1)(a^{(p-1)/2} + 1) \equiv 0 \pmod{p}$. Hence, either $a^{(p-1)/2} \equiv 1$ or $a^{(p-1)/2} \equiv -1 \pmod{p}$. ■

Theorem 6.6 (Euler's criterion) *If p is an odd prime and* $\gcd(a, p) = 1$, *then*

$$\left(\frac{a}{p}\right) \equiv a^{(p-1)/2} \pmod{p}.$$

Proof Suppose p is an odd prime, $\gcd(a, p) = 1$, and $1 \leqslant r \leqslant p - 1$. Since $rx \equiv a$ has a unique solution modulo p there is exactly one element s, $1 \leqslant s \leqslant p - 1$, such that $rs \equiv a \pmod{p}$. If a is a QNR modulo p, $(\frac{a}{p}) = -1$, then $r \not\equiv s \pmod{p}$ and the elements $1, 2, \ldots, p - 1$ can be grouped into pairs $r_i s_i$, such that $r_i s_i \equiv a \pmod{p}$, for $i = 1, 2, \ldots, (p - 1)/2$. Thus, from Wilson's Theorem,

$$-1 \equiv (p-1)! \equiv \prod_{i=1}^{\frac{p-1}{2}} r_i s_i \equiv a^{(p-1)/2} \pmod{p}.$$

If a is a QR modulo p, $(\frac{a}{p}) = 1$, there exists an integer b such that $b^2 \equiv a \pmod{p}$. By Fermat's Little Theorem, $a^{(p-1)/2} \equiv b^{p-1} \equiv 1 \pmod{p}$. Therefore, in either case, it follows that

$$\left(\frac{a}{p}\right) \equiv a^{(p-1)/2} \pmod{p}. \quad ■$$

Corollary *If p is an odd prime with* $\gcd(a, p) = 1$, $\gcd(b, p) = 1$, *and* $a \equiv b \pmod{p}$, *then*

$$\left(\frac{a}{p}\right) = \left(\frac{b}{p}\right).$$

For example, according to Euler's criterion,

$$\left(\frac{3}{31}\right) \equiv 3^{(31-1)/2} \equiv 3^{15} \equiv -1 \pmod{31}.$$

Hence, the equation $x^2 \equiv 3 \pmod{31}$ has no solution. Since

$$\left(\frac{6}{29}\right) \equiv 6^{(29-1)/2} \equiv 6^{14} \equiv 1 \pmod{29},$$

the equation $x^2 \equiv 6 \pmod{29}$ has a solution. The next result was conjectured by Fermat around 1630 and proven by Euler in 1750.

Theorem 6.7 *If p is an odd prime,*

$$\left(\frac{-1}{p}\right) = \begin{cases} 1 & \textit{if } p \equiv 1 \pmod{4}, \\ -1 & \textit{if } p \equiv 3 \pmod{4}. \end{cases}$$

Proof If $p = 4k + 1$, then

$$\left(\frac{-1}{p}\right) = (-1)^{(p-1)/2} = (-1)^{2k} = 1.$$

If $p = 4k + 3$, then

$$\left(\frac{-1}{p}\right) = (-1)^{(p-1)/2} = (-1)^{2k+1} = -1. \blacksquare$$

The next result can be used to simplify computations with Legendre symbols.

Theorem 6.8 *If p is an odd prime and p does not divide ab, then*

$$\left(\frac{ab}{p}\right) = \left(\frac{a}{p}\right)\left(\frac{b}{p}\right).$$

Proof We have

$$\left(\frac{ab}{p}\right) \equiv (ab)^{(p-1)/2} \equiv a^{(p-1)/2}b^{(p-1)/2} \equiv \left(\frac{a}{p}\right)\left(\frac{b}{p}\right) \pmod{p}.$$

Since the only possible values for $(\frac{a}{p})$, $(\frac{b}{p})$, and $(\frac{ab}{p})$ modulo p are ± 1, an examination of the various cases establishes that

$$\left(\frac{ab}{p}\right) = \left(\frac{a}{p}\right)\left(\frac{b}{p}\right). \blacksquare$$

Corollary *If p is an odd prime with* $\gcd(n, p) = 1$ *and* $n = \prod_{i=1}^{r} p_i^{\alpha_i}$, *then*

$$\left(\frac{n}{p}\right) = \prod_{i=1}^{r}\left(\frac{p_i^{\alpha_i}}{p}\right) = \prod_{i=1}^{r}\left(\frac{p_i}{p}\right)^{\alpha_i}.$$

For example, since

$$\left(\frac{24}{31}\right)=\left(\frac{2}{31}\right)^3\left(\frac{3}{31}\right)=1^3(-1)=-1,$$

24 is a quadratic nonresidue of 31. Hence, the equation $x^2 \equiv 24 \pmod{31}$ has no solution. The next result was obtained by Gauss in 1808. It leads to the third proof of his celebrated quadratic reciprocity law, an extremely efficient method for determining whether an integer is a quadratic residue or not of an odd prime p.

Theorem 6.9 (Gauss's Lemma) *If p is an odd prime with* $\gcd(a, p) = 1$, *then* $(\frac{a}{p}) = (-1)^s$, *where s denotes the number of elements* $\{a, 2a, 3a, \ldots, \frac{1}{2}(p-1)a\}$ that exceed $p/2$.

Proof Let S denote the set of least positive residues modulo p of the set $\{a, 2a, 3a, \ldots, \frac{1}{2}(p-1)a\}$. Let s denote the number of elements of S that exceed $p/2$ and $r = (p-1)/2 - s$. Relabel the elements of S as $a_1, a_2, \ldots, a_r, b_1, b_2, \ldots, b_s$, where $a_i < p/2$, for $i = 1, 2, \ldots, r$, and $b_j > p/2$, for $j = 1, 2, \ldots, s$. Since the elements are the least positive residues of a, $2a, \ldots, \frac{1}{2}(p-1)a$,

$$\left(\prod_{i=1}^{r} a_i\right)\left(\prod_{j=1}^{s} b_j\right) \equiv n^{(p-1)/2}\left(\frac{p-1}{2}\right)! \pmod{p}.$$

Consider the set T consisting of the $(p-1)/2$ integers $a_1, a_2, \ldots, a_r$, $p-b_1, p-b_2, \ldots, p-b_s$. Since $p/2 < b_j < p$, for $j = 1, 2, \ldots, s$, $0 < p - b_j < p/2$, all the elements of T lie between 1 and $(p-1)/2$. In addition, if $a_i \equiv p - b_j \pmod{p}$, for any $1 \leqslant i \leqslant r$ and $1 \leqslant j \leqslant r$, then $0 \equiv p \equiv a_i + b_j = ha + ka = (h+k)a \pmod{p}$, for $1 \leqslant h < k \leqslant (p-1)/2$. Hence, p divides $(h+k)a$. Since $\gcd(p, a) = 1$, p must divide $h + k$, but that is impossible since $0 < h + k < p$. Thus the elements of T are distinct and, hence, must consist precisely of the integers $1, 2, \ldots, (p-1)/2$. Thus,

$$\left(\frac{p-1}{2}\right)! \equiv \left(\prod_{i=1}^{r} a_i\right)\left(\prod_{j=1}^{s}(p-b_j)\right) \equiv (-1)^s\left(\prod_{i=1}^{r} a_i\right)\left(\prod_{j=1}^{s} b_j\right)$$

$$\equiv (-1)^s n^{(p-1)/2}\left(\frac{p-1}{2}\right)! \pmod{p}.$$

Cancelling $((p-1)/2)!$ from both sides of the congruence yields $1 \equiv (-1)^s n^{(p-1)/2} \pmod{p}$. Therefore, from Euler's criterion,

$$\left(\frac{n}{p}\right) \equiv (-1)^s \pmod{p}. \blacksquare$$

For example, if $p = 31$ and $a = 3$, then, with respect to the multiples of 3, we have 3, 6, 9, 12, 15, 18, 21, 24, 27, 30, 33, 36, 39, 42, 45 which are congruent to 3, 6, 9, 12, 15, 18, 21, 24, 27, 30, 2, 5, 8, 11, 14 modulo 31, respectively. Hence, $s = 5$ and

$$\left(\frac{3}{31}\right) = (-1)^5 = -1.$$

Therefore, the congruence $x^2 \equiv 3 \pmod{31}$ has no solution. The next result, established by Legendre in 1775, gives an efficient way to evaluate the Legendre symbol when the numerator equals 2.

Theorem 6.10 *If p is an odd prime, then*

$$\left(\frac{2}{p}\right) = \begin{cases} 1 & \text{if } p \equiv \pm 1 \pmod{8}, \\ -1 & \text{if } p \equiv \pm 3 \pmod{8}. \end{cases}$$

Proof Let s denote the number of elements $2, 4, 6, \ldots, 2((p-1)/2)$ that exceed $p/2$. A number of the form $2k$ is less than $p/2$ whenever $k \leqslant p/4$. Hence, $s = (p-1)/2 - [\![p/4]\!]$. If $p = 8k + 1$, then $s = 4k - [\![2k + \frac{1}{4}]\!] = 4k - 2k \equiv 0 \pmod{0}$. If $p = 8k + 3$, then $s = 4k + 1 - [\![2k + \frac{3}{4}]\!] = 4k + 1 - 2k \equiv 1 \pmod{2}$. If $p = 8k + 5$, then $s = 4k + 2 - [\![2k + 1 + \frac{1}{4}]\!] = 2k + 1 \equiv 1 \pmod{2}$. If $p = 8k + 7$, then $s = 4k + 3 - [\![2k + 1 + \frac{3}{4}]\!] = 2k + 2 \equiv 0 \pmod{2}$. $\blacksquare$

Since $(p^2 - 1)/8$ satisfies the same congruences as does s in the proof of Theorem 6.10, we obtain the following formula which can be used to determine for which primes 2 is a QR and for which it is a QNR.

Corollary *If p is an odd prime, then*

$$\left(\frac{2}{p}\right) = (-1)^{(p^2-1)/8}.$$

It is often difficult and sometimes nearly impossible to credit a mathematical result to just one person, often because there is a person who first stated the conjecture, one who offered a partial proof of the conjecture, one who proved it conclusively, and one who generalized it. The quadratic reciprocity law,

$$\text{if } p \text{ and } q \text{ are odd primes then } \left(\frac{p}{q}\right) = \pm\left(\frac{q}{p}\right),$$

is no exception. It was mentioned in 1744 by Euler who estabished several special cases of the law in 1783. In 1785, Legendre stated and attempted to prove the *Loi de réciprocité* in *Recherches d'analyse indéterminée* and again in a 1798 paper. In both attempts, he failed to show that for each prime $p \equiv 3 \pmod 4$ there exists a prime $q \equiv 3 \pmod 4$ such that $(\frac{p}{q}) \equiv -1$.

Gauss gave the first complete proof in 1795 just prior to his 18th birthday and remarked that the problem had tormented him for a whole year. In 1801, he published his first proof of the quadratic reciprocity law in *Disquisitiones*. He wrote, 'engaged in other work I chanced upon an extraordinary arithmetic truth ... since I considered it to be so beautiful in itself and since I suspected its connections with even more profound results, I concentrated on it all my efforts in order to understand the principles on which it depends and to obtain a rigorous proof'. Gauss eventually devised eight proofs for the quadratic reciprocity law.

Theorem 6.11 will allow us to efficiently determine whether or not an integer is a quadratic residue modulo a prime. In essence, the quadratic reciprocity law states that if p and q are prime then, unless both are congruent to 3 modulo 4, $x^2 \equiv p \pmod q$ and $x^2 \equiv q \pmod p$ are solvable. In the case that $p \equiv 3 \pmod 4$ and $q \equiv 3 \pmod 4$, one of the equations is solvable and the other is not. The geometric proof offered below is due to Ferdinand Eisenstein, Gauss's pupil, who published it in 1840. Eisenstein discovered a cubic reciprocity law as well.

Theorem 6.11 (Gauss's quadratic reciprocity law) *If p and q are distinct odd primes, then*

$$\left(\frac{p}{q}\right)\left(\frac{q}{p}\right) = (-1)^{\frac{1}{2}(p-1)\frac{1}{2}(q-1)}.$$

Proof Let p and q be distinct odd primes. Consider the integers q_k and r_k, where $kp = pq_k + r_k$, and $1 \leqslant r_k \leqslant p-1$, for $k = 1, 2, \ldots, (p-1)/2$. Hence, $q_k = [\![kq/p]\!]$, and r_k is the least residue of kq modulo p. As in the proof of Gauss's Lemma, we let $a_1, a_2, \ldots, a_r$ denote those values of r_k which are less than $p/2$, and $b_1, b_2, \ldots, b_s$ denote those values of r_k which are greater than $p/2$. Hence, $a_1, a_2, \ldots, a_r, p-b_1, p-b_2, \ldots, p-b_s$ are just the integers $1, 2, \ldots, (p-1)/2$ in some order and

$$\left(\frac{q}{p}\right) = (-1)^s.$$

Let

$$a = \sum_{i=1}^{r} a_i \text{ and } b = \sum_{j=1}^{s} b_j,$$

so

$$a + b = \sum_{k=1}^{\frac{p-1}{2}} r_k.$$

Therefore,

$$(^*) \qquad a + sp - b = \sum_{i=1}^{r} a_i + \sum_{j=1}^{s}(p - b_j) = \sum_{k=1}^{\frac{p-1}{2}} k = \frac{p^2 - 1}{8}.$$

Moreover, if we let

$$u = \sum_{k=1}^{\frac{p-1}{2}} q_k = \sum_{k=1}^{\frac{p-1}{2}} \left[\!\left[\frac{kq}{p}\right]\!\right]$$

and sum the equations $kq = pq_k + r_k$, for $1 \leqslant k \leqslant (p-1)/2$, we have

$$(^{**}) \qquad pu + a + b = p\left(\sum_{k=1}^{\frac{p-1}{2}} q_k\right) + a + b$$
$$= \sum_{k=1}^{\frac{p-1}{2}}(pq_k + r_k) = \sum_{k=1}^{\frac{p-1}{2}} kq = \left(\frac{p^2 - 1}{8}\right)q.$$

Subtracting $(^*)$ from $(^{**})$, we obtain

$$pu + 2b - sp = \left(\frac{p^2 - 1}{8}\right)(q - 1).$$

Since $p \equiv q \equiv 1 \pmod 2$, $u \equiv s \pmod 2$. Therefore,

$$\left(\frac{q}{p}\right) = (-1)^s = (-1)^u.$$

Repeating the above process with the roles of p and q interchanged and with

$$v = \sum_{j=1}^{\frac{q-1}{2}} \left[\!\left[\frac{jp}{q}\right]\!\right],$$

we obtain

$$\left(\frac{p}{q}\right) = (-1)^v.$$

Therefore,

$$(***) \qquad \left(\frac{p}{q}\right)\left(\frac{q}{p}\right) = (-1)^{u+v}.$$

We need only show that

$$u + v = \left(\frac{p-1}{2}\right)\left(\frac{q-1}{2}\right).$$

Consider all the lattice points (i, j), in the Cartesian plane, such that $1 \leqslant i \leqslant (p-1)/2$ and $1 \leqslant j \leqslant (q-1)/2$. If the lattice point (i, j) lies on the line l: $py = qx$, then $pj = qi$ (see Figure 6.1). However, p and q are coprime implying that p divides i, which is impossible since $1 \leqslant i \leqslant (p-1)/2$. Thus, each such lattice point lies either above l or below l. If (i, j) is a lattice point below l, then $pj < qi$, so $j < qi/p$. Thus, for each fixed value for i, $1 \leqslant j \leqslant [\![qi/p]\!]$ whenever (i, j) is below l. Therefore, the total number of lattice points below l is given by

$$\sum_{i=1}^{\frac{p-1}{2}} \left[\!\!\left[\frac{qi}{p}\right]\!\!\right] = u.$$

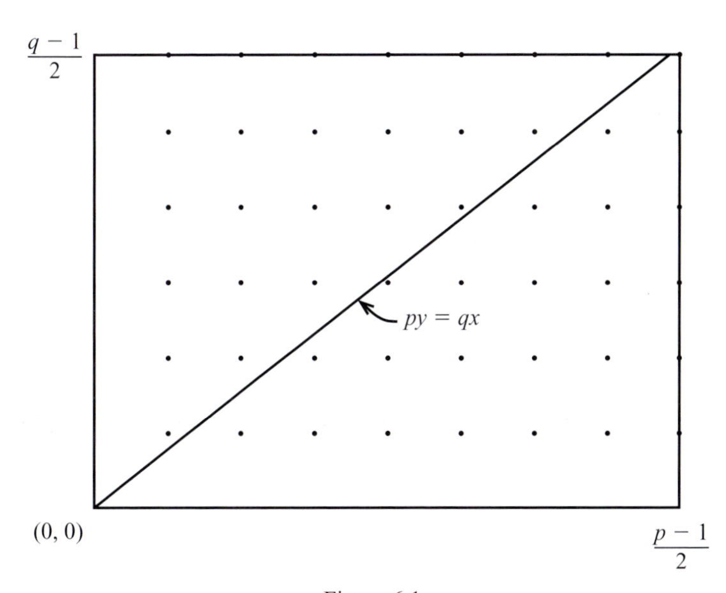

Figure 6.1

Similarily, the total number of lattice points above l is given by

$$\sum_{j=1}^{\frac{q-1}{2}} \left[\!\left[\frac{jp}{q} \right]\!\right] = v.$$

Since each one of the $(\frac{p-1}{2})(\frac{q-1}{2})$ points must lie above or below l,

$$u + v = \left(\frac{p-1}{2}\right)\left(\frac{q-1}{2}\right).$$

Therefore, from (***), it follows that

$$\left(\frac{p}{q}\right)\left(\frac{q}{p}\right) = (-1)^{\frac{1}{2}(p-1)\frac{1}{2}(q-1)}. \quad \blacksquare$$

For example, $283 \equiv 17 \pmod{19}$, and $19 \equiv 2 \pmod{17}$; using Theorem 6.8 and the quadratic reciprocity law in the form

$$\left(\frac{p}{q}\right) = \left(\frac{q}{p}\right)(-1)^{\frac{1}{2}(p-1)\frac{1}{2}(q-1)},$$

we obtain

$$\left(\frac{19}{283}\right) = \left(\frac{283}{19}\right)(-1)^{\frac{18}{2}\frac{282}{2}}$$

or

$$-\left(\frac{17}{19}\right) = -\left(\frac{19}{17}\right)(-1)^{\frac{16}{2}\frac{18}{2}}$$

or

$$-\left(\frac{2}{17}\right) = -1.$$

Therefore, $x^2 \equiv 19 \pmod{283}$ has no solutions.

Quartic and higher order reciprocity laws have been developed. The construction of such criteria now belongs to the branch of number theory called class field theory which was introduced by David Hilbert in 1898. A general law of reciprocity was established by Emil Artin in 1927.

It is possible to generalize Legendre's symbol for cases in which the denominator is composite. If we let $a \neq 0$ and m be positive integers with canonical representation $m = \prod_{i=1}^{r} p_i^{\alpha_i}$, then the Jacobi symbol $(\frac{a}{m})$, which first appeared in Crelle's *Journal* in 1846, is defined by

$$\left(\frac{a}{m}\right) = \prod_{i=1}^{r} \left(\frac{a}{p_i}\right)^{\alpha_i},$$

where $(\frac{a}{p_i})$, with p_i prime, represents the Legendre symbol.

Unlike the Legendre symbol the Jacobi symbol may equal unity without the numerator being a quadratic residue modulo m. For example,

$$\left(\frac{2}{9}\right) = \left(\frac{2}{3}\right)^2 = 1,$$

but $x^2 \equiv 2 \pmod 9$ is not solvable! However, if

$$\left(\frac{a}{m}\right) = -1,$$

for a composite positive integer m, then the equation $x^2 \equiv a \pmod m$ has no solution. For example,

$$\left(\frac{21}{997}\right) = \left(\frac{3}{997}\right)\left(\frac{7}{997}\right) = (1)(-1) = -1.$$

Hence, $x^2 \equiv 21 \pmod{997}$ has no solution.

Important properties of the Jacobi symbol, whose proofs follow from the definition and properties of the Legendre symbol, include the following:

(a) $a \equiv b \pmod p$ implies that $\left(\frac{a}{m}\right) = \left(\frac{b}{m}\right)$,

(b) $\left(\frac{a}{mn}\right) = \left(\frac{a}{m}\right)\left(\frac{a}{n}\right)$,

(c) $\left(\frac{ab}{m}\right) = \left(\frac{a}{m}\right)\left(\frac{b}{m}\right)$,

(d) $\left(\frac{-1}{m}\right) = (-1)^{(m-1)/2}$, if m is odd,

(e) $\left(\frac{2}{m}\right) = (-1)^{(m^2-1)/8}$,

(f) $\left(\frac{n}{m}\right)\left(\frac{m}{n}\right) = (-1)^{\frac{1}{2}(n-1)\frac{1}{2}(m-1)}$, for m and n odd and $\gcd(m, n) = 1$.

Exercises 6.2

1. Find all the quadratic residues modulo 29.
2. Evaluate the following Legendre symbols:

   (a) $\left(\frac{2}{29}\right)$, (b) $\left(\frac{-1}{29}\right)$, (c) $\left(\frac{5}{29}\right)$, (d) $\left(\frac{11}{29}\right)$,

   (e) $\left(\frac{2}{127}\right)$, (f) $\left(\frac{-1}{127}\right)$, (g) $\left(\frac{5}{127}\right)$, (h) $\left(\frac{11}{127}\right)$.

3. Which of the following quadratic congruences have solutions?

   (a) $x^2 \equiv 2 \pmod{29}$, (e) $x^2 \equiv 2 \pmod{127}$,

(b) $x^2 \equiv 28 \pmod{29}$,
(c) $x^2 \equiv 5 \pmod{29}$,
(d) $x^2 \equiv 11 \pmod{29}$,
(f) $x^2 \equiv 126 \pmod{127}$,
(g) $x^2 \equiv 5 \pmod{127}$,
(h) $x^2 \equiv 11 \pmod{127}$.

4. Determine whether or not the following quadratic congruences are solvable. If solvable find their solutions.
(a) $5x^2 + 4x + 7 \equiv 0 \pmod{19}$.
(b) $7x^2 + x + 11 \equiv 0 \pmod{17}$.
(c) $2x^2 + 7x - 13 \equiv 0 \pmod{61}$.
5. Evaluate the following Jacobi symbols:

(a) $\left(\frac{21}{221}\right)$, (b) $\left(\frac{215}{253}\right)$, (c) $\left(\frac{631}{1099}\right)$,

(d) $\left(\frac{1050}{1573}\right)$, (e) $\left(\frac{89}{197}\right)$.

6. If p is an odd prime show that
$$\sum_{a=1}^{p-1}\left(\frac{a}{p}\right) = 0.$$
7. If p is an odd prime and $\gcd(a, p) = \gcd(b, p) = 1$, show that at least one of a, b and ab is a quadratic residue of p.
8. If p is an odd prime use Euler's criterion to show that -1 is a quadratic residue of p if and only if $p \equiv 1 \pmod 4$.
9. If p and q are odd primes with $p = 2q + 1$, use the quadratic reciprocity law to show that
$$\left(\frac{p}{q}\right) = \left(\frac{-1}{p}\right).$$
10. If p and q are distinct primes with $p \equiv 3 \pmod 4$ and $q \equiv 3 \pmod 4$, then use the quadratic reciprocity law to show that p is a quadratic residue modulo q if and only if q is a quadratic nonresidue modulo p.
11. Prove that 19 does not divide $4n^2 + 4$ for any integer n.
12. If p is a prime and $h + k = p - 1$, show that $h! \cdot k! \equiv (-1)^{k+1} \pmod p$.
13. If p is an odd prime with $p = 1 + 4r$ use the previous exercise, with $h = k = 2r$, to show that $2r!$ is a solution to $x^2 \equiv -1 \pmod p$.
14. Prove that if $p > 3$ is a prime then
$$1 + \frac{1}{2^2} + \frac{1}{3^2} + \cdots + \frac{1}{(p-1)^2} \equiv 0 \pmod p.$$
[J. Wolstenholme 1862].

6.3 Primitive roots

We now describe a general method to solve polynomial congruences of higher degree modulo a prime. We begin by considering fundamental congruences of the type $x^m \equiv a \pmod{p}$, where p is an odd prime, $a > 2$, and $\gcd(a, p) = 1$. If $x^m \equiv a \pmod{p}$ is solvable, we say that a is an mth power residue of p. If n is a positive integer and $\gcd(a, n) = 1$, the least positive integer k such that $a^k \equiv 1 \pmod{n}$ is called the order of a modulo n and is denoted by $\text{ord}_n(a)$. For any positive integer n, $a^{\phi(n)} \equiv 1$. Thus, the Euler–Fermat Theorem, implies that $\text{ord}_n(a)$ is well-defined and always less than $\phi(n)$.

Theorem 6.12 *If* $\text{ord}_n(a) = k$ *then* $a^h \equiv 1 \pmod{n}$ *if and only if k divides h.*

Proof Suppose $\gcd(a, n) = 1$, $\text{ord}_n(a) = k$, and $a^h \equiv 1 \pmod{n}$. The division algorithm implies there exist integers q and s such that $h = kq + s$, with $0 \leqslant s < k$. Thus $a^h = a^{kq+s} = (a^k)^q a^s$. Since $a^k \equiv 1 \pmod{n}$, it follows that $a^s \equiv 1 \pmod{n}$, so $s \neq 0$ would contradict the fact that k is the least positive integer with the property that $a^k \equiv 1 \pmod{n}$. Hence, $s = 0$ and k divides h. Conversely, if $k|h$, then there is an integer t such that $kt = h$. Since $\text{ord}_n(a) = k$, $a^h \equiv a^{kt} \equiv (a^k)^t \equiv 1 \pmod{n}$. ∎

If we know the order of a modulo n, with a little more effort, we can determine the order of any power of a modulo n as illustrated in the next result.

Theorem 6.13 *If* $\text{ord}_n(a) = k$ *then* $\text{ord}_n(a^m) = k/\gcd(m, k)$.

Proof Let $\text{ord}_n(a) = k$, $\text{ord}_n(a^m) = r$, $d = \gcd(m, k)$, $m = bd$, $k = cd$, and $\gcd(b, c) = 1$. Hence, $(a^m)^c = (a^{bd})^c = (a^{cd})^b = (a^k)^b \equiv 1 \pmod{n}$. Theorem 6.12 implies that $r|c$. Since $\text{ord}_n(a) = k$, $(a^{mr}) = (a^m)^r \equiv 1 \pmod{n}$. Hence, Theorem 6.12 implies that $k|mr$. Thus, $cd|(bd)r$, inplying that $c|br$. Since b and c are coprime, c divides r. Hence, c equals r. Therefore, $\text{ord}_n(a^m) = r = c = k/d = k/\gcd(m, k)$. ∎

From Theorem 6.12, it follows that the order of every element modulo a prime p is a divisor of $p - 1$. In addition, Theorem 6.13 implies that if d is a divisor of $p - 1$ then there are exactly $\phi(d)$ incongruent integers modulo p having order d. For example, if $p = 17$, 8 is a divisor of $p - 1$. Choose an element, say 3, that has order 16 modulo 17. In Theorem 6.17, we show that this can always be done. For example, the $\phi(8) = 4$ elements k with

$1 \leqslant k \leqslant 16$ such that $\gcd(k, 16) = 2$ are 2, 6, 10, 14. The four elements of order 8 modulo 17 are 3^2, 3^6, 3^{10}, and 3^{14}.

The following corollaries follow directly from the previous two theorems and the definition of the order of an element. We state them without proof.

Corollary 6.1 *If* $\mathrm{ord}_n(a) = k$, *then* k *divides* $\phi(n)$.

Corollary 6.2 *If* $\mathrm{ord}_n(a) = k$, *then* $a^r \equiv a^s \pmod n$ *if and only if* $r \equiv s \pmod k$.

Corollary 6.3 *If* $k > 0$ *and* $\mathrm{ord}_n(a) = hk$, *then* $\mathrm{ord}_n(a^h) = k$.

Corollary 6.4 *If* $\mathrm{ord}_n(a) = k$, $\mathrm{ord}_n(b) = h$, *and* $\gcd(h, k) = 1$, *then* $\mathrm{ord}_n(ab) = hk$.

We use the order of an element to establish the following primality test devised by the nineteenth century French mathematician J.F.T. Pepin.

Theorem 6.14 (Pepin's primality test) *For* $n \geqslant 1$, *the nth Fermat number* F_n *is prime if and only if* $3^{(F_n-1)/2} \equiv -1 \pmod{F_n}$.

Proof If F_n is prime, for $n \geqslant 1$, $F_n \equiv 2 \pmod 3$. Hence, from the quadratic reciprocity law,

$$\left(\frac{3}{F_n}\right)\left(\frac{F_n}{3}\right) = \left(\frac{3}{F_n}\right)\left(\frac{2}{3}\right) = \left(\frac{3}{F_n}\right)(-1) = 1.$$

Thus,

$$\left(\frac{3}{F_n}\right) = -1.$$

From Euler's criterion, $3^{(F_n)/2} \equiv -1 \pmod{F_n}$. Conversely, suppose that $3^{(F_n-1)/2} \equiv -1 \pmod{F_n}$. If p is any prime divisor of F_n then $3^{(F_n-1)/2} \equiv -1 \pmod p$. Squaring both sides of the congruence, we obtain $3^{F_n-1} \equiv 1 \pmod p$. If m is the order of 3 modulo p, according to Theorem 6.12, m divides $F_n - 1$. That is, m divides 2^{2^n}. Hence, $m = 2^r$, with $0 \leqslant r \leqslant 2^n$. If $r = 2^n - s$, where $s > 0$, then $3^{(F_n-1)/2} = 3^{2^{2^n-1}} = 3^{2^{r+s-1}} = (3^{2^r})^{2^{s-1}} = 1$. A contradiction, since we assumed $3^{(F_n-1)/2} \equiv -1 \pmod p$. Thus, $s = 0$ and 3 has order 2^{2^n} modulo p. From Theorem 6.12, 2^{2^n} divides $p - 1$. Hence, $2^{2^n} \leqslant p - 1$ implying that $F_n \leqslant p$. Therefore, if p is a prime divisor of F_n, then $F_n = p$. That is, F_n is prime. ■

For some positive integers n, there is a number q, $1 < q \leqslant n - 1$, such that powers of q generate the reduced residue system modulo n. That is, for each integer r, $1 \leqslant r \leqslant n - 1$, with $\gcd(r, n) = 1$ there is a positive integer k for which $q^k = r$. In this case, q can be used to determine the order of an element in Z_n^* and to determine the QRs and NQRs of n as well. The existence of such a number is crucial to the solutions of polynomial congruences of higher degree. We call a positive integer q a primitive root of n if $\mathrm{ord}_n(q) = \phi(n)$. We now show that primitive roots of n generate the reduced residue system modulo n.

Theorem 6.15 *If q is a primitive root of n, then $q, q^2, \ldots, q^{\phi(n)}$ form a reduced residue system modulo n.*

Proof Since q is a primitive root of n, $\mathrm{ord}_n(q) = \phi(n)$, implying that $\gcd(q, n) = 1$. Hence, $\gcd(q^i, n) = 1$, for $i = 1, 2, \ldots, \phi(n)$. The elements $q, q^2, \ldots, q^{\phi(n)}$ consist of $\phi(n)$ mutually incongruent positive integers. If $q^i \equiv q^j \pmod{n}$, for $1 \leqslant i < j \leqslant \phi(n)$, then, from Corollary 6.2, $i \equiv j \pmod{\phi(n)}$. Hence, $\phi(n)$ divides $j - i$, which is impossible since $0 < j - i < \phi(n)$. Hence, $q^i \not\equiv q^j \pmod{\phi(n)}$, for $1 \leqslant i < j \leqslant \phi(n)$, and $q, q^2, \ldots, q^{\phi(n)}$ form a reduced residue system modulo n. ∎

Theorem 6.16 (Lambert) *If p is an odd prime, h a positive integer, and q a prime such that q^h divides $p - 1$, then there exists a positive integer b such that* $\mathrm{ord}_p(b) = q^h$.

Proof By Lagrange's Theorem and the fact that $p \geqslant 3$, the equation $x^{(p-1)/q} \equiv 1 \pmod{p}$ has at most $(p-1)/q$ solutions where

$$\frac{p-1}{q} \leqslant \frac{p-1}{2} \leqslant p - 2.$$

Therefore, at least one element, say a, with $1 \leqslant a \leqslant p - 1$, and $\gcd(a, p) = 1$, is not a solution. Hence, $a^{(p-1)/q} \not\equiv 1 \pmod{p}$. Let $b = a^{(p-1)/q^h}$ and suppose that $\mathrm{ord}_p(b) = m$. Since $b^{q^h} \equiv a^{p-1} \pmod{p}$, Theorem 6.12 implies that m divides q^h. Suppose $m < q^h$. Since q is prime, m divides q^{h-1} and there is an integer k such that $mk = q^{h-1}$. Thus, $a^{(p-1)/q} = b^{q^{h-1}} = (b^m)^k \equiv 1^k \equiv 1 \pmod{p}$, contradicting our assumption. Hence, $q^h = m = \mathrm{ord}_p(b)$. ∎

In 1769, in connection with his work on decimal expansions of $1/p$, where p is an odd prime, J.H. Lambert established Theorem 6.16 and claimed that primitive roots of p exist for every prime p. Euler introduced the term 'primitive root' in 1773 when he attempted to establish Lambert's conjecture. Euler proved that there are exactly $\phi(p-1)$ primitive roots of p. At age 11, Gauss began working with primitive roots attempting to determine their relation to decimal expansions of fractions. He was able to show that if 10 is a primitive root of a prime p, then the decimal expansion of $1/p$ has period $p-1$. Gauss showed that if $m = 2^{\alpha}5^{\beta}$, then the period of the decimal expansion for m/p^n is of the order of 10 modulo p^n. He also showed that primitive roots exist modulo n for $n = 2$, 4, p, p^k, and $2p^k$, where p is an odd prime and k is a positive integer. In addition, he proved that if q is a primitive root of an odd prime p, then $q^p - p$, $q^p - qp$, and at least one of q and $q+p$ is a primitive root of p^2; if r is a primitive root of p^2, then r is a primitive root of p^k, for $k \geqslant 2$; and if s is a primitive root of p^k and s is odd then s is a primitive root of $2p^k$, if s is even then $s + p^k$ is a primitive root of $2p^k$. In addition, he proved that if m and n are coprime positive integers both greater than 3 there are no primitive roots of mn. For a positive integer $n > 2$, there are no primitive roots of 2^n, as shown in the next result.

Theorem 6.17 *There are no primitive roots of* 2^n, *for* $n > 2$.

Proof We use induction to show that if $\gcd(a, 2^n) = 1$, for $n > 2$, then $\text{ord}_{2^n}(a) = 2^{n-2}$. Hence, a cannot be a primitive root of 2^n. If $n = 3$ and $\gcd(a, 2^3) = 1$, then $a \equiv 1, 3, 5, 7, \pmod 8$. In addition, $1^2 \equiv 3^2 \equiv 5^2 \equiv 7^2 \equiv 1 \pmod 8$. Hence, if $\gcd(a, 2^3) = 1$, then $\text{ord}_8(a) = 2 = 2^{3-2}$. Let $k > 3$ and suppose that if $\gcd(m, 2^k) = 1$, for some positive integer m, then $\text{ord}_{2^k}(m) = 2^{k-2}$. That is, $m^{2^{k-2}} \equiv 1 \pmod{2^k}$ with $m^s \not\equiv 1 \pmod{2^k}$ for $1 \leqslant s < 2^{k-2}$. Let b be such that $\gcd(b, 2^{k+1}) = 1$. Hence, $\gcd(b, 2^k) = 1$ and, from the induction assumption, it follows that $\text{ord}_{2^k}(b) = 2^{k-2}$. Thus there is an integer r such that $b^{2^{k-2}} = 1 + r \cdot 2^k$. In addition $b^{2^{k-1}} = (b^{2^{k-2}})^2 = (1 + 2r \cdot 2^k + r^2 \cdot 2^{2k}) \equiv 1 \pmod{2^{k+1}}$. Suppose there is an integer s such that $b^s \equiv 1 \pmod{2^{k+1}}$ for $1 \leqslant s < 2^{k-1}$. We have $b^s = 1 + t \cdot 2^{k+1} = 1 + 2t \cdot 2^k$, implying that $b^s \equiv 1 \pmod{2^k}$, a contradiction . Therefore, $\gcd(b, 2^{k+1}) = 1$ implies that $\text{ord}_{2^{k+1}}(b) = 2^{k-1}$ and the result is established. ■

Finding primitive roots even of a prime is not an easy task. In 1844, A.L. Crelle devised an efficient scheme to determine whether an integer is a primitive root of a prime. The method works well for small primes. It uses the property that, if $1 \leqslant a \leqslant p-1$, s_i is the least residue of $a \cdot i$ modulo p, and t_j is the least residue of a^j, for $1 \leqslant i, j \leqslant p-1$, then $t_k \equiv s_{t_{k-1}}$

Table 6.1.

k	0	1	2	3	4	5	6	7	8	9	10	11	12	13	14	15	16
$3k$	0	3	6	9	12	15	1	4	7	10	13	16	2	5	8	11	14
3^k	1	3	9	10	13	5	15	11	16	14	8	7	4	12	2	6	1

(mod p), for $1 \leqslant k \leqslant p-1$. Crelle's algorithm follows since $a^{j-1} \cdot a \equiv a^j$ (mod p), for $1 \leqslant a \leqslant p-1$.

Example 6.3 If $p = 17$, and $a = 3$, then we generate the powers of 3 using the multiples of 3, as shown in Table 6.1. In particular, suppose the rows for k and $3k$ have been completed and we have filled in $3^0 = 1$, $3^1 = 3$ and $3^2 = 9$ on the bottom row. In order to determine 3^3 modulo 17, go to column 9 (since $3^2 \equiv 9 \pmod{17}$), to find that $3^3 \equiv 3 \cdot 9 \equiv 10 \pmod{17}$. Hence, $3^3 \equiv 10 \pmod{17}$. To determine 3^4 modulo 17, go to column 10 (since $3^3 \equiv 10 \pmod{17}$) to find that $3^4 \equiv 3 \cdot 10 \equiv 13 \pmod{17}$. Hence, $3^4 \equiv 13 \pmod{17}$. To determine $3^5 \pmod{17}$, go to column 13 (since $3^4 \equiv 3 \cdot 10 \equiv 13 \pmod{17}$) to find that $3^5 = 3 \cdot 13 \equiv 5 \pmod{17}$. Hence, $3^5 \equiv 5 \pmod{17}$, and so forth. The smallest value for k, $1 \leqslant k \leqslant 16$, for which $3^k \equiv 1 \pmod{17}$ is 16. Hence, 3 is primitive root modulo 17.

Theorem 6.18 *If p is an odd prime, then there exist $\phi(p-1)$ primitive roots modulo p.*

Proof If $p - 1 = \prod_{i=1}^{r} p_i^{\alpha_i}$, where $\alpha_i \geqslant 1$, for $i = 1, 2, \ldots, r$, is the canonical representation for $p-1$, by Theorem 6.16, there exist integers n_i such that $\text{ord}_p(n_i) = p_i$, for $1 \leqslant i \leqslant r$. By a generalization of Corollary 6.4, if $m = \prod_{i=1}^{r} n_i$, then $\text{ord}_p(m) = \prod_{i=1}^{r} p_i^{\alpha_i} = p - 1$, and m is the desired primitive root. From Theorem 6.13, if q is a primitive root of p and $\gcd(r, p-1) = 1$ then q^r is a primitive root of p. Therefore, there are $\phi(p-1)$ primitive roots of p. ■

Hence if q is a primitive root of p, then the $\phi(p-1)$ incongruent primitive roots of p are given by $q^{\alpha_1}, q^{\alpha_2}, \ldots, q^{\alpha_{\phi(p-1)}}$, where $\alpha_1, \alpha_2, \ldots, \alpha_{\phi(p-1)}$ are the $\phi(p-1)$ integers less than $p-1$ and coprime to $p-1$. For example, in order to determine all the primitive roots of 17, we use the fact that 3 is a primitive root of 17 and $\phi(16) = 8$. The eight integers less than 16 and coprime to 16 are 1, 3, 5, 7, 9, 11, 13, and 15. In addition, $3^1 \equiv 3$, $3^3 \equiv 10$, $3^5 \equiv 5$, $3^7 \equiv 11$, $3^9 \equiv 14$, $3^{11} \equiv 7$, $3^{13} \equiv 12$, and $3^{15} \equiv 6 \pmod{17}$. Therefore, the primitive roots of 17 are 3, 5, 6, 7, 10, 11, 12, and

14. If $\gcd(q, m) = 1$, then q is a primitive root of m if and only if $q^{\phi(m)/p} \not\equiv 1 \pmod{m}$ for all prime divisors p of $\phi(m)$. In general, if a primitive root exists for m, then there are $\phi(\phi(m))$ incongruent primitive roots of m.

Theorem 6.19 *If q is a primitive root of a prime p, the quadratic residues of p are given by q^{2k} and the quadratic nonresidues by q^{2k-1}, where $0 \leqslant k \leqslant (p-1)/2$.*

Proof Using Euler's criterion, if $\gcd(a, p) = 1$, then

$$(q^{2k})^{(p-1)/2} = (q^{p-1})^k \equiv 1 \pmod{p}$$

and

$$(q^{2k-1})^{(p-1)/2} = (q^{p-1})^k \cdot (q^{(p-1)/2})^{-1} \equiv (q^{(p-1)/2})^{-1} \equiv -1 \pmod{p}.$$

Conversely, if a is a QR of p then $a = (q^k)^2 = q^{2k}$ and if a is a QNR of p then $a = (q^2)^k \cdot q = q^{2k+1}$ where $0 \leqslant k \leqslant (p-1)/2$. ■

For example, since 3 is a primitive root of 17, the quadratic residues of 17 are $3^0, 3^2, 3^4, 3^6, 3^8, 3^{10}, 3^{12}, 3^{14}$, and 3^{16}.

Gauss thought that 10 was a primitive root for infinitely many primes. In 1920, Artin conjectured that there are infinitely many primes p with the property that 2 is a primitive root. Artin's conjecture has been generalized to state that if n is not a kth power then there exist infinitely many primes p such that n is a primitive root. In 1927, Artin conjectured further that every positive nonsquare integer is a primitive root of infinitely many primes. There are infinitely many positive integers for which Artin's conjecture is true and a few for which it fails.

According to Euler's criterion $x^2 \equiv a \pmod{p}$ is solvable if and only if $a^{(p-1)/2} \equiv 1 \pmod{p}$. A necessary condition that $x^m \equiv a \pmod{p}$ be solvable is that $a^{(p-1)/d} \equiv 1 \pmod{p}$, with $d = \gcd(m, p-1)$. In order to see this, suppose $\gcd(a, b) = 1$ and b is a solution of $x^m \equiv a \pmod{p}$. Fermat's Little Theorem implies that $a^{(p-1)/d} \equiv b^{(p-1)m/d} \equiv (b^{p-1})^r \equiv 1 \pmod{p}$, with $r = m/d$. The next result generalizes Euler's criterion for mth power residues of a prime. The proof is constructive and will enable us to determine where polynomial congruences of the form $x^m \equiv a \pmod{p}$ have solutions.

Theorem 6.20 *Let p be an odd prime with $\gcd(a, p) = 1$, then $x^m \equiv a \pmod{p}$ is solvable if and only if $a^{(p-1)/d} \equiv 1 \pmod{p}$, where $d = \gcd(m, p-1)$.*

Proof We need only establish the necessity. Suppose that $a^{(p-1)/d} \equiv 1 \pmod{p}$ where $\gcd(a, p) = 1$, $d = \gcd(m, p-1)$, and q is a primitive root modulo p. There exists an integer s such that $a = q^s$. Hence, $q^{s(p-1)/d} \equiv a^{(p-1)/d} \equiv 1 \pmod{p}$. Since q is a primitive root of p, $\text{ord}_p(q) = p - 1$. Thus, $s/d = k$ is an integer and $a \equiv q^{kd} \pmod{p}$. Since $d = \gcd(m, p-1)$, there are integers u and v such that $d = um + v(p-1)$. Thus, $a = q^{kd} = q^{kum+kv(p-1)} = q^{kum}q^{(p-1)kv} = q^{(ku)m} \cdot 1 = q^{(ku)m}$. Therefore, q^{ku} is a solution to $x^m \equiv a \pmod{p}$. ■

For example, the equation $x^7 \equiv 15 \pmod{29}$ is not solvable since $15^{28/7} \equiv 15^4 \equiv -9 \not\equiv 1 \pmod{29}$. The equation $x^{16} \equiv 8 \pmod{73}$ has a solution since $\gcd(16, 72) = 8$ and $8^{72/8} = 8^9 \equiv 1 \pmod{73}$.

Given a fixed value for m, it is possible to find that all the mth power residues modulo a prime as illustrated in the next result.

Theorem 6.21 *If p is an odd prime, q is a primitive root of p, and $d = \gcd(m, p-1)$, then the mth power residues of p are given by q^d, q^{2d}, $\ldots$, $q^{d(p-1)/d}$.*

Proof Let p be an odd prime, q a primitive root of p, and $d = \gcd(m, p-1)$. From the proof of Theorem 6.20, each element in the set $\{q^d, q^{2d}, \ldots, q^{d(p-1)/d}\}$ is an mth power residue of p. In addition, they are incongruent modulo p, for if $q^{id} \equiv q^{jd} \pmod{p}$, for some $1 \leqslant i < j \leqslant (p-1)/d$, from Corollary 6.2 $p - 1$ divides $d(j - i)$, which is impossible since $0 < d(j - i) < p - 1$. Suppose a is an mth power residue of p. Hence, there is an element b, $1 \leqslant b \leqslant p - 1$, such that $b^m \equiv a \pmod{p}$. There is an integer k, $1 \leqslant k \leqslant p - 1$, such that $b \equiv q^k \pmod{p}$, hence, $a \equiv b^m \equiv q^{kd} \pmod{p}$. Let, r, s, t, u, be such that $ud = m$, $td = p - 1$, $uk = st + r$ with $0 \leqslant r < t$. So $a \equiv q^{mk} \equiv q^{ukd} \equiv q^{(st+r)d} \equiv q^{(p-1)s}q^{rd} \equiv q^{rd} \pmod{p}$. Therefore, a is included in the set $\{q^d, q^{2d}, \ldots, q^{d(p-1)/d}\}$. ■

For example, in order to find the 12th power residues of 17, we use $p - 1 = 16$, $m = 12$, $d = \gcd(12, 16) = 4$ and the fact that 3 is a primitive root of 17. Hence, the 12th power residues of 17 are $3^4 \equiv 13$, $3^8 \equiv 16$, $3^{12} \equiv 4$, and $3^{16} \equiv 1$. Therefore, $x^{12} \equiv a \pmod{17}$ is solvable if and only if $a = 1, 4, 13$, or 16.

There is a relationship between primitive roots and quadratic nonresidues of odd primes. In particular, if p is an odd prime and a is a quadratic residue of p, then there exists an element b, $1 \leqslant b \leqslant p - 1$, such that

Table 6.2.

k	1	2	3	4	5	6	7	8	9	10	11	12	13	14	15	16
$I(k)$	16	14	1	12	5	15	11	10	2	3	7	13	4	9	6	8

$b^2 \equiv a \pmod{p}$, Hence, $a^{(p-1)/2} \equiv b^{((p-1)/2)^2} \equiv b^{p-1} \equiv 1 \pmod{p}$. Therefore, for an odd prime p, every primitive root of p is a quadratic nonresidue of p. In addition, we have the following result.

Theorem 6.22 *If p is an odd prime, every quadratic nonresidue of p is a primitive root of p if and only is $p = 2^k + 1$, for k a positive integer.*

Proof There are $(p-1)/2$ quadratic nonresidues of p and $\phi(p-1)$ primitive roots of p. Every quadratic nonresidue of p is a primitive root of p if and only if $\phi(p-1) = (p-1)/2$, but $\phi(n) = n/2$ if and only if $n = 2^k$. Therefore, $p = 2^k + 1$. ■

Gauss introduced a method to solve a number of polynomial congruences of higher degree modulo a prime. In particular, if p is an odd prime and q is a primitive root of p, we say that r is an index of n to the base q modulo p and write $r = I_q(n) \pmod{p}$ if and only if $n \equiv q^r \pmod{p}$ and $0 \leqslant r < p-1$. Note that $q^{I_q(n)} \equiv n \pmod{p}$. If p and q are known and the context is clear we simply write $I(n)$ to denote the index of n to the base q modulo p. In 1839, Jacobi published a table of indices for all primes less than a thousand in his *Canon arithmeticus*. In 1968, A.E. Western and J.C.P. Miller published a table of indices for all primes less than 50 021. A table of indices for the primitive root 3 modulo 17 can be generated from Table 6.1 by dropping the second row, rewriting the third row in ascending order, and interchanging the third row with the first row as shown in Table 6.2.

Indices are not additive but act like and play a role similar to that of logarithms. The next result provides us with enough machinery to solve a number of polynomial congruences of higher degree as well as other problems in modular arithmetic.

Theorem 6.23 *If p is an odd prime, q a primitive root of p, m and n integers such that* $\gcd(m, p) = \gcd(n, p) = 1$, *and r and k are positive integers, then*

(a) $m \equiv n \pmod{p}$ *if and only if* $I(m) \equiv I(n) \pmod{p-1}$,

(b) $I(q^r) \equiv r \pmod{p-1}$,
(c) $I(1) = 0$ *and* $I(q) = 1$,
(d) $I(mn) \equiv I(m) + I(n) \pmod{p-1}$,
(e) $I(n^k) \equiv k \cdot I(n) \pmod{p-1}$.

Proof Since q is a primitive root modulo p, $\text{ord}_p(q) = p - 1$. Let $r = I(m)$ and $s = I(n)$; hence, $q^r \equiv m \pmod{p}$ and $q^s \equiv n \pmod{p}$.

(a) $m \equiv n \pmod{p}$ if and only if $q^r \equiv q^s \pmod{p}$ if and only if $r \equiv s \pmod{p-1}$ if and only if $I(m) \equiv I(n) \pmod{p-1}$. ∎

(b) Since $q^r \equiv m \pmod{p}$ it follows from (a) that $I(q^r) \equiv I(m) \equiv r \pmod{p-1}$. ∎

(c) $1 \equiv q^0 \pmod{p}$ and $q \equiv q^1 \pmod{p}$. Hence, $I(1) = 0$ and $I(q) = 1$. ∎

(d) $q^{r+s} = q^r q^s \equiv mn \pmod{p}$. Hence, from (a), we have $I(mn) \equiv r + s \equiv I(m) + I(n) \pmod{p-1}$. ∎

(e) $q^{st} \equiv n^t \pmod{p}$. Hence, from (a), we have $I(n^t) \equiv ts \equiv t \cdot I(n) \pmod{p-1}$. ∎

Example 6.4 We use indices, the fact that 3 is a primitive root of 17, and Table 6.2, to solve $11x \equiv 9 \pmod{17}$.

$$\begin{aligned} I(11x) &\equiv I(9) \pmod{16}, \\ I(11) + I(x) &\equiv I(9) \pmod{16}, \\ 7 + I(x) &\equiv 2 \pmod{16}, \\ I(x) &\equiv 11 \pmod{16}, \\ x &\equiv 7 \pmod{17}. \end{aligned}$$

Example 6.5 Solve $x^3 + 6 \equiv 0 \pmod{17}$. We have

$$\begin{aligned} x^3 &\equiv -6 \equiv 11 \pmod{17}, \\ I(x^3) &\equiv I(11) \pmod{16}, \\ 3(I(x)) &\equiv 7 \pmod{16}, \\ I(x) &\equiv 77 \equiv 13 \pmod{16}, \\ x &\equiv 12 \pmod{17}, \end{aligned}$$

and

$$x \equiv 8,\ 11,\ \text{or } 7 \pmod{13}.$$

Example 6.6 Evaluate $1134^{729} \cdot 432^{97}$ modulo 17. We have $x \equiv 1134^{729} \cdot$

$432^{97} \equiv 12^{729} \cdot 7^{97} \pmod{17}$. In addition, $I(x) \equiv 729 \cdot I(12) + 97 \cdot I(7) \equiv 9 \cdot 13 + 1 \cdot 11 \equiv 0 \pmod{16}$. Therefore, $x \equiv 1 \pmod{17}$.

Exercises 6.3

1. Determine all positive integers that have exactly one primitive root.
2. Show that $3^{(F_3-1)/2} \equiv -1 \pmod{F_3}$, where $F_3 = 2^{2^3} + 1$.
3. Use Crelle's method to show that 2 is a primitive root modulo 29.
4. Use Crelle's method to show that 5 is not a primitive root modulo 29.
5. Find all $\phi(28)$ primitive roots modulo 29.
6. Construct a table of indices modulo 29.
7. Find the fourth and seventh power residues modulo 29.
8. Find all noncongruent solutions to $x^7 \equiv 12 \pmod{29}$.
9. Find all solutions to $x^9 \equiv 12 \pmod{29}$.
10. Use Table 6.2 to solve the following congruences.
    (a) $7x \equiv 5 \pmod{17}$.
    (b) $x^7 \equiv 5 \pmod{17}$.
    (c) $x^8 \equiv 8 \pmod{17}$.
11. Construct a table of indices modulo 11 and use it to solve the following congruences.
    (a) $7x^3 \equiv 3 \pmod{11}$.
    (b) $3x^4 \equiv 5 \pmod{11}$.
    (c) $x^8 \equiv 10 \pmod{11}$.
12. Use indices to find the remainder when $3^{24} \cdot 5^{13}$ is divided by 17.
13. Use indices to find the remainder when $x = 434\,421^{919} \cdot 3415^{783}$ is divided by 29.
14. Prove that the product of all the primitive roots of a prime $p > 3$ is congruent to 1 modulo p.
15. Prove that if $p \equiv 3 \pmod{28}$, then $(\frac{7}{p}) = 1$.
16. Show that $(\frac{3}{p})$ equals 1 if $p \equiv \pm 1 \pmod{12}$ and -1 if $p \equiv \pm 5 \pmod{12}$.
17. Show that $(\frac{5}{p})$ equals 1 if $p \equiv \pm 1 \pmod{10}$ and -1 if $p \equiv \pm 3 \pmod{10}$.

6.4 Miscellaneous exercises

1. In 1879, in *The Educational Times*, Christine Ladd showed that no power of 3 is of the form $13n - 1$ and found the lowest power of 3 of the form $29n - 1$. Duplicate her feat. Ladd received a PhD from Johns Hopkins in 1926, 44 years after she completed the requirements for the

Table 6.3.

					p				
q	3	5	7	11	13	17	19	23	29
3									
5									
7									
11									
13									
17									
19									
23									
29									

degree. Her advisors were J.J. Sylvester and C.S. Peirce. She is the only person ever to have received an honorary degree from Vassar College.

2. If p is an odd prime and d divides $p - 1$, show that $x^d - 1 \equiv 0 \pmod{p}$ has exactly d incongruent solutions modulo p.
3. If p is an odd prime and d divides $p - 1$, determine the d incongruent solutions to $x^d - 1 \equiv 0 \pmod{p}$.
4. If p is a Sophie Germain prime of the form $2q + 1$, where q is a prime of the form $4k + 1$, show that 2 is a primitive root of p.
5. If p is a Sophie Germain prime of the form $2q + 1$, where q is a prime of the form $4k + 3$, show that -2 is a primitive root of p.
6. If $p = 4q + 1$ and $q = 3r + 1$ are prime then show that 3 is a primitive root of p.
7. If p is a prime show that the sum of the primitive roots is 0.
8. Fill in the values of $(\frac{p}{q})$ in Table 6.3, where p and q are distinct odd primes with $3 \leqslant p \leqslant q \leqslant 29$.
9. A group G is called cyclic if it contains an element a, called a generator, such that for every element g in G there is an integer k such that $g = a^k$. That is, every element of G can be represented as a power of a. Show that, for p a prime, Z_p^* is a cyclic of order $p - 1$.
10. Find all the generators of Z_{13}^*.
11. Every subgroup of a cyclic group is cyclic. Determine all the subgroups of Z_{13}^*.

6.5 Supplementary exercises

1. Solve for x if $3x^7 + 2x^6 + x^5 + 5x^4 - 2x^3 + x^2 + x + 2 \equiv 0 \pmod{7}$.
2. Solve for x if $x^2 + 6x - 15 \equiv 0 \pmod{23}$.

3. Solve for x if $x^2 + 7x + 6 \equiv 0 \pmod{37}$.
4. Solve for x if $x^4 + x + 1 \equiv 0 \pmod{15}$.
5. Solve for x if $3x^3 + 2x + 1 \equiv 0 \pmod{21}$.
6. Solve for x if $5x^2 + 7x - 3 \equiv 0 \pmod{35}$.
7. Solve for x if $2x^3 + x^2 + 2x + 1 \equiv 0 \pmod{25}$.
8. Solve for x if $5x^2 + 2x + 5 \equiv 0 \pmod{49}$.
9. Solve for x if $3x^4 + 2x^3 + 2x^2 + 1 \equiv 0 \pmod{27}$.
10. Find all solutions to $x^2 + x + 7 \equiv 0 \pmod{27}$.
11. Solve for x if $x^2 + 6x - 31 \equiv 0 \pmod{72}$.
12. Determine the remainder when $18! + 25!$ is divided by 23.
13. Determine the remainder when $35! + 42!$ is divided by 37.
14. Determine the remainder when $28! + 37!$ is divided by 31.
15. For what value of c does $3x^2 - 3x + c \equiv 0 \pmod{11}$ have solutions?
16. Evaluate $(\frac{7}{23})$, $(\frac{11}{31})$, $(\frac{19}{37})$, and $(\frac{113}{307})$.
17. Evaluate $(\frac{15}{77})$, $(\frac{21}{65})$, $(\frac{100}{143})$, and $(\frac{91}{165})$.
18. Which of the following equations have solutions?
    (a) $x^2 \equiv 211 \pmod{233}$
    (b) $x^2 \equiv 73 \pmod{79}$
    (c) $x^2 \equiv 37 \pmod{53}$
    (d) $x^2 \equiv 71 \pmod{79}$
    (e) $x^2 \equiv 31 \pmod{641}$
19. Which of the following equations have solutions?
    (a) $x^2 \equiv 713 \pmod{1009}$
    (b) $x^2 \equiv 2663 \pmod{3299}$
    (c) $x^2 \equiv 109 \pmod{385}$
    (d) $x^2 \equiv 20964 \pmod{1987}$
    (e) $x^2 \equiv 60 \pmod{379}$
20. For any prime p with $p \equiv 3 \pmod{28}$ show that $(\frac{7}{p}) = 1$.
21. Find a primitive root modulo 23.
22. Find all primitive roots modulo 23.
23. Fine all quadratic residues modulo 23.
24. Solve for x if $17x \equiv 21 \pmod{23}$.
25. Determine the remainder when $1234^{5678} \cdot 5678^{1234}$ is divided by 23.
26. Solve for x if $17x^3 \equiv 19 \pmod{23}$.
27. Solve for x if $x^6 \equiv 4 \pmod{23}$.
28. Use Crelle's method to show that 2 is not a primitive root modulo 31.
29. Use Crelle's method to show that 3 is a primitive root modulo 31.
30. Find all the primitive roots modulo 31.
31. Does $x^2 \equiv 7 \pmod{31}$ have solutions? If so, find them.
32. Find all the quadratic residues modulo 31.

33. Use Crelle's method to show that 2 is a primitive root modulo 37.
34. Find all primitive roots modulo 37.
35. Does $x^2 \equiv 19 \pmod{37}$ have solutions? If so, find them.
36. Find all quadratic residues modulo 37.
37. Determine all sixth power residues modulo 37.
38. Solve for x if $51x \equiv 29 \pmod{37}$.
39. Determine all the solutions to $5x^8 \equiv 11 \pmod{37}$.
40. Determine the remainder when $1234^{5678} \cdot 5678^{1234}$ is divided by 37.

7

Cryptology

> I have resumed the study of mathematics with great avidity. It was ever my favourite one . . . where no uncertainties remain on the mind; all is demonstration and satisfaction.
>
> *Thomas Jefferson*

7.1 Monoalphabetic ciphers

Crypto is from the Greek *kryptos*, meaning hidden or secret. Cryptology is the study of secrecy systems, cryptography, the design and implementation of secrecy systems, and cryptanalysis, the study of systems or methods of breaking ciphers. The message to be altered into secret form, the message we want to send, is called the plaintext. The message we actually send is called the ciphertext. The device used to transform the plaintext into the ciphertext is called a cipher. Plaintext and ciphertext may be composed of letters, numbers, punctuation marks, or other symbols. Encryption or enciphering is the process of changing plaintext into ciphertext. Decryption or deciphering is the process of changing ciphertext back into plaintext. In order to make decryption more difficult, plaintext and ciphertext are often broken up into message units of a fixed number of characters. The enciphering transformation can be thought of as a one-to-one function that takes plaintext message units into corresponding ciphertext message units. The process or method used in going from the plaintext to ciphertext and back to the plaintext is called a cryptosystem. A cipher is called monoalphabetic if it uses only one cipher alphabet.

Encryption or decryption is often mistaken for encoding or decoding, respectively. A code, however, is a system used for brevity or secrecy of communication, in which arbitrarily chosen words, letters, or phrases are assigned definite symbols. In most cases a code book is necessary to decode coded messages.

The demand for and use of cryptography are directly proportional to the literacy and paranoia of the peoples involved. The history of cryptology has Babylonian, Egyptian, and Hindu roots. A Babylonian cuneiform tablet, dating from about 1500 BC, contains an encrypted recipe for making pottery glaze. Al-Khalil, an eighth century philologist, wrote the

Book of Secret Language, in which he mentions decoding Greek cryptograms. Homer's works were originally passed on from generation to generation orally. One of the earliest references to Greek writing is found in Book 6 of the *Iliad* when King Proetus sends Bellerophon to Lycia with a document containing secret writing. In Book 5 of *The History*, Herodotus remarked that Histiaeus, the despot of Miletus who was being held by Darius, shaved and tattooed a message to revolt against the Persians on the head of a trusted slave. After waiting for the hair to grow in again, Histiaeus sent the slave to his son-in-law Aristagoras in Miletus who shaved the head and found the message. *The History* also includes an account of a very subtle secret message. Thrasybulus, despot of Miletus, gives no written or verbal message to a messenger from Periander, tyrant of Corinth and one of the seven sages of the ancient world, but while walking through a field of corn with him, cuts down any corn that was growing above the rest. This act of removing the fairest and strongest is related to Periander by the messenger and he interprets it as having to murder the most eminent citizens of Corinth.

The Spartans are credited with the first system of military cryptography. They enciphered some messages by wrapping a strip of papyrus or parchment helically around a long cylindrical rod called a *skytale*. The message was written lengthwise down the cylinder. The paper was unwound and sent. Given a rod of the same radius and length, the strip could be wound around it helically and the message deciphered. One of the earliest known works on cryptanalysis was Aeneas the Tactician's *On the Defense of Fortified Places* which includes a clever method of hidden writing whereby holes are pricked in a document or page of a book directly above the letters in the secret message to be sent. A variation of this method was used by the Germans in World War II.

Polybius, the second century BC Greek politician, diplomat, and historian, devised a cryptographic system that replaced plaintext letters with a pair of symbols as shown in Table 7.1, where we have used the English alphabet and the numerals 1, 2, 3, 4, 5. According to Polybius's method, the message

LET NONE ENTER IGNORANT OF GEOMETRY

would be sent as

31 15 44 33 34 33 15 15 33 44 15 42 24 22 33
34 42 11 33 44 34 21 22 15 34 32 15 44 42 54,

where the first numeral indicates the location of the row and the second the column of the plaintext letter.

Table 7.1.

	1	2	3	4	5
1	A	B	C	D	E
2	F	G	H	IJ	K
3	L	M	N	O	P
4	Q	R	S	T	U
5	V	W	X	Y	Z

Table 7.2.

A	B	C	D	E	F	G	H	I	J	K	L	M	N	O	P	Q	R	S	T	U	V	W	X	Y	Z
D	E	F	G	H	I	J	K	L	M	N	O	P	Q	R	S	T	U	V	W	X	Y	Z	A	B	C

Character ciphers are systems based on transforming each letter of the plaintext into a different letter to produce the ciphertext, that is, each letter is changed by substitution. Character ciphers can be traced back to the Romans. Valerius Probus, a grammarian, wrote a treatise on the ciphers used by Julius Caesar. Suetonius, the Roman historian, wrote that Caesar used a cipher which simply replaced each letter in the alphabet by the letter three letters to the right, with the stipulation that X, Y, and Z were replaced by A, B, and C respectively, as shown in Table 7.2, where we use the English rather than the Latin alphabet and have preserved the natural lengths of words. The plaintext message

BOUDICCA HAS BURNED LONDINIUM

would be enciphered using Caesar's cipher into the ciphertext

ERXGLFFD KDV EXUQHG ORQGLQLXP.

Augustus Caesar (Octavian) used a much simplified version of his uncle's cipher in which he transformed plaintext to ciphertext by merely substituting, with the exception of writing AA for X, the next letter of the alphabet. One can hardly fail to get a feeling for the dearth of literacy during this period of Roman history.

We can generalize character ciphers mathematically by translating the letters of the alphabet of any plaintext into numerical equivalents, for example, using Table 7.3. Let the letter P denote the numerical equivalent of a letter in the plaintext and the letter C denote the numerical equivalent

Table 7.3.

A	B	C	D	E	F	G	H	I	J	K	L	M	N	O	P	Q	R	S	T	U	V	W	X	Y	Z
0	1	2	3	4	5	6	7	8	9	10	11	12	13	14	15	16	17	18	19	20	21	22	23	24	25

Table 7.4.

A	B	C	D	E	F	G	H	I	J	K	L	M	N	O	P	Q	R	S	T	U	V	W	X	Y	Z
7	1	3	4	13	2	2	6	8	$*$	1	4	2	8	7	2	$*$	8	6	9	3	1	2	$*$	2	$*$

of the corresponding letter in the ciphertext. Caesar's cipher would then be represented by the transformation $C \equiv P + 3 \pmod{26}$ and its inverse by $P \equiv C - 3 \pmod{26}$. Any cipher of the form $C \equiv P + k \pmod{26}$, with $0 \leqslant k \leqslant 25$, is called a shift transformation, where k, the key, represents the size of the shift. Accordingly, the corresponding deciphering transformation is given by $P \equiv C - k \pmod{26}$. If we include the case where $k = 0$, where the letters of the plaintext are not altered at all, there are 26 possible shift transformations. For example, consider the shift transformtion with key $k = 17$ and the plaintext message

THOMAS JEFFERSON LIVES.

We use the cipher $C \equiv P + 17 \pmod{26}$ to transform the numerical plaintext

19 7 14 12 0 18 9 4 5 5 4 17 18 14 13 11 8 21 4 18

into the ciphertext

10 24 5 3 17 9 0 21 22 22 21 8 9 5 4 2 25 12 21 9

and send the message as

KYFDRJ AVWWVIJFE CZMVJ.

The major difficulty with shift transformations is their vulnerability to being deciphered easily using the relative frequency of the letters. In a relatively long sample of English text, the most frequently occurring letter will normally be E, followed by T, N, I, R, O and A, respectively. Table 7.4 exhibits the percent frequency of the occurrence of letters in a standard English text, where an asterisk is used to denote that the normal occurrence of the letter is less than one percent.

Similar tables exist for most major languages. However, we cannot always assume that the natural frequency prevails in the plaintext, for it is not impossible to circumvent the natural frequencies of a language as well.

Table 7.5.

A	B	C	D	E	F	G	H	I	J	K	L	M	N	O	P	Q	R	S	T	U	V	W	X	Y	Z
6	0	0	4	8	9	2	2	0	2	0	0	7	1	1	2	11	3	1	7	7	0	0	3	0	4

La Disparition, a novel written in 1969 by George Perec, included over 85 000 words and not one of them contained the letter 'e'. Nevertheless, using the information in Table 7.4, we may be able to decipher a long ciphertext which has been encoded using a shift transformation by frequency analysis as illustrated in the next example.

Example 7.1 Suppose that we wish to decipher the ciphertext

URUTM HQEQQ ZMXUF FXQRM DFTQD FTMZA FTQDE UFUEN QOMGE

QUTMH QEFAA PAZFT QETAG XPQDE ARSUM ZFEJJ

given that a shift transformation was used to encipher the plaintext message. The encipherer has divided the ciphertext into a uniform set of letters, quintuplets in this case, to disguise any natural lengths that may be apparent in the plaintext. The frequency of letters for our ciphertext is given in Table 7.5. Since the letter that occurs most frequently is Q, we assume that E was sent as Q. Hence, $k = 12$. The plaintext message expressed in quintuplets would read

IFIHA VESEE NALIT TLEFA RTHER THANO THERS ITISB ECAUS EIHAV

ESTOO DONTH ESHOU LDERS OFGIA NTSXX,

or with natural word length

IF I HAVE SEEN A LITTLE FARTHER THAN OTHERS IT IS BECAUSE I

HAVE STOOD ON THE SHOULDERS OF GIANTS,

a quote attributed to Isaac Newton.

Ciphers of the form $C \equiv aP + b \pmod{26}$, where $0 \leqslant a,b \leqslant 25$, and $\gcd(a, 26) = 1$, are called affine ciphers. Shift ciphers are affine ciphers with $a = 1$. There are $\phi(26) = 12$ choices for a and 26 choices for b, hence, 312 possible affine ciphers. The deciphering transformation for an affine cipher is given by $P \equiv a^{-1}(C - b) \pmod{26}$, where $0 \leqslant P \leqslant 25$ and $aa^{-1} \equiv 1 \pmod{26}$. For convenience, Table 7.6 gives the inverses of positive integers less than and coprime to 26 modulo 26.

Table 7.6.

a	1	3	5	7	9	11	15	17	19	21	23	25
a^{-1}	1	9	21	15	3	19	7	23	11	5	17	25

Example 7.2 We encode the plaintext

SHAKESPEARE WAS A PEN NAME FOR EDWARD DE VERE THE
EARL OF OXFORD,

using the affine transformation $C \equiv aP + b$, with $a = 5$ and $b = 8$. From Table 7.3, the numerical equivalent of the plaintext is given by

18 7 0 10 4 18 15 4 0 17 4 22 0 18 0 15 4 13
13 0 12 4 6 14 17 4 3 22 0 17 3 3 4 21 4 17
5 19 7 4 4 0 17 11 14 5 14 23 5 14 17 3.

Applying the cipher $C \equiv 5P + 8 \pmod{26}$, we obtain

20 17 8 6 2 20 5 2 8 15 2 14 8 20 8 5 2 21 21
8 16 2 12 0 15 2 23 0 8 15 23 23 2 9 2 15 2 25
17 2 2 8 15 11 0 7 0 19 7 0 15 23.

Transforming from numerical to alphabetic quintuplet ciphertext we obtain

URIGC UFCIP COIUI FCVVI QCMAP CXAIP XXCJC PCZRC CIPLA
HATHA PXTTT,

where we have added XXX to the end of the plaintext message to preserve the quintuplicate nature of the ciphertext and to make the message more difficult to decipher.

Nevertheless, a deciphering technique using the relative frequency of letters can be used to decipher most affine transformations as illustrated in the next example.

Example 7.3 Albeit the message

FJJIF JLIIO JFLIH YJJYJ GINJQ YJPQL ZGZGZ

is relatively short, we can use frequency analysis to decipher it. From Table 7.7, we see that the letter J appears nine times and the letter I five times. Suppose E corresponds to J and T corresponds to I. Let $C = aP + b \pmod{26}$. With $C = 9$ when $P = 4$ and $C = 8$ when $P = 19$, we obtain

$$9 \equiv 4a + b \pmod{26}$$

Table 7.7.

A	B	C	D	E	F	G	H	I	J	K	L	M	N	O	P	Q	R	S	T	U	V	W	X	Y	Z
0	0	0	0	0	3	3	1	5	9	0	3	0	1	1	1	2	0	0	0	0	0	0	0	3	3

and

$$8 \equiv 19a + b \pmod{26}.$$

Subtracting the first equation from the second, we obtain $15a \equiv -1 \equiv 25 \pmod{26}$. Multiplying both sides of the congruence by 7, the inverse of 15 modulo 26, we get $a \equiv 19 \pmod{26}$. Substituting the value $a = 19$ into the first equation, we find that $b \equiv 11 \pmod{26}$. Thus, the message was enciphered using the affine transformation $C \equiv 19P + 11 \pmod{26}$. Applying the inverse transformation, $P \equiv 11C + 9 \pmod{26}$, to the numerical ciphertext, we recover the plaintext message:

MEET ME AT THE MATINEE NEXT WEDNESDAY.

In Europe, the period from about 400 to about 800, following the collapse of the Roman Empire, is referred to by many historians as the Dark Ages. The barbarians were at the gates, culture and literacy went seriously into decline, and with them went cryptography. In 529, after existing for over nine centuries, Plato's Academy was closed. Almost singlehandedly, Benedictine monasteries continued to serve as effective educational institutions throughout the Dark Ages. According to conservative estimates over 90 percent of the literate men between 600 and 1100 received their instruction in a monastic order. Very few scientific commentaries appeared and many of those that did were woefully primitive. People had a rough time just making ends meet. Most began looking for a better life in the hereafter.

As with mathematics and science, cryptology developed in India and Islamic countries during the European Dark Ages. The *Kamasutra*, written sometime between the third and fifth centuries and attributed to Vatsyayana, lists secret writing as one of the arts a woman should understand and practice. One ancient Hindu cipher consisted of substituting a set of letters of the Hindu alphabet in the plaintext for each other and leaving the remaining letters unaltered. In 855, Abu Bakr Ahmad included several ciphers in *Book of the Frenzied Devotee's Desire to Learn about the Riddles of Ancient Scripts*. Ibn Khaldun's *Muqaddimah* describes several codes used by Islamic tax and military bureaucrats. A compilation of Islamic knowledge of cryptography was included in a compendium of all branches of knowledge useful to civil servants written by al-Qulqashandi in 1412. As with Euclid's *Elements*, much of the content of al-Qulqashandi's book was based on works of his predecessors, the chapter on cryptology being no exception for much of it came from a fourteenth century treatise by al-Duraihim. Many of the cryptographic methods mentioned in al-Duraihim's work were quite sophisticated, for example, letter substitution using numeric as well as symbolic substitution and a method whereby vowels were deleted and the letters of each word were reversed.

For example, let us look at some of the more fundamental ways a message can be altered using transpositions. We could send the plaintext

BURN ALL YOUR CODES

using a simple transposition cipher as follows:

B R A L O R O E
 U N L Y U C D S

and send it as

BRAL OROE UNLY UCDS.

We could have written the plaintext in columns–

B O
U U
R R
N C

A O
L D
L E
Y S

–and sent the message as

BOUU RRNC AOLD LEYS.

We could have written the message in a matrix as

B A O O
U L U D
R L R E
N Y C S

and sent it as

BAOO ULUD RLRE NYCS.

The few European ciphertext manuscripts that exist from the period from 400 to 1400 employ very primitive encryption systems, for example, transposition ciphers with $k = 1$, simple letter substitution using foreign alphabets or symbols, dots substituted for vowels, and phrases written backwards or vertically. There were a few notable exceptions. Gerbert, Pope Sylvester II, used a shorthand encryption system to record important notes and messages. Hildegard von Bingen, a twelfth century Benedictine abbess and composer of liturgical music, used a cipher alphabet consisting of a mixture of German and Latin, which came to her in a vision. Roger Bacon, an English Franciscan scholar, wrote a treatise, *Secret Works of Art and the Nullity of Magic*, in the mid thirteenth century, in which he listed a number of primitive encryption systems. Geoffrey Chaucer, using a simple alphabet substitution, enciphered a few lines of the *The Equatorie of the Planets*. The earliest known manuscript devoted entirely to cryptanalysis, including rules for deciphering simple substitution ciphers where word order has been preserved, was written in 1747 by Cicco Simonetta, a Milanese civil servant.

Exercises 7.1

1. Use Polybius's method to encipher the message

   NO MAN IS AN ISLAND.

2. Decipher the message

   24 44 43 22 42 15 15 25 44 34 32 15

   given that it was enciphered using Table 7.1 and Polybius's method.
3. Using the Caesar cipher, encipher the following messages:
   (a) I HAVE A SECRET;
   (b) SIC SEMPER TYRANNIS;
   (c) SEND HELP.
4. Decipher the following messages assuming that each has been enciphered using the Caesar cipher.
   (a) DOOPH QDUHP RUWDO;
   (b) SHULF XOXPL QPRUD;
   (c) LQYLWR SDWUH VLGHUD YHUVR.
5. Use frequency analysis and the knowledge that the message was enciphered using a shift transformation to decipher

   PXAHE WMAXL XMKNM ALMHU XLXEY XOBWX GMMAT MTEEF XGTKX
   VKXTM XWXJN TEMAT MMAXR TKXXG WHPXW URMAX BKVKX TMHKP
   BMAVX KMTBG NGTEB XGTUE XKBZA MLMAT MTFHG ZMAXL XTKXE
   BYXEB UXKMR TGWMA XINKL NBMHY ATIIB GXLLQ

6. Encipher the message

THERE IS A MOLE IN THE OFFICE

using the affine transformation $C \equiv 7P + 4 \pmod{26}$.

7. Decipher the message

   WHSNK FGLNJ ELHFY JQTGX YZGI,

   which was enciphered using the affine transformation $C \equiv 11P + 6 \pmod{26}$.
8. If the most common letter in a long ciphertext, enciphered by a shift transformation $C \equiv P + k \pmod{26}$, is S, what is the most likely value for k?
9. Decipher the ciphertext

   YFXMP CESPZ CJTDF DPQFW QZCPY NTASP CTYRX PDDLR PD,

   given that it was enciphered using a shift transformation.
10. If the two most common letters in a long ciphertext enciphered by an affine transformation $C \equiv aP + b \pmod{26}$ are V and A respectively, then what are the two most likely values for a and b?
11. Decipher the following ciphertext given that the message was enciphered using an affine transformation in which E and T were enciphered as L and U, respectively.

    BSLGU SLRGL HYLTU JPRYL YPRVL JURVT YZTHT

    DGJUX RFYGT VLUSL VTYZD JGRUW RYBSL USLYX

    RRFBRG RYKRJ UEFUS RBXRF CKTXL AUSLH TVLMM

12. Decipher the following cipher given that the message was enciphered using a simple transposition cipher.

    D E S D O H N N S R S E T T E I E

13. Decipher the following cipher given that the message was enciphered using a simple transposition column cipher.

    T E W E E S N K T I Y D T D H O R O

14. Decipher the following cipher given that the message was enciphered using a simple transposition matrix cipher.

    NACBNF ESHYYE VSOSTW EOWOOW RMEMSS WUDAOC

7.2 Polyalphabetic ciphers

In an attempt to hinder decryption by frequency analysis, a method was introduced in the early fifteenth century whereby simple substitution is used to alter consonants and multiple substitution to alter vowels. Around 1467, Leon Battista Alberti, the Italian artist and author of the first printed book on architecture, wrote a treatise on cryptanalysis, which was published posthumously in 1568. The treatise included instruction on how to construct a cipher disk. This was the earliest appearance of a polyalphabetic cipher, one involving two or more cipher alphabets, and forms the basis for modern cryptograms, and secret decoder rings as well. Alberti's cipher disk was made from two copper disks of unequal size with a pin through their centers to hold them together. A cipher disk using modern English letters, where the letters Y and Z have been omitted, is shown in Figure 7.1. Alberti divided each disk into 24 equal parts listing the plaintext consisting of 20 letters of the Italian alphabet and the first 4 natural numbers on the larger outer disk. The numbers on the outer disk were used in pairs, triples, or taken 4 at a time to represent encoded words or phrases which he inserted into the ciphertext. After enciphering (and/or encoding) part of the plaintext, the inner disk was rotated and another part of the message enciphered using a different cipher. The process was repeated until the complete message was enciphered. Besides the ability to encode as well as to encipher messages, the main advantage of Alberti's cipher disk was that the word THE in the plaintext may be encoded as PWR in one part of the message and as UVA in another.

Example 7.4 Suppose we wished to encipher the message

EAT MORE BROCCOLI.

One option would be to encipher the first two words using the cipher disk as shown in Figure 7.1, where A is encoded as Q, then rotating the inner disk countcrclockwisc scvcn positions so A is encoded as W, as shown in Figure 7.2. The ciphertext message would appear as

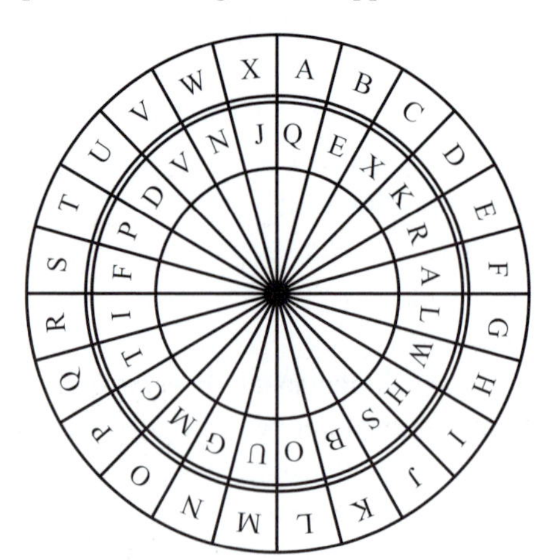

Figure 7.1.

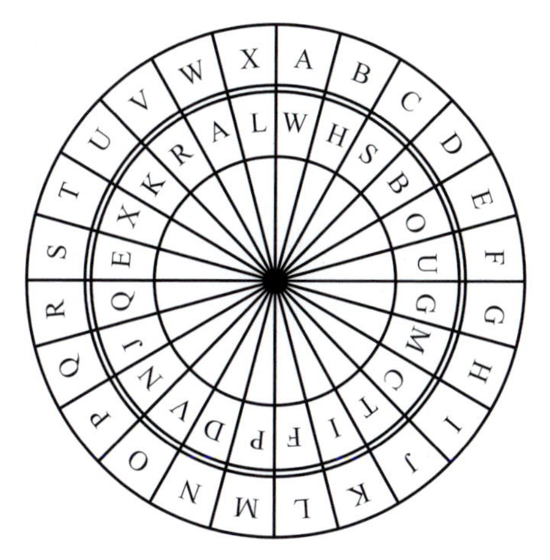

Figure 7.2.

RQPUM IRHQV SSVFC.

In 1499, Johannes Trithemius wrote a trilogy on communicating with spirits called *Steganographia*, Greek for ‘hidden writing’. The text was criticized by Protestants. It was included, along with the works of Copernicus, Kepler, and Galileo, on the *Index librorum prohibitorum*, a list of books Roman Catholics were forbidden to read “until corrected”. The third volume, on occult astrology, consisted mainly of tables of numbers that many believed contained secret incantations for conjuring up spirits. In 1676, Wolfgang Ernst Heidel, a lawyer from Mainz, claimed to have deciphered Trithemius’s passages, but he wrote his solution with a secret cipher that no one could decipher. In 1996, Thomas Ernst of La Roche College in Pittsburgh and, independently two years later, Jim Reeds of AT&T Labs in Florham Park, New Jersey, deciphered Johannes Trithemius’s third volume. Disappointingly, the messages turned out to be mainly trite sayings. Ernst turned his attention to Heidel and deciphered his manuscript. He found that Heidel had in fact deciphered the secret passages in Trithemius’s third volume.

The first printed book on cryptography, *Polygraphia*, appeared in 1518. It had been written by Trithemius about 10 years earlier. The bulk of the text is taken up with hundreds of columns of Latin words each preceded by a letter. The book’s most important innovation in cryptology was the transformation of the wheel cipher into an alphabetic square to encode plaintext shown in Table 7.8. Rows corresponded to key letters and columns to plaintext letters. Ciphertext letters are found at the intersections

Table 7.8.

A	B	C	D	E	F	G	H	I	J	K	L	M	N	O	P	Q	R	S	T	U	V	W	X	Y	Z
A	B	C	D	E	F	G	H	I	J	K	L	M	N	O	P	Q	R	S	T	U	V	W	X	Y	Z
B	C	D	E	F	G	H	I	J	K	L	M	N	O	P	Q	R	S	T	U	V	W	X	Y	Z	A
C	D	E	F	G	H	I	J	K	L	M	N	O	P	Q	R	S	T	U	V	W	X	Y	Z	A	B
D	E	F	G	H	I	J	K	L	M	N	O	P	Q	R	S	T	U	V	W	X	Y	Z	A	B	C
E	F	G	H	I	J	K	L	M	N	O	P	Q	R	S	T	U	V	W	X	Y	Z	A	B	C	D
F	G	H	I	J	K	L	M	N	O	P	Q	R	S	T	U	V	W	X	Y	Z	A	B	C	D	E
G	H	I	J	K	L	M	N	O	P	Q	R	S	T	U	V	W	X	Y	Z	A	B	C	D	E	F
H	I	J	K	L	M	N	O	P	Q	R	S	T	U	V	W	X	Y	Z	A	B	C	D	E	F	G
I	J	K	L	M	N	O	P	Q	R	S	T	U	V	W	X	Y	Z	A	B	C	D	E	F	G	H
J	K	L	M	N	O	P	Q	R	S	T	U	V	W	X	Y	Z	A	B	C	D	E	F	G	H	I
K	L	M	N	O	P	Q	R	S	T	U	V	W	X	Y	Z	A	B	C	D	E	F	G	H	I	J
L	M	N	O	P	Q	R	S	T	U	V	W	X	Y	Z	A	B	C	D	E	F	G	H	I	J	K
M	N	O	P	Q	R	S	T	U	V	W	X	Y	Z	A	B	C	D	E	F	G	H	I	J	K	L
N	O	P	Q	R	S	T	U	V	W	X	Y	Z	A	B	C	D	E	F	G	H	I	J	K	L	M
O	P	Q	R	S	T	U	V	W	X	Y	Z	A	B	C	D	E	F	G	H	I	J	K	L	M	N
P	Q	R	S	T	U	V	W	X	Y	Z	A	B	C	D	E	F	G	H	I	J	K	L	M	N	O
Q	R	S	T	U	V	W	X	Y	Z	A	B	C	D	E	F	G	H	I	J	K	L	M	N	O	P
R	S	T	U	V	W	X	Y	Z	A	B	C	D	E	F	G	H	I	J	K	L	M	N	O	P	Q
S	T	U	V	W	X	Y	Z	A	B	C	D	E	F	G	H	I	J	K	L	M	N	O	P	Q	R
T	U	V	W	X	Y	Z	A	B	C	D	E	F	G	H	I	J	K	L	M	N	O	P	Q	R	S
U	V	W	X	Y	Z	A	B	C	D	E	F	G	H	I	J	K	L	M	N	O	P	Q	R	S	T
V	W	X	Y	Z	A	B	C	D	E	F	G	H	I	J	K	L	M	N	O	P	Q	R	S	T	U
W	X	Y	Z	A	B	C	D	E	F	G	H	I	J	K	L	M	N	O	P	Q	R	S	T	U	V
X	Y	Z	A	B	C	D	E	F	G	H	I	J	K	L	M	N	O	P	Q	R	S	T	U	V	W
Y	Z	A	B	C	D	E	F	G	H	I	J	K	L	M	N	O	P	Q	R	S	T	U	V	W	X
Z	A	B	C	D	E	F	G	H	I	J	K	L	M	N	O	P	Q	R	S	T	U	V	W	X	Y

of rows and columns. For example, to encode the word DEUS with Trithemius's cipher, we leave the 1st letter unaltered. We replace the 2nd letter by F, the letter under E in the 3rd row. We replace the 3rd letter by W, the letter under U in the 4th row. Finally, we replace the 4th letter by V, the letter under S in the 5th row. The ciphertext obtained is DFWV. For long messages, the 26th row is followed by the 1st row and the process cycles.

In 1553, Giovan Batista Belaso introduced a polyalphabetic cipher similar to Trithemius's cipher where a key phrase is used to indicate the column by which successive letters are enciphered. For example, using the key phrase

SIC SEMPER TYRANNIS ET MURES XX,

using Table 7.8, we encipher

THE TREE OF LIBERTY

as

LPG LVQT SF MGSEEGG.

The 'S' column is used to encipher T as L, the 'I' column is used to encipher H as P, the 'C' column is used to encipher E as G, and so forth.

In 1550, Girolamo Cardano, the physician–mathematician and author of the first text on probability, devised a technique whereby a mask with windows was placed over a piece of paper and the message written in the windows. The mask was then removed and the rest of the paper filled with words and phrases. When the mask was placed over the document the message was revealed. Several sixteenth and seventeenth century diplomats made use of Cardano's system.

Cardano described an innovative but incomplete autokey cipher system, where the message itself is used as the key phrase. The earliest valid autokey system was formulated in 1563 by Giovanni Battista Porta who invented the camera obscura. In *De furtivis literarum notis*, Porta included the cryptographic contributions of Alberti, Trithemius, Belaso, and Cardano. He described numerous cipher systems and suggested making deliberate misspellings, transposing letters, and using nonsense words as keys in enciphering plaintext. *De furtivis* included a pair of cipher disks and a cipher whereby a 26 by 26 matrix consisting of 676 distinct symbols was used to encipher and decipher messages. Each symbol in the matrix represented a pair of letters. For example, if the symbol □ in the 3rd column and 9th row represented the letter pair CI and the symbol ✿ in the 1st row and 14th column represented the pair AO, then □✿ stands for CIAO.

Giordano Bruno, a peripatetic Dominican friar, resided at the home of the French ambassador in London from 1583 to 1585. He used the alias Henry Fagot when he sent messages back to France. He devised a cipher where each vowel is exchanged with the next letter of the alphabet. Hence,

ALLISWELL

would be sent as

BLLJSWFLL.

Bruno was the first modern European to profess belief that the universe is infinite and that the stars are suns. Bruno was brought before the Inquisition for his beliefs, not his espionage, and burned at the stake in 1600. Ironically, the English and French term for the bundles of wood used to kindle the flames when Bruno and other heretics were burned at the stake is fagots.

In the early seventeenth century, Matteo Argenti, a cryptologist for several popes, wrote a primer on Renaissance ciphers, many of which he

Table 7.9.

E	N	R	I	C	O	A	B	D	F	G	H	L	M	P	Q	S	T	U	V	Z
10	11	12	13	14	15	16	17	18	19	20	21	22	23	24	25	26	27	28	29	30

and his uncle, also a papal cryptologist, had devised. They were the first to use a mnemonic key to encipher an alphabet. One, for example, is shown in Table 7.9 using the modern Italian alphabet with key ENRICO. To discourage decryption by frequency analysis the Argentis suggested using several numbers, for example 5, 7, and 9, interspersed frequently throughout the text, representing nulls. They also stressed multiple vowel substitution and deleting the second member of a double letter consonant combinations. For example, *mezo* for *mezzo* and *mile* for *mille*. As a further hindrance to would-be cryptanalysis, they used other numbers to represent often used words such as 'and', 'this', 'that', 'which' and 'what'. Cryptanalysts had their hands full when attempting to decipher an Argenti ciphertext.

Philip II of Spain used both multiple vowel and multiple consonant substitutions in his ciphers. François Viète, a lawyer by profession whose mathematical work revolutionized algebra, worked as a cryptanalyst at the court of Henry IV, King of France. The Cambridge educated mathematician, John Wallis, deciphered messages for Charles I, Charles II, and William and Mary. In 1641, John Wilkins, first secretary of the Royal Society, introduced the words cryptographia (secret writing) and cryptologia (secret speech) into the English language.

In 1586, using an array similar to that shown in Table 7.8, Blaise de Vigenère [VEE zhen AIR], a French author, diplomat, and cryptanalyst for Charles IX of France, devised a number of polyalphabetic ciphers that appear in his *Traicté des chiffres*. Two of the autokey ciphers he devised deserve note. In one, the plaintext is the key and in another the ciphertext is the key, where the first key letter is known to both the encipher and the decipher. For example, suppose we received the ciphertext message

CWRQPAFVQABRC,

and were told the first key letter was K, and that the first letter of the plaintext message was S. According to this cipher, S would be the second letter in the key, as shown in Table 7.10. Using Table 7.8, now known as the Vigenère tableau, the second letter in the plaintext is E, and it becomes the third letter in the key, and so forth. Thus, the plaintext message is

Table 7.10.

Key	K	S	E	N	D	M	O	R	E	M	O	N	E
Plaintext	S	E	N	D	M	O	R	E	M	O	N	E	Y
Ciphertext	C	W	R	Q	P	A	F	V	Q	A	B	R	C

Table 7.11.

Key	R	C	Q	E	O	C	W
Plaintext	L	O	O	K	O	U	T
Ciphertext	C	Q	E	O	C	W	P

SEND MORE MONEY.

Suppose we are given the ciphertext

CQEOCWP

and know that R is the first letter of the key and the ciphertext has been used as the key. We fill in the rest of the code as shown in Table 7.11 and use Table 7.8 to recover the plaintext message,

LOOK OUT.

Vigenère ciphers employ an alphabetic matrix, as shown in Table 7.8, and use a simple key word that is repeated. For example, let us encipher the message

OH TO BE IN ENGLAND NOW THAT APRILS THERE

using the key word VOILA. Use column 'V' to encipher O as J. Use column 'O' to encipher H as V. Use column 'I' to encipher T as B, and so forth.

VOILA VOILA VOILA VOILA VOILA VOLIA VOILA

OHTOB EINEN GLAND NOWTH ATAPR ILSTH EREXX

The ciphertext, in quintuplets, would appear as

JVBZB ZWVPN BZIYD ICEEH VHIAR DZAEH ZFMIX

This Vigenère cipher can be broken much more easily than his autokey ciphers using a method developed successfully by F. W. Kasiski in 1863. In our example, once the cryptanalyst knows that the key has five letters, frequency analysis may be often employed on successive sets containing every fifth letter. In 1925, the American cryptanalyst, William Friedman, developed a method that would determine the length of the key word in any Vigenère cipher.

Unfortunately, Vigenère's work had relatively little influence on his contemporaries. Vigenère tableaux were rediscovered by a number of

Table 7.12.

C	A	M	B	R
IJ	D	G	E	F
H	K	L	N	O
P	Q	S	T	U
V	W	X	Y	Z

cryptanalysts including the English mathematician and author Charles Dodgson (Lewis Carroll). A similar array, known as a Beaufort tableau, was published in 1857 by Sir Francis Beaufort, Rear-Admiral of the Royal Navy, inventor of the Beaufort wind scale ranging from 0 (calm) to 12 (hurricane). Beaufort's alphabetic array contains 22 rows and 22 columns, key letters and plaintext were denoted on the rows, the ciphertext letters on the columns. Nevertheless, Beaufort's tableaux had been used to encipher plaintext by Giovanni Sestri as early as 1710.

Charles Babbage, whose analytic engine was the precursor of our modern computers, constructed a 26-volume code breaking dictionary. He deciphered a message sent by Henrietta Maria, consort of Charles I, personal advertisements found in *The Times*, and a number of Vigenère ciphers. He served as a cryptographical advisor to Beaufort during the Crimean War. Babbage wrote that deciphering is a fascinating art and one which he had wasted more time on than it deserved. He thought, as did many cryptanalysts, that he was capable of constructing a cipher that no one else could break. Unfortunately, the cleverer the person the more deep-seated was the conviction. He was particularly adept at deciphering digraphic ciphers, where letters are paired and encoded together. These ciphers were devised in 1854 by Charles Wheatstone, inventor of the Wheatstone bridge, a circuit used in physics. There are many variations of Wheatstone's cipher, one, in particular, with keyword CAMBRIDGE is shown in Table 7.12.

To encipher messages using Wheatstone's cipher, letters were paired up. Paired letters on the same row or column were encoded cyclically. Hence, AQ and WZ would be enciphered as DW and XV, respectively. Similarly, NO and EN would be enciphered as OH and NT, respectively. If the two letters are not on the same row or column then they form opposite vertices of a rectangle and are replaced by the two letters forming the other two vertices of the rectangle with the proviso that letters on the same row replace each other. For example, DN and OS were replaced by EK and LU, respectively.

The letters I and J were considered identical and double letters were separated by an X. Hence,

WILLIAM TELL

would be enciphered as if it were

WILXLIAM TELXL.

Using the Wheatstone cipher shown in Table 7.12, the message

ALWAYS DENY IT

would be sent as

MKADXT GFTB EP.

When serving as Secretary of State for George Washington in the 1790s, Thomas Jefferson devised a wheel cipher. His cipher was about six inches long and consisted of 36 wooden disks each about $\frac{1}{6}$ of an inch thick held together with a bolt and nuts on each end, similar to that shown in Figure 7.3. The outer rim of each disk was divided into 26 equal parts where the letters of the alphabet appeared in random order. To encipher a message, the wheels were rotated until the message to be sent appeared and then one of the remaining 25 jumbled lines sent as the ciphertext. Jefferson did not recommend his method to his successors and it was forgotten. Several years later, when he was President, he chose a Vigenère cipher as the

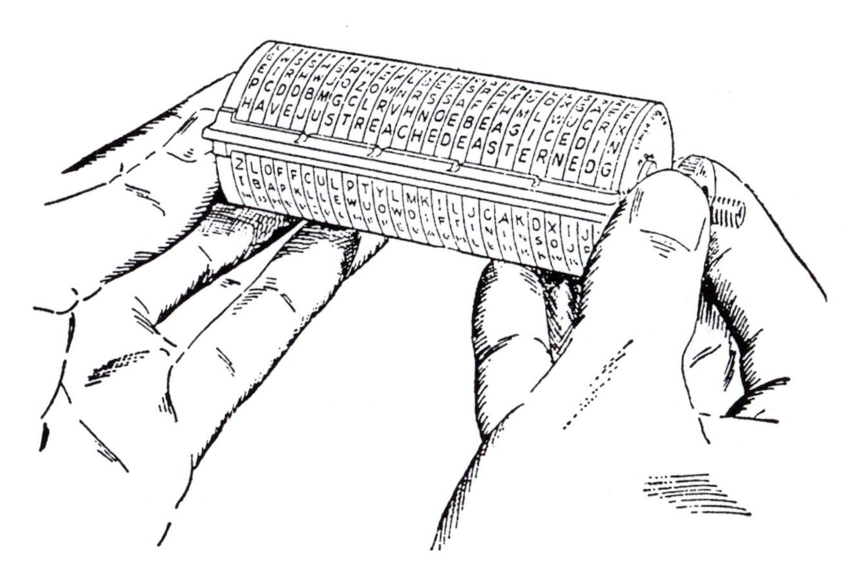

Figure 7.3.

official method for Meriweather Lewis and William Clark to encipher their messages to him during their expedition to explore the Louisiana Territory in 1802–4. Wheel ciphers were rediscovered by the US Army in 1922 and were used by the US Navy up to 1960.

Exercises 7.2

1. Use the disk cipher shown in Figure 7.1 to encipher

   RUMPLESTILTSKIN.

2. Decipher

   FDIIRGKRI WX VDSO

   assuming the first word was enciphered using Figure 7.1 and the last words using Figure 7.2.
3. Encipher

   MAKE MY DAY

   using Trithemius's cipher.
4. Decipher

   HFTHA JGYMW YEMSF PYULH ZJHIM VTIGW VZZOE QOCQI SGYKE TSGAX
   BMOSV RPJXH BNOCD BFPIC HGAVO OCZSP PLWKX LMCYN AZPPJZL

   given that it was enciphered using Trithemius's cipher.
5. Encipher

   MEET ME TONIGHT AT MIDNIGHT

   using Belaso's cipher with key phrase

   ARIVEDERCI ROMA ARIVEDERCI.

6. Decipher

   FFYPGWZFWT

   given it was enciphered using Belaso's cipher with key

   FOURSCORE AND SEVEN YEARS AGO.

7. Decipher

   JLJMP ORTFD FCHFR CHFRT PVJPV RSLBV FRJTF

   given it was enciphered using Bruno's cipher.
8. Decipher the ciphertext

   YVKZR WTZJZ XALIP PXFQG QIHGM ALAWYQ

   given it was enciphered with a Vigenère autocipher using the plaintext as key and R as the first letter of the key.
9. Decipher the ciphertext

FULTE MEXEI KBVZK OGZLZ MFM

given that it was enciphered with Vigenère autocipher using the ciphertext as key and F as the first letter of the key.

10. Encipher

THE WHOLE NINE YARDS

using the standard Vigenère cipher (Table 7.8) with key word MATHS.

11. Decipher the ciphertext

VLMNR FMPSM ITASB HSUTV NTMJP

given that it was enciphered with a standard Vigenère cipher with key SHAZAM.

12. Encipher

ARE WE HAVING FUN

using the standard Vigenère cipher with key word KEY.

13. Decipher the message

SSAHXYOO

given that it was enciphered using standard Vigenère cipher with key word ME.

14. Devise a Wheatstone cipher with keyword KELVIN and encipher

GRANTCHESTER.

15. Use Wheatstone's cipher shown in Table 7.12 to decipher

DIGPK HSOGG DFPNE HLHON BYPHU LGRNY KCYYN IBFGP ULBGR HBOFB UORDU FFDEL IDGNE QKCPE UGUFI YFBNI LSBLGW.

7.3 Knapsack and block ciphers

Knapsack ciphers, like character ciphers, are based on modular arithmetic. However, numbers not letters are transmitted with knapsack ciphers. Knapsack ciphers originated from an ancient problem in which a knapsack's weight was given together with the weights of the individual objects before they were placed in the knapsack. The problem was to determine how many of each type of object were in the knapsack. Modern knapsack ciphers use superincreasing sequences and binary representations for letters of the alphabet. Recall that a superincreasing sequence is a sequence $a_1, a_2, \ldots, a_n$, with $a_{k+1} > \sum_{i=1}^{k} a_i$, for $k = 1, 2, \ldots$. For example, 1, 2, 4, 8, 16, 32, 64 and 2, 12, 16, 32, 65, 129, 275 are superincreasing sequences.

Knapsack ciphers can be constructed as follows. Given a superincreasing sequence $a_1, a_2, \ldots, a_{10}$ of length 10, choose an integer n such that

Table 7.13.

A	00000	J	01001	S	10010
B	00001	K	01010	T	10011
C	00010	L	01011	U	10100
D	00011	M	01100	V	10101
E	00100	N	01101	W	10110
F	00101	O	01110	X	10111
G	00110	P	01111	Y	11000
H	00111	Q	10000	Z	11001
I	01000	R	10001		

$n > 2a_{10}$ and an integer w such that $\gcd(w, n) = 1$. Form the superincreasing sequence $wa_1, wa_2, \ldots, wa_{10}$, where the terms are taken modulo n. To encipher the message, group adjacent letters in pairs and use Table 7.13 to partition the message into blocks of 10 binary digits. Use vector multiplication on the decadal binary blocks and the modified superincreasing sequence. Knapsack ciphers can be made even more difficult to decipher by multiplying the decadal binary block by a nonzero scalar before the vector multiplication.

Example 7.5 Given the superincreasing sequence 2, 7, 11, 31, 58, 117, 251, 482, 980, 1943, let us encipher the message

SEND HELP.

Choose $n = 3891 > 3886 = 2 \cdot 1943$ and $w = 1001$, where $\gcd(1001, 3891) = 1$. Multiplying each term by w and reducing modulo 3891, we transform the given superincreasing sequence into the sequence 2002, 3116, 3229, 3794, 3584, 387, 2227, 3889, 448, 3334. Partition the message into blocks of 10 binary digits using Table 7.13.

S E	N D	H E	L P
10010 00100	01101 00011	00111 00100	01011 01111

We now transform the block corresponding to the combination SE under vector multiplication into $1 \cdot 2002 + 0 \cdot 3116 + 0 \cdot 3229 + 1 \cdot 3794 + \cdots + 0 \cdot 3334 = 9685$. The block corresponding to ND under vector multiplication is transformed into $0 \cdot 2002 + 1 \cdot 3116 + 1 \cdot 3229 + 0 \cdot 3794 + \cdots + 1 \cdot 3334 = 13\,711$. Thus, the resulting ciphertext is given by

$$9685 \quad 13\,711 \quad 12\,926 \quad 18\,822.$$

To decipher the message, we use Sanderson's algorithm to determine, 3650, the inverse of 1001 modulo 3891. Since $3650 \cdot 9685 \equiv 515 \pmod{3891}$ and $515 = 2 + 31 + 482$, from Table 7.13, we find that 515, in our original superincreasing sequence, corresponds to 10010 00100. That is, to the pair SE.

A block cipher is a polygraphic cipher that substitutes for each block of plaintext of a specified length a block of ciphertext of the same length. Such ciphers act on blocks of letters, and not on individual letters, and, hence, are not as vulnerable to cryptanalysis based on letter frequency. Block ciphers were devised in 1929 by Lester Hill at Hunter College.

Hill cipher systems are obtained by splitting the plaintext into blocks of n letters, translating the letters into their numerical equivalents, and then forming the ciphertext using the relationship $C = AP \pmod{26}$, where A is an n by n matrix with determinant coprime to 26, C is the 1 by n column matrix with entries $C_1, C_2, \ldots, C_n$, and P is the 1 by n column matrix with entries $P_1, P_2, \ldots, P_n$, where the C_i are the ciphertext blocks corresponding to the plaintext blocks P_i, for $i = 1, 2, \ldots, n$. The ciphertext numbers are then translated back into letters. To decipher a Hill cipher encoded message use A^{-1}, the inverse of the matrix A, taken modulo 26, since $A^{-1}C \equiv A^{-1}(AP) \equiv (A^{-1}A)P \equiv P \pmod{26}$. A Hill cipher is called digraphic if $n = 2$, trigraphic if $n = 3$, and polygraphic if $n > 3$.

Example 7.6 In order to encipher the plaintext

GAUSS WAS VERY BRIGHT

using a Hill cipher with

$$A = \begin{pmatrix} 1 & 2 \\ 4 & 3 \end{pmatrix},$$

we partition the plaintext into blocks of length 2 and use Table 7.3 to translate the blocks into their numerical equivalents:

G A	U S	S W	A S	V E	R Y	B R	I G	H T
6 0	20 18	18 22	0 18	21 4	17 24	1 17	8 6	7 19

We have added XX to the end of the message so that the cipher text is composed of quintuplets. Perform the matrix calculations:

$$\begin{pmatrix} 1 & 2 \\ 4 & 3 \end{pmatrix}\begin{pmatrix} 6 \\ 0 \end{pmatrix} \equiv \begin{pmatrix} 6 \\ 24 \end{pmatrix} \pmod{26}, \quad \begin{pmatrix} 1 & 2 \\ 4 & 3 \end{pmatrix}\begin{pmatrix} 20 \\ 18 \end{pmatrix} \equiv \begin{pmatrix} 4 \\ 4 \end{pmatrix} \pmod{26},$$

$$\begin{pmatrix} 1 & 2 \\ 4 & 3 \end{pmatrix}\begin{pmatrix} 18 \\ 22 \end{pmatrix} \equiv \begin{pmatrix} 10 \\ 8 \end{pmatrix} \pmod{26}, \quad \begin{pmatrix} 1 & 2 \\ 4 & 3 \end{pmatrix}\begin{pmatrix} 0 \\ 18 \end{pmatrix} \equiv \begin{pmatrix} 10 \\ 2 \end{pmatrix} \pmod{26},$$

$$\begin{pmatrix} 1 & 2 \\ 4 & 3 \end{pmatrix}\begin{pmatrix} 21 \\ 4 \end{pmatrix} \equiv \begin{pmatrix} 3 \\ 18 \end{pmatrix} \pmod{26}, \quad \begin{pmatrix} 1 & 2 \\ 4 & 3 \end{pmatrix}\begin{pmatrix} 17 \\ 24 \end{pmatrix} \equiv \begin{pmatrix} 13 \\ 10 \end{pmatrix} \pmod{26},$$

$$\begin{pmatrix} 1 & 2 \\ 4 & 3 \end{pmatrix}\begin{pmatrix} 1 \\ 17 \end{pmatrix} \equiv \begin{pmatrix} 9 \\ 3 \end{pmatrix} \pmod{26}, \quad \begin{pmatrix} 1 & 2 \\ 4 & 3 \end{pmatrix}\begin{pmatrix} 8 \\ 6 \end{pmatrix} \equiv \begin{pmatrix} 20 \\ 24 \end{pmatrix} \pmod{26},$$

$$\begin{pmatrix} 1 & 2 \\ 4 & 3 \end{pmatrix}\begin{pmatrix} 7 \\ 19 \end{pmatrix} \equiv \begin{pmatrix} 19 \\ 7 \end{pmatrix} \pmod{26}, \quad \begin{pmatrix} 1 & 2 \\ 4 & 3 \end{pmatrix}\begin{pmatrix} 23 \\ 23 \end{pmatrix} \equiv \begin{pmatrix} 17 \\ 16 \end{pmatrix} \pmod{26}.$$

6 24	4 4	10 8	10 2	3 18	13 10	9 3	20 24	19 7	17 16
G Y	E E	K I	K C	D S	N K	J D	U Y	T H	R Q

Hence, the resulting ciphertext is

GYEEK IKCDS NKJDU YTHRQ.

To decipher the message, the cryptanalyst must determine the inverse of the enciphering matrix A modulo 26. In general, the inverse of a 2 by 2 matrix

$$M = \begin{pmatrix} a & b \\ c & d \end{pmatrix}$$

is given by

$$M^{-1} = \frac{1}{ad - bc}\begin{pmatrix} d & -b \\ -c & a \end{pmatrix}.$$

Hence, in our example, we find that

$$A^{-1} = \begin{pmatrix} 1 & 2 \\ 4 & 3 \end{pmatrix}^{-1} \equiv \frac{1}{-5}\begin{pmatrix} 3 & -2 \\ -4 & 1 \end{pmatrix} \equiv \frac{1}{21}\begin{pmatrix} 3 & 24 \\ 22 & 1 \end{pmatrix}$$
$$\equiv 5\begin{pmatrix} 3 & 24 \\ 22 & 1 \end{pmatrix} \equiv \begin{pmatrix} 15 & 16 \\ 6 & 5 \end{pmatrix} \pmod{26}.$$

In digraphic ciphers, there are $26^2 = 676$ possible blocks of length 2. However, studies on the relative frequencies of typical English text have led to methods for deciphering digraphic Hill ciphers. The most common pair of juxtaposed letters in the English language is TH followed closely by HE. In addition, 10 words–the, of, and, to, a, in, that, it, is and I-make up a quarter of a typical English text.

Example 7.7 Suppose a Hill digraphic cipher system has been employed and the most common pair of letters in the ciphertext is JX followed by TM; it is likely that JX corresponds to TH and TM corresponds to HE. Therefore, the block $\begin{pmatrix} 19 \\ 7 \end{pmatrix}$ corresponds to the block $\begin{pmatrix} 9 \\ 23 \end{pmatrix}$ and the block $\begin{pmatrix} 7 \\ 4 \end{pmatrix}$ corresponds to the block $\begin{pmatrix} 19 \\ 12 \end{pmatrix}$. Let A denote the enciphering matrix; then

$$A \cdot \begin{pmatrix} 19 & 7 \\ 7 & 4 \end{pmatrix} \equiv \begin{pmatrix} 9 & 19 \\ 23 & 12 \end{pmatrix} \pmod{26}.$$

Since

$$\begin{pmatrix} 19 & 7 \\ 7 & 4 \end{pmatrix}^{-1} \equiv \begin{pmatrix} 4 & 19 \\ 19 & 19 \end{pmatrix} \pmod{26},$$

we have

$$A \equiv \begin{pmatrix} 9 & 19 \\ 23 & 12 \end{pmatrix} \begin{pmatrix} 4 & 19 \\ 19 & 19 \end{pmatrix} \equiv \begin{pmatrix} 7 & 12 \\ 8 & 15 \end{pmatrix} \pmod{26}.$$

Hence,

$$A^{-1} \equiv \begin{pmatrix} 19 & 16 \\ 2 & 21 \end{pmatrix},$$

and we use $P = A^{-1} \cdot C$ to decipher the message.

Exercises 7.3

1. Use the superincreasing sequence and n and w from Example 7.5 to encode the message

   NUTS.

2. Decode the message

   3564 9400 16 703,

   given that it was encoded using the superincreasing sequence and n and w from Example 7.5.
3. Show that

$$\begin{pmatrix} 7 & 12 \\ 8 & 15 \end{pmatrix}^{-1} \equiv \begin{pmatrix} 19 & 16 \\ 2 & 21 \end{pmatrix} \pmod{26}.$$

4. Use the digraphic cipher that sends the plaintext blocks P_1 and P_2 to the cipherblocks C_1 and C_2, such that

$$C_1 \equiv 3P_1 + 5P_2 \pmod{26},$$
$$C_2 \equiv 4P_1 + 7P_2 \pmod{26},$$

   that is,

$$\begin{pmatrix} C_1 \\ C_2 \end{pmatrix} \equiv \begin{pmatrix} 3 & 5 \\ 4 & 7 \end{pmatrix} \begin{pmatrix} P_1 \\ P_2 \end{pmatrix} \pmod{26},$$

   to encipher the message

   BUT WHO WILL GUARD THE GUARDS.

5. Decipher the ciphertext message

Table 7.14.

A	B	C	D	E	F	G	H	I	J	K	L	M	N	O	P	Q	R	S	T	U	V	W	X	Y	Z
00	01	02	03	04	05	06	07	08	09	10	11	12	13	14	15	16	17	18	19	20	21	22	23	24	25

RR QB IQ IT UV QO HW ZI,

which was enciphered using the digraphic cipher

$$C_1 \equiv 7P_1 + 2P_2 \pmod{26},$$
$$C_2 \equiv 8P_1 + 3P_2 \pmod{26}.$$

6. The two most common digraphs in a ciphertext are ZI and UG and these pairs correspond to the two most common pairs in the English text, TH and HE. The plaintext was enciphered using a Hill digraphic cipher. Determine a, b, c, and d if

$$C_1 \equiv aP_1 + bP_2 \pmod{26},$$
$$C_2 \equiv cP_1 + dP_2 \pmod{26}.$$

7. The three most common triples of letters in a ciphertext are AWG, FMD, and RXJ. Suppose these triples correspond to the common triples: THE, AND, and THERE. If the plaintext was enciphered using a Hill trigraphic cipher described by $C \equiv AP \pmod{26}$, then determine the 3 by 3 enciphering matrix A.

7.4 Exponential ciphers

Exponential ciphers are a type of polygraphic cipher developed in 1978 by Martin Hellman at Stanford. So far, they are relatively resistant to cryptanalysis. To encipher a plaintext using a digraphic exponential cipher we first transform pairs of the letters of the plaintext into their numerical equivalents in sets of four digits using Table 7.14. For example,

SEND HELP

would be represented digraphically as

1804 1303 0704 1115.

Choose a prime p such that $2525 < p < 252\,525$ and a positive integer e, called the enciphering key, such that $\gcd(e, p-1) = 1$. Encipher each block P of plaintext into a cipher block C using the exponential congruence $C \equiv P^e \pmod{p}$, where $0 \leqslant C \leqslant p$. If the enciphering key e and the prime p are known, then the plaintext P is easily recovered. Since $\gcd(e, p-1) = 1$ there exists an integer f such that $ef \equiv 1 \pmod{p-1}$, so that for some integer k $ef = 1 + k(p-1)$ and, from Fermat's Little Theorem, $C^f \equiv (P^e)^f \equiv P^{1+k(p-1)} \equiv P(P^{(p-1)k}) \equiv P \pmod{p}$.

In general, with n-ary exponential ciphers, we group the resulting numerical equivalent of the plaintext into blocks of length $2n$, with n chosen so that the largest integer formed by adjoining n decimal equivalents of plaintext letters is less than p.

Example 7.8 We send the message

WAIT UNTIL THE SUN SHINES NELLIE

using $p = 2819$ and $e = 23$. The letters of the plaintext are converted into their numerical equivalents and then grouped into blocks of length 4 to obtain

2200	0819	2013	1908	1119
0704	1820	1318	0708	1318
1813	0411	1108	0423	

where the letter X has been added at the end of the plaintext to fill out the final block of four digits. Encoding the numerical plaintext using the formula $C \equiv P^{23} \pmod{2819}$, we obtain

602	2242	1007	439	2612
280	1303	1981	1511	1981
233	1013	274	540	

Since $\gcd(2818, 23) = 1$, to decipher the ciphertext, we use the Euclidean algorithm to obtain $23 \cdot 2573 - 21 \cdot 2818 = 1$. Hence, 2573 is the inverse of 23 modulo 2818. The deciphering congruence $C^{2573} \equiv P \pmod{2819}$ will return the message to the plaintext. For example, $602^{2573} \equiv 2200 \pmod{2819}$.

Exponential ciphers discourage cryptanalysis since the cryptanalyst needs to determine the prime and exponent involved in enciphering the message, a formidable task even with a high-speed computer. In a public-key encryption system, we are given a number of individuals who wish to communicate with each other. Each person chooses an enciphering key E, which is published in a book of keys and made available to all users of the system, and a deciphering key D, whose inverse is E and which is kept secret. In order to be a secure system, each deciphering key should be essentially impossible to discover or compute even though the enciphering key is public knowledge. Suppose individuals A and B wish to communicate using the system. Since E_A and E_B are known to all users of the system, A can send a message M to B by transmitting $E_B(M)$, that is, by applying E_B to M. Since $D_B(E_B(M)) = (D_B E_B)M = M$ and only B knows D_B, the deciphering key, only B can compute M and read the message. To respond to A with message N, B would transmit $E_A(N)$ to A, who would decipher it using D_A. That is, A would compute $D_A(E_A(N)) = (D_A E_A)N = N$.

If the composition of enciphering and deciphering is commutative, that is $(ED)M = (DE)M = M$, for all messages M, then it is possible to send signed messages, important in such matters as the electronic transfer of large sums of money. For example, if A wished to send a signed message M to B, then A, using B's published enciphering key and A's deciphering key, would compute and send $E_B(D_A(M))$. To decipher the message, B would compute $E_A(D_B(E_B(D_A(M)))) = (E_A D_A)(E_B D_B)M$ and obtain M. Moreover, if the deciphered message were legible B would know that the message could only come from someone who knew A's deciphering key, E_A. This does not affect the security of the message since only A knows D_A and only B knows D_B. The practicality of such a system eventually depends on the ability of all parties to be able to calculate efficiently with the enciphering and deciphering keys.

In 1976, a very useful and practical public-key encryption system based on exponential ciphers was devised independently by W. Diffie and M.E. Hellman at Stanford and R.C. Merkle at Berkeley, and implemented at MIT in 1978 by Ronald L. Rivest, Adi Shamir, and Leonard M. Adleman. The RSA system, as it is known, works as follows. Each individual in the system chooses two very large primes p and q, say of approximately 100 digits each and calculates $r = pq$. Each person determines a positive integer s, such that $\gcd(s, \phi(r)) = 1$, and integers t and k such that $st = 1 + k\phi(r)$. Hence, $st \equiv 1 \pmod{\phi(r)}$. The pair (r, s) forms the enciphering key and is published in the public register of such keys, but t, the deciphering key, is kept secret by the individual.

In the RSA system, a message, M, is altered into its numerical equivalent using Table 7.14 and grouped into blocks of length $2n$, as with exponential ciphers. The successive numerical blocks obtained from the plaintext are enciphered using s, the receiver's encryption key, and the equation $C = E(P) \equiv P^s \pmod{r}$, where $0 \leqslant C < r$, and the numerical ciphertext is sent. From the Euler–Fermat Theorem, $P^{\phi(r)} \equiv 1 \pmod{r}$. Hence, $D(C) = C^t \equiv (P^s)^t \equiv P^{st} \equiv P^{1+k\cdot\phi(r)} \equiv P \cdot P^{k\cdot\phi(r)} \equiv P \pmod{r}$ with

$0 \leqslant P < r$. Therefore, the receiver applies the inverse operator and deciphers the message.

We may choose s to be any prime greater than pq, such that $2^s > r = pq$, and it would be virtually impossible to recover the plaintext block P by simply calculating the sth root of C. Knowledge of the enciphering key (r, s) does not lead to the deciphering key (t, r). To determine t, the inverse of s modulo $\phi(r)$, one must first determine $\phi(r) = \phi(pq) = (p-1)\cdot(q-1)$, which requires the decipherer to know the factorization of r, which is virtually impossible without knowing p and q. For example, when p and q contain 100 decimal digits, $r = pq$ has around 200 decimal digits. Using the fastest factorization techniques known would require approximately 3.8×10^9 years of computer time to factor $\phi(r)$. Nevertheless, if r and $\phi(r)$ are known then p and q can be determined using the identity $(p-q)^2 - (p+q)^2 = -4pq$, since $p + q = pq - \phi(r) + 1 = r - \phi(r) + 1$ and $p - q = [(p+q)^2 - 4pq]^{1/2} = [(p+q)^2 - 4r]^{1/2}$:

$$p = \frac{(p+q) + (p-q)}{2}$$

and

$$q = \frac{(p+q) - (p-q)}{2}.$$

Example 7.9 Suppose we wish to send the message

VEE IS FOR VICTORY

using the RSA system, where $p = 61$, $q = 47$, $r = pq = 2867$, and $\phi(r) = 60 \cdot 46 = 2760$. If we let $s = 17$, from the Euclidean algorithm, we find that t, the inverse of 17 modulo 2760, equals 2273. We publish the key (2867, 17) and keep 2273 hidden. We change the plaintext into its numerical equivalent, and group the numbers into blocks of size 4 to obtain

2104 0408 1805 1417 2108 0219 1417 2423,

where we have added a 23, an X, at the end of the message to fill out the final block of digraphic plaintext. We use the congruence $C \equiv P^{17} \pmod{2867}$ to encipher the numerical plaintext. For example, $2104^{17} \equiv 2458 \pmod{2867}$. We obtain

2458 0300 0778 2732 1827 2608 2732 0129.

To decipher the ciphertext the receiver would use the deciphering congruence $C^{2273} \equiv P \pmod{2867}$. In particular, $2458^{2273} \equiv 2104 \pmod{2867}$.

Diffie and Hellman devised a technique whereby two participants in a public-key cipher system are able to share the same key. In particular, suppose a prime p and a positive integer $s < p$ with $\gcd(s, p-1) = 1$ are known to both participants. Let the participants, say A and B, choose positive integers $a < p$ and $b < p$, respectively. A and B compute $u = s^a \pmod{p}$ and $v = s^b \pmod{p}$, respectively. A sends u to B and B sends v to A. A and B, respectively, compute $v^a \pmod{p}$ and $u^b \pmod{p}$. Since, modulo p, $k = u^b \equiv (s^a)^b \equiv s^{ab} \equiv (s^b)^a \equiv v^a$, both A and B use k as their common key. For example, if $p = 9199$, $s = 13$, $a = 10$ and $b = 23$, then their common key would be $k = 13^{230} \equiv 7999 \pmod{9199}$.

Exercises 7.4

1. Using an exponential cipher with $p = 2591$, $e = 5$, and $n = 2$, encipher

   HAVE A GOOD DAY.

2. Using an exponential cipher with $p = 3307$, $e = 17$, and $n = 2$, encipher

   HAPPY DAYS ARE HERE AGAIN.

3. Using an exponential cipher with $p = 7193$, $e = 97$, and $n = 2$, encipher

   SEND HELP.

4. Decipher the ciphertext message

   2771 1794 3187 1013 3228 1259,

   given it was enciphered digraphically using an exponential cipher with $p = 3373$ and $e = 95$.
5. Decipher the ciphertext message

   1843 0288 2142 2444,

   given it was enciphered digraphically using an exponential cipher with $p = 2591$ and $e = 157$.
6. Decipher the ciphertext message

   1391 1958 1391 2558 0709 1425 2468
   1311 1123 0079 2468 1774 0993 1915
   1123 0846

   given it was enciphered digraphically using an exponential cipher with $p = 2671$ and $e = 49$.
7. Determine primes p and q used in an RSA cipher given that $r = 4\,386\,607$ and $\phi(r) = 4\,382\,136$. If $s = 5$ determine t.

8. Determine primes p and q used in an RSA cipher given that $r = 4\,019\,651$ and $\phi(r) = 4\,015\,632$. If $s = 17$ determine t.
9. If $p = 8461$, $s = 61$, A chooses $a = 17$, and B chooses $b = 31$, determine a public key k that would be common to A and B.

7.5 Supplementary exercises

1. Use Polybius's method to decipher the following message: 44 23 15 42 11 13 15 24 43 33 34 44 11 52 11 54 43 52 34 33 12 54 44 23 15 43 52 24 21 44 34 42 44 23 15 12 42 11 51 15
2. Decipher the following Caesar cipher: WKLVP HVVDJ HLVWR SVHFU HWZZZ.
3. Decipher the following Caesar cipher: LFDPH LVDZL FRQTX HUHG.
4. Decipher the following Caesar Augustus cipher: BWF DBFTBS.
5. Decipher LWWRO OVIRU WKHH given that it was enciphered with a shift transformation with key $k = 3$.
6. Decipher the message WDVKN ACQNX AHVJT NBXWN CQRWT given that it was enciphered with a shift transformation.
7. Decipher the message JCFYG DWVQY TNFGP QWIJC PFVKO GVJKU EQAPG UUNCF AYGTG PQETK OGYGY QWNFU KVFQY PCPFV JKPMY JKESY CAVQY CNMCP FRCUU QWTNQ PINQX GUFCA given that it was enciphered with a shift transformation.
8. Decipher the message NYVEZ EKYVT FLIJV FWYLO REVMV EKJ given the it was enciphered with a shift transformation.
9. Decipher the message IEXXK FZKXC UUKZC STKJW that was enciphered using the affine transformation $C \equiv 11P + 18 \pmod{26}$.
10. Decipher the message MBNJH MEPNK KPHMS CJDPG, that was enciphered using the affine transformation $C = 5P + 21 \pmod{26}$.
11. Decipher ANLLL OISEF OARIW RAIR.
12. Decipher ATITE RIETO RTHSH IFNSH UX.
13. Decipher WISIE ALUNL ITNEL THSSY IEHNX.
14. Decipher YRAEAL EGTINA SIHSTU VNEAAS IIRSCX.
15. If a frequency count of a message enciphered with a shift transformation gives R as the most frequent cipher letter and J as the second most frequent letter, determine the most likely values for a and b such that $C \equiv aP + b \pmod{26}$.
16. If a frequency count of a message enciphered with a shift transformation gives X as the most frequent cipher letter and M as the second

most frequent letter, determine the most likely values for a and b such that $C \equiv aP + b \pmod{26}$.

17. Use the Vigenère method to decipher the message OCCU AVOH NQXP ALTD PMWB given that it was enciphered using the key word WILD.
18. Use the Vigenère method to decipher the message OSNXF UNLZE HPYDH HYZVA AGQBZ FDACV DUKWH FMDVG PSQNY KGMCU SLSLH XFK given that it was enciphered using the key phrase BOSTONMASSACHUSETTS.
19. Use the Vigenère method to decipher the message DELCY MZFBR WTSBW IKUJE CUEGZ VRGQS CLWGJ YEBEO YELWH YNVDC LWGJB VRLTR KYWB given that it was enciphered using the key word CALIFORNIA.
20. Decipher the ciphertext CBATF XMVKH VHVZP ZJZHD NQOZG LMAHN GHVHV ZRRZZ VYHVZ PZGLT TTLW given that it was enciphered with a Vigenère autocipher using the plaintext as key and J as the first letter of the key.
21. Decipher the ciphertext GIIXQ QYLXV XXMFF NAOIZ DHHYD XIBTB JBESFJ given that it was enciphered with a Vigenère autocipher using the ciphertext as key and S as the first letter of the key.
22. Use the Wheatstone cipher with key word KELVIN to decipher QVANGLKR.
23. Use the Wheatstone cipher with key word CAMBRIDGE to decipher SHBW EPMD CDLTMB.
24. Decode VUJIR WHMYV given that it was enciphered using a Hill cipher system and encoding matrix

$$A = \begin{pmatrix} 4 & 11 \\ 1 & 22 \end{pmatrix}.$$

25. Decode RJHMQO given that it was enciphered using a Hill cipher system and encoding matrix

$$A = \begin{pmatrix} 5 & 2 \\ 1 & 7 \end{pmatrix}.$$

26. Decode the message 13587 4724 2614 given that it was encoded using the superincreasing sequence and n and w from Example 7.5.
27. Given the superincreasing sequence 3, 5, 10, 20, 40, 90, 171, 361, 701, 1500, $n = 3001$ and $w = 1111$, use the knapsack cipher to encode LET'S GO.

28. Encipher 07 10 24 13 02 06 given that it was enciphered with an exponential cipher with $p = 29$ and $e = 19$:
29. Decipher 2767 2320 3151 2690 1399 2174 given that it was enciphered biographically using an exponential cipher with $p = 3373$ and $e = 95$.
30. Decipher 795 647 480 2710, given that it was enciphered bigraphically using an exponential cipher with $p = 2819$ and $e = 1691$.

8

Representations

When you have eliminated the impossible, whatever remains, *however improbable*, must be the truth.
Sherlock Holmes, in The Sign of Four, by Sir Arthur Conan Doyle

8.1 Sums of squares

In this chapter, we make use of several number theoretic tools established earlier to determine which integers may be represented as sums of squares, cubes, triangular numbers, and so forth. The branch of number theory dealing with such integral representation has led to the advances in the theory of sphere packing, the theory of unique factorization domains, and ideal theory.

Being able to express a positive integer as the sum of two squares of nonnegative integers is a problem that had intrigued ancient as well as modern mathematicians. In an earlier section dealing with Pythagorean triples, we were able to express certain square numbers as the sum of two integral squares. Diophantus, in Book II of *Arithmetica* gave

$$x = \frac{2am}{m^2+1} \text{ and } y = \frac{a(m^2-1)}{m^2+1},$$

where a is a nonnegative integer and m a positive integer constant, as rational solutions to the equation $x^2 + y^2 = a^2$. In 1225, Fibonacci devoted a good part of *Liber quadratorum* to such problems. The specific problem of determining exactly which positive integers can be represented as the sum of two integral squares was posed first by the Dutch mathematician, Albert Girand, in 1627 and independently by Fermat a few years later. Methods for solving Girand's problem can be straightforward but tedious. For example, given an integer n, we can determine whether or not it can be represented as the sum of two integral squares be calculating $n - 1^2$, $n - 2^2$, $n - 3^2$, ..., $n - [\![\sqrt{n}/2]\!]^2$ until we either obtain a square or exhaust all possibilities. The process may be started from the other direction by subtracting the square of the greatest integer not greater than the square root of n. For example, if the number is 7522, the greatest integer not greater than $\sqrt{7522}$ is 86. Hence,

$$7522 - 86^2 = 126,$$
$$7522 - 85^2 = 297,$$
$$7522 - 84^2 = 466,$$
$$7522 - 83^2 = 633,$$
$$7522 - 82^2 = 798,$$
$$7522 - 81^2 = 961 = 31^2.$$

Therefore, $7522 = 81^2 + 31^2$.

Example 8.1 According to Theorem 2.13, in order to determine if a number z is the z-component of a primitive Pythagorean triple (x, y, z), the hypotenuse of a Pythagorean triangle, we need only express z as the sum of two coprime squares of opposite parity. That is, $z = s^2 + t^2$, $y = s^2 - t^2$, and $x = 2st$, $s > t$, $\gcd(s, t) = 1$, where one of s and t is even and the other is odd. For example, if $z = 10\,394 = 95^2 + 37^2$, then $y = 95^2 - 37^2$ and $x = 2 \cdot 37 \cdot 95$. Thus (7030, 7656, 10 394) is a primitive Pythagorean triple and, accordingly, 10 394 is the hypotenuse of a Pythagorean triangle.

For each positive integer n, let the function $h(n)$ equal 1 if n can be represented as the sum of two integral squares and 0 otherwise. The values of $h(n)$, for $1 \leqslant n \leqslant 100$, are given in Table 8.1. It appears, from Table 8.1, that there are an infinite number of values for which $h(n) = 0$. This indeed is the case and is implied by either of the next two results.

Theorem 8.1 *If* $n \equiv 3 \pmod 4$, *then* $h(n) = 0$.

Proof If $h(n) = 1$, then there exist integers x and y such that $n = x^2 + y^2$. The integers x and y are congruent to either 0 or 1 modulo 2. Hence, $x^2 + y^2$ can only be congruent to 0, 1, or 2 modulo 4 and the result follows by contraposition. ■

Theorem 8.2 *If* $h(n) = 0$, *then* $h(4n) = 0$.

Proof The result is established by contraposition. If $h(4n) = 1$, then $4n = x^2 + y^2$, for some values of x and y. In this case, x and y must both be even, say $x = 2r$ and $y = 2s$. We obtain $4n = 4r^2 + 4s^2$ or $n = x^2 + y^2$, hence, $h(n) = 1$. ■

Table 8.1.

n	$h(n)$	n	$h(n)$	n	$h(n)$	n	$h(n)$
1	1	26	1	51	0	76	0
2	1	27	0	52	1	77	0
3	0	28	0	53	1	78	0
4	1	29	1	54	0	79	0
5	1	30	0	55	0	80	1
6	0	31	0	56	0	81	1
7	0	32	1	57	0	82	1
8	1	33	0	58	1	83	0
9	1	34	1	59	0	84	0
10	1	35	0	60	0	85	1
11	0	36	1	61	1	86	0
12	0	37	1	62	0	87	0
13	1	38	0	63	0	88	0
14	0	39	0	64	1	89	1
15	0	40	1	65	1	90	1
16	1	41	1	66	0	91	0
17	1	42	0	67	0	92	0
18	1	43	0	68	1	93	0
19	0	44	0	69	0	94	0
20	1	45	1	70	0	95	0
21	0	46	0	71	0	96	0
22	0	47	0	72	1	97	1
23	0	48	0	73	1	98	1
24	0	49	1	74	1	99	0
25	1	50	1	75	0	100	1

Theorem 8.3 *If an odd prime p can be expressed as the sum of two integral squares then* $p \equiv 1 \pmod 4$.

Proof Suppose p is an odd prime and $p = x^2 + y^2$. Since p is odd, we have a contradiction if either x and y are even or x and y are odd. Suppose that one is even, the other odd, say $x = 2r$ and $y = 2s + 1$. Hence, $p = 4r^2 + 4s^2 + 4s + 1$. Therefore, $p \equiv 1 \pmod 4$. ■

In 1202, Fibonacci included the identity $(a^2 + b^2)(c^2 + d^2) = (ad + bc)^2 + (ac - bd)^2 = (ac + bd)^2 + (ad - bc)^2$ in *Liber abaci*. The identity had been used implicity by Diophantus in *Arithmetica*. In 1749, Euler used the identity to establish the next result.

Theorem 8.4 *If* $h(m) = 1$ and $h(n) = 1$, *then* $h(mn) = 1$.

Proof Suppose $h(m) = 1$ and $h(n) = 1$, then there exist integers a, b, c, d

such that $m = a^2 + b^2$ and $n = c^2 + d^2$. Hence, from Fibonacci's identity, $mn = (a^2 + b^2)(c^2 + d^2) = (ad + bc)^2 + (ac - bd)^2$. Thus, $h(mn) = 1$ and the result is established. ∎

In Theorem 8.1, we showed that a number of the form $4n + 3$ cannot be written as the sum of two integral squares. Using Fermat's method of descent, we now establish a much stronger result.

Theorem 8.5 *An integer n can be expressed as a sum of two squares if and only if every prime divisor of n of the form $4k + 3$ has even exponent in the canonical representation of n.*

Proof Suppose that $n = x^2 + y^2$ and p is a prime divisor of n. Hence, $x^2 \equiv -y^2 \pmod{p}$. That is, $-y^2$ is a quadratic residue modulo p. It follows from the theory of quadratic residues that

$$1 = \left(\frac{-y^2}{p}\right) = \left(\frac{-1}{p}\right)\left(\frac{y}{p}\right)^2 = \left(\frac{-1}{p}\right) = (-1)^{(p-1)/2}.$$

If $p \equiv 3 \pmod{4}$, $(-y^2/2) = -1$, a contradiction, unless $x \equiv y \equiv 0 \pmod{p}$. In that case, $x = pr$, $y = ps$, and $n = p^2m$ with $m = r^2 + s^2$. Continuing the process, we find that $n = p^{2t}w$, for some positive integer t. Therefore, if $p \equiv 3 \pmod{4}$ is prime, it appears in the canonical representation of n to an even power. Conversely, let p be a prime of the form $4k + 1$. Hence,

$$\left(\frac{-1}{p}\right) = (-1)^{(p-1)/2} = (-1)^{2k} = 1.$$

Thus, the equation $x^2 \equiv -1 \pmod{p}$ has a solution, say a, with $1 \leqslant a < p/2$. Hence, there exists an integer m such that $mp = a^2 + 1$. Since $0 < mp = a^2 + 1 < p^2/4 + 1 < p^2/4 + 3p^2/4 = p^2$, m is a positive integer such that $mp = a^2 + 1$, with $p > m$. Let t be the least positive integer such that tp is the sum of two integral squares. That is, there exist integers x and y such that $tp = x^2 + y^2$, with $0 < t \leqslant m < p$, and t is the least positive integer for which this is the case. If $t > 1$, from the Corollary to Theorem 2.2, it follows that $x = qt + r$ and $y = ut + v$, with $-|t|/2 < r \leqslant |t|/2$ and $-|t|/2 < v \leqslant |t|/2$. Thus, $tp = x^2 + y^2 = (q^2t^2 + 2qrt + r^2) + (u^2t^2 + 2tuv + v^2)$. If we let $w = p - q^2t - 2qr - u^2t - 2uv$, $wt = r^2 + v^2 \leqslant (t/2)^2 + (t/2)^2 < t^2$. Hence, wt is a multiple of t and $0 \leqslant w < t$. If $w = 0$, then $r = v = 0$, implying that $x = qt$ and $y = vt$. Hence, $tp + x^2 + y^2 = t^2(q^2 + v^2)$. Thus, t divides p, a contradiction since $1 < t < p$ and p is prime. Hence, $w \neq 0$ and wp is a multiple of p

with $0 < w < t$. Since $p = t(q^2 + u^2) + 2(qr + uv) + w$, it follows that $wp = wt(q^2 + u^2) + 2w(qr + uv) + w^2 = (w + qr + uv)^2 + (qv - ru)^2$. However, this contradicts the assumption that tp was the least positive multiple of p expressible as the sum of two integral squares. Therefore, $t = 1$ and $p = a^2 + 1$. That is, p can be expressed as the sum of two integral squares. Since $(4n + 3)^{2k} = ((4n + 3)^k)^2 + 0^2$, the result follows from Theorem 8.4. ■

Theorem 8.4 and Theorem 8.5 enable us to completely determine which positive integers can be expressed as a sum of two integral squares. For example, the only primes of the form $4k + 3$ in the canonical representation of 8820 are 3 and 7 and each appears to an even power. Hence, according to Theorem 8.5, 8820 can be represented as the sum of two squares. One useful technique to accomplish this is to factor 8820 into two components, represent each component as the sum of two squares, and use Fibonacci's identity. We have $8820 = 2^2 \cdot 3^2 \cdot 5 \cdot 7^2 = (2^2 \cdot 7^2)(3^2 \cdot 5) = 196 \cdot 45 = (14^2 + 0^2)(6^2 + 3^2) = 84^2 + 42^2$.

In 1747, in a letter to Goldbach, Euler claimed that every prime divisor of the sum of two coprime squares is itself the sum of two squares. The result is implied by the next theorem.

Theorem 8.6 *If p is an odd prime that divides $a^2 + b^2$, with* $\gcd(a, b) = 1$, *then* $p \equiv 1 \pmod 4$.

Proof Suppose that p divides $(a^2 + b^2)$ where $\gcd(a, b) = 1$. If $p|a$, then $p|a^2$ implying that $p|b^2$ and, hence, $p|b$, a contradiction. Thus, p divides neither a nor b. Since p divides $a^2 + b^2$, $-a^2 \equiv b^2 \pmod p$. Thus, $(-a^2)^{(p-1)/2} \equiv (b^2)^{(p-1)/2} \pmod p$ or $(-1)^{(p-1)/2} a^{p-1} \equiv b^{p-1} \pmod p$. Since $\gcd(a, p) = \gcd(b, p) = 1$, it follows from Fermat's Little Theorem that $a^{p-1} \equiv b^{p-1} \equiv 1 \pmod p$. Hence, $(-1)^{(p-1)/2} \equiv 1 \pmod p$. Therefore, $p \equiv 1 \pmod 4$. ■

There are an infinite number of integers that may be expressed as a sum of two integral squares in more than one way. For example, $50 = 7^2 + 1^2 = 5^2 + 5^2$. In 1621, Bachet noted that $5525 = 55^2 + 50^2 = 62^2 + 41^2 = 70^2 + 25^2 = 71^2 + 22^2 = 73^2 + 14^2 = 74^2 + 7^2$. According to Theorem 8.3, since $1073 = 32^2 + 7^2 = 28^2 + 17^2 = 7^2 + 32^2 = 17^2 + 28^2$, $5\,928\,325 = 5525 \cdot 1073$ can be expressed as the sum of two squares in at least 24 ways, albeit they all might not be distinct.

Disregarding order and signs, that is, not counting $(-2)^2 + 3^2$,

$(-3)^2 + (-2)^2$, or $3^2 + 2^2$ as being distinct from $2^2 + 3^2$, 13 can be represented uniquely as the sum of two squares. Let p be a prime of the form $4k + 1$ having two distinct representations as a sum of integral squares, $p = a^2 + b^2 = c^2 + d^2$. From the proof of Theorem 8.5, -1 is a quadratic residue of p. Hence, there is a solution, say w, to the equation $x^2 = -1 \pmod{p}$. From our assumption, $a^2 \equiv -b^2 \equiv w^2 b^2 \pmod{p}$ and $c^2 \equiv -d^2 \equiv w^2 d^2 \pmod{p}$, hence, $a \equiv \pm wb$ and $c \equiv \pm wd \pmod{p}$. Thus, $ac + bd \equiv w^2 bd + bd \equiv 0$ and $ad - bc \equiv \pm w(bd - bd) \equiv 0 \pmod{p}$. Hence, there exist integers m and n such that $ac + bd = mp$ and $ad - bc = np$. From Theorem 8.4, $p^2 = (a^2 + b^2)(c^2 + d^2) = (ac + bd)^2 + (ad - bc)^2 = (mp)^2 + (np)^2$. Hence, $1 = m^2 + n^2$, but this is the case only if m or n equals 0, that is, only if $ac + bd = 0$ or $ad - bc = 0$. Since $\gcd(a, b) = \gcd(c, d) = 1$, $ac + bd = 0$ or $ad - bc = 0$ if and only if $a = \pm c$ and $b = \pm d$ or $a = \pm d$ and $b = \mp c$. In either case, the representation is unique and we have established Theorem 8.7, a solution to Girand's problem. The first published proof of the result, due to Euler, appeared in 1754.

Theorem 8.7 (Girand–Euler Theorem) *Disregarding order and signs, any prime of the form* $4k + 1$ *can be represented uniquely as the sum of two integral squares.*

Let us generalize the square representation function $h(n)$ to the function $f(n)$ which denotes the number of different representations of n as the sum of two integral squares, taking signs and order into account. For example, $f(2) = 4$, since $2 = 1^2 + 1^2 = 1^2 + (-1)^2 = (-1)^2 + 1^2 = (-1)^2 + (-1)^2$. Table 8.2 illustrates values of $f(n)$ for $1 \leqslant n \leqslant 100$. From a casual glance at Table 8.2 it appears that $f(n)$ is always a multiple of 4. This indeed is the case and follows from the fact that solutions of the form $(a, 0)$ and (a, a) each contribute 4 to the multiplicity of $f(n)$, and solutions of the form (b, c), where b and c are distinct, contribute 8 to the value of $f(n)$. In 1829, at age 25, Jacobi established the following result which is offered without proof; for a proof see [Niven, Zuckerman, and Montgomery].

Theorem 8.8 (Jacobi) *If* $\tau(m, n)$ *denotes the number of positive divisors of* n *which are congruent to* m *modulo* 4, *then* $f(n) = 4[\tau(1, n) - \tau(3, n)]$.

For example, $234 = 2 \cdot 3^2 \cdot 13$, $\tau(1, 234) = 4$, and $\tau(3, 234) = 2$. Hence, $f(n) = 4[4 - 2] = 8$. Taking order and signs into consideration, the eight

Table 8.2.

n	$f(n)$	n	$f(n)$	n	$f(n)$	n	$f(n)$
1	4	26	8	51	0	76	0
2	4	27	0	52	8	77	0
3	0	28	0	53	8	78	0
4	4	29	8	54	0	79	0
5	8	30	0	55	0	80	8
6	0	31	0	56	0	81	4
7	0	32	4	57	0	82	8
8	4	33	0	58	8	83	0
9	4	34	8	59	0	84	0
10	8	35	0	60	0	85	16
11	0	36	4	61	8	86	0
12	0	37	8	62	0	87	0
13	8	38	0	63	0	88	0
14	0	39	0	64	4	89	8
15	0	40	8	65	16	90	8
16	4	41	8	66	0	91	0
17	8	42	0	67	0	92	0
18	4	43	0	68	8	93	0
19	0	44	0	69	0	94	0
20	8	45	8	70	0	95	0
21	0	46	0	71	0	96	0
22	0	47	0	72	4	97	8
23	0	48	0	73	8	98	8
24	0	49	4	74	8	99	0
25	12	50	12	75	0	100	12

representations of 234 are $15^2 + 3^2$, $(-15)^2 + 3^2$, $15^2 + (-3)^2$, $(-15)^2 + (-3)^2$, $3^2 + 15^2$, $(-3)^2 + 15^2$, $3^2 + (-15)^2$, and $(-3)^2 + (-15)^2$.

There are a number of geometric interpretations involving the representation function. For example, the sum $\sum_{i=1}^{n} f(i)$ represents the number of lattice points in the Cartesian plane satisfying the inequality $x^2 + y^2 \leqslant n$. It also represents the area, in square units, of the region K formed by all unit squares whose centers (x, y) lie inside or on the circle $x^2 + y^2 = n$. If we denote the average value of $f(n)$ by $F(n)$, then

$$F(n) = \frac{1}{n+1} \sum_{i=1}^{n} f(i) = \frac{1}{n+1} \cdot (\text{area of region } K).$$

Since the diagonal of a unit square equals $\sqrt{2}$, the region K is completely contained in the circular disk centered at the origin having radius $\sqrt{n} + \sqrt{2}/2$ and completely contains the circular disk centered at the origin having radius $\sqrt{n} - \sqrt{2}/2$, as shown in Figure 8.1. Hence,

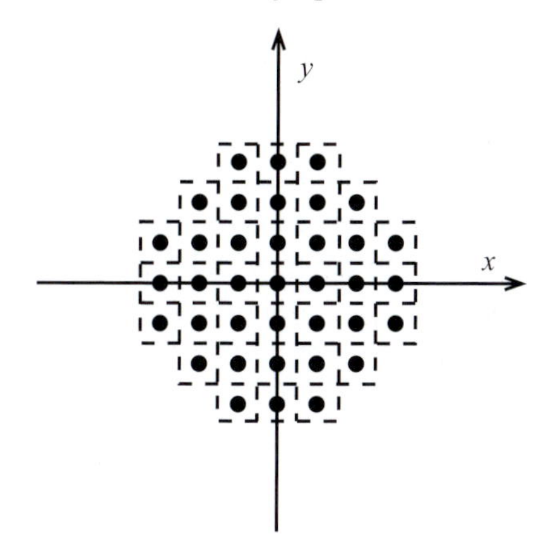

Figure 8.1

$$\frac{\pi}{n+2}\left(\sqrt{n}-\frac{\sqrt{2}}{2}\right)^2 < F(n) < \frac{\pi}{n+1}\left(\sqrt{n}+\frac{\sqrt{2}}{2}\right)^2.$$

Letting n approach infinity, we obtain one of the most elegant results (Theorem 8.9) concerning the sums of two squares. It was originally established by Gauss and was discovered among his unpublished manuscripts after his death in 1855.

Theorem 8.9 *If*

$$F(n) = \frac{1}{n+1}\sum_{i=1}^{n} f(i)$$

then $\lim_{n\to\infty} F(n) = \pi$.

Diophantus considered the problem of representing integers as sums of more than two squares. In the process, he realized that it is not possible to express all integers as the sum of three integral squares. In 1636, in a letter to Mersenne, Fermat conjectured that no number of the form $8k + 7$ can be expressed as a sum of three integral squares. Two years later, Descartes verified Fermat's conjecture. The result follows from the fact that the square of any integer is congruent to 0, 1, or 4 modulo 8.

The conjecture was generalized in the eighteenth century to state that any positive integer can be expressed as a sum of three nonzero integral squares if and only if it is not of the form $4^n(8k + 7)$, where n and k are nonnegative integers. A proof of the conjecture was offered by Legendre in 1798 assuming that if $\gcd(a, b) = 1$, then infinitely many terms of the sequence $a, a + b, a + 2b, \ldots$ were prime. In 1837, Dirichlet completed

Legendre's proof. Gauss offered a different proof in 1801. The sufficiency is difficult to establish and is beyond the scope of this text, but the necessity follows from the fact that if $4^m(8k+7)$, where $n = m - 1$ and k are nonnegative integers, is expressible as the sum of three integral squares then so is $4^{m-1}(8k+7)$ and, hence, so is $8k+7$, a contradition . Hence, no integer of the form $4^n(8k+7)$, where n and k are positive integers, can be represented as the sum of three nonzero integral squares. In 1785, Legendre was able to show that if a, b, c are squarefree, not all positive or all negative, $abc \neq 0$, and $\gcd(a, b) = \gcd(a, c) = \gcd(b, c) = 1$, then $ax^2 + by^2 + cz^2 = 0$, has a nontrivial solution, that is, with $(a, b, c) \neq (0, 0, 0)$, if and only if, using the Jacobi symbol,

$$\left(\frac{-ab}{|a|}\right) = \left(\frac{-bc}{|b|}\right) = \left(\frac{-ca}{|c|}\right) = 1,$$

where, a, b, c are not equal to 1.

Using the previous results, we can show that the equation $x^2 + y^2 + z^2 + x + y + z = 1$ has no integral solution (x, y, z). If (r, s, t) were a solution, then we could multiply both sides of the equation by 4 and complete the square, to obtain $(2r+1)^2 + (2s+1)^2 + (2t+1)^2 = 8 \cdot 0 + 7$, a contradiction.

Since each number of the form $8n+3$ can be written as the sum of three integral squares and each summand must be the square of an odd number, we have $8n+3 = (2r+1)^2 + (2s+1)^2 + (2t+1)^2$. Expanding and collecting terms, we obtain $n = r(r+1)/2 + s(s+1)/2 + t(t+1)/2$, establishing Gauss's result that every positive integer is the sum of three or fewer triangular numbers. For example, $59 = 8 \cdot 7 + 3 = 7^2 + 3^2 + 1^2 = (2 \cdot 3 + 1)^2 + (2 \cdot 1 + 1)^2 + (2 \cdot 0 + 1)^2$. Hence, $7 = 3(3+1)/2 + 1(1+1)/2 + 0(0+1)/2 = 6 + 1 + 0$.

A number of interesting identities occur when numbers are represented as sums of squares. For example, if $n > 1$ and $m = n(2n+1)$, then $m^2 + (m+1)^2 + \cdots + (m+n)^2 = (m+n+1)^2 + (m+n+2)^2 + \cdots + (m+2n)^2$. If $n = 1$, we obtain $3^2 + 4^2 = 5^2$. If $n = 3$, we have $21^2 + 22^2 + 23^2 + 24^2 = 25^2 + 26^2 + 27^2$.

There are many unanswered questions regarding sums of squares, in particular, whether there are infinitely many primes that can be represented as the sum of squares of consecutive positive integers. For example, $5 = 1^2 + 2^2$, $13 = 2^2 + 3^2$, $61 = 5^2 + 6^2$, and so forth. It is an open question whether there are an infinite number of primes p such that $p = n^2 + (n+1)^2 + (n+2)^2$, where n is a positive integer. For example, $29 = 2^2 + 3^2 + 4^2$ and $149 = 6^2 + 7^2 + 8^2$.

Bachet wrote, in 1621, that since there is no explicit reference in *Arithmetica*, Diophantus must have assumed that every positive integer can be represented as the sum of at most four nonzero integral squares. Bachet added that he would welcome a proof of the result. Fermat using the method of descent sketched a proof of Bachet's conjecture. Euler worked on the problem for almost 25 years and in the process established a number of crucial results. In particular, he discovered the identity

$$(a^2 + b^2 + c^2 + d^2)(e^2 + f^2 + g^2 + h^2) = (ae + bf + cg + dg)^2$$
$$+ (af - be + ch - dg)^2 + (ag - bh - ce + df)^2 + (ah + bg - cf - de)^2$$

and the fact that there are integral solutions to $x^2 + y^2 + 1 \equiv 0 \pmod{p}$, where p is prime. For example, $3 = 1^2 + 1^2 + 1^2 + 0^2$ and $17 = 4^2 + 1^2 + 0^2 + 0^2$. Hence, $459 = 3^2 \cdot 3 \cdot 17 = 3^2(1^2 + 1^2 + 1^2 + 0^2)(4^2 + 1^2 + 0^2 + 0^2) = 3^2[(4 + 1 + 0 + 0)^2 + (1 - 4 + 0 + 0)^2 + (1 - 0 - 4 - 0)^2 + (0 + 0 - 1 - 0)^2] = 3^2[5^2 + 3^2 + 4^2 + 1^2] = 15^2 + 9^2 + 12^2 + 3^2$. Building on Euler's work, in 1770 Lagrange gave the first proof of the four-square theorem. We state Lagrange's result without proof; for a proof see [Strayer].

Theorem 8.10 (Lagrange) *Every positive integer can be represented as the sum of four or fewer integral squares.*

In 1829, Jacobi proved that the number of representations of an integer of the form $2^\alpha m$, taking order and signs into consideration, where m is odd, is $8 \cdot \sigma(m)$ if $\alpha = 0$, and $24 \cdot \sigma(m)$ if $\alpha > 1$. For example, 13 has $8 \cdot 14 = 112$ representations. 64 derive from the representation $13 = 3^2 + 2^2 + 0^2 + 0^2$ and 48 from the representation $13 = 2^2 + 2^2 + 2^2 + 1^2$. The number $36 = 2^2 \cdot 3^2$ has $24 \cdot 13 = 312$ representations: 192 derive from $5^2 + 3^2 + 1^2 + 1^2$, 96 from $4^2 + 4^2 + 2^2 + 0^2$, 16 from $3^2 + 3^2 + 3^2 + 3^2$, and 8 from $6^2 + 0^2 + 0^2 + 0^2$.

In 1884, at age 18, Hermann Minkowski proved that all numbers of the form $8n + 5$ are sums of five odd squares. Einstein's theory of general relativity, where gravity is treated as a warping of space and not as a force, is based on results in tensor calculus developed by Minkowski.

The above results all lead naturally to Waring's problem. Edward Waring, sixth Lucasian professor of mathematics at Cambridge, had lots of problems, but the ones we are interested in are mathematical in nature. For example, is there a least positive integer $g(k)$ such that every positive integer can be expressed as the sum of at most $g(k)$ kth powers of nonnegative integers? That is, can any positive integer n be represented in at least one way as $a_1^k + a_2^k + \cdots + a_{g(k)}^k$, where $a_i \geqslant 0$ are not necessarily distinct? From Theorem 8.10 we know that $g(2) = 4$. Several cubic and quartic representations are quite intriguing. For example,

$$\begin{aligned}
153 &= 1^3 + 5^3 + 3^3, \\
370 &= 3^3 + 7^3 + 0^3, \\
371 &= 3^3 + 7^3 + 1^3, \\
407 &= 4^3 + 0^3 + 7^3, \\
1634 &= 1^4 + 6^4 + 3^4 + 4^4, \\
8208 &= 8^4 + 2^4 + 0^4 + 8^4, \\
9474 &= 9^4 + 4^4 + 7^4 + 4^4.
\end{aligned}$$

In addition, 635 381 657 is the smallest number that can be written as the sum of two fourth powers in two distinct ways, namely as $133^4 + 134^4$ and $59^4 + 158^4$.

In 1770, in *Meditationes algebraicae*, Waring stated, without proof, as was his nature, that $g(k) = [\![(\frac{3}{2})^k]\!] + 2^k - 2$, where $k \geqslant 2$ is a positive integer and $[\![\cdot]\!]$ denotes the greatest integer function. That is, every positive integer can be expressed as the sum of 4 or fewer squares, 9 or fewer cubes, 19 or fewer fourth powers, 37 or fewer fifth powers, and so forth. Since 7 requires exactly 4 squares, 23 requires exactly 9 cubes, 79 requires exactly 19 fourth powers, and 223 requires exactly 37 fifth powers, Waring's problem has been shown to be the best estimate for squares, cubes, fourth powers, and fifth powers.

It is important, when dealing with odd exponents, that the solutions are required to be nonnegative integers. For example, if n is a positive integer, then since $n^3 \equiv n \pmod 6$ there is an integer k such that $n^3 = n + 6k$ and we have $n = n^3 - 6k = n^3 + k^3 + k^3 + (-k-1)^3 + (1-k)^3$. Therefore, if no restrictions are placed on the integral solutions then any positive integer may be represented as the sum of five cubes.

In 1772, Euler's son, Johannes Albert, showed that for any positive integer n, $g(k) \geqslant [\![(\frac{3}{2})^k]\!] + 2^k - 2$, for $k \geqslant 1$. His result follows from the fact that for a given positive integer k, the number $n = 2^k \cdot [\![(\frac{3}{2})^k]\!] - 1$ cannot be written by a sum of fewer than $[\![(\frac{3}{2})^k]\!] + 2^k - 2$ kth powers. Since $n \leqslant 2^k(\frac{3}{2})^k - 1 < 3^k$, only summands of the forms 1^k and 2^k can be used to represent n as a sum of kth powers. In addition, the maximum number of summands of form 2^k that we can use to represent n without exceeding n is $[\![(\frac{3}{2})^k]\!] - 1$. Thus, the number of summands of the form 1^k is given by $n - 2 \cdot ([\![(\frac{3}{2})^k]\!] - 1) = (2^k \cdot [\![(\frac{3}{2})^k]\!] - 1) - 2 \cdot ([\![(\frac{3}{2})^k]\!] - 1) = 2^k - 1$. Therefore, the minimal number of summands needed to represent n is $[\![(\frac{3}{2})^k]\!] + 2^k - 2$. For example, if $k = 3$, then $n = 23$, $g(3) = 9$, $[\![(\frac{3}{2})^3]\!] - 1 = 2$, and $23 - 2^3 \cdot [[\![(\frac{3}{2})^3]\!] - 1] = 23 - 8 \cdot 2 = 7$. Thus, 23 expressed as a sum of cubes requires exactly nine summands, namely, $2^3 + 2^3 + 1^3 + 1^3 + 1^3 + 1^3 + 1^3 + 1^3 + 1^3$. In 1909, David Hilbert solved Waring's problem when he proved that for every positive integer $k \geqslant 2$ there is a number $g(k)$ such that every positive integer can be represented as the sum of at most $g(k)$ kth powers. Currently, it is known that $g(4) = 19$, $g(5) = 37$, $g(6) = 73$, $g(7) = 143$, $g(8) = 279$, $g(9) = 548$, and $g(10) = 1079$.

We can generalize Waring's problem in another direction, by defining $G(k)$ to be the least positive integer such that all integers from some point on can be represented as the sum of at most $G(k)$ kth powers. That is, all but a finite number of integers can be represented as the sum of $G(k)$ kth powers and infinitely many positive integers cannot be written as the sum of fewer that $G(k)$ kth powers. From the definitions of the functions g and G, it follows that $G(k) \leqslant g(k)$. Since an infinite number of positive integers cannot be written as the sum of three squares, it follows from Theorem 8.10 that $G(2) = 4$. All positive integers except 23 and 239 can be represented as the sum of eight or fewer cubes. In addition, all positive integers greater than 454 can be expressed as the sum of seven or fewer cubes. In fact, 8042 is the largest positive integer requiring seven cubes. Hence, $G(3) < 7$. In 1908, E. Maillet and A. Hurwitz showed that $G(k) \geqslant k + 1$, and in 1920, G.H. Hardy and J.E. Littlewood showed that $G(k) \leqslant 2^{k-1}(k - 2) + 5$. Up to now, the following results concerning the function $G(k)$ are known: $G(4) = 16$, $6 \leqslant G(5) \leqslant 21$, $9 \leqslant G(6) \leqslant 31$, $8 \leqslant G(7) \leqslant 45$, $32 \leqslant G(8) \leqslant 62$, $13 \leqslant G(9) \leqslant 82$, and $12 \leqslant G(10) \leqslant 102$.

The number 325 is the smallest positive integer that can be represented three essentially different ways as the sum of two squares, namely, $325 = 1^2 + 18^2 = 6^2 + 17^2 = 10^2 + 15^2$. The story of Hardy's visit to Ramanujan in a London hospital illustrates a stellar property of the number 1729. Ramanujan was suffering from tuberculosis, a disease that would end his short but enormously mathematically productive life a few years later. Hardy told Ramanujan that he had arrived in a taxi having the quite undistinguished number 1729. Whereupon Ramanujan replied that it was not so dull a number as Hardy thought for it is the smallest number which can be represented as the sum of two cubes in two essentially different ways.

Fermat showed that a cube cannot be expressed as the sum of two cubes.

However, generalizing Pythagorean triples, it is possible to find cubic quadruples, that is, 4-tuples (x, y, z, w) such that $x^3 + y^3 + z^3 = w^3$. For example, (1, 6, 8, 9) and (3, 4, 5, 6) are examples of cubic quadruples. In 1769, Euler conjectured that no nth power could be represented as the sum of fewer than n nth powers, that is, $x_1^n + x_2^n + \cdots + x_k^n = z^n$ has nontrivial integral solutions if and only if $k \geqslant n$. However, in 1968, L.J. Lander and T.R. Parkin discovered that $144^5 = 27^5 + 84^5 + 110^5 + 133^5$. In 1988, N.J. Elkies found an infinite number of counterexamples to Euler's conjecture for the case when $n = 4$ including $20\,615\,673^4 = 2\,682\,440^4 + 15\,365\,639^4 + 18\,796\,760^4$ and the smallest counterexample known, namely $422\,481^4 = 95\,800^4 + 217\,519^4 + 414\,560^4$. In 1936, K. Mahler discovered the identity $(1 + 9n^3)^3 + (3n - 9n^4)^3 + (9n^4)^3 = 1$. Hence, 1 can be written infinitely many ways as the sum of three cubes.

There have been conjectures as to whether each natural number n can be expressed as the sum of finite number of kth powers of primes. In 1937, I.M. Vinogradov showed that for every $k \geqslant 1$ there exists a natural number $V(k)$ such that every sufficiently large natural number is the sum of at most $V(k)$ kth powers of prime numbers. In 1987, K. Thanigasalam showed that $V(5) \leqslant 23$, $V(6) \leqslant 33$, $V(7) \leqslant 47$, $V(8) \leqslant 63$, $V(9) \leqslant 83$, and $V(10) \leqslant 107$.

Fermat's Last Theorem states that the equation $x^n + y^n = z^n$ has no integral solution (a, b, c) with $abc \neq 0$, if $n \geqslant 3$. In 1637, Fermat claimed to have a proof but the margins of his copy of Bachet's version of Diophantus's *Arithmetica* were too narrow to sketch the proof. In his correspondence, Fermat showed that there are no integral solutions for the case when $n = 4$ (Theorem 2.14) and he probably had a proof for the case when $n = 3$. Euler considered the factorization $a^3 - b^3 = (a - b)(a - b\omega)(a + b\omega)$, where $\omega = (-1 + \sqrt{3})/2$ is a cube root of unity. However, Euler assumed that all numbers of the form $a + b\omega$, where a and b are integers, factor uniquely and used the method of descent to establish the case when $n = 3$. It was Gauss who showed that factorization of numbers of the form $a + b\omega$ is indeed unique. In 1820, Sophie Germain showed that if p and $2p + 1$ are prime then $x^p + y^p = z^p$ has no solution when xyz is not divisible by p. The case where the exponents are prime is crucial for if p is any prime that divides n, say $n = pm$, then $(x^m)^p + (y^m)^p = (z^m)^p$. Hence, if Fermat's Last Theorem is true for primes then it is true for all positive integers.

In 1825, Dirichlet and Legendre established the theorem for the case when $n = 5$. In 1832, Dirichlet showed the result is true for the case when $n = 14$. In 1839, Lamé proved it for the case when $n = 7$, but ran into difficulty for the general case when he assumed unique factorization for more general algebraic number fields. In particular, Lamé was interested in working with cyclotomic integers, numbers of the form $a_0 + a_1\zeta + a_2\zeta^2 + \cdots + a_p\zeta^p$, where a_i is an integer, for $0 \leqslant i \leqslant p$, p is an odd prime, and $\zeta \neq 1$ is a complex pth root of unity, that is $\zeta = \alpha + \beta i$, where α and β are real numbers and $\zeta^p = 1$. Liouville and Dirichlet remarked that unique factorization may fail to hold in such general number systems. In addition, while 1 and -1 are the only two integers that have multiplicative inverses, many nontrivial cyclotomic integers have multiplicative inverses. From 1844 to 1847, Ernst Eduard Kummer, after encountering the same problems as Lamé, attacked the problem and in the process founded the theory of ideals, a theory that was developed by Richard Dedekind in the nineteenth century. In 1849, Kummer showed that except possibly for $n = 37$, 59, and 67, the theorem was true for all positive integer exponents less than 100. After receiving his degree from the University of Halle and before assuming a position at the University of Breslau, Kummer spent 10 years as a high school mathematics teacher. When Dirichlet replaced Gauss at Göttingen in 1855, Kummer was chosen to replace Dirichlet at the University of Berlin. After teaching at the University of Zürich for 5 years, he spent 50 years teaching high school mathematics in Brunswick, Germany.

There have been more than a thousand alleged proofs of Fermat's Last Theorem. In 1984, Gerd Faltings, a German mathematician from the University of Wuppertal, was awarded the Fields Medal, considered by many to be the Nobel Prize in mathematics, for solving the Mordell conjecture. Faltings proved an auxiliary result, first posed by the Cambridge mathematician, Louis J. Mordell, in 1922, namely, for each integer $n > 2$ the equation $x^n + y^n = z^n$ has finitely many solutions. Most attempts at proving the theorem relied on devising original factoring techniques. Much progress has been made in this direction by Ken Ribet, Jean-Pierre Serre, Goro Shimura, Yutaka Taniyama, Barry Mazur, and Richard Taylor. In 1985, L.M. Adleman, D.R. Heath-Brown, and E. Fouvry showed that there are infinitely many non-Sophie Germain primes p such that $x^p + y^p = z^p$ has no solutions where p does not divide xyz. In 1986, Gerhard Frey suggested that there was a correspondence between the theorem and elliptic curves. By 1990, Fermat's Last Theorem had been established for all positive integers less than 10^8. In 1994, more than 350 years after Fermat proposed the question, Cambridge-educated Andrew Wiles of Princeton, working virtually by himself for six years and building on the work of his predecessors and colleagues, proved Fermat's Last

Theorem. Wiles presented a flawed version of his proof at a conference at the Isaac Newton Institute for Mathematical Research at Cambridge University in June 1993. After a year of intensive work back at Princeton using a different approach, he published a valid proof of Fermat's Last Theorem in the fall of 1994.

Exercises 8.1

1. In *Arithmetica*, Diophantus gave
$$x = \frac{2am}{m^2 + 1}$$
and
$$y = \frac{a(m^2 - 1)}{m^2 + 1},$$
where m is a nonzero constant, as a rational solution to the equation $x^2 + y^2 = a^2$ for a given value of a. Verify that it is a valid rational solution to the equation.
2. Use Fermat's method to express 8650 as the sum of two squares.
3. Determine the values of $h(n)$ and $f(n)$, for $101 \leqslant n \leqslant 200$.
4. If n is a multiple of 4 and can be represented as the sum of two squares, such as $n = x^2 + y^2$, then show that both x and y must be even.
5. If $n \equiv 12 \pmod{16}$, then show that n cannot be written as a sum of two squares.
6. If $n \equiv 6 \pmod 8$, then show that n cannot be written as the sum of two squares.
7. Prove that if $n \equiv 7 \pmod 8$, then n cannot be written as the sum of two squares.
8. Show that if 3 does not divide n, then $6n$ cannot be written as a sum of two squares.
9. Prove that if n can be written as the sum of two squares then $2n$ can also be written as the sum of two squares. [Charles Dodgson]
10. Determine the smallest positive number which can be written in two different ways as the sum of two positive squares and exhibit the two distinct representations.
11. Use Fibonacci's identity to find a positive integer with three distinct representations as the sum of two integral squares. Exhibit the three distinct representations.

12. Determine a representation of 2^{2k+1} as the sum of two integral squares, where k is a positive integer.
13. Prove that 2^{2k} cannot be represented as the sum of two integral nonzero squares, where k is a positive integer.
14. Use Theorem 8.8 to determine $f(3185)$, $f(7735)$, $f(72\,581)$, $f(226\,067)$.
15. Show that 6525 is the hypotenuse of a Pythagorean triangle and determine the legs of the triangle.
16. Show that 6370 is the hypotenuse of a triangle with integral sides.
17. If n is a positive integer, show that either n or $2n$ can be expressed as the sum of three integral squares.
18. Find two different representations for 1729 as the sum of two cubes.
19. Find two different representations for 40 033 as the sum of two cubes.
20. In how many ways can n appear as the hypotenuse of a Pythagorean triangle where
    (a) $n = 16\,120$,
    (b) $n = 56\,144$?
21. Given that $30 = 1^2 + 2^2 + 3^2 + 4^2$ and $29 = 2^2 + 5^2 + 0^2 + 0^2$ express $870 = 29 \cdot 30$ as the sum of four squares.
22. Show that (3, 4, 5, 6) is a cubic quadruple.
23. Find the missing integer in the following cubic quadruples: $(2, 12, 16, a)$, $(9, 12, 15, b)$, $(3, 10, c, 19)$, and $(d, 14, 17, 20)$.
24. What identity results when $n = 2$ ($n = 4$) in the equation $m^2 + (m + 1)^2 + \cdots + (m + n)^2 = (m + n + 1)^2 + (m + n + 2)^2 + \cdots + (m + 2n)^2$, where $m = n(2n + 1)$?
25. Prove that a positive integer n can be written as the difference of two squares if and only if $n \not\equiv 2 \pmod 4$.
26. Prove that every Fermat number, $F_n = 2^{2^n} + 1$, where $n \geqslant 1$, can be expressed as the difference of two squares.
27. Prove that every odd prime can be expressed as the difference of two squares.
28. Find three primes p, other than 5, 13, and 41, such that $p = n^2 + (n + 1)^2$, where n is a positive integer.
29. Find two primes p, other than 29 and 149, such that $p = n^2 + (n + 1)^2 + (n + 2)^2$, where n is a positive integer.
30. Express 459 as the sum of three integral squares.
31. Show that there are no integer solutions to the equation $y^3 = x^2 + (x + 1)^2$.
32. If $3n$ is a sum of four squares, show that n is the sum of four squares, where n is a positive integer. [Sylvester 1847]

33. Represent 192 as the sum of three triangular numbers.
34. Show that if x, y, z are integers such that $x^{p-1} + y^{p-1} = z^{p-1}$, where p is prime, then p divides xyz.
35. Show that if x, y, z are integers such that $x^p + y^p = z^p$, then p divides $x + y - z$.
36. Can 1999^{1999} be expressed as the sum of two squares? Justify your answer.
37. Can 5 941 232 be expressed as the sum of three integral squares?
38. Show that $8(k+1) = (2a+1)^2 + (2a-1)^2 + (2b+1)^2 + (2b-1)^2 + (2c+1)^2 + (2c-1)^2 + (2d+1)^2 + (2d-1)^2$, where $k = a^2 + b^2 + c^2 + d^2$. The identity implies that any multiple of 8 can be expressed as the sum of the square of eight odd integers.
39. In 1844, E.C. Catalan conjectured that 8, 9 are the only consecutive integers that are powers. That is, $3^2 - 2^3 = 1$. Positive integers not congruent to 2 modulo 4 can be represented as the difference of two powers each greater than the first. Note that $2 = 3^3 - 5^2$ and $3 = 2^7 - 5^3$. Express 4, 5, 7, 8, 9, 10, 11, 12, 13 as differences of two powers each greater than the first.

8.2 Pell's equation

Euler, after a cursory reading of Wallis's *Opera Mathematica*, mistakenly attributed the first serious study of nontrivial solutions to equations of the form $x^2 - dy^2 = 1$, where $x \neq 1$ and $y \neq 0$, to John Pell, mathematician to Oliver Cromwell. However, there is no evidence that Pell, who had taught at the University of Amsterdam, had ever considered solving such equations. They would be more aptly called Fermat's equations, since Fermat first investigated properties of nontrivial solutions of each equations. Nevertheless, Pellian equations have a long history and can be traced back to the Greeks. Theon of Smyrna used x/y to approximate $\sqrt{2}$, where x and y were integral solutions to $x^2 - 2y^2 = 1$. In general , if $x^2 = dy^2 + 1$, then $x^2/y^2 = d + 1/y^2$. Hence, for y large, x/y is a good approximation of $\sqrt{d}$, a fact well known to Archimedes.

Archimedes's *problema bovinum* took two thousand years to solve. According to a manuscript discovered in the Wolfenbüttel library in 1773 by Gotthold Ephraim Lessing, the German critic and dramatist, Archimedes became upset with Apollonius of Perga for criticizing one of his works. He divised a cattle problem that would involve immense calculation to solve and sent it off to Apollonius. In the accompanying correspondence, Archimedes asked Apollonius to compute, if he thought he was smart enough, the number of the oxen of the sun that grazed once upon the plains of the Sicilian isle Trinacria and that were divided according to color into four herds, one milk white, one black, one yellow and one dappled, with the following constraints:

$$\text{white bulls} = \text{yellow bulls} + \left(\frac{1}{2} + \frac{1}{3}\right) \text{black bulls},$$

$$\text{black bulls} = \text{yellow bulls} + \left(\frac{1}{4} + \frac{1}{5}\right) \text{dappled bulls},$$

$$\text{dappled bulls} = \text{yellow bulls} + \left(\frac{1}{6} + \frac{1}{7}\right) \text{white bulls},$$

$$\text{white cows} = \left(\frac{1}{3} + \frac{1}{4}\right) \text{black herd},$$

$$\text{black cows} = \left(\frac{1}{4} + \frac{1}{5}\right) \text{dappled herd},$$

$$\text{dappled cows} = \left(\frac{1}{5} + \frac{1}{6}\right) \text{yellow herd, and}$$

$$\text{yellow cows} = \left(\frac{1}{6} + \frac{1}{7}\right) \text{white herd.}$$

Archimedes added, if you find this number, you are pretty good at numbers, but do not pat yourself on the back too quickly for there are two more conditions, namely:

white bulls plus black bulls is square and
dappled bulls plus yellow bulls is triangular.

Archimedes concluded, if you solve the whole problem then you may 'go forth as conqueror and rest assured that thou art proved most skillful in the science of numbers'.

The smallest herd satisfying the first seven conditions in eight unknowns, after some simplifications, lead to the Pellian equation $x^2 - 4\,729\,494y^2 = 1$. The least positive solution, for which y has 41 digits, was discovered by Carl Amthov in 1880. His solution implies that the number of white bulls has over 2×10^5 digits. The problem becomes much more difficult when the eighth and ninth conditions are added and the first complete solution was given in 1965 by H.C. Williams, R.A. German, and C.R. Zarnke of the University of Waterloo.

In *Arithmetica*, Diophantus asks for rational solutions to equations of the type $x^2 - dy^2 = 1$. In the case where $d = m^2 + 1$, Diophantus offered the integral solution $x = 2m^2 + 1$ and $y = 2m$. Pellian equations are found in Hindu mathematics. In the fourth century, the Indian mathematican

Baudhayana noted that $x = 577$ and $y = 408$ is a solution of $x^2 - 2y^2 = 1$ and used the fraction $\frac{577}{408}$ to approximate $\sqrt{2}$. In the seventh century Brahmagupta considered solutions to the Pellian equation $x^2 - 92y^2 = 1$, the smallest solution being $x = 1151$ and $y = 120$. In the twelfth century the Hindu mathematician Bhaskara found the least positive solution to the Pellian equation $x^2 - 61y^2 = 1$ to be $x = 226\,153\,980$ and $y = 1\,766\,319\,049$.

In 1657, Fermat stated without proof that if d was positive and nonsquare, then Pell's equation had an infinite number of solutions. For if (x, y) is a solution to $x^2 - dy^2 = 1$, then $1^2 = (x^2 - dy^2)^2 = (x^2 + dy^2) - (2xy^2)d$. Thus, $(x^2 + dy^2, 2xy)$ is also a solution to $x^2 - dy^2 = 1$. Therefore, if Pell's equation has a solution, it has infinitely many.

In 1657 Fermat challenged William Brouncker, of Castle Lynn in Ireland, and John Wallis to find integral solutions to the equations $x^2 - 151y^2 = 1$ and $x^2 - 313y^2 = 1$. He cautioned them not to submit rational solutions for even 'the lowest type of arithmetician' could devise such answers. Wallis replied with (1 728 148 040, 140 634 693) as a solution to the first equation. Brouncker replied with (126 862 368, 7 170 685) as a solution to the second. Lord Brouncker claimed that it only took him about an hour or two to find his answer. Samuel Pepys, secretary of the Royal Society, had a low opinion of Brouncker's moral character but thought that his mathematical ability was quite adequate. In the section on continued fractions, in this chapter, we will demonstrate the method Wallis and Brouncker used to generate their answers.

In 1770, Euler showed that no triangular number other than unity was a cube and none but unity was a fourth power. He devised a method, involving solutions to Pellian equations, to determine natural numbers that were both triangular and square. In particular, he was looking for positive integers m and n such that $n(n+1)/2 = m^2$. To accomplish this, he multiplied both sides of the latter equation by 8 and added 1 to obtain $(2n+1)^2 = 8m^2 + 1$. He let $x = 2n + 1$ and $y = 2m$ so that $x^2 - 2y^2 = 1$. Solutions to this Pellian equation produce square–triangular numbers since

$$\left(\frac{x-1}{2}\right)\left(\frac{x-1}{2}+1\right)\Big/2 = \left(\frac{y}{2}\right)^2.$$

That is, the $((x-1)/2)$th triangular number equals the $(y/2)$th square number. Using notation introduced in Chapter 1, $t_{x-1/2} = s_{y/2}$. For example, from the solution $x = 3$ and $y = 2$, it follows that $m = n = 1$,

Table 8.3.

x	y	m	n	sqr–tri #
3	2	1	1	1
17	12	6	8	36
99	70	35	49	1 225
577	408	204	288	41 616

yielding the square–triangular number 1. Table 8.3 lists several solutions (x, y) to $x^2 - 2y^2 = 1$ and their associated square–triangular numbers. A natural question arises. Does the method generate all squarc–triangular numbers? If one is more methodical about how one obtains the solutions, one can see that it does.

Since $1 = x^2 - 2y^2 = (x - y\sqrt{2})(x + y\sqrt{2})$, it follows that $1 = 1^2 = (x - y\sqrt{2})^2(x + y\sqrt{2})^2 = ((2y^2 + x^2) - 2xy\sqrt{2})((2y^2 + x^2) + 2xy\sqrt{2}) = (2y^2 + x^2)^2 - 2(2xy)^2$. Thus, if (x, y) is a solution to $1 = x^2 - 2y^2$, so is $(2y^2 + x^2, 2xy)$. For example, the solution $(3, 2)$ generates the solution $(2 \cdot 2^3 + 3^2, 2 \cdot 2 \cdot 3) = (17, 12)$. The solution $(17, 12)$ generates the solution $(2 \cdot 12^2 + 17^2, 2 \cdot 12 \cdot 17) = (577, 408)$. The square–triangular number generated by the solution $(2y^2 + x^2, 2xy)$ to $1 = x^2 - 2y^2$ is distinct from the square–triangular number generated by the solution (x, y). Therefore, there exist an infinite number of square–triangular numbers. Lagrange in a series of papers presented to the Berlin Academy between 1768 and 1770 showed that a similar procedure will determine all solutions to $x^2 - dy^2 = 1$, where d is positive and nonsquare. By the fundamental or least positive solution of $x^2 - dy^2 = 1$, we mean the solution (r, s) such that for any other solution (t, u) $r < t$ and $s < u$. In 1766, Lagrange proved that the equation $x^2 - dy^2 = 1$ has an infinite number of solutions whenever d is positive and not square.

Theorem 8.11 (Lagrange) *If (r, s) is the fundamental solution of $x^2 - dy^2 = 1$, where d is positive and nonsquare, then every solution to $x^2 - dy^2 = 1$ is given by (x_n, y_n) where $x_n + y_n\sqrt{d} = (r + s\sqrt{d})^n$ for $n = 1, 2, 3, \ldots$.*

Proof Let (r, s) be a fundamental solution of $x^2 - dy^2 = 1$, where d is positive and nonsquare, and $x_n + y_n\sqrt{d} = (r + s\sqrt{d})^n$, for $n = 1, 2, 3, \ldots$. It follows that $x_n^2 - dy_n^2 = (x_n + y_n\sqrt{d})(x_n - y_n\sqrt{d}) = (r + s\sqrt{d})^n(r - s\sqrt{d})^n = (r^2 - s^2d)^n = 1^n = 1$. Hence, (x_n, y_n) is a solution to $x^2 - dy^2 = 1$, where $x_n + y_n\sqrt{d} = (r + s\sqrt{d})^n$ for $n = 1, 2, 3, \ldots$. We show that if (a, b) is a solution to $x^2 - dy^2 = 1$, where a and b are positive, there is a positive integer n such that $(a, b) = (x_n, y_n)$. Suppose that is not the case. Hence there is a positive integer k such that $(r + s\sqrt{d})^k < a + b\sqrt{d} < (r + s\sqrt{d})^{k+1}$. Since $(r + s\sqrt{d})^{-k} = (r - s\sqrt{d})^k$, dividing by $(r + s\sqrt{d})^k$, we obtain $1 < (a + b\sqrt{d})(r - s\sqrt{d})^k < (r + s\sqrt{d})$. Let $u + v\sqrt{d} = (a + b\sqrt{d})(r - s\sqrt{d})^k$; hence, $u^2 - v^2 d = (u + v\sqrt{d})(u - v\sqrt{d}) = (a + b\sqrt{d})(r - s\sqrt{d})^k(a - b\sqrt{d})(r + s\sqrt{d})^k = (a^2 - b^2 d)(r^2 - s^2 d)^k = 1$. Thus, (u, v) is a solution to $x^2 - dy^2 = 1$. However, since $u + v\sqrt{d} > 1$, $0 < u - v\sqrt{d} < 1$. Hence, $2u = (u + v\sqrt{d}) + (u - v\sqrt{d}) > 1 + u > 0$ and $2v\sqrt{d} = (u + v\sqrt{d}) - (u - v\sqrt{d}) > 1 - 1 = 0$. Therefore, $u > 0$, $v > 0$, and $u + v\sqrt{d} < r + s\sqrt{d}$, contradicting the assumption that (r, s) is the fundamental solution, and the result is established. ■

In particular, if (x_k, y_k) is the solution to $x^2 - 2y^2 = 1$ generating the square–triangular number E_k, then (x_{k+1}, y_{k+1}), the solution generating the next square–triangular number E_{k+1}, is obtained as follows: $1 = 9 - 8 = (3 + 2\sqrt{2})(3 - 2\sqrt{2})$ and $1 = x_k^2 - 2y_k^2 = (x_k + \sqrt{2}y_k)(y_k - \sqrt{2}x_k)$. Hence, $1 = 1 \cdot 1 = (x_k + \sqrt{2}y_k)(x_k - \sqrt{2}y_k)(3 + 2\sqrt{2})(3 - 2\sqrt{2}) = [(3x_k + 4y_k) + (2x_k + 3y_k)\sqrt{2}][(3x_k + 4y_k) - (2x_k + 3y_k)\sqrt{2}] = (3x_k + 4y_k)^2 - 2(2x_k + 3y_k)^2$. Therefore, $x_{k+1} = 3x_k + 4y_k$ and $y_{k+1} = 2x_k + 3y_k$, in a sense, is the 'next' solution to $x^2 - 2y^2 + 1$. If we represent the kth square–triangular number by $E_k = y_k^2/4 = (x_k^2 - 1)/2$, it follows that $x_k = 2\sqrt{8E_k + 1}$ and $y_k = 2\sqrt{E_k}$. Hence, the next square–triangular number is given by

$$E_{k+1} = \frac{(y_{k+1})^2}{4} = \frac{(2x_k + 3y_k)^2}{4} = \frac{4x_k^2 + 12x_k y_k + 9y_k^2}{4}$$
$$= 17E_k + 1 + 6\sqrt{8E_k^2 + E_k}.$$

For example, the square–triangular number after 41 616 is $17 \cdot 41\,616 + 1 + 6\sqrt{8 \cdot 41\,616^2 + 41\,616} = 1\,413\,721$.

Frenicle compiled a table of least positive solutions to $x^2 - dy^2 = 1$, where d is nonsquare and $1 \leqslant d \leqslant 150$. A brief version of Frenicle's table is shown in Table 8.4. The *Canon Pellianus* computed by C.F. Degenin, 1817, gave least positive solutions to Pell's equation for all positive nonsquare values of $d \leqslant 1000$.

Pell's equation is of considerable importance in number theory and can be used to find optimal rational approximations to square roots of positive integers. In particular, if $x^2 - dy^2 = 1$, then x/y gives a good approximation to $\sqrt{d}$. This follows since if $x > y\sqrt{d}$ then

Table 8.4.

d	x	y	d	x	y
1	—	—	26	51	10
2	3	2	27	26	5
3	2	1	28	127	24
4	—	—	29	9 801	1 820
5	9	4	30	11	2
6	5	2	31	1 520	273
7	8	3	32	17	3
8	3	1	33	23	4
9	—	—	34	35	6
10	19	6	35	6	1
11	10	3	36	—	—
12	7	2	37	73	12
13	649	180	38	37	6
14	15	4	39	25	4
15	3	1	40	19	3
16	—	—	41	2 049	320
17	33	8	42	13	2
18	17	4	43	3 482	531
19	170	39	44	199	30
20	9	2	45	161	24
21	55	12	46	24 335	3 588
22	197	42	47	48	7
23	24	5	48	1	7
24	5	1	49	—	—
25	—	—	50	99	14

$$\left|\frac{x}{y} - \sqrt{d}\right| = \left|\frac{1}{y}\right|\left|\frac{x^2 - dy^2}{x + y\sqrt{d}}\right| = \left|\frac{1}{y}\right|\left|\frac{1}{x + y\sqrt{d}}\right| < \frac{1}{y(2y\sqrt{d})} < \frac{\sqrt{d}}{2y^2\sqrt{d}}$$
$$= \frac{1}{2y^2}.$$

Elliptic curves, Diophantine equations of form $y^2 = x^3 + ax^2 + bx + c$, are more general than Pellian equations. In 1621, Bachet studied elliptic equations of the form $y^2 = x^3 + c$. He claimed correctly that the only solution (x, y) to the equation $y^2 = x^3 - 2$ is (3, 5). In 1657, Fermat claimed that the only solutions to $y^2 = x^3 - 4$ were (2, 2) and (5, 11). Euler showed that the only solution to $y^2 = x^3 + 1$ is (2, 3). In 1922, Louis J. Mordell, Sadlerian Professor of Mathematics at Cambridge, proved that, for a fixed value of c, Bachet's equation has only a finite number of solutions. In 1965 Alan Baker of Cambridge was awarded the Fields Medal for devising a finite procedure for determining solutions to Bachet's equation.

Exercises 8.2

1. Find the square–triangular number generated by the solution $x = 19\,601$ and $y = 13\,860$ to the equation $x^2 - 2y^2 = 1$.
2. Find positive solutions (x, y) to the following Pellian equations.
   (a) $x^2 - 3y^2 = 1$,
   (b) $x^2 - 5y^2 = 1$,
   (c) $x^2 - 6y^2 = 1$.
3. Find a Pellian formula to generate square–pentagonal numbers.
4. Find two square–pentagonal numbers.
5. Find two triangular–pentagonal numbers.
6. Find the next two square–triangular numbers following 1 413 721.
7. Why is it necessary, in determining a solution to Pell's equation $x^2 - dy^2 = 1$, that d not be a square?
8. Prove that if the Bachet equation $y^2 = x^3 + 2$ has a solution then x and y must both be odd.
9. Show that $3x^2 + 2 = y^2$ has no integral solutions.

8.3 Binary quadratic forms

Fermat considered the representation of integers by Diophantine polynomials of the form $x^2 \pm cy^2$ and in 1761 Euler those of the form $x^2 + xy + y^2$ or $x^2 + cy^2$. In 1763, Euler showed that every prime of the form $6n + 1$ can be represented by $x^2 + 3y^2$ and every prime of the form $8n + 1$ can be represented by $x^2 + 2y^2$. Representing integers as sums of squares and Pellian problems are special cases of a more general problem, namely, representing integers by integral expressions of the form $ax^2 + bxy + cy^2 + dx + ey + f$. Gauss devoted almost 60 percent of *Disquisitiones* to deriving properties of such expressions.

In general, an integral expression $f(x, y)$ consisting of a finite number of terms of the form $ax^r y^s$, with a an integer and x and y indeterminates, is called a Diophantine polynomial in two variables. We say that $f(x, y)$ represents the integer n if there exist integers x and y such that $f(x, y) = n$. In addition, the integer n is said to be properly represented by $f(x, y)$ if $f(x, y) = n$ with $\gcd(x, y) = 1$. We say that $f(x, y)$ is universal if it represents every integer, and positive definite if it represents only nonnegative integers. One of the first problems that arised is that of equivalence. Two Diophantine polynomials $f(x, y)$ and $g(u, v)$ are said to be equivalent, denoted by $f \sim g$, if there is a linear transformation $x = au + bv$ and $y = cu + dv$, with $ad - bc = \pm 1$, such that $f(x, y) = g(u, v)$. For example, the Diophantine polynomials $f(x, y) = x^3 + xy + y^2 + x - 2$ and $g(u, v) = u^3 + v^3 + 2u^2 + 3u^2v + 3uv^2 + 7uv + 6v^2 + u + v - 2$ are equivalent under the transformation $x = u + v$, $y = u + 2v$. If two Diophantine polynomials are equivalent they represent the same numbers. For example, under the transformation $x = -u + 3v$, $y = -u + 2v$, $f(x, y) = 3x^2 - 10xy + 8y^2$ is equivalent to $g(u, v) \equiv u^2 - v^2$. Both polynomials represent all integers not of the form $4k + 2$.

In 1773, Lagrange made the first investigations of binary quadratic forms, which are Diophantine equations of the type $f(x, y) = ax^2 + bxy + cy^2$. The term $b^2 - 4ac$ is called the discriminant of the binary quadratic form. Equivalent binary quadratic forms have the same discriminant. However, binary quadratic forms with the same discriminant need not be equivalent. It can be shown that there exists a binary quadratic form with discriminant d if and only if $d \equiv 0$ or $1 \pmod 4$. In particular if $d \equiv 0 \pmod 4$, then $x^2 - (d/4)y^2$ has discriminant d. If $d \equiv 0 \pmod 4$, then $x^2 + xy - ((d - 1)/4)y^2$ has discriminant d. Gauss showed that the number of binary quadratic forms with a given discriminant is finite. It can be shown that the integer n can be properly represented by $ax^2 + bxy + cy^2$ if and only if $x^2 \equiv d \pmod{4n}$ has a solution, for a proof see [Baker].

Theorem 8.12 *A binary quadratic form $ax^2 + bxy + cy^2$ is positive definite if and only if $a \geqslant 0$, $c \geqslant 0$, $a^2 + c^2 > 0$, and $b^2 - 4ac \leqslant 0$.*

Proof Suppose that $f(x, y) = ax^2 + bxy + cy^2$ is a positive definite binary quadratic form. Since $f(1, 0) = a$ and $f(0, 1) = c$, neither $a < 0$ nor $c < 0$. Hence $a \geqslant 0$ and $c \geqslant 0$. If $a = b = c = 0$, then the binary quadratic form represents only 0 and hence cannot be positive definite. If $a = c = 0$ and $b \neq 0$, then $f(x, y) = bxy$. So $f(1, 1) = b$ and $f(1, -1) = -b$, hence, the form cannot be positive definite. Thus, at least one of a and c must be nonzero. Therefore, $a^2 + c^2 > 0$. If $a > 0$, then $f(b, -2a) = -a(b^2 - 4ac) > 0$. If $a = 0$ and $c > 0$, then $f(-2c, d) = -c(b^2 - 4ac) > 0$. In either case, $b^2 - 4ac \leqslant 0$. Conversely, suppose $a \geqslant 0$, $c \geqslant 0$, $a^2 + c^2 > 0$, and $b^2 - 4ac \leqslant 0$. If $a > 0$, then $f(x, y)$ represents at least one positive integer since $f(1, 0) = a$, and $4af(x, y) = 4a(ax^2 + bxy + cy^2) = (2ax + by)^2 - (b^2 - 4ac)y^2 \geqslant 0$. If $a = 0$, then $b = 0$ and $c > 0$, so $f(x, y) = cy^2 > 0$. In any case, $f(x, y)$ is positive definite. ■

In 1996, John Conway and William Schneeberger showed that if a positive definite quadratic form represents all integers from 1 to 15, then it represents all positive integers. Given two integers n and d, with $n \neq 0$, there is a binary quadratic form with discriminant d that represents n properly if and only if the quadratic equation $x^2 \equiv d \pmod{4|n|}$ has a solution. Therefore, if $d \equiv 0$ or $1 \pmod 4$ and if p is an odd prime then there is a binary quadratic form of discriminant d that represents n if and only if $(\frac{d}{p}) = 1$; for a proof see [Niven, Zuckerman, and Montgomery].

Exercises 8.3

1. Show that equivalence between binary quadratic forms is an equivalence relation.
2. Show that $f(x, y) = -x^2 + 2y^2$ and $g(u, v) = 14u^2 + 20uv + 7v^2$ are equivalent under the transformation $x = 2u + v$, $y = 3u + 2v$.
3. Given the binary quadratic forms in the preceding exercise, $f(3, 2) = -1$, what values of u and v yield $g(u, v) = -1$?
4. Show that $2x^2 - y^2$ and $2u^2 - 12uv - v^2$ are equivalent.
5. Show that $x^3 + y^3$ and $35u^3 - 66u^2v + 42uv^2 - 9v^3$ are equivalent.
6. Use the transformation $x = 5u + 2v$, $y = 7u + 3v$ to find a binary quadratic form equivalent to $2x^2 + 5xy - y^2$.
7. Show that equivalent binary quadratic forms have the same discriminant.
8. Show that if d is the discriminant of the binary quadratic form $ax^2 + bxy + cy^2$, then $d \equiv 0 \pmod 4$ or $d \equiv 1 \pmod 4$.
9. Find a binary quadratic form with discriminant 12.
10. Which of the following binary quadratic forms are positive definite?
    (a) $f_1(x, y) = 6xy$;
    (b) $f_2(x, y) = x^2 + 3xy + 2y^2$;
    (c) $f_3(x, y) = -x^2 + 3xy - 12y^2$;
    (d) $f_4(x, y) = x^2 + 3xy + 3y^2$;
    (e) $f_5(x, y) = x^2 + xy - y^2$.
11. Are $f(x, y) = 2x^2 + 3xy + 3y^2$ and $g(x, y) = x^2 + y^2$ equivalent?
12. Can $x^2 + 6y^2$ ever represent 31 or 415?
13. Use the second derivative test for functions of several variables to show if $a \geqslant 0$, $c \geqslant 0$, $a^2 + c^2 > 0$, and $b^2 - 4ac \leqslant 0$, then the surface $f(x, y) = ax^2 + bxy + cy^2$ lies on or above the xy-plane in Euclidean 3-space.
14. Show that the equations $x^2 + 3y^2 = 1$ and $7u^2 + 10uv + 4v^2 = 1$ have corresponding solutions under the transformation $x = 2u + v$ and $y = u + v$.

8.4 Finite continued fractions

An iterated sequence of quotients of the form

$$a_1 + \cfrac{1}{a_2 + \cfrac{1}{a_3 + \cfrac{1}{\cdots}}} \quad \ddots \quad + \cfrac{1}{a_{n-1} + \cfrac{1}{a_n}},$$

denoted by $[a_1, a_2, \ldots, a_n]$, where a_i are real numbers and $a_i > 0$, for $2 \leqslant i \leqslant n$, is called a finite continued fraction. The notation $[a_1, a_2, \ldots, a_n]$ was introduced by Dirichlet in 1854. If the a_i are required to be integers, the resulting expression is called a simple finite continued fraction. For example,

$$\frac{34}{79} = 0 + \cfrac{1}{2 + \cfrac{1}{3 + \cfrac{1}{11}}}$$

is a simple finite continued fraction and is denoted by [0, 2, 3, 11]. References to continued fractions can be found in Indian mathematical works, in particular, in those of Aryabhata in the sixth century and Bhaskara in the twelfth century. Both employed continued fractions to solve linear equations. Fibonacci uses and attempts a general definion of continued fractions in *Liber abaci*. In 1572, Bombelli employed simple continued fractions to approximate the values of square roots as did Cataldi before him. It was, however, Cataldi who first developed a symbolism and properties of continued fractions. The term 'continued fraction' first appeared in the 1653 edition of John Wallis's *Arithmetica infinitorum*. In a posthumous paper, *Descriptio automati planetarii*, Christiaan Huygens used continued fraction expansions to determine the number of teeth on the gears of a planetarium he was constructing. A continued fraction expansion appears on the first page of Gauss's diary for the year 1796. The modern theory of continued fractions began, in 1737, with Euler's *De fractionibus continuis*. In 1882, Carl Lindemann used continued fractions to prove that π was a transcendental number, that is, not the solution to a polynomial equation with rational coefficients.

A straightforward inductive argument shows that every finite simple continued fraction represents a rational number. A finite simple continued fraction of length one is an integer and, hence, rational. Suppose that every finite simple continued fraction with k terms is rational and consider $[a_1, a_2, \ldots, a_k, a_{k+1}]$. We have $[a_1, a_2, \ldots, a_k, a_{k+1}] = a_1 + 1/[a_2,$

$\ldots, a_k, a_{k+1}]$, the sum of two rational numbers. Hence, $[a_1, a_2, \ldots, a_k, a_{k+1}]$ is rational. The converse is also true, namely, every rational number can be expressed as a finite simple continued fraction. Let a/b be any rational number with $b > 0$. From the Euclidean algorithm, we obtain

$$\begin{aligned}
a &= bq_1 + r_1, \text{ where } 0 \leqslant r_1 < b, \\
b &= r_1 q_2 + r_2, \text{ where } 0 \leqslant r_2 < r_1, \\
r_1 &= r_2 q_3 + r_3, \text{ where } 0 \leqslant r_3 < r_2, \\
&\cdots \\
r_{n-2} &= r_{n-1} q_n + r_n, \text{ where } 0 \leqslant r_n < r_{n-1}, \\
r_{n-1} &= r_n q_{n+1}.
\end{aligned}$$

Hence, dividing, we have

$$\begin{aligned}
\frac{a}{b} &= q_1 + \frac{r_1}{b} = q_1 + \frac{1}{\frac{b}{r_1}}, \\
\frac{b}{r_1} &= q_2 + \frac{r_2}{r_1} = q_2 + \frac{1}{\frac{r_1}{r_2}}, \\
\frac{r_1}{r_2} &= q_3 + \frac{r_3}{r_2} = q_3 + \frac{1}{\frac{r_2}{r_3}}, \\
&\cdots \\
\frac{r_{n-1}}{r_n} &= q_{n+1}.
\end{aligned}$$

The multiplicative inverse of the fraction at the end of the kth row is the first term in the $(k+1)$th row. By substitution, we obtain

$$\frac{a}{b} = q_1 + \cfrac{1}{q_2 + \cfrac{1}{q_3 + \cdots}} \quad \ddots \quad + \cfrac{1}{q_n + \cfrac{1}{q_{n+1}}}.$$

That is, $a/b = [q_1, q_2, \ldots, q_{n+1}]$ and we have established the following result.

Theorem 8.13 *Every rational number can be expressed as a finite simple continued fraction and every finite simple continued fraction represents a rational number.*

For example,

$$[7, 4, 2, 5] = 7 + \cfrac{1}{4 + \cfrac{1}{2 + \cfrac{1}{5}}} = 7 + \cfrac{1}{4 + \cfrac{5}{11}} = 7 + \frac{11}{49} = \frac{354}{49}.$$

In order to represent the fraction $\frac{73}{46}$ as a finite simple continued fraction, we first employ the Euclidean algorithm to obtain

$$\begin{aligned}
73 &= 46 \cdot 1 + 27, & \frac{73}{46} &= 1 + \frac{27}{46}, \\
46 &= 27 \cdot 1 + 19, & \frac{46}{27} &= 1 + \frac{19}{27}, \\
27 &= 19 \cdot 1 + 8, & \frac{27}{19} &= 1 + \frac{8}{19}, \\
19 &= 8 \cdot 2 + 3, & \frac{19}{8} &= 2 + \frac{3}{8}, \\
8 &= 3 \cdot 2 + 2, & \frac{8}{3} &= 2 + \frac{2}{3}, \\
3 &= 2 \cdot 1 + 1, & \frac{3}{2} &= 1 + \frac{1}{2}, \\
2 &= 1 \cdot 2.
\end{aligned}$$

Substituting, we obtain

$$\frac{73}{46} = 1 + \cfrac{1}{1 + \cfrac{1}{1 + \cfrac{1}{2 + \cfrac{1}{2 + \cfrac{1}{1 + \frac{1}{2}}}}}}, \text{ or } [1, 1, 1, 2, 2, 1, 2].$$

Since $[a_1, a_2, \ldots, a_{n-1}, a_n] = [a_1, a_2, \ldots, a_{n-2}, a_{n-1} + 1/a_n]$ and $[a_1, a_2, \ldots, a_n] = [a_1, a_2, \ldots, a_n - 1, 1]$, the representation for a finite continued fraction is not unique. However, $[a_1, a_2, \ldots, a_n]$ and $[a_1, a_2, \ldots, a_n - 1, 1]$ are the only two finite simple continued fractional representations for a rational number.

Let $[a_1, a_2, \ldots, a_n]$ be a finite continued fraction. The terms

$$c_1 = a_1, \; c_2 = a_1 + \frac{1}{a_2},$$

$$c_3 = a_1 + \cfrac{1}{a_2 + \cfrac{1}{a_3}}, \ldots,$$

$$c_k = a_1 + \cfrac{1}{a_2 + \cfrac{1}{a_3 + \cfrac{1}{\ddots + \cfrac{1}{a_{k-1} + \cfrac{1}{a_k}}}}},$$

and so forth are called the convergents of $[a_1, a_2, \ldots, a_n]$. In general, the kth convergent of $[a_1, a_2, \ldots, a_n]$, denoted by c_k, is given by $[a_1, a_2, \ldots, a_k]$. Convergents were first described by Daniel Schwenter, Professor of Hebrew, Oriental Languages, and Mathematics at the University of Altdorf, who included the convergents of $\frac{177}{233}$ in his *Geometrica practica* in 1618. The recursive formulas for convergents first appeared in Wallis's *Arithmetica infinitorum* in 1665.

Given a rational number a/b, with $a \geqslant b > 0$, we can apply Saunderson's algorithm, albeit with different initial conditions, to develop a practical method to determine the kth convergents of a/b. Suppose $a/b = [q_1, q_2, \ldots, q_n, q_{n+1}]$, $x_i = x_{i-2} + x_{i-1}q_i$, $y_i = y_{i-2} + y_{i-1}q_i$, for $i = 1, 2, \ldots, n+1$, $x_{-1} = 0$, $x_0 = 1$, $y_{-1} = 1$, and $y_0 = 0$. Hence,

$$\frac{x_1}{y_1} = \frac{x_{-1} + x_0 q_1}{y_{-1} + y_0 q_1} = \frac{0 + 1 \cdot q_1}{1 + 0 \cdot q_1} = q_1 = c_1,$$

$$\frac{x_2}{y_2} = \frac{x_0 + x_1 q_2}{y_0 + y_1 q_2} = \frac{1 + q_1 \cdot q_2}{0 + 1 \cdot q_2} = \frac{1 + q_1 q_2}{q_2} = q_1 + \frac{1}{q_2} = c_2,$$

and

$$\frac{x_3}{y_3} = \frac{x_1 + x_2 q_3}{y_1 + y_2 q_3} = \frac{q_1 + (1 + q_1 \cdot q_2)q_3}{1 + q_2 \cdot q_3} = q_1 + \frac{q_3}{1 + q_2 q_3}$$
$$= q_1 + \cfrac{1}{q_2 + \cfrac{1}{q_3}} = c_3.$$

Suppose that for integer m, with $2 < m \leqslant n$,

$$c_m = [q_1, q_2, \ldots, q_{m-1}, q_m] = \frac{x_m}{y_m} = \frac{x_{m-2} + x_{m-1} q_m}{y_{m-2} + y_{m-1} q_m}.$$

Consider

Table 8.5.

i	-1	0	1	2	3	$\dots$	n
q_i			q_1	q_2	q_3	$\dots$	q_n
x_i	0	1	x_1	x_2	x_3	$\dots$	x_n
y_i	1	0	y_1	y_2	y_3	$\dots$	y_n
c_i			$\frac{x_1}{y_1}$	$\frac{x_2}{y_2}$	$\frac{x_3}{y_3}$	$\dots$	$\frac{x_n}{y_n}$

Table 8.6.

i	-1	0	1	2	3	4	5	6
q_i			1	2	2	3	4	2
x_i	0	1	1	3	7	24	103	230
y_i	1	0	1	2	5	17	73	163
c_i			$\frac{1}{1}$	$\frac{3}{2}$	$\frac{7}{5}$	$\frac{24}{17}$	$\frac{103}{73}$	$\frac{230}{163}$

$$c_{m+1} = [q_1, q_2, \dots, q_{m-1}, q_m, q_{m+1}] = \left[q_1, q_2, \dots, q_{m-1}, q_m + \frac{1}{q_{m+1}}\right]$$

$$= c_{m+1} = \frac{x_{m-2} + x_{m-1}\left(q_m + \dfrac{1}{q_{m+1}}\right)}{y_{m-2} + y_{m-1}\left(q_m + \dfrac{1}{q_{m+1}}\right)}$$

$$= \frac{x_{m-1} + (x_{m-2} + x_{m-1}q_m)q_{m+1}}{y_{m-1} + (y_{m-2} + y_{m-1}q_m)q_{m+1}}$$

$$= \frac{x_{m-1} + x_m q_{m+1}}{y_{m-1} + y_m q_{m+1}} = \frac{x_{m+1}}{y_{m+1}}$$

and we have established the following result.

Theorem 8.14 *If $a/b = [q_1, q_2, \dots, q_n]$, $x_i = x_{i-2} + x_{i-1}q_i$, $y_i = y_{i-2} + y_{i-1}q_i$, for $i = 1, 2, \dots, n$, $x_{-1} = 0$, $x_0 = 1$, $y_{-1} = 1$, and $y_0 = 0$, then the kth convergent, c_k, is given by $c_k = x_k/y_k$, for $0 \leqslant k \leqslant n$.*

For example, let us determine the convergents of $\frac{230}{163} = [1, 2, 2, 3, 4, 2]$. Using the algorithm described in the proof of Theorem 8.14, we fill in Table 8.5 with our data to obtain Table 8.6 and find that the convergents are $c_1 = 1$, $c_2 = \frac{3}{2}$, $c_3 = \frac{7}{5}$, $c_4 = \frac{24}{17}$, $c_5 = \frac{103}{73}$, and $c_6 = \frac{230}{163}$.

Table 8.7.

i	-1	0	1	2	3
q_i			29	2	3
x_i	0	1	29	59	206
y_i	1	0	1	2	7
c_i			$\frac{29}{1}$	$\frac{59}{2}$	$\frac{206}{7}$

Example 8.2 The sidereal period of Saturn, the time it takes Saturn to orbit the Sun, is 29.46 years; in Huygens's day it was thought to be 29.43 years. In order to simulate the motion of Saturn correctly, he needed to efficiently construct two gears, one with p teeth, the other with q teeth, such that p/q is approximately 26.43. To be efficient, p and q were required to be relatively small. The convergents of 29.43 are given in Table 8.7. A reasonably close approximation of 29.43 is given by $\frac{206}{7} = 29.4285$. Thus, to simulate the motion of Saturn with respect to the Earth's motion, Huygens made one gear with 7 teeth and the other with 206 teeth.

Theorem 8.15 *If $c_k = x_k/y_k$ is the kth convergent of $a/b = [q_1, q_2, \ldots, q_n]$, then $y_k \geqslant k$, for $1 \leqslant k \leqslant n$.*

Proof Recall that $y_i = y_{i-2} + y_{i-1}q_i$, for $i = 1, 2, \ldots, y_{-1} = 1$ and $y_0 = 0$. It follows that $y_1 = 1 \geqslant 1$ and $y_2 = 0 + 1 \cdot q_2 \geqslant 1$ since $q_2 > 0$. Suppose that $y_i \geqslant i$, for $2 \leqslant i \leqslant k$, and consider y_{k+1}. We have $y_{k+1} = y_{k-1} + y_k y_{k+1} \geqslant k - 1 + k \cdot 1 = 2k - 1 \geqslant k + 1$. The result follows by induction. ■

Theorem 8.16 *If $c_k = x_k/y_k$ is the kth convergent of $a/b = [q_1, q_2, \ldots, q_n]$, then x_k and y_k are coprime.*

Proof We claim that $x_k y_{k-1} - y_k x_{k-1} = (-1)^k$, for $1 \leqslant k \leqslant n$. If $k = 1$, using the notation of Theorem 8.14, $x_1 y_0 - y_1 x_0 = q_1 \cdot 0 - 1 \cdot 1 = (-1)^1$. If $k = 2$, $x_2 y_1 - y_2 x_1 = (1 + q_1 q_2) \cdot 1 - q_2 q_1 = (-1)^2$. Suppose that for some m, with $1 < m \leqslant n$, $x_m y_{m-1} - y_m x_{m-1} = (-1)^m$. Hence, $x_{m+1} y_m - y_{m+1} x_m = (x_{m-1} + x_m q_m) y_m - (y_{m-1} + y_m q_m) x_m = x_{m-1} y_m - y_{m-1} x_m = -(x_m y_{m-1} - y_m x_{m-1}) = -(-1)^m = (-1)^{m+1}$. The result follows from induction and Theorem 2.7. ■

Table 8.8.

i	-1	0	1	2	3	4
q_i			2	2	2	3
x_i	0	1	2	5	12	41
y_i	1	0	1	2	5	17
c_i			$\frac{2}{1}$	$\frac{5}{2}$	$\frac{12}{5}$	$\frac{41}{17}$

The method outlined in the next result was used by Bhaskara in the twelfth century. It offers a practical way to solve linear Diophantine equations of the form $ax - by = 1$ using convergents of finite simple continued fractions. With a little ingenuity, the method can be adapted to solve Diophantine equations of the form $ax + by = c$ and $ax - by = c$, where there are no restrictions placed on the integers a, b and c.

Theorem 8.17 *If* $\gcd(a, b) = 1$, $a > b > 0$, *and* $c_{n-1} = x_{n-1}/y_{n-1}$ *is the penultimate convergent of* a/b, *then* $x = (-1)^n y_{n-1}$, $y = (-1)^n x_{n-1}$ *is a solution to the equation* $ax - by = 1$.

Proof From the proof of Theorem 8.16, $x_n y_{n-1} - y_n x_{n-1} = (-1)^n$. Thus, $a((-1)^n y_{n-1}) - b((-1)^n x_{n-1}) = (-1)^{2n} = 1$ and the result is established. ∎

Corollary *If* $\gcd(a, b) = 1$, $a > b > 0$, *and* $c_{n-1} = x_{n-1}/y_{n-1}$ *is the penultimate convergent of* a/b, *then the equation* $ax - by = c$ *has solution* $x = (-1)^n \cdot c \cdot y_{n-1}$, $y = (-1)^n \cdot c \cdot x_{n-1}$.

For example, consider the equation $230x - 163y = 1$. We have $\gcd(230, 163) = 1$, $230 > 163 > 0$, and from Table 8.6, the penultimate convergent of $\frac{230}{163} = [1, 2, 2, 3, 4, 2]$ is $c_5 = x_5/y_5 = 103/73$. From the corollary to Theorem 8.17, $x = (-1)^6 73 = 73$ and $y = (-1)^6 103 = 103$. Hence, $230x - 163y = 1$.

Consider the equation $41x - 17y = 13$. We have $\gcd(41, 17) = 1$, $41 > 17 > 0$, and $\frac{41}{17} = [2, 2, 2, 3]$. From Table 8.8 the penultimate convergent of $\frac{41}{17}$ is $c_3 = x_3/y_3 = 12/5$. Thus, a solution to the equation $41x - 17y = 13$ is given by $x = (-1)^4 \cdot 5 \cdot 13 = 65$ and $y = (-1)^4 \cdot 13 \cdot 12 = 156$.

With the few examples we have considered, you may have noticed that the odd convergents, c_{2k+1}, of a/b seem to be monotonically increasing and always less than a/b, while the even convergents, c_{2k}, of a/b seem to be monotonically decreasing and always greater than a/b. This indeed is the case as illustrated by the next result.

Theorem 8.18 *If* $a/b = [q_1, q_2, \ldots, q_n]$, $a > b > 0$, *and* $c_k = x_k/y_k$ *denotes the kth convergent of* a/b, *then* $c_1 < c_3 < c_5 < \cdots \leqslant a/b \leqslant \cdots c_4 < c_2 < c_0$.

Proof Using the notation of Theorem 8.14 and Theorem 8.15, we have

$$\begin{aligned} c_k - c_{k-2} &= \frac{x_k}{y_k} - \frac{x_{k-2}}{y_{k-2}} = \frac{x_k y_{k-2} - y_k x_{k-2}}{y_k y_{k-2}} \\ &= \frac{(x_{k-2} + x_{k-1} q_k) y_{k-2} - (y_{k-2} + y_{k-1} q_k) x_{k-2}}{y_k y_{k-2}} \\ &= \frac{q_k (x_{k-1} y_{k-2} - y_{k-1} x_{k-2})}{y_k y_{k-2}} = \frac{q_k (-1)^{k-1}}{y_k y_{k-2}}. \end{aligned}$$

Since q_i and y_i are positive for $1 \leqslant i \leqslant n$, if k is even, say $k = 2r$, then $c_{2r} - c_{2r-2} = q_{2r}(-1)^{2r-1}/y_{2r}y_{2r-2} < 0$. Hence, $c_{2r} < c_{2r-2}$ and the sequence of even convergents is decreasing. Similarly, if k is odd, say $k = 2r + 1$, then $c_{2r+1} - c_{2r-1} = q_{2r+1}(-1)^{2r}/y_{2r+1}y_{2r-1} > 0$. Hence, $c_{2r+1} > c_{2r-1}$ and the sequence of odd convergents is increasing. Consider the difference of two consecutive convergents. We have

$$c_k - c_{k-1} = \frac{x_k}{y_k} - \frac{x_{k-1}}{y_{k-1}} = \frac{x_k y_{k-1} - y_k x_{k-1}}{y_k y_{k-1}} = \frac{(-1)^k}{y_k y_{k-1}}.$$

If k is even and m is odd and less than k, then m (odd) $< k - 1$ (odd) $< k$ (even). Hence, $c_m \leqslant c_{k-1}$ and $c_k - c_{k-1} > 0$. Thus, $c_m \leqslant c_{k-1} < c_k$. If k is odd and m is even and greater than k, then k (odd) $< k - 1$ (even) $\leqslant m$ (even). Hence, $c_{k-1} \leqslant c_m$ and $c_k - c_{k-1} < 0$. Thus, $c_k < c_{k-1} \leqslant c_m$. In any case, the odd convergents are bounded above by all the even convergents and the even convergents are bounded below by all the odd convergents. Since the ultimate convergent $c_n = a/b$ is either the smallest even convergent or the largest odd convergent the result is established. ■

Exercises 8.4

1. Determine the rational number represented by [1, 2, 3, 2, 1].
2. Determine the rational number represented by [1, 2, 3, 4, 5, 6].
3. Show that if $x = [a_1, a_2, \ldots, a_n]$, then $1/x = [0, a_1, a_2, \ldots, a_n]$.
4. Determine the convergents of $\frac{177}{233}$.

5. Devise a formula to solve equations of the form $ax + by = c$.
6. Determine a necessary and sufficient condition that $[a_1, a_2, \ldots, a_n]$ be palindromic.
7. Given the continued fraction expansion $[a_1, a_2, \ldots, a_n]$, show that the convergents $c_k = x_k/y_k$ may be obtained by matrix multiplication. That is,

$$\begin{pmatrix} a_0 & 1 \\ 1 & 0 \end{pmatrix}\begin{pmatrix} a_1 & 1 \\ 1 & 0 \end{pmatrix}\cdots\begin{pmatrix} a_k & 1 \\ 1 & 0 \end{pmatrix} = \begin{pmatrix} x_k & x_{k-1} \\ y_k & y_{k-1} \end{pmatrix}, \text{ for } 1 \leqslant k \leqslant n.$$

8.5 Infinite continued fractions

An expression of the form

$$a_1 + \cfrac{1}{a_2 + \cfrac{1}{a_3 + \cfrac{1}{a_4 + \cdots}}},$$

where the a_i, for $i = 1, 2, \ldots,$ except possibly a_1 which may be negative, are positive real numbers, is called an infinite continued fraction and is denoted by $[a_1, a_2, a_3, \ldots]$. If the a_i, for $i \geqslant 1$, are required to be integers then the expression is called a simple infinite continued fraction. Whereas simple finite continued fractions represent rational numbers. simple infinite continued fractions represent irrational numbers. In particular, if c_n denotes the nth convergent of $[a_1, a_2, a_3, \ldots]$, we define the value of the simple infinite continued fraction $[a_1, a_2, a_3, \ldots]$ to be the real number $\lim_{n\to\infty} c_n$, whenever the limit exists. It can be shown that if the values of two simple infinite continued fractions $[a_1, a_2, a_3, \ldots]$ and $[b_1, b_2, b_3, \ldots]$ are equal then $a_i = b_i$, for $i \geqslant 1$.

Lambert used a continued fraction expansion for $\tan(x)$ to show that π is irrational. Since $\tan(x)$ is irrational if x is a nonzero rational number and $\tan(\pi/4) - 1$, it follows that π is an irrational number.

Recall, from mathematical analysis, that every bounded monotonic (either increasing or decreasing) sequence converges. The odd convergents $(c_1, c_3, c_5, \ldots)$ form an increasing sequence bounded above by c_2 and the even convergents $(c_2, c_3, c_6, \ldots)$ form a decreasing sequence bounded below by c_1, hence, both sequences converge. Let $\lim_{n\to\infty} c_{2n+1} = L$ and $\lim_{n\to\infty} c_{2n} = M$. From Theorem 8.14 and Theorem 8.15, it follows that

$$0 < |c_{2n+1} - c_n| = \left|\frac{x_{2n+1}}{y_{2n+1}} - \frac{x_{2n}}{y_{2n}}\right| = \left|\frac{x_{2n+1}y_{2n} - y_{2n+1}x_{2n}}{y_{2n+1}y_{2n}}\right|$$
$$= \left|\frac{(-1)^{2n+1}}{y_{2n+1}y_{2n}}\right| = \frac{1}{y_{2n+1}y_{2n}} \leqslant \frac{1}{(2n+1)(2n)}.$$

Since $\lim_{n\to\infty}|c_{2n+1} - c_{2n}| = 0$, $L = M$. If $L = M = \gamma$ is rational, say $\gamma = a/b$, then since $\lim_{n\to\infty} c_n = \gamma = a/b$, $|c_n - \gamma|$ can be made as small as we please. In addition,

$$0 < |c_n - \gamma| = \left|\frac{x_n}{y_n} - \frac{a}{b}\right| = \left|\frac{bx_n - ay_n}{by_n}\right|.$$

Therefore, let n be such that

$$0 < \left|\frac{bx_n - ay_n}{by_n}\right| < \frac{1}{by_n}.$$

Hence, $0 < |bx_n - ay_n| < 1$. However, since a, b, x_n, y_n are integers, this implies that there is an integer between 0 and 1, a contradiction. Therefore, we have established the following result.

Theorem 8.19 *A simple infinite continued fraction represents an irrational number.*

We carry our reasoning one step further to show that if $\gamma = [a_1, a_2, a_3, \ldots]$ and $c_n = x_n/y_n$ denotes the nth convergent of γ, then

$$0 < |\gamma - c_n| < |c_{n+1} - c_n| < \left|\frac{x_{n+1}}{y_{n+1}} - \frac{x_n}{y_n}\right| = \left|\frac{x_{n+1}y_n - x_n y_{n+1}}{y_{n+1}y_n}\right|$$
$$= \left|\frac{(-1)^{n+1}}{y_{n+1}y_n}\right| = \frac{1}{y_{n+1}y_n} < \frac{1}{y_n^2}.$$

The latter inequality follows from the nature of the y_i for $i > -1$, as noted in the proof of Theorem 8.11. Therefore,

$$\left|\gamma - \frac{x_n}{y_n}\right| < \frac{1}{y_n^2}$$

and we have established the following result.

Theorem 8.20 *Given any irrational number γ and positive integer n there is a rational number a/b such that $|\gamma - a/b| < 1/n$.*

In 1753, Robert Simpson derived the Fibonacci numbers as components of terms in successive convergents of the irrational number $(1 + \sqrt{5})/2$. In 1891, A. Hurwitz showed that if γ is irrational then there exist infinitely many rational numbers a/b such that $|\gamma - a/b| < 1/\sqrt{5}b^2$. In addition, $\sqrt{5}$ is best possible in the sense that given any real number $\alpha > \sqrt{5}$ there is an irrational number γ such that there exist only a finite number of rational numbers a/b with the property that $|\gamma - a/b| < 1/\alpha b^2$.

Given an irrational number γ, we may represent γ as a simple infinite continued fraction $[a_1, a_2, a_3, \ldots]$ in the following manner. Let $\gamma_1 = \gamma$ and $a_i = [\![\gamma_i]\!]$, where $\gamma_{i+1} = 1/(\gamma_i - a_i)$, for $i \geqslant 1$. Rewriting the last equation, we have $\gamma_i = a_i + 1/\gamma_{i+1}$, for $i \geqslant 1$. Hence,

$$\begin{aligned}
\gamma &= a_1 + \frac{1}{\gamma_2} \\
&= a_1 + \cfrac{1}{a_2 + \cfrac{1}{\gamma_3}} \\
&= a_1 + \cfrac{1}{a_2 + \cfrac{1}{a_3 + \cfrac{1}{\gamma_4}}} \\
&\cdots \\
&= a_1 + \cfrac{1}{a_2 + \cfrac{1}{a_3 + 1 \ddots + \cfrac{1}{a_n + \cfrac{1}{\gamma_{n+1}}}}} \\
&= [a_1, a_2, a_3, \ldots, a_n, \gamma_{n+1}].
\end{aligned}$$

In addition, γ equals the $(n+1)$st convergent c_{n+1}. That is,

$$\gamma = \frac{\gamma_{n+1}x_n + x_{n-1}}{\gamma_{n+1}y_n + y_{n-1}},$$

where, since γ is irrational, γ_i is irrational and greater than 1, for $i \geqslant 1$. Thus,

$$\begin{aligned}
0 \leqslant |\gamma - c_n| &= \left|\gamma - \frac{x_n}{y_n}\right| = \left|\frac{\gamma_{n+1}x_n + x_{n-1}}{\gamma_{n+1}y_n + y_{n-1}} - \frac{x_n}{y_n}\right| = \left|\frac{x_{n-1}y_n - y_{n-1}x_n}{y_n(\gamma_{n+1}y_n + y_{n-1})}\right| \\
&= \left|\frac{(-1)^{n-1}}{y_n(\gamma_{n+1}y_n + y_{n-1})}\right| < \frac{1}{y_n} < \frac{1}{n}.
\end{aligned}$$

Hence, $\lim_{n\to\infty} c_n = \gamma$. See [Niven, Zuckerman, and Montgomery] for the proof of uniquenes of the representation.

For example, consider the irrational number $\pi = 3.141\,59\ \ldots$. Let

Table 8.9.

i	-1	0	1	2	3	4	5
q_i			3	7	15	1	292
x_i	0	1	3	22	333	355	103 933
y_i	1	0	1	7	106	113	33 102
c_i			$\frac{3}{1}$	$\frac{22}{7}$	$\frac{333}{106}$	$\frac{355}{113}$	$\frac{103\,993}{33\,102}$

$\pi_1 = \pi$, so $a_1 = [\![\pi_1]\!] = [\![\pi]\!] = 3$, $\pi_2 = 1/(\pi - 3) = 1/0.141\,59\ldots = 7.062\,513\ldots$. Hence, $a_2 = [\![\pi_2]\!] = [\![7.062\,513\ldots]\!] = 7$. Further, $\pi_3 = 1/(\pi_2 - a_2) = 1/0.062\,513\ldots = 15.996\ldots$, hence, $a_3 = [\![\pi_3]\!] = [\![15.996\ldots]\!] = 15$. From Table 8.9, we find that π can be represented as $[3, 7, 15, 1, 292, 1, \ldots]$. Thus $3, \frac{22}{7}, \frac{333}{106}, \frac{355}{113}, \frac{103\,993}{33\,102}, \ldots$ are successive approximations to π. The continued fraction expansion for e appears in *Logometria*, a treatise written by Roger Cotes, Plumian Professor of Astronomy and Experimental Philosophy at Cambridge University, in 1714. It was from discussions with Cotes on continued fractions that Saunderson devised his practical algorithm (Theorem 2.12). Cotes's work so impressed Newton that upon his death, at age 34, Newton said, 'If he had lived we might have known something.'

We have $\sqrt{17} = 4.123\,10\ldots = [4, 8, 8, 8, \ldots]$, $\sqrt{23} = 4.795\,83\ldots = [4, 1, 3, 1, 8, 1, 3, 1, 8, \ldots]$, and $\sqrt{35} = 5.91607\ldots = [5, 1, 10, 1, 10, \ldots]$. The length of thc pcriod of $\sqrt{17}$ is 1, of $\sqrt{23}$ is 4, and of $\sqrt{35}$ is 2. In 1770, Lagrange showed that every expression of the form $(a + b\sqrt{c})/d$, where a, b, c, d are positive integers and c is nonsquare, has a periodic simple infinite continued fractional representation.

In particular, if α has a periodic continued fraction expansion $[a_1, a_2, \ldots, \overline{a_k, a_{k+1}, \ldots, a_{k+r}}]$, where the bar indicates that the sequence $a_k, a_{k+1}, \ldots, a_{k+r}$ repeats indefinitely, then it can be shown, see [Olds], that there exist positive integers a, b, c, d, with c nonsquare, such that $\alpha = (a + b\sqrt{c})/d$. For the sufficiency, let $\beta = [\overline{a_k, \ldots, a_{k+r}}]$, then β is an infinite continued fraction and, thus, from Theorem 8.19, it is irrational. From the proof of Theorem 8.19, $\beta = (\beta u_r + u_{r+1})/\beta v_r + v_{r-1})$, where u_{r-1}/v_{r-1} and u_r/v_r are the last two convergents of β. Hence, $\beta^2 v_r + \beta(v_{r-1} - u_r) - u_{r-1} = 0$. Thus $\beta = (r + s\sqrt{t})/w$, where r, s, t, w are positive integers and t is nonsquare. In addition, $\alpha = [a_1, \ldots, a_k, \beta]$, hence, by rationalizing the denominator we obtain

Table 8.10.

	$\sqrt{26} = [5, \overline{10}]$
$\sqrt{2} = [1, \overline{2}]$	$\sqrt{27} = [5, \overline{5, 10}]$
$\sqrt{3} = [1, \overline{1, 2}]$	$\sqrt{28} = [5, \overline{3, 2, 3, 10}]$
	$\sqrt{29} = [5, \overline{2, 1, 1, 2, 10}]$
$\sqrt{5} = [2, \overline{4}]$	$\sqrt{30} = [5, \overline{2, 10}]$
$\sqrt{6} = [2, \overline{2, 4}]$	$\sqrt{31} = [5, \overline{1, 1, 3, 5, 3, 1, 1, 10}]$
$\sqrt{7} = [2, \overline{1, 1, 1, 4}]$	$\sqrt{32} = [5, \overline{1, 1, 1, 10}]$
$\sqrt{8} = [2, \overline{1, 4}]$	$\sqrt{33} = [5, \overline{1, 2, 1, 10}]$
	$\sqrt{34} = [5, \overline{1, 4, 1, 10}]$
$\sqrt{10} = [3, \overline{6}]$	$\sqrt{35} = [5, \overline{1, 10}]$
$\sqrt{11} = [3, \overline{3, 6}]$	
$\sqrt{12} = [3, \overline{2, 6}]$	$\sqrt{37} = [6, \overline{12}]$
$\sqrt{13} = [3, \overline{1, 1, 1, 6}]$	$\sqrt{38} = [6, \overline{6, 12}]$
$\sqrt{14} = [3, \overline{1, 2, 1, 6}]$	$\sqrt{39} = [6, \overline{4, 12}]$
$\sqrt{15} = [3, \overline{1, 6}]$	$\sqrt{40} = [6, \overline{3, 12}]$
	$\sqrt{41} = [6, \overline{2, 2, 12}]$
$\sqrt{17} = [4, \overline{8}]$	$\sqrt{42} = [6, \overline{2, 12}]$
$\sqrt{18} = [4, \overline{4, 8}]$	$\sqrt{43} = [6, \overline{1, 1, 3, 1, 5, 1, 3, 1, 1, 12}]$
$\sqrt{19} = [4, \overline{2, 1, 3, 1, 2, 8}]$	$\sqrt{44} = [6, \overline{1, 1, 1, 2, 1, 1, 1, 12}]$
$\sqrt{20} = [4, \overline{2, 8}]$	$\sqrt{45} = [6, \overline{1, 2, 2, 2, 1, 12}]$
$\sqrt{21} = [4, \overline{1, 3, 1, 8}]$	$\sqrt{46} = [6, \overline{1, 3, 1, 1, 2, 6, 2, 1, 1, 3, 1, 12}]$
$\sqrt{22} = [4, \overline{1, 2, 4, 2, 1, 8}]$	$\sqrt{47} = [6, \overline{1, 5, 1, 12}]$
$\sqrt{23} = [4, \overline{1, 3, 1, 8}]$	$\sqrt{48} = [6, \overline{1, 12}]$
$\sqrt{24} = [4, \overline{1, 8}]$	
	$\sqrt{50} = [7, \overline{14}]$

$$\alpha = \frac{\beta x_k + x_{k-1}}{\beta y_k + y_{k-1}} = \frac{\left(\dfrac{r + s\sqrt{t}}{w}\right) x_k + x_{k-1}}{\left(\dfrac{r + s\sqrt{t}}{w}\right) y_k + y_{k-1}} = \frac{a + b\sqrt{c}}{d}.$$

In the above example β was a special type of infinite continued fraction called purely periodic. More precisely, an infinite continued fraction is called purely periodic if it is of the form $[\overline{a_k, \ldots, a_n}]$. It can be shown that if $-1 < (a - b\sqrt{c})/d < 0$, the infinite continued fraction for $(a + b\sqrt{c}/d$ is purely periodic, for a proof see [Niven, Zuckerman, and Montgomery]. Let $\sqrt{n} = [b_0, b_1, \ldots, b_n, \ldots]$, then $b_0 = [\![\sqrt{n}]\!]$. If $\alpha = [\![\sqrt{n}]\!] + \sqrt{n}$, then the conjugate of α, namely $[\![\sqrt{n}]\!] - \sqrt{n}$, is such that $-1 < [\![\sqrt{n}]\!] - \sqrt{n} < 0$. Hence, α is purely periodic and $\alpha = [\![\sqrt{n}]\!] + \sqrt{n} = [\overline{2b_0, b_1, \ldots, b_n}]$.

Subtracting $[\![\sqrt{n}]\!]$ from both sides of the equation we find that $\sqrt{n} = [b_0, \overline{b_1, \ldots, b_n, 2b_0}]$.

If $\alpha = [\overline{a_1, a_2, \ldots, a_{n-1}, a_n}]$ is purely periodic, then the continued fraction expansion of the negative reciprocal of the conjugate of α, $-1/\overline{\alpha}$, is given by $[\overline{a_n, a_{n-1}, \ldots, a_2, a_1}]$. In addition, if $\alpha > 1$, then $1/\alpha = [0, \overline{a_1, a_2, \ldots, a_{n-1}, a_n}]$. Hence, if n is positive and nonsquare, then the infinite continued fraction expansion of $\sqrt{n}$ is given by $[[\![\sqrt{n}]\!], \overline{a_1, a_2, a_3, \ldots, a_3, a_2, a_1, 2[\![\sqrt{n}]\!]}]$. Periodic infinite continued fractional expansions for square roots of nonsquare integers n, for $1 < n < 50$, are illustrated in Table 8.10.

Let us determine a representation for the infinite periodic continued fraction

$$[\overline{1, 3, 5}] = 1 + \cfrac{1}{3 + \cfrac{1}{5 + \cfrac{1}{[\overline{1, 3, 5}]}}}.$$

If $x = [\overline{1, 3, 5}]$, then

$$x = 1 + \cfrac{1}{3 + \cfrac{1}{5 + x}}.$$

Hence, $8x^2 - 9x - 2 = 0$. Using the quadratic formula, we find that $x = (9 + \sqrt{145})/16$.

If $x = [a, b, a, b, a, b, \ldots]$, $a = bc$, where c is an integer, then $x = a \pm \sqrt{a^2 + 4c}/2$. Hence, $[1, 1, 1, 1 \ldots] = (\sqrt{5} + 1)/2 = \tau$, and $[2, 1, 2, 1, 2, 1, \ldots] = \sqrt{3} + 1$. In addition,

$$\sqrt{2} = 1 + (\sqrt{2} - 1) = 1 + \frac{1}{\sqrt{2} + 1} = 1 + \frac{1}{2 + (\sqrt{2} - 1)},$$

hence, $\sqrt{2} = [1, 2, 2, 2, \ldots]$. Bombelli showed that $\sqrt{a^2 + b} = [a, 2a/b, 2a/b, \ldots]$, which leads to a number of straightforward continued fraction representations for square roots of integers. The first publication of Evariste Galois, in 1828, dealt with periodic continued fractions. Galois, who died in a duel at the age of 20, had an exceptionally brilliant mathematical mind. His work, as a teenager, founded the theory of solvability of algebraic equations by radicals.

We state the following important result without proof. The interested reader can find the proof in [Robbins].

Theorem 8.21 *If* $\gcd(a, b) = 1$, $b > 0$ *and* γ *is irrational with* $|\gamma - a/b| < 1/2b^2$, *then* a/b *is a convergent of* γ.

Table 8.11.

i	-1	0	1	2	3	4	5	6	7
a_i				1	2	2	2	2	2
x_i	0	1	1	3	7	17	41	99	239
y_i	1	0	1	2	5	12	29	70	169

Let (a, b) be a solution to $x^2 - dy^2 = 1$. Since

$$a - b\sqrt{d} = \frac{1}{a + b\sqrt{d}} \text{ implies that } \frac{a}{b} - \sqrt{d} = \frac{1}{b(a + b\sqrt{d})},$$

it follows that if $a > b\sqrt{d}$, then $a + b\sqrt{d} > 2b\sqrt{d}$. Hence, $0 < a/b - \sqrt{d} < 1/2b^2\sqrt{d} < 1/2d^2$. Therefore, from Theorem 8.21, it follows that if (a, b) is a solution to $x^2 - dy^2 = 1$ then it is one of the convergents of $\sqrt{d}$. On the other hand, however, not every convergent of $\sqrt{d}$ is a solution to $x^2 - dy^2 = 1$. In particular, any positive solution $x = x_k$, $y = y_k$ to $x^2 - dy^2 = 1$ has the property that $c_k = x_k/y_k$ is a convergent of $\sqrt{d}$. The next result outlines the method devised independently by Bhaskara and Brouncker to solve Pell's equations. It is offered without proof: for a proof see [Robbins]. In 1907, this result was generalized by Major Percy MacMahon who showed that integral solutions to $x^n - dy^n = 1$, where n is a positive integer, can be obtained from the convergents of $\sqrt[n]{a}$.

Theorem 8.22 (Bhaskara–Brouncker) *Let $c_k = x_k/y_k$ denote the kth convergent of $\sqrt{d}$ and n the length of the period of $\sqrt{d}$. If n is even, every positive solution to $x^2 - dy^2 = 1$ is given by $x = x_{kn-1}$, $y = y_{kn-1}$, for $k \geqslant 1$. If n is odd, every positive solution to $x^2 - dy^2 = 1$ is given by $x = x_{2kn-1}$, $y = y_{2kn-1}$ for $k \geqslant 1$.*

For example, in order to find solutions to the Pellian equation $x^2 - 2y^2 = 1$, where $\sqrt{2} = [1, 2, 2, \ldots]$, we construct Table 8.11. Hence, (3, 2), (17, 12), (99, 70), ..., are solutions (x, y) to $x^2 - 2y^2 = 1$, and (7, 5), (41, 29), (239, 169), ..., are solutions to $x^2 - 2y^2 = -1$.

Exercises 8.5

1. Use the process outlined in the section to determine the continued fraction expansions for
   (a) $\sqrt{3}$ (use 1.732 050 81),
   (b) $\sqrt{5}$ (use 2.236 067 98),

(c) $\sqrt{7}$ (use 2.645 751 31),
(d) $\sqrt{10}$ (use 3.162 277 66).

2. Determine the first 12 terms of the continued fraction for e. [Cotes]
3. Determine the first 12 terms of the continued fraction for $(e+1)/(e-1)$.
4. Determine five solutions of the equation $x^2 - 3y^2 = \pm 1$.
5. If n is a positive integer, then determine the number represented by the periodic infinite continued fraction $[\overline{n}]$.
6. A more generalized form of continued fraction was used by the ancients to approximate square roots. In particular,

$$\sqrt{a^2+b} = a + \cfrac{b}{2a + \cfrac{b}{2a + \cfrac{b}{2a + \cdots}}}.$$

Use the formula to approximate $\sqrt{13}$ and $\sqrt{18}$.

7. Suppose that $a/b < \gamma < c/d$, where γ is irrational, a, b, c, d are positive, and $bc - ad = 1$. Prove that either a/b or c/d is a convergent of γ.

8.6 *p*-Adic analysis

A field is a nonempty set F with two operations, called addition and multiplication, that is distributive, an Abelian group under addition with identity 0, and the nonzero elements of F form an Abelian group under multiplication. A function υ from a field F to the nonnegative real numbers is called a valuation or norm on F if for all x and y in F the following properties hold:

(1) $\upsilon(x) \geqslant 0$, and $\upsilon(x) = 0$ if and only if $x = 0$,
(2) $\upsilon(x, y) = \upsilon(x)\upsilon(y)$, and
(3) $\upsilon(x+y) \leqslant \upsilon(x) + \upsilon(y)$.

From the first properties it follows that if e denotes the multiplicative identity of the field F then $\upsilon(e) = 1$ and $\upsilon(-e) = 1$. Hence, for any element a in F, $\upsilon(-a) = \upsilon(a)$. The third condition is just the triangle inequality. Two examples of valuations over the reals are the trivial valuation given by

$$|x|_0 = \begin{cases} 1, & \text{if } x \neq 0 \\ 0, & \text{otherwise,} \end{cases}$$

and the familiar absolute value function

$$|x| = \begin{cases} x, & \text{if } x \geqslant 0 \\ -x, & \text{if } x < 0. \end{cases}$$

A valuation is called non-Archimedean if it satisfies the ultrametric inequality,

(4) $v(x + y) \leqslant \max\{v(x), v(y)\}$, for all x and y in F,

otherwise it is called Archimedean. The ultrametric inequality implies the triangle inequality. The trivial metric is an example of a non-Archimedean valuation and the absolute value is an example of an Archimedean valuation.

Given any prime p, every rational number q can be written uniquely as $(a/b)p^\alpha$, where $\gcd(a, b) = 1$, and $b > 0$. That is, $p^\alpha \| q$. The p-adic valuation, denoted by $|\cdot|_p$, is defined over the rational numbers in the following manner:

$$|q|_p = \begin{cases} p^{-\alpha} & \text{if } p^\alpha \| q, \\ 0 & \text{if } q = 0. \end{cases}$$

For example, since $450 = 2 \cdot 3^2 \cdot 5^2$, $|450|_2 = 1/2$, $|450|_3 = 1/3^2$, $|450|_5 = 1/5^2$, and $|450|_p = 1$, for any other prime p. We leave the proof that $|\cdot|_p$ is a valuation over the rationals as an exercise. Properties of p-adic valuations were first investigated by Kurt Hensel in 1908.

There are a number of interesting p-adic properties. For example, $|q|_p \leqslant 1$ for any integer q and any prime p. For any p-adic valuation, $\prod_p |q|_p = 1/|q|$, where p runs through all primes and q is nonzero. If r and s are integers, then r divides s if and only if $|s|_p \leqslant |r|_p$ for every prime p. In 1918, A. Ostrowski showed that every nontrivial valuation in the rational numbers is equivalent to either the absolute value or a p-adic valuation.

A distance function or metric d is a nonnegative real valued function defined on ordered pairs of elements of a set such that

(1) $d(x, y) \geqslant 0$, and $d(x, y) = 0$ if and only if $x = y$,
(2) $d(x, y) = d(y, x)$,
(3) $d(x, y) \leqslant d(x, z) + d(z, y)$.

The third condition is the familiar triangle inequality. Each valuation on a field generates a metric or distance function, namely, $d(x, y) = v(x - y)$. The ordinary metric in Euclidean space is generated by the absolute value. The trivial valuation gives rise to the trivial metric $d_0(x, y)$ which equals 1 if $x \neq y$ and equals 0 otherwise. Non-Archimedean metrics can generate strange properties. If d is the metric generated by the non-Archimedean valuation υ, then, since $x - y = (x - z) + (z - y)$, $d(x, y) = \upsilon(x - y) = \upsilon((x - z) + (z - y)) \leqslant \max\{\upsilon((x - z), \upsilon(z - y)\} = \max\{d(x, z), d(z, y)\}$.

Example 8.3 Let d be the metric determined by the non-Archimedean valuation υ. Consider three points x, y, and 0 where, without loss of generality, we have let one of the points be the origin. The three distances determined by the points are $d(x, 0) = \upsilon(x)$, $d(y, 0) = \upsilon(y)$, and $d(x, y) = \upsilon(x - y)$. We have $d(x, y) = \upsilon(x - y) = \upsilon(x + (-y)) \leqslant \max\{\upsilon(x), \upsilon(y)\}$. If $\upsilon(x) \neq \upsilon(y)$, say $\upsilon(x) > \upsilon(y)$, then $d(x, y) \leqslant \upsilon(x)$. However, $\upsilon(x) = \upsilon((x - y) + y) \leqslant \max\{\upsilon(x - y), \upsilon(y)\}$ and since $\upsilon(x) > \upsilon(y)$, $\upsilon(x) \leqslant \upsilon(x - y) = d(x, y)$, implying that $\upsilon(x) = d(x, y)$. Thus in a non-Archimedean geometry, $\upsilon(x, y) = \max\{\upsilon(x), \upsilon(y)\}$ whenever $\upsilon(x) \neq \upsilon(y)$. Therefore, every triangle in a non-Archimedean geometry has the property that its two longest sides are of equal length.

We say that the sequence $a_1, a_2, a_3, \ldots$ converges p-adically to the real number L, if the sequence $|a_1 - L|_p, |a_2 - L|_p, \ldots$ converges in the usual sense. That is, given any real positive number ϵ there is a natural number N such that $|a_n - L|_p < \epsilon$ whenever $n > N$. Similarly, we say that S is the sum of the series $\sum_{n=1}^{\infty} a_n$ if and only if the sequence of partial sums $s_1, s_2, \ldots$, where $s_k = \sum_{i=1}^{k} a_i$, for $k \geqslant 1$, converges to S. It follows that if p is prime the sequence $p, p^2, p^3, \ldots$ converges to 0 p-adically. Another interesting consequence of the definition of p-adic convergence is that, 7-adically speaking, $-1 = 6 + 6 \cdot 7 + 6 \cdot 7^2 + 6 \cdot 7^3 + 6 \cdot 7^4 + \cdots$. To see why this is the case, add 1 to both sides of the equation and continue to combine terms to obtain

$$\begin{aligned}
0 &= 7 + 6 \cdot 7 + 6 \cdot 7^2 + 6 \cdot 7^3 + 6 \cdot 7^4 + \cdots \\
&= 0 + 7 \cdot 7 + 6 \cdot 7^2 + 6 \cdot 7^3 + 6 \cdot 7^4 + \cdots \\
&= 0 + 0 \quad + 7 \cdot 7^2 + 6 \cdot 7^3 + 6 \cdot 7^4 + \cdots \\
&= 0 + 0 \quad + 0 \quad + 7 \cdot 7^3 + 6 \cdot 7^4 + \cdots \\
&= 0 + 0 \quad + 0 \quad + 0 \quad + 7 \cdot 7^4 + \cdots \\
&= 0 + 0 \quad + 0 \quad + 0 \quad + 0 \quad + \cdots \\
&= \cdots .
\end{aligned}$$

In addition, 5-adically speaking, to evaluate $x = 2 + 5 + 5^2 + 5^3 + 5^4 + 5^5 + \cdots$ we multiply both sides of the equation by 4 and combine terms to obtain

$$\begin{aligned}4x &= 8 + 4\cdot 5 + 4\cdot 5^2 + 4\cdot 5^3 + 4\cdot 5^4 + \cdots\\ &= 3 + 5\cdot 5 + 4\cdot 5^2 + 4\cdot 5^3 + 4\cdot 5^4 + \cdots\\ &= 3 + 0 \quad + 5\cdot 5^2 + 4\cdot 5^3 + 4\cdot 5^4 + \cdots\\ &= 3 + 0 \quad + 0 \quad + 5\cdot 5^3 + 4\cdot 5^4 + \cdots\\ &= 3 + 0 \quad + 0 \quad + 0 \quad + 0 \quad + \cdots\\ &= \cdots.\\ &= 3.\end{aligned}$$

Hence, $x = \frac{3}{4}$. More formally, a sequence (a_n) of rational numbers is called a p-adic Cauchy sequence if for every positive number ϵ there is an integer N such that whenever m and n are greater than N, $|a_n - a_m|_p < \epsilon$. Two p-adic Cauchy sequences (a_n) and (b_n) are called equivalent if $\lim_{n\to\infty}|a_n - b_n|_p = 0$. This is an equivalence relation and, hence, partitions the p-adic Cauchy sequences into equivalence classes, denoted by Q_p. If we define the operations of addition and multiplication on Q_p to be componentwise addition and multiplication, that is $(a_n) + (b_n) = (a_n + b_n)$ and $(a_n)\cdot(b_n) = (a_n \cdot b_n)$, then Q_p becomes a field. Any nonzero element r of Q_p can be represented uniquely as $r = p^n(a_0, a_0 + a_1 p, a_0 + a_1 p + a_2 p^2, \ldots)$, where n and a_i are integers such that $0 \leqslant a_0 < p$ and $a_0 \neq 1$, for $i = 1, 2, 3 \ldots$. Equivalently, we could take the sequence of partial sums above and represent r in the form of a series where, in that case, $r = a_0 p^n + a_1 p^{n+1} + a_2 p^{n+2} + \cdots$. For example,

$$1, 6, 31, 156, \ldots = 1 + 1\cdot 5 + 1\cdot 5^2 + 1\cdot 5^3 + \cdots,$$
$$3, 3, 3, \ldots = 3 + 0\cdot 5 + 0\cdot 5^2 + 0\cdot 5^3 + \cdots,$$

and

$$75, 275, 1525, 7775, \ldots = 5^2(3, 1 + 2\cdot 5, 1 + 2\cdot 5 + 2\cdot 5^2, 1 + 2\cdot 5 + 2\cdot 5^2 + 2\cdot 5^3 + \ldots).$$

p-Adic analysis is a useful tool. However, most of its important applications are outside our present scope. For its use in establishing polynomial congruences see [Edgar]. To see how it may be applied to the analysis of binary quadratic forms see [Cassels].

Exercises 8.6

1. Show that the absolute value is an Archimedean valuation.
2. If v is a valuation on the field F show that

(a) $v(e) = 1$,
(b) $v(-e) = 1$, and
(c) $v(-a) = v(a)$ for any a in F.

3. Prove that the trivial valuation satisfies the three properties of a valuation.
4. Show that the ultrametric inequality implies the triangle inequality.
5. Determine $|600|_p$, for any prime p.
6. Determine $|q^k|_p$, where p and q are prime and k is an integer.
7. Prove that the p-adic valuation satisfies the ultrametric inequality.
8. If p is prime, show that the p-adic valuation satisfies the three conditions for a valuation.
9. Prove that if r and s are rational numbers, then r divides s if and only if $|s|_p \leqslant |r|_p$ for every prime p.
10. Determine a 2-adic value for $1 + 2 + 2^2 + 2^3 + \cdots$.
11. Determine a 3-adic value for $5 + 2 \cdot 3 + 2 \cdot 3^2 + 2 \cdot 3^3 + \cdots$.
12. Determine a 7-adic series expansion for $\frac{5}{6}$.
13. Show that $-1 = 6 + 6 \cdot 7 + 6 \cdot 7^2 + 6 \cdot 7^3 + 6 \cdot 7^4 + \cdots$.
14. Find the p-adic distance between 48 and 36 for any prime p.
15. In Q_7 determine the first four terms of the series represented by (3, 31, 227, 1599, ...).
16. In Q_7 determine the first four terms of the sequence represented by $2 \cdot 7^2 + 2 \cdot 7^3 + 2 \cdot 7^4 + \cdots$.
17. Define the unit disk U in the 2-dimensional Cartesian plane to be the set of all points where distance from the origin is at most one. Describe the unit disk geometrically if the distance from $P = (x_1, y_1)$ to $Q = (x_2, y_2)$ is given by
    (a) the trivial metric,
    (b) $d(P, Q) = \sqrt{(x_2 - x_1)^2 + (y_2 - y_1)^2}$,
    (c) $d(P, Q) = |x_1 - x_2| + |y_1 - y_2|$,
    (d) $d(P, Q) = \max\{|x_1 - x_2|, |y_1 - y_2|\}$,
    (e) a p-adic valuation on points whose coordinates are both rational.
18. Show that if v is a non-Archimedean valuation on the field F, then every point of $D(a, r) = \{x \text{ in } F\colon v(x - a) < r)$, the disk centered at a with radius r, can be considered as being at the center.

8.7 Supplementary exercises

1. Show that if n can be represented as the sum of two squares then n^2 can be represented as the sum of two squares.
2. Express $36^2 + 37^2 + 38^2 + 39^2 + 40^2$ as the sum of four squares.

3. Express $55^2 + 56^2 + 57^2 + 58^2 + 59^2 + 60^2$ as the sum of five squares.
4. From Theorem 8.9, determine $F(100)$, $F(200)$, and $F(250)$.
5. Can 31747100 be represented as a sum of two squares.
6. Taking signs and order into consideration, how many ways can 164 be represented as the sum of two squares.
7. Taking signs and order into consideration, how many ways can 2011 be represented as the sum of two squares.
8. Can 1001 be expressed as the sum of three squares?
9. Use Jacobi's theorem to determine $f(42)$, $f(2861)$, and $f(3003)$.
10. In how many ways can 4, 16, 195, and 1386 be written as a sum of four squares taking order and signs into consideration.
11. Express 21, 37, and 85 as the sum of five odd squares.
12. Express 3 as the sum of five cubes.
13. Express 4 as the sum of five cubes.
14. Determine a lower bound for $g(11)$.
15. Determine lower and upper bounds for $G(11)$.
16. Find x and y so as to represent 31, 61, and 79 in the form $x^2 + 3y^2$.
17. Find x and y so as to represent 41, 72, and 113 in the form $x^2 + 2y^2$.
18. Determine the rational number represented by [1, 1, 1, 1, 1, 1].
19. Determine the rational number represented by the finite continued fraction consisting of n ones.
20. Determine the rational number represented by [1, 5, 7, 2, 4, 1].
21. Express $\frac{356}{301}$ as a finite simple continued fraction.
22. Show that $[1, 2, 3, 4, 2, 3] = [1, 2, 3, 4, 2, 2, 1]$.
23. Determine the convergents of $\frac{177}{233}$.
24. Use the penultimate convergent in the answer to Exercise 23 to find a solution to $177x - 233y = 1$.
25. Find a solution to $177x - 233y = 3$.
26. Find a representation for the infinite continued fraction $[\overline{3, 2, 1}]$.
27. Use the decimal expansion of $\sqrt{53}$ to generate a solution to the Pell equation $x^2 - 53y^2 = -1$.
28. Use the decimal expansion of $\sqrt{51}$ to generate two solutions to the Pell equation $x^2 - 51y^2 = 1$.
29. Given that $x = 8$ and $y = 3$ is a solution to the Pell equation $x^2 - 7y^2 = 1$, generate a sequence of rational numbers of length 5 converging to $\sqrt{7}$.
30. Use the continued fraction expansion for $\sqrt{11}$ to obtain a fraction that differs from $\sqrt{11}$ by less than 0.0001.

9

Partitions

Say, is there Beauty yet to find?
And Certainty? and Quiet kind?
Deep meadows yet, for to forget
The lies, and truths, and pain? . . . oh! yet
Stands the Church clock at ten to three?
And is there honey still for tea?

Rupert Brooke

9.1 Generating functions

Given a sequence $a_0, a_1, a_2, \ldots$ of integers, the expression $G(x) = a_0 + a_1x + a_2x^2 + \cdots$ is called the generating function for the sequence. More generally, if $f(n)$ is an integral valued function defined on the nonnegative integers, then the generating function for $f(n)$ is given by $G(x) = \sum_{n=0}^{\infty} f(n)x^n$. In this chapter, our main concern is with the algebraic manipulation of the coefficients of generating functions. We are not interested in the convergence or divergence of generating functions considered as infinite series.

Generating functions were introduced in 1748 by Euler in his *Introductio in analysin infinitorum*. He used generating functions as a tool to discover a number of interesting properties concerning partitions. Several straightforward generating functions for familiar sequences can be derived by simple polynomial division. The generating function for the sequence 1, 1, 1, 1, . . . or equivalently for the constant function $f(n) = 1$, for n a positive integer, is given by $1/(1 - x)$. Since $1 + x + 2x^2 + 3x^3 + 4x^4 + \cdots = 1/(1 - x)^2$, the sequence of natural numbers is generated by $1/(1 - x)^2$. The sequence of triangular numbers is generated by $1/(1 - x)^3 = 1 + 3x + 6x^2 + 10x^3 + 15x^4 + 21x^5 + \cdots$. The sequence of even positive integers is generated by $1/(1 - 2x)$.

Suppose $G(x) = a_0 + a_1x + a_2x^2 + \cdots$ represents the generating function for the Fibonacci sequence 1, 1, 2, 3, 5, 8, $\ldots$, u_n, $\ldots$, where $u_0 = u_1 = 1$ and $u_n = u_{n-1} + u_{n-2}$, for $n \geqslant 3$. Hence, $xG(x) = u_0x + u_1x^2 + u_2x^3 + \cdots$, and $x^2G(x) = u_0x^2 + u_1x^3 + u_2x^4 + \cdots$. Thus, $G(x) - xG(x) - x^2G(x) = u_0 + (u_1 - u_0)x + (u_2 - u_1 - u_0)x^2 + (u_3 - u_2 - u_1)x^3 + \cdots + (u_n - u_{n-1} - u_{n-2})x^n + \cdots = 1 + 0 \cdot x + 0 \cdot x^2 +$

$0 \cdot x^3 + \cdots + 0 \cdot x^n + \cdots = 1$. Therefore, $G(x) = 1/(1 - x - x^2)$ is the generating function for the Fibonacci sequence.

If $G(x) = a_0 + a_1x + a_2x^2 + \cdots$ represents the generating function for the sequence 0, 1, 5, 18, 55, ..., a_n, ..., where $a_n = 5a_{n-1} - 7a_{n-2}$, then $G(x) - 5xG(x) + 7x^2G(x) = a_0 + (a_1 - 5a_0)x + (a_2 - 5a_1 + 7a_0)x^2 + \cdots + (a_n - 5a_{n-1} + 7a_{n-2})x^n + \cdots = x$. Hence, $G(x) = x/(1 - 5x + 7x^2)$.

Many other number theoretic functions we have encountered have nontrivial generating functions. In a paper dated 1747, but published posthumously, Euler noted that the generating function for $\sigma(n)$ is given by

$$\sum_{n=1}^{\infty} \frac{nx^n}{1 - x^n}.$$

In 1771, Johann Lambert discovered that the generating function for $\tau(n)$ is

$$\sum_{n=1}^{\infty} \frac{x^n}{1 - x^n}.$$

Exercises 9.1

1. Identify the sequence for which $1/(1 - x)^4$ is the generating function.
2. Identify the sequence for which $1/(1 - x)^5$ is the generating function.
3. Identify the sequence for which $1/(1 - x)^n$ is the generating function.
4. Describe the sequence for which $x/(1 - x)^4$ is the generating function.
5. Describe the sequence for which $x/(1 - x)^5$ is the generating function.
6. Describe the sequence for which $x/(1 - x)^n$ is the generating function.
7. Describe the sequence for which $x^2/(1 - x)^2$ is the generating function.
8. Identify the sequence for which $(1 + x)/(1 - x)^2$ is the generating function.
9. Identify the sequence for which $(x + x^2)/(1 - x)^3$ is the generating function.
10. Identify the sequence for which $x(x^2 + 4x + 1)/(1 - x)^4$ is the generating function.
11. Determine the generating function for the sequence of fourth powers of nonnegative integers 0, 1, 16, 81, 256, 625, 1296, 2401, ...
12. Determine the generating function for σ_k, the sum of the kth powers of the divisors of n.

13. Determine a generating function for the Lucas sequence 1, 3, 4, 7, 11, ...
14. Determine a generating function for the difference equation $a_n = 3a_{n-1} - 7a_{n-2}$, where $a_0 = 0$ and $a_1 = 1$.

9.2 Partitions

By a partition of a positive integer n we mean an expression of n as a sum of positive integers. For any positive integer n, there are 2^{n-1} ordered partitions of n. Consider a linear array of n ones. In each of the $n-1$ spaces between two of the ones, we may or may not put a slash. From the multiplication principle, there are 2^{n-1} choices for all the slashes and each choice generates an ordered partition of n. For example, if $n = 7$,

$$1\ 1\ /\ 1\ 1\ /\ 1\ /\ 1\ 1$$

represents the partition $2 + 2 + 1 + 2$, and

$$1\ /\ 1\ /\ 1\ /\ 1\ 1\ 1\ 1$$

represents the partition $1 + 1 + 1 + 4$. Consider the representation of a partition of n using n ones and $k+1$ slashes, where two slashes are external and the remaining $k-1$ are internal. For example, / 1 1 1 / 1 / 1 1 / represents the partition $3 + 1 + 2$ of 6. Since there are

$$\binom{n-1}{k-1}$$

ways of placing the $k-1$ slashes in the $n-1$ slots between the ones, the number of ordered partitions of the positive integer n into exactly k parts equals

$$\binom{n-1}{k-1}.$$

Summing over all possible cases, we obtain

$$\sum_{k=1}^{n}\binom{n-1}{k-1} = 2^{n-1}.$$

For the remainder of the chapter, we restrict ourselves to partitions of the positive integer n where the order of the summands is ignored and repetitions are allowed. That is, we consider only the partitions of n which are expressions of n as a sum of positive integers in descending order. We denote the number of such partitions by $p(n)$. For convenience, we set $p(0) = 1$ and use the convention that if $n = x_1 + x_2 + \cdots + x_k$ represents a partition of n, the terms are written in descending order, $x_1 \geqslant$

Table 9.1.

n	$p(n)$
1	1
2	2
3	3
4	5
5	7
6	11
7	15

$x_2 \geqslant \cdots \geqslant x_k \geqslant 1$. Values of $p(n)$, for $1 \leqslant n \leqslant 7$, are given in Table 9.1. For example, the partitions of 1, 2, 3, 4, 5, 6 and 7 are given by

1	2	3	4	5	6	7
	1+1	2+1	3+1	4+1	5+1	6+1
		1+1+1	2+2	3+2	4+2	5+2
			2+1+1	3+1+1	4+1+1	5+1+1
			1+1+1+1	2+2+1	3+3	4+3
				2+1+1+1	3+2+1	4+2+1
				1+1+1+1+1	3+1+1+1	4+1+1+1
					2+2+2	3+3+1
					2+2+1+1	3+2+2
					2+1+1+1+1	3+2+1+1
					1+1+1+1+1+1	3+1+1+1+1
						2+2+2+1
						2+2+1+1+1
						2+1+1+1+1+1
						1+1+1+1+1+1+1

The origin of partition theory can be traced back to 1669 when Gottfried Leibniz wrote Johann Bernoulli asking him if he had ever considered determining the number of ways a given positive integer may be separated into parts. Leibniz commented that the problem seemed difficult but important. In 1740, Philipp Naudé, a Berlin mathematician originally from Metz, France, proposed the following two questions to Euler.

(1) Find the number of ways a number is a sum of a given number of distinct parts.
(2) Find the number of ways a number is a sum of a given number of equal or distinct parts.

Euler realized that the coefficient of $x^n z^m$ in the expression $(1 + xz)(1 + x^2z)(1 + x^3z)(1 + x^4z) \cdots$ represented the number of ways n can be written as a sum of m distinct positive integers. For example, the coefficient of x^9z^3 is 3 and it results from summing the terms $x^6z \cdot x^2z \cdot$

x^1z, $x^5z \cdot x^3z \cdot x^1z$, and $x^4z \cdot x^3z \cdot x^2z$, that is, the terms corresponding respectively to the partitions $6+2+1$, $5+3+1$, and $4+3+2$ of 9. If we let $z=1$ in the expression, we find that the coefficient of x^n in $(1+x)(1+x^2)(1+x^3)(1+x^4)(1+x^5)\cdots$ represents the number of ways n can be written as a sum of distinct positive integers, which we denote by $p_{\rm d}(n)$. That is, the generating function for $p_{\rm d}(n)$, the number of ways n can be written as a sum of distinct positive integers, is given by $\prod_{n=1}^{\infty}(1+x^n)$, solving Naudé's first problem. Generalizing Euler's argument, we find that $\prod_{n=1}^{\infty}(1+x^{2n+1})$, $\prod_{n=1}^{\infty}(1+x^{2n})$, and $\prod_{n=1}^{\infty}(1+x^{n^2})$ represent respectively the generating functions for the number of ways the positive integer n can be written as a sum of distinct odd positive integers, even positive integers, and squares.

With respect to Naudé's second problem, Euler realized that

$$\begin{aligned}
&\frac{1}{(1-xz)(1-x^2z)(1-x^3z)(1-x^4z)\cdots}\\
&=\left(\frac{1}{1-xz}\right)\left(\frac{1}{1-x^2z}\right)\left(\frac{1}{1-x^3z}\right)\left(\frac{1}{1-x^4z}\right)\cdots\\
&=(1+xz+x^2z^2+x^3z^3+\cdots)(1+x^2z+x^4z^2+x^6z^3+\cdots)\\
&\quad\times(1+x^3z+x^6z^2+x^9z^3+\cdots)(1+x^4z+x^8z^2+x^{12}z^3+\cdots)\cdots.
\end{aligned}$$

Hence the coefficient of x^nz^m in the expression represents the number of ways that n can be written as a sum of m not necessarily distinct positive integers. For example, the coefficient of x^8z^3 is 5 and it results from summing the terms $(x^6z)(x^2z)$, $(x^5z)(x^2z)(xz)$, $(x^4z)(x^3z)(xz)$, $(x^4z)(x^4z^2)$, and $(x^6z^2)(x^2z)$. They are the terms corresponding respectively to the partitions $6+1+1$, $5+2+1$, $4+3+1$, $4+2+2$, and $3+3+2$ of 8. If we let $z=1$ in the above expression, we obtain

$$\begin{aligned}
&\frac{1}{(1-x)(1-x^2)(1-x^3)(1-x^4)\cdots}\\
&=\left(\frac{1}{1-x}\right)\left(\frac{1}{1-x^2}\right)\left(\frac{1}{1-x^3}\right)\left(\frac{1}{1-x^4}\right)\cdots\\
&=(1+x+x^2+\cdots)(1+x^2+x^4+\cdots)(1+x^3+x^6+\cdots)\\
&\quad\times(1+x^4+x^8+\cdots)\cdots\\
&=1+x+2x^2+3x^3+5x^4+7x^5+11x^6+15x^7+22x^8+\cdots,
\end{aligned}$$

where the coefficient of x^n represents the number of ways n can be written as the sum of not necessarily distinct positive integers. For example, the partition $3+2+2+2+1$ of 10 corresponds, in the previous expression, to $x \cdot x^6 \cdot x^3 \cdot 1 \cdot 1 \cdot 1 \cdots$. That is, in the product of sums, it corresponds to choosing x from the first sum, $x^6 = x^{2+2+2}$ from the second sum, x^3 from the third sum, and 1 from the remaining sums. The partition $2 + 2 + 2 + 2 + 1 + 1$ of 10 corresponds to $x^2 \cdot x^8 \cdot 1 \cdot 1 \cdot 1 \cdots$. That is, in the product of sums, it corresponds to choosing x^2 from the first sum, $x^8 = x^{2+2+2+2}$ from the second sum, and 1 from the remaining sums. In addition, the terms $x^3x^6x^1$ and x^8x^2 each contribute exactly 1 to the coefficient of x^{10}. In general, each partition of 10 contributes exactly once to the coefficient of x^{10}. Therefore, the generating function for $p(n)$, the number of ways n can be written as a sum of not necessarily distinct positive integers, is given by

$$G(x) = \prod_{n=1}^{\infty} \frac{1}{1 - x^n},$$

solving Naudé's second problem.

In general, the coefficient of x^n in

$$\begin{aligned} &\frac{1}{(1 - x^a)(1 - x^b)(1 - x^c)(1 - x^d)\cdots} \\ &= \left(\frac{1}{1 - x^a}\right)\left(\frac{1}{1 - x^b}\right)\left(\frac{1}{1 - x^c}\right)\left(\frac{1}{1 - x^d}\right)\cdots \\ &= (1 + x^a + x^{2a} + \cdots)(1 + x^b + x^{2b} + \cdots)(1 + x^c + x^{2c} + \cdots) \\ &\quad \times (1 + x^d + x^{2d} + \cdots)\cdots \end{aligned}$$

is of the form $x^{k_1a}x^{k_2b}x^{k_3c}x^{k_4d}\cdots$, where $n = k_1a + k_2b + k_3c + k_4d + \cdots$. Hence, the term $x^{k_1a}x^{k_2b}x^{k_3c}x^{k_4d}\cdots$ represents writing n as the sum of k_1 as, k_2 bs, k_3 cs, k_4 ds, and so forth. Therefore,

$$G(x) = \frac{1}{(1 - x^a)(1 - x^b)(1 - x^c)(1 - x^d)\cdots}$$

is the generating function for expressing the positive integer n as a sum of as, bs, cs, ds, and so forth. Thus,

$$\prod_{n=1}^{\infty} \frac{1}{1 - x^{2n}}, \quad \prod_{n=1}^{\infty} \frac{1}{1 - x^{2n+1}}, \quad \prod_{n=1}^{\infty} \frac{1}{1 - x^{n^2}}$$

represent, respectively, the generating functions for the number of ways of representing the positive integer n as a sum of not necessarily distinct positive even integers, positive odd integers, and squares. In addition, $1/(1 - x^6)(1 - x^8)(1 - x^{10})\cdots$ represents the generating function for the number of partitions of the positive integer n into even integers greater than 6. Analogously, the generating function for $p_k(n)$, the number of ways of partitioning the positive integer n using only positive integers less than or equal to k, is $1/(1 - x)(1 - x^2)(1 - x^3)\cdots(1 - x^k)$.

Let $p_o(n)$ and $p_e(n)$ denote the number of partitions of the positive integer n using only odd or only even positive integers, respectively. For example, the only ways to partition 7 into odd positive integers are 7, $5+1+1$, $3+3+1$, $3+1+1+1+1$, and $1+1+1+1+1+1+1$. Therefore, $p_o(7)=5$. The only ways to partition 6 into even positive integers are 6, $4+2$, and $2+2+2$. Therefore, $p_e(6)=3$. The elegant proof of the next result is due to Euler.

Theorem 9.1 (Euler's parity law) *For any positive integer n, the number of partitions of n using only odd positive integers equals the number of partitions of n into distinct parts.*

Proof The generating function for $p_o(n)$ is

$$\frac{1}{(1-x)(1-x^3)(1-x^5)\cdots} = \frac{(1-x^2)(1-x^4)(1-x^6)\cdots}{(1-x)(1-x^2)(1-x^3)\cdots}$$
$$= (1+x)(1+x^2)(1+x^3)\cdots,$$

which is the generating function for $p_d(n)$, the number of partitions of n into distinct parts. Therefore, $p_o(n) = p_d(n)$. ■

Exercises 9.2

1. Determine all the ordered partitions of 4 and 5.
2. Write out the partitions for $n = 8$ and 9.
3. What does the coefficient of $x^n z^m$ in the expression $(1+x^a z)(1+x^b z)(1+x^c z)(1+x^d z)(1+x^e z)\cdots$ represent?
4. What does the coefficient of $x^n z^m$ in the expression
$$\frac{1}{(1-x^a z)(1-x^b z)(1-x^c z)(1-x^d z)(1-x^e z)\cdots}$$
represent?
5. Determine the generating function for the number of ways the positive integer n can be written as a distinct sum of cubes.
6. Determine the generating function for the number of ways the positive integer n can be written as a distinct sum of triangular numbers.
7. Determine the generating function for the number of ways the positive integer n can be written as a distinct sum of prime numbers.
8. Determine the generating function for the number of ways the postive integer n can be written as a sum of cubes.
9. Determine the generating function for the number of ways the positive integer n can be written as a sum of triangular numbers.

10. Determine the generating function for the number of ways the positive integer n can be written as a sum of prime numbers.
11. Determine the generating function for the numbers of ways of representing the positive integer n as a sum of primes each greater than 7.
12. Determine the generating function for the number of ways of representing the positive integer n as a sum of odd numbers greater than 11.
13. Determine the generating function for the number of ways of representing the positive integer n as a sum of even numbers between 6 and 20 inclusive.
14. Determine the first 10 coefficients of $(1+x)(1+x^2)(1+x^4)(1+x^8)(1+x^{16})\cdots$, the generating function for the number of ways of representing the positive integer n as a sum of powers of 2.
15. Determine all the odd partitions of 9 and all the partitions of 9 into distinct parts.
16. Find all the even partitions of 10.
17. Find all the partitions of 10 using only the integers 3, 4, 5, 6, 7.
18. Show that the number of partitions of n into at most two parts is given by $[\![n/2]\!]$.
19. For $1 \leqslant n \leqslant 9$, construct a table with columns $p(n)$, the number of partitions of n; $p_{\mathrm{e}}(n)$, the number of partitions of n using only even positive integers; $p_{\mathrm{o}}(n)$, the number of partitions of n using only odd positive integers; $p_{\mathrm{d}}(n)$, the number of partitions of n using distinct positive integers; $p_{\mathrm{ed}}(n)$, the number of partitions of n into an even number of distinct parts; $p_{\mathrm{od}}(n)$, the number of partitions of n into an odd number of distinct parts; and $p_1(n)$, the total number of 1s that appear in the partitions of n.

9.3 Pentagonal Number Theorem

In 1853, Norman Macleod Ferrers communicated to J.J. Sylvester an ingenious method for representing partitions. Ferrers, an Etonian, was Senior Wrangler and First Smith's Prizeman at Cambridge in 1851. He edited *The Mathematical Papers of George Green* and served as Master of Gonville and Caius College and Vice-Chancellor of Cambridge University. His geometric representation is useful in establishing a number of results concerning partitions. Given a partition $n_1 + n_2 + n_3 + n_4 + \cdots$ of the positive integer n, the Ferrers diagram associated with the partition is an array with n_k dots in the kth row. If we interchange the rows and columns of a Ferrers diagram, we obtain the conjugate Ferrers diagram.

For example, in Figure 9.1, the partition $8 + 4 + 3 + 3 + 2 + 1 + 1$ of 22 is represented by a Ferrers diagram. The Ferrers diagram of its congugate partition, $7 + 5 + 4 + 2 + 1 + 1 + 1 + 1$, is shown in Figure 9.2.

Using our convention of expressing each partition of a positive integer with terms in descending order, the longest row of each Ferrers diagram will be at the top and the longest column will be the first. Any Ferrers diagram identical with its conjugate is called a selfconjugate Ferrers diagram. For example, the partition $5 + 3 + 2 + 1 + 1$ of 12 is self-conjugate. Its Ferrers diagram is shown in Figure 9.3.

In 1882, J.J. Sylvester and William Pitt Durfee, a graduate student at Johns Hopkins, noted that in any selfconjugate partition the shells outlined in the selfconjugate Ferrers diagrams, shown in Figure 9.4, contain an odd number of dots. Thus, the Ferrers diagrams represent the partition of a positive integer into a sum of odd parts as for selfcongugate partitions of 12 and 24. Conversely, any partition of a positive integer into a sum of odd

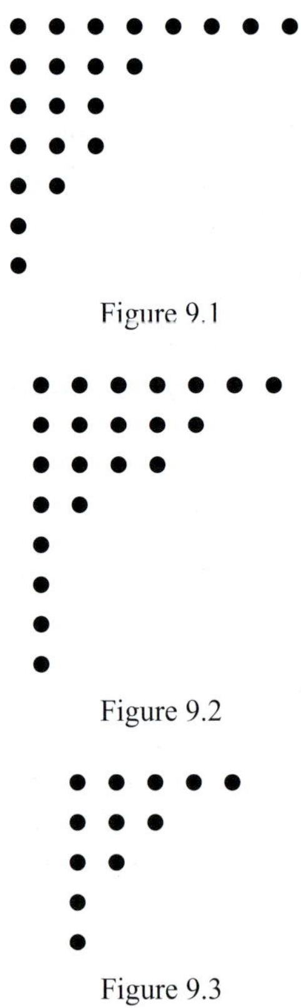

Figure 9.1

Figure 9.2

Figure 9.3

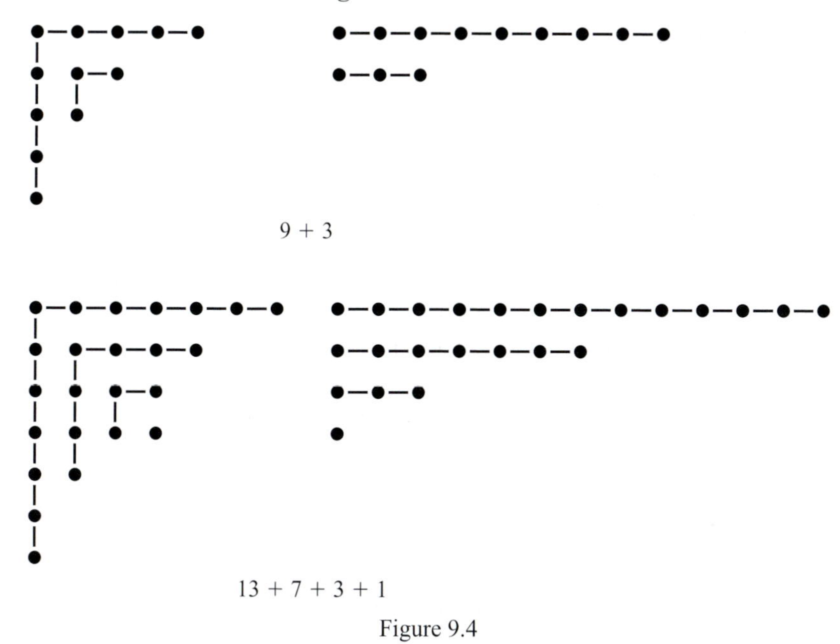

Figure 9.4

parts yields a selfconjugate partition of that positive integer. The result is stated as the next theorem. After receiving his degree from Johns Hopkins, Durfee taught mathematics at Hobart College, now Hobart–William Smith College, in Geneva, New York.

Theorem 9.2 (Durfee–Sylvester) *The number of partitions of a positive integer n into odd distinct parts equals the number of partitions of n whose Ferrers diagrams are selfconjugate.*

Let $p_k(n)$ represent the number of partitions of n into parts none of which exceeds k and $p(n, k)$ the number of partitions of n into exactly k parts. Hence, $p_k(n) - p_{k-1}(n)$ represents the number of partitions of n into parts the largest of which is k. For each partition for which the largest part is k, the conjugate partition has k parts and vice versa. Hence, the number of partitions of n into k parts equals the number of partitions of n into parts the largest of which is k. Similarly, the number of partitions of n into at most k parts equals the number of partitions of n into parts which do not exceed k. Hence, we have established the next result.

Theorem 9.3 (Ferrers) *For any positive integer n, $p(n, k) = p_k(n) - p_{k-1}(n)$.*

Table 9.2.

							k									
n	1	2	3	4	5	6	7	8	9	10	11	12	13	14	15	16
1	1															
2	1	1														
3	1	1	1													
4	1	2	1	1												
5	1	2	2	1	1											
6	1	3	3	2	1	1										
7	1	3	4	3	2	1	1									
8	1	4	5	5	3	2	1	1								
9	1	4	7	6	5	3	2	1	1							
10	1	5	8	9	7	5	3	2	1	1						
11	1	5	10	11	10	7	5	3	2	1	1					
12	1	6	12	15	13	11	7	5	3	2	1	1				
13	1	6	14	18	18	14	11	7	5	3	2	1	1			
14	1	7	16	23	23	20	15	11	7	5	3	2	1	1		
15	1	7	19	27	30	26	21	15	11	7	5	3	2	1	1	
16	1	8	21	34	37	35	28	22	15	11	7	5	3	2	1	1

Let $p_m(n, k)$ denote the number of partitions of n into k parts none of which is larger than m. Consider a Ferrers diagram of the positive integer $a - c$ with $b - 1$ parts none of which is larger than c and adjoin a new top row of length c to obtain a Ferrers diagram representing a partition of a into b parts the largest of which is c. The conjugate of the revised Ferrers diagram represents a partition of a into c parts the largest of which is b. Deleting the top row of the conjugate Ferrers diagram we obtain a Ferrers diagram representing a partition of $a - b$ into $c - 1$ parts the largest of which is b. The operations are reversible, hence, we have established the next result, first established by Sylvester in 1853.

Theorem 9.4 (Sylvester) *If a, b, c are positive integers such that $a > b$ and $b > c$, then $p_c(a - c, b - 1) = p_b(a - b, c - 1)$.*

For a given positive integer n, there is no elementary formula for determining $p(n)$. However, the next result, due to Euler, can be used to evaluate $p(n)$. Since $p(n, k)$ denotes the number of partitions of n into exactly k parts, it follows that $p(n) = \sum_{k=1}^{n} p(n, k)$. Some values of $p(n, k)$ are shown in Table 9.2. For convenience, we denote the order of a set A, that is the number of elements in A, by $|A|$.

Theorem 9.5 (Euler) *For positive integers* n *and* k *with* $k \leqslant n$, $p(n, k) = p(n-1, k-1) + p(n-k, k)$.

Proof From Theorem 9.3, the number of partitions of n into exactly k parts, $p(n, k)$, is also the number of partitions of n into parts the largest of which is k. Let S represent the set of partitions of n into parts the largest of which is k. Hence, $|S| = p(n, k)$. Let T represent the union of the set A of partitions of $n-1$ whose largest term is $k-1$ and the set B of partitions of $n-k$ whose largest term is k. Since A and B are disjoint, $|T| = |A| + |B| = p(n-1, k-1) + p(n-k, k)$. Any partition in S has the form $x_1 + \cdots + x_{r-1} + x_r = n$, where $k = x_1 \geqslant \cdots \geqslant x_r$. If $k = x_1 > x_2$, we associate it with $(x_1 - 1) + x_2 + \cdots + x_r = n - 1$, an element of A. If $x_1 = x_2 = k$, associate it with $x_2 + x_3 + \cdots + x_r = n - x_1 = n - k$, an element of B. The association is a one-to-one mapping from S into T, hence $|S| \leqslant |T|$. Any partition in T is of the form $\alpha = u_1 + u_2 + \cdots + u_r = n - 1$, where $k - 1 = u_1 \geqslant \cdots \geqslant u_r$, or $\beta = v_1 + v_2 + \cdots + v_s = n - k$, where $k = v_1 \geqslant \cdots \geqslant v_s$. Any partition of the form α, we associate with the partition $(u_1 + 1) + \cdots + u_r = n$. Since $u_1 + 1 = k$, this partition is in S. Any partition of the form β, we associate with the partition $k + v_1 + \cdots + v_s = n$, which is an element of S. This association is a one-to-one mapping from T into S, hence, $|S| \geqslant |T|$. Therefore, $|S| = |T|$ and the result is established. ■

The next result, first proven in 1881 by Fabian Franklin, a professor of mathematics at Johns Hopkins University, is instrumental in deriving Euler's Pentagonal Number Theorem. Franklin was the husband of the mathematician–psychologist, Christine Ladd Franklin. When he left Johns Hopkins to begin a career in journalism in New York, Ladd taught at Columbia.

Consider a Ferrers diagram for n, with b dots on the bottom row and s dots on the rightmost NE–SW diagonal. If $b < s$ remove the b dots on the bottom row and adjoin one each to the end of each of the first b rows of the diagram. For example, in Figure 9.5, where $b = 2$ and $s = 3$, the partition $6 + 5 + 4 + 2 + 2$ of 19 is transformed into the partition $7 + 6 + 4 + 2$ of 19. This process transforms a partition of n with an even number of distinct parts into a partition of n with an odd number of distinct parts and vice versa.

If $b > s + 1$ remove the rightmost diagonal and adjoin it to the bottom of the diagram making it the new bottom row. For example, in Figure 9.6, where $b = 4$ and $s = 2$, the partition $7 + 6 + 4$ of 17 is transformed into

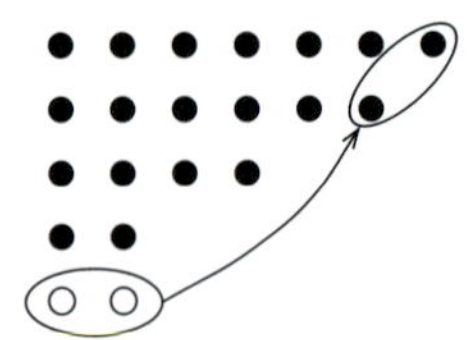

Figure 9.5

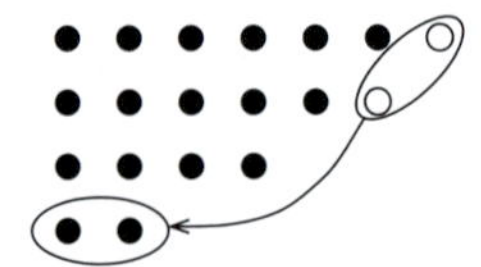

Figure 9.6

the partition $6 + 5 + 4 + 2$ of 17. This process transforms a partition of n with an even number of distinct parts into a partition of n with an odd number of distinct parts and vice versa.

In the two remaining cases, where $b = s$ or $b = s + 1$, no similar process can be carried out, for example, in Figure 9.7, where $b = s = 3$ in the partition $5 + 4 + 3$ of 12, or in Figure 9.8, where $b = 3$ and $s = 2$ and the bottom row and rightmost diagonal have a point in common.

If $b = s$ then $n = b + (b + 1) + \cdots + (2b - 1) = b(3b - 1)/2$ and if $b = s + 1$ then $n = (s + 1) + (s + 2) + \cdots + 2s = s(3s + 1)/2$. If b does not equal s or $s + 1$, then exactly one of the above operations can be carried out. Hence, there is a one-to-one correspondence between partitions of n into an even number of distinct parts and partitions of n into an odd number of distinct parts, and for these values of n, $p_{\text{ed}}(n) - p_{\text{od}}(n) = 0$. In the two exceptional cases, when $n = k(3k \pm 1)/2$, the difference is $(-1)^k$ and we have established Theorem 9.6.

Theorem 9.6 (Franklin) *If n is a positive integer, and $p_{\text{ed}}(n)$ and $p_{\text{od}}(n)$ represent, respectively, the number of partitions of the positive integer n into even and odd distinct parts, then*

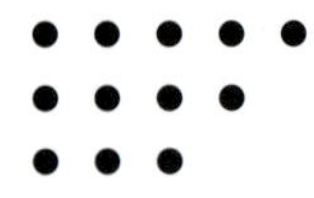

Figure 9.7

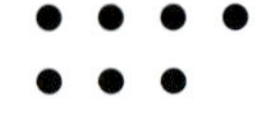

Figure 9.8

$$p_{\mathrm{ed}}(n) - p_{\mathrm{od}}(n) = \begin{cases} 0 & \text{if } n \neq \dfrac{3k \pm 1}{2}, \\ (-1)^k & \text{if } n = \dfrac{3k \pm 1}{2}. \end{cases}$$

Recall that the generating function for $p_{\mathrm{d}}(n)$, the number of ways n can be written as a sum of distinct positive integers, is $\prod_{n=1}^{\infty}(1 + x^n)$. Substituting $-x$ for x, we account for the contribution of a plus or minus 1 to each coefficient depending on whether the number of distinct parts in the partition is even or odd respectively. Hence, it follows from Theorem 9.6 that

$$\begin{aligned}\prod_{n=1}^{\infty}(1 - x^n) &= 1 + \prod_{n=1}^{\infty}(p_{\mathrm{ed}}(n) - p_{\mathrm{od}}(n))x^n \\ &= 1 + \sum_{n=1}^{\infty}(-1)^n x^{n(3n+1)/2} + \sum_{n=1}^{\infty}(-1)^n x^{n(3n-1)/2}\end{aligned}$$

and we have established the Pentagonal Number Theorem.

Theorem 9.7 (Euler's Pentagonal Number Theorem) *For any positive integer n,*

$$\prod_{n=1}^{\infty}(1 - x^n) = 1 + \sum_{n=1}^{\infty}(-1)^n x^{n(3n+1)/2} + \sum_{n=1}^{\infty}(-1)^n x^{n(3n-1)/2}.$$

Euler used the Pentagonal Number Theorem in 1750 to develop a formula to determine values of $p(n)$ recursively as illustrated in the next result.

Theorem 9.8 (Euler) *For any positive integer n, $p(n)$, the number of partitions of n, is given by*

$$\begin{aligned}&p(n-1) + p(n-2) - p(n-5) - p(n-7) + p(n-12) + p(n-15) \\ &+ \cdots + (-1)^{k+1}\left[p\left(n - \frac{3k^2 - k}{2}\right) + p\left(n - \frac{3k^2 + k}{2}\right)\right].\end{aligned}$$

Proof Recall that the generating function of $p(n)$ is given by

$$\prod_{n=1}^{\infty}\frac{1}{1 - x^n}.$$

From Theorem 9.7,

$$1 = \frac{1}{\prod_{n=1}^{\infty}(1-x^n)} \cdot \prod_{n=1}^{\infty}(1-x^n)$$

$$= \left[\sum_{n=0}^{\infty} p(n)x^n\right] \cdot [1 - x - x^2 + x^5 + x^7 - x^{12} - \cdots$$

$$+ (-1)^k(x^{(3k^2-k)/2} + x^{(3k^2+k)/2})].$$

Expanding and collecting terms, we obtain

$$1 = p(0) - [p(0) - p(1)]x + [p(2) - p(0) - p(1)]x^2 + \cdots$$

$$+ \Bigg[p(n) - p(n-1) - p(n-2) + p(n-5) + p(n-7)$$

$$- p(n-12) - p(n-15) + \cdots$$

$$+ (-1)^k\left[p\left(n - \frac{3k^2-k}{2}\right) + p\left(n - \frac{3k^2+k}{2}\right)\right] + \cdots\Bigg]x^n + \cdots .$$

Cancelling $p(0) = 1$ from both sides of the equations and equating the coefficients of x^n, for $n \geqslant 1$, to 0, we obtain Euler's partition formula

$$p(n) = p(n-1) + p(n-2) - p(n-5) - p(n-7) + p(n-12)$$

$$+ p(n-15) - \cdots$$

$$+ (-1)^{k+1}\left[p\left(n - \frac{3k^2-k}{2}\right) + p\left(n - \frac{3k^2+k}{2}\right)\right]. \quad \blacksquare$$

Major Percy Alexander MacMahon used Euler's result to calculate the value of $p(200)$, which he found to be 3 972 999 029 388. After a distinguished career with the Royal Artillery in Madras and as an instructor at the Royal Military Academy, Woolwich, MacMahon at age 58 went up to Cambridge University to pursue research in combinatorial number theory. He was elected a member of St John's College and served as president of the London Mathematical Society and of the Royal Astronomical Society.

About the same time that he derived the partition formula, Euler devised an analogous formula for $\sigma(n)$, the sum of the divisors of n.

Theorem 9.9 (Euler) *If n is a positive integer, then*

$$\sigma(n) = \sigma(n-1) + \sigma(n-2) - \sigma(n-5) - \sigma(n-7) + \sigma(n-12)$$

$$+ \sigma(n-15) + \cdots + (-1)^{k+1}\left[\sigma\left(n - \frac{3k^2-k}{2}\right) + \sigma\left(n - \frac{3k^2+k}{2}\right)\right],$$

where $\sigma(k) = 0$ *if* $k < 0$ *and* $\sigma(0) = k$.

Proof Let

$$G(x) = \sum_{n=1}^{\infty} \sigma(n)x^n = \sum_{n=1}^{\infty} \frac{nx^n}{1-x^n}.$$

Assume that $|x| < 1$. Divide both sides by x and integrate with respect to x, and use Theorem 9.7 to obtain

$$\int \frac{G(x)\,dx}{x} = \int \sum_{n=1}^{\infty} \frac{nx^{n-1}\,dx}{1-x^n} = -\sum_{n=1}^{\infty} \ln(1-x^n) = -\ln\left(\prod_{n=1}^{\infty}(1-x^n)\right)$$

$$= -\ln(1 - x - x^2 + x^5 + x^7 - x^{12} - x^{15} + \cdots).$$

Differentiate both sides with respect to x, we obtain

$$\frac{G(x)}{x} = \frac{-1 - 2x + 5x^4 + 7x^6 - \cdots}{1 - x - x^2 + x^5 + x^7 - \cdots}.$$

Hence,

$$G(x) = \frac{-x - 2x^2 + 5x^5 + 7x^7 - \cdots}{1 - x - x^2 + x^5 + x^7 - \cdots} = \sum_{n=1}^{\infty} \sigma(n)x^n.$$

By crossmultiplying and equating coefficients of x^n the result follows. ∎

In 1829 Jacobi established the triple product identity $\prod_{n=1}^{\infty}(1 - x^{2n})(1 + x^{2n-1}z)(1 + x^{2n-1}z^{-1}) = \sum_{n=-\infty}^{\infty} x^{n^2} z^n$, where $z \neq 0$ and $|x| < 1$. He used it to established the following results.

(a) $\displaystyle\prod_{n=0}^{\infty}(1 - x^{2n+2})(1 + x^n) = \sum_{n=-\infty}^{\infty} x^{n(n+1)/2}$,

(b) $\displaystyle\prod_{n=0}^{\infty}\frac{1 - x^{2n}}{1 - x^{2n+1}} = \sum_{n=0}^{\infty} x^{n(n+1)/2}$, and

(c) $\displaystyle\prod_{n=0}^{\infty}(1 - x^{2n})^3 = \sum_{n=0}^{\infty}(-1)^n(2n+1)x^{n(n+1)/2}$.

For example if we let $x = z = u^{1/2}$ in the triple product identity on the left we obtain $\prod_{n=1}^{\infty}(1 - u^n)(1 + u^n)(1 + u^{n-1}) = \sum_{n=-\infty}^{\infty} u^{n(n+1)/2}$. However,

$$\prod_{n=1}^{\infty}(1 - u^n)(1 + u^n)(1 + u^{n-1}) = \prod_{n=1}^{\infty}(1 - u^{2n})(1 + u^{n-1})$$

$$= \prod_{n=0}^{\infty}(1 - u^{2n+2})(1 + u^n)$$

and (a) is established.

In 1878, Franklin considered the partitions of n which contain at most one 1. If a partition contained exactly one 1, he counted it as 1. If it contained no 1s he counted the number of distinct elements in it. He found

8	1
7 + 1	1
6 + 2	2
5 + 3	2
5 + 2 + 1	1
4 + 4	1
4 + 3 + 1	1
4 + 2 + 2	2
3 + 3 + 2	2
3 + 2 + 2 + 1	1
2 + 2 + 2 + 2	1
	15 = $p(7)$

Figure 9.9

the total sum to be $p(n-1)$, as illustrated in Figure 9.9 for the case when $n = 8$.

MacMahon discovered an interesting relationship concerning partitions which he included in *Combinatorial Analysis*. MacMahon defined a partition of n to be perfect if every integer from 1 to $n-1$ can be represented in a unique way as a sum of parts from the partition.

For example, the partition $7 = 4 + 2 + 1$ is a perfect partition of 7 since

$$\begin{aligned} 1 &= 1, \\ 2 &= 2, \\ 3 &= 2 + 1, \\ 4 &= 4, \\ 5 &= 4 + 1, \\ 6 &= 4 + 2. \end{aligned}$$

The other perfect partitions of 7 are $4 + 1 + 1 + 1$, $2 + 2 + 2 + 1$, and $1 + 1 + 1 + 1 + 1 + 1 + 1$.

Theorem 9.10 (MacMahon) *The number of perfect partitions of n equals the number of ways of factoring $n + 1$, where the order of the factors counts and factors of* 1 *are not counted.*

Proof There must be at least one 1 in any perfect partition of $n + 1$ and if there are x 1s then the next smallest element in the partition must be $x + 1$ since all smaller integers can be written as the sum of 1s alone. If there are y parts of $x + 1$ the next smallest number in the partition must be $x + y(x + 1) + 1 = (x + 1)(y + 1)$. Hence, if the different parts of the partition occur $x, y, z, \ldots$ times then $(x + 1)(y + 1)(z + 1) \cdots = n + 1$, and the number of perfect partitions of n is the same as the number of ways

Table 9.3.

Partitions	Number of distinct parts in each partition
6	1
5 + 1	2
4 + 2	2
4 + 1 + 1	2
3 + 3	1
3 + 2 + 1	3
3 + 1 + 1 + 1	2
2 + 2 + 2	1
2 + 2 + 1 + 1	2
2 + 1 + 1 + 1 + 1	2
1 + 1 + 1 + 1 + 1 + 1	1
Total	19

of factoring $n + 1$ where the order of the factors counts and factors of 1 are not counted. ■

Ramanujan proved that the number of partitions of n with unique smallest part (it occurs only once) and largest part at most twice the smallest part is equal to the number of partitions of n in which the largest part is odd and the smallest part is larger than half the largest part. George Andrews of Pennsylvania State University proved that the number of partitions of n in which only odd parts may be repeated equals the number of partitions of n in which no part appears more than three times. In 1958, R.K. Guy showed that the numbers of partitions of a positive integer into (a) odd parts greater than unity, (b) unequal parts such that the greatest two parts differ by unity, and (c) unequal parts which are not powers of 2, are all equal.

Let $p_1(n)$ denote the 1s number of a positive integer n, that is, the total number of 1s that appear in all the partitions of n. Richard Stanley, a combinatorialist at MIT, defined the parts number of n, denoted by $p_\text{p}(n)$, to be the sum of distinct parts in each partition of n. For example, if $n = 6$, $p_\text{p}(6) = p_1(6) = 19$, as illustrated in Table 9.3.

Theorem 9.11 (Stanley) *For any positive integer n, $p_1(n) = p_\text{p}(n)$.*

Proof If we add a 1 to any partition of $n - 1$, we obtain a partition of n with at least one 1. Hence, the number of partitions of n which have at least one 1 is $p(n - 1)$. The number of partions of n which have two or more 1s is $p(n - 2)$, and so forth. Hence, the 1s number of n equals

$p(n) + p(n-1) + \cdots + p(1) + 1$. Since k occurs in exactly $p(n-k)$ partitions of n, the parts number of n also equals $p(n) + p(n-1) + \cdots + p(1) + 1$. ■

It can be shown that

$$\lim_{n\to\infty} \frac{p(n+1)}{p(n)} > 1$$

(see [Grosswald]). Finding reasonable bounds for $p(n)$ is a difficult task. However, we are able to derive the following upper bound for $p(n)$.

Theorem 9.12 *For any positive integer n,* $p(n) < e^{3\sqrt{n}}$.

Proof Let $G(x) = \sum_{i=0}^{\infty} p(n)x^n = \prod_{k=1}^{\infty}(1-x^k)^{-1}$ be the generating function for $p(n)$. Hence,

$$\begin{aligned}
\ln G(x) &= -\ln(1-x) - (\ln(1-x^2) - \ln(1-x^3) - \cdots \\
&= \left(x + \frac{x^2}{2} + \frac{x^3}{3} + \cdots\right) + \left(x^2 + \frac{x^4}{2} + \frac{x^6}{3} + \cdots\right) \\
&\quad + \left(x^3 + \frac{x^6}{2} + \frac{x^9}{3} + \cdots\right) + \cdots \\
&= (x + x^2 + x^3 + \cdots) + \left(\frac{x^2}{2} + \frac{x^4}{2} + \frac{x^6}{2} + \cdots\right) \\
&\quad + \left(\frac{x^3}{3} + \frac{x^6}{3} + \frac{x^9}{3} + \cdots\right) + \cdots \\
&= \left(\frac{x}{1-x}\right) + \frac{1}{2}\left(\frac{x^2}{1-x^2}\right) + \frac{1}{3}\left(\frac{x^3}{1-x^3}\right) + \cdots .
\end{aligned}$$

If $0 < x < 1$, then $x^{n-1} < x^{n-2} < \cdots < x^2 < x < 1$. Since the average of a set of numbers is bigger than the smallest number in the set,

$$x^{n-1} < \frac{1 + x + x^2 + \cdots + x^{n-1}}{n}$$

or

$$\frac{x^{n-1}}{1 + x + x^2 + \cdots + x^{n-1}} < \frac{1}{n}.$$

Thus,

$$\frac{x^n}{1-x^n} = \frac{x^{n-1}}{1 + x + x^2 + \cdots + x^{n-1}} \cdot \frac{x}{1-x} < \frac{1}{n} \cdot \frac{x}{1-x}.$$

Hence,

$$\begin{aligned}\ln G(x) &< \left(\frac{x}{1-x}\right) + \left(\frac{1}{2}\right)^2\left(\frac{x}{1-x}\right) + \left(\frac{1}{3}\right)^2\left(\frac{x}{1-x}\right) + \cdots \\ &= \left(\frac{x}{1-x}\right)\left(1 + \frac{1}{2^2} + \frac{1}{3^2} + \cdots\right) \\ &< \left(\frac{x}{1-x}\right)\left(1 + \int_1^\infty \frac{1}{x^2}\,\mathrm{d}x\right) \\ &= \frac{2x}{1-x}.\end{aligned}$$

Thus, $G(x)$, a sum of positive terms, is bigger than any one of its terms, in particular, $G(x) > p(n)x^n$. Therefore, $\ln p(n) < \ln G(x) - n \cdot \ln(x)$. That is,

$$\ln p(n) < \frac{2x}{1-x} - n \ln x < 2\left(\frac{x}{1-x}\right) + n\left(\frac{1-x}{x}\right).$$

If we now let $x = \sqrt{n}/\sqrt{n+1}$, we obtain $\ln p(n) < 3\sqrt{n}$ and the result is established. ■

Hardy and Ramanujan were able to show that

$$p(n) \approx \frac{1}{4\sqrt{3}n}\,\mathrm{e}^{\pi\sqrt{2\sqrt{n/3}}},$$

a result made exact by Hans Rademacher, a number theorist at the University of Pennsylvania. Rademacher found an expression that, when rounded to the nearest integer, equaled $p(n)$. In 1919 Ramanujan discovered a number of modular identities concerning partition numbers. In particular, for any positive integer n, he showed the following.

(a) $p(5n+4) \equiv 0 \pmod 5$,
(b) $p(7n+5) \equiv 0 \pmod 7$, and
(c) $p(11n+6) \equiv 0 \pmod{11}$.

One may generalize partitions to Young tableaux, whose properties were developed by Alfred Young, a Fellow of Clare College, Cambridge, who served for many years as the rector at Birdbrook in Essex, England. Given a positive integer n, a Young tableau for n of shape $(n_1, n_2, \ldots, n_m)$ is a Ferrers diagram for the partition $n_1 + n_2 + \cdots + n_m$ of n, where adjacent boxes are employed rather than dots, the ith row contains n_i elements, the integers from 1 to n are distributed in the boxes in such a way that all rows and columns are strictly increasing and each of the numbers $1, 2, 3, \ldots, n$ occurs exactly once. For example, a Young tableau for the partition $5+4+2+1$ of 12, that is a Young tableau of shape $(5, 4, 2, 1)$, is illustrated in Table 9.4. Young tableaux can be used to generate symmetric groups in group representation theory a topic that is beyond the scope of this book.

Table 9.4.

1	3	4	7	11
2	5	10	12	
6	9			
8				

Exercises 9.3

1. Determine the 17th row of Table 9.2.
2. Show that $p_{\mathrm{p}}(7) = p_{\mathrm{I}}(7)$.
3. Construct Ferrers diagrams for all 15 partitions of 7. Which of them are selfconjugate?
4. Show that the partition $2 + 2 + 2 + 1$ of 7 is perfect.
5. Use Jacobi's triple product identity with $x = u^{3/2}$ and $z = -u^{1/2}$ to establish Euler's Pentagonal Number Theorem.
6. In 1944, F.J. Dyson defined the rank of a partition to be the largest part minus the number of parts. Prove that the ranks of a partition and its conjugate differ only in sign.
7. Determine the sum of the ranks of the five partitions of 4 modulo 5.
8. Determine the sum of the ranks of the 30 partitions of 9 modulo 5.
9. Show that, in general, if $n \equiv 4 \pmod 5$ there are an equal number of ranks in each least positive residue class modulo 5. Hence, $p(5k + 4) \equiv 0 \pmod 5$.
10. Determine at least two Young tableaux of shape (5, 4, 2, 1).
11. Determine all 16 Young tableaux of shape (3, 2, 1).

9.4 Supplementary exercises

1. Identify the sequence for which $G(x) = 1/(1 - x^2)$ is the generating function.
2. Determine a generating function for the difference equation $a_n = 6a_{n-1} - 5a_{n-2}$, where $a_0 = 0$ and $a_1 = 1$.
3. Determine a generating function for the difference equation $a_n = 6a_{n-1} - 5a_{n-2}$, where $a_0 = 1$ and $a_1 = 3$.
4. Determine the generating function for the sequence of fifth powers of nonnegative integers 0, 1, 32, 243, 1024, 3125, ...
5. Identify the sequence for which

$$G(x) = \frac{x(x^5 + 57x^4 + 302x^3 + 302x^2 + 57x + 1)}{(1 - x)^7}$$

is the generating function.

6. Determine the generating function for the number of ways the positive integer n can be written as a sum of fourth powers.
7. Determine the generating function for the number of ways the positive integer n can be written as a sum of primes each less than or equal to 23.
8. Determine the generating function for the number of ways the positive integer n can be written as a sum of even numbers less than 30.
9. Determine the generating function for the number of ways the positive integer n can be written as a sum of numbers from seven to eleven.
10. Find all the partitions of 11 with exactly three parts.
11. Find all the partitions of 11 with exactly four parts.
12. Let $p^m(n, k)$ denote the number of partitions of n having exactly k parts with each part greater than or equal to m. Show that $p(n - k, k) = p^2(n, k)$, with the convention that $p(n, k) = 0$ if $n < k$.
13. Show that $p(n, k) = p(n - 1, k - 1) + p^2(n, k)$.
14. Determine r such that $p^m(n, k) = p(r, k)$.
15. Find r such that $\sum_{i=1}^{k} p(n, i) = p(r, k)$.
16. Find r such that $p_d(n, 3) = p(r, 3)$.
17. Find r such that $p_d(n, k) = p(r, k)$.
18. Determine the number of partitions of 19 using distinct parts and the number of partitions of 19 using only odd parts.
19. List all the perfect partitions of 11 and indicate the one-to-one correspondence with the factorizations of 12.
20. Use Euler's pentagonal number theorem to calculate the number of partitions of 20.
21. Characterize the partition(s) $n_1 + n_2 + \cdots + n_k$ of n such that $n_1 n_2 \cdots n_k$ is maximal.
22. Determine the generating function for the number of ways the positive integer n can be written as the sum of distinct Fibonacci numbers.
23. For any positive integer $n > 1$, let $P^+(n)$ denote the greatest prime divisor of n and let $P^+(1) = 1$. We call a partition $n_1 + n_2 + \cdots + n_k = n$ m-smooth if $P^+(n_i) \leq m$ for $1 \leq i \leq k$. Let $p^+(n, m)$ denote the number of m-smooth partitions of n. Determine $p^+(7, 2)$, $p^+(8, 4)$, and $p^+(9, 3)$.

Tables

Table T.1. *List of symbols used*

$(a)_m$	$\{a + km\colon\ k \in \mathbb{Z}\}$
$A(n)$	arithmetic mean of the divisors of n
B_n	nth Bernoulli number
$d_k(n)$	number of distinct solutions to the equation $x_1 \cdot x_2 \cdots x_k = n$
$D(n)$	smallest positive integer with n divisors
$E(n)$	excess of the number of divisors of n of the form $4k + 1$ over the number of divisors of n of the form $4k + 3$
E_n	nth square–triangular number
f_n	nth fortunate number
$F_r(n)$	least number of rs to represent n
$f^m{}_n$	nth mth order figurate number
F_n	nth Fermat number
$\mathscr{F}_n$	Farey fractions of order n
$G(n)$	geometric mean of the divisors of n
H	set of Hilbert numbers
H_n	nth harmonic number
$H(n)$	harmonic mean of the divisors of n
$I(n)$	index of n
K_n	n-digit Kaprekar constant
K_{a_1}	Kaprekar sequence with first term a_1
M_n	nth Monica set
M_p	Mersenne prime
n_b	n written to the base b
o_n	nth oblong number
O_n	nth octahedral number
$p^m{}_n$	nth m-gonal number
$p(n)$	number of partitions of n
$p_e(n)$	number of partitions of n using only even integers
$p_{ed}(n)$	number of partitions of n into even distinct parts
$p_k(n)$	number of ways of partitioning n using only integers less than or equal to k
$p_o(n)$	number of partitions of n using only odd positive integers
$p_{od}(n)$	number of partitions of n into odd distinct parts
$p_1(n)$	ones number of a positive integer n

Table T.1. (*cont.*)

$p_p(n)$	parts numbers of n
$p(n, k)$	number of partitions of n into exactly k parts
$p_m(n, k)$	number of partitions of n into k parts none of which is larger than m
$p^{\#}(n)$	nth primorial number
$P^{+}(n)$	the greatest prime divisor of n
$P^3{}_n$	nth tetrahedral number
$P^4{}_n$	nth pyramidal number
$P^*(n)$	product of the unitary divisors of n
$p^+(n, m)$	the number of m-smooth partitions of n
$\mathscr{P}_n$	nth prime
$Q(n)$	number of squarefull numbers less than n
$r_{n,k}$	rectangular number of the form $n(n+k)$
R_n	nth repunit
$s_d(n)$	sum of the digits of n
$s_d(n, b)$	sum of the digits of n base b
$s_p(n, b)$	prime digital sum of n expressed in base b
s_n	nth square number
$s(n)$	sum of divisors of n that are less than n
$s^*(n)$	Chowla's function, $\sigma(n) - n - 1$
$\overline{S}$	complement of set S
$S^*(n)$	psuedo-Smarandache function
S_n	nth Suzanne set
$S_r(n)$	sum of the rth powers of the digits of n
$S_{r,b}(n)$	sum of the rth powers of digits of n in base b
$S(n)$	sum of the squarefree positive integers less than n
$S_d(n, b)$	extended digital sum of n
t_n	nth triangular number
$t(n-k, k)$	number of divisors of $n-k$ greater than k
$T(n)$	Trigg operator
$T^m{}_n$	nth m-triangular number
u_n	nth Fibonacci number
$u(m, n)$	Ulam (m, n)-numbers
v_n	nth Lucas number
$v(m, n)$	non-Ulam (m, n)-numbers
$V(n)$	number of perfect numbers less than n
$Z(n)$	Zeckendorf representation function
Z_m	$\{0, 1, 2, \ldots, m-1\}$
Z^*_m	$\{1, 2, \ldots, m-1\}$
γ	Euler–Mascheroni number
Δ_n	nth differences of a sequence
$\zeta(n)$	Riemann zeta-function
$\eta(n)$	primeness of a positive integer
$\theta(n)$	excess of the sum of odd divisors of n over the even divisors of n
$\lambda(n)$	Liouville lambda-function
$\Lambda(n)$	Von Mangolt's function
$\Lambda_c(n)$	Carmichael's lambda function
$\mu(n)$	Möbius function

Table T.1. (*cont.*)

$\nu(n)$	sum of the Möbius function over the divisors of n
$\xi(n)$	number of positive integers k, $1 \leqslant k \leqslant n$, such that k is not a divisor of n and $\gcd(k, n) \neq 1$
$\pi(x)$	number of primes less than or equal to x
$\rho(n)$	digital root of the positive integer n
$\sigma(n)$	sum of the positive divisors of n
$\sigma^*(n)$	sum of the unitary divisors of n
$\sigma_e(n)$	sum of the even divisors of n
$\sigma_k(n)$	sum of the kth powers of the divisors of n
$\sigma_o(n)$	sum of the odd divisors of n
τ	golden ratio
$\tau(n)$	number of positive divisors of n
$\tau_e(n)$	number of even divisors of n
$\tau_k(n)$	generalized number of divisors of n
$\tau_o(n)$	number of odd divisors of n
$\tau(m, n)$	number of positive divisors of n which are congruent to m modulo 4
$\phi(n)$	number of positive integers less than n and coprime to n
$\chi(n)$	$\xi(d)$ summed over the divisors of n
$\Psi(n)$	inner product of primes and powers in the canonical representation of n
$\omega(n)$	number of distinct prime factors of n
$\Omega(n)$	the degree of the positive integer n
$a \mid b$	a 'divides' b
$a \nmid b$	a 'does not divide' b
$p^\alpha \| m$	p^α 'exactly divides' m
$\lvert x \rvert$	absolute value of x
$\lvert n \rvert_p$	p-adic valuation of n
$[\![x]\!]$	greatest integer not greater than x
$(a_0a_1 \cdots a_n)_b$	base b expansion of $a_0a_1 \cdots a_n$
$[a_1, a_2, \ldots, a_n]$	simple continued fraction
$\gcd(a, b)$	greatest common divisor of a and b
$\operatorname{lcm}(a, b)$	least common multiple of a and b
$\operatorname{ord}_n(a)$	order of a modulo n
$a \equiv b \pmod{n}$	a is 'congruent to' b modulo n
$\binom{n}{r}$	binomial coefficient
$\left(\frac{n}{p}\right)$	Legendre symbol
$\prod_{d \mid n}$	product of the divisors of n
$\sum_{d \mid n}$	summation over the divisors of n
$*_n$	nth star number
$n!$	n factorial
$!n$	$0! + \cdots + (n-1)!$
$\approx$	approximately equal
$\sim$	equivalence of binary quadratic forms
$f * g$	Dirichlet product of f and g

Table T.2. *Primes less than* 10 000

2	151	353	577	811	1049	1297	1559
3	157	359	587	821	1051	1301	1567
5	163	367	593	823	1061	1303	1571
7	167	373	599	827	1063	1307	1579
11	173	379	601	829	1069	1319	1583
13	179	383	607	839	1087	1321	1597
17	181	389	613	853	1091	1327	1601
19	191	397	617	857	1093	1361	1607
23	193	401	619	859	1097	1367	1609
29	197	409	631	863	1103	1373	1613
31	199	419	641	877	1109	1381	1619
37	211	421	643	881	1117	1399	1621
41	223	431	647	883	1123	1409	1627
43	227	433	653	887	1129	1423	1637
47	229	439	659	907	1151	1427	1657
53	233	443	661	911	1153	1429	1663
59	239	449	673	919	1163	1433	1667
61	241	457	677	929	1171	1439	1669
67	251	461	683	937	1181	1447	1693
71	257	463	691	941	1187	1451	1697
73	263	467	701	947	1193	1453	1699
79	269	479	709	953	1201	1459	1709
83	271	487	719	967	1213	1471	1721
89	277	491	727	971	1217	1481	1723
97	281	499	733	977	1223	1483	1733
101	283	503	739	983	1229	1487	1741
103	293	509	743	991	1231	1489	1747
107	307	521	751	997	1237	1493	1753
109	311	523	757	1009	1249	1499	1759
113	313	541	761	1013	1259	1511	1777
127	317	547	769	1019	1277	1523	1783
131	331	557	773	1021	1279	1531	1787
137	337	563	787	1031	1283	1543	1789
139	347	569	797	1033	1289	1549	1801
149	349	571	809	1039	1291	1553	1811

Table T.2. (*cont.*)

1823	2131	2437	2749	3083	3433	3733	4073
1831	2137	2441	2753	3089	3449	3739	4079
1847	2141	2447	2767	3109	3457	3761	4091
1861	2143	2459	2777	3119	3461	3767	4093
1867	2153	2467	2789	3121	3463	3769	4099
1871	2161	2473	2791	3137	3467	3779	4111
1873	2179	2477	2797	3163	3469	3793	4127
1877	2203	2503	2801	3167	3491	3797	4129
1879	2207	2521	2803	3169	3499	3803	4133
1889	2213	2531	2819	3181	3511	3821	4139
1901	2221	2539	2833	3187	3517	3823	4153
1907	2237	2543	2837	3191	3527	3833	4157
1913	2239	2549	2843	3203	3529	3847	4159
1931	2243	2551	2851	3209	3533	3851	4177
1933	2251	2557	2857	3217	3539	3853	4201
1949	2267	2579	2861	3221	3541	3863	4211
1951	2269	2591	2879	3229	3547	3877	4217
1973	2273	2593	2887	3251	3557	3881	4219
1979	2281	2609	2897	3253	3559	3889	4229
1987	2287	2617	2903	3257	3571	3907	4231
1993	2293	2621	2909	3259	3581	3911	4241
1997	2297	2633	2917	3271	3583	3917	4243
1999	2309	2647	2927	3299	3593	3919	4253
2003	2311	2657	2939	3301	3607	3923	4259
2011	2333	2659	2953	3307	3613	3929	4261
2017	2339	2663	2957	3313	3617	3931	4271
2027	2341	2671	2963	3319	3623	3943	4273
2029	2347	2677	2969	3323	3631	3947	4283
2039	2351	2683	2971	3329	3637	3967	4289
2053	2357	2687	2999	3331	3643	3989	4297
2063	2371	2689	3001	3343	3659	4001	4327
2069	2377	2693	3011	3347	3671	4003	4337
2081	2381	2699	3019	3359	3673	4007	4339
2083	2383	2707	3023	3361	3677	4013	4349
2087	2389	2711	3037	3371	3691	4019	4357
2089	2393	2713	3041	3373	3697	4021	4363
2099	2399	2719	3049	3389	3701	4027	4373
2111	2411	2729	3061	3391	3709	4049	4391
2113	2417	2731	3067	3407	3719	4051	4397
2129	2423	2741	3079	3413	3727	4057	4409

Table T.2. (*cont.*)

4421	4759	5099	5449	5801	6143	6481	6841
4423	4783	5101	5471	5807	6151	6491	6857
4441	4787	5107	5477	5813	6163	6521	6863
4447	4789	5113	5479	5821	6173	6529	6869
4451	4793	5119	5483	5827	6197	6547	6871
4457	4799	5147	5501	5839	6199	6551	6883
4463	4801	5153	5503	5843	6203	6553	6899
4481	4813	5167	5507	5849	6211	6563	6907
4483	4817	5171	5519	5851	6217	6569	6911
4493	4831	5179	5521	5857	6221	6571	6917
4507	4861	5189	5527	5861	6229	6577	6947
4513	4871	5197	5531	5867	6247	6581	6949
4517	4877	5209	5557	5869	6257	6599	6959
4519	4889	5227	5563	5879	6263	6607	6961
4523	4903	5231	5569	5881	6269	6619	6967
4547	4909	5233	5573	5897	6271	6637	6971
4549	4919	5237	5581	5903	6277	6653	6977
4561	4931	5261	5591	5923	6287	6659	6983
4567	4933	5273	5623	5927	6299	6661	6991
4583	4937	5279	5639	5939	6301	6673	6997
4591	4943	5281	5641	5953	6311	6679	7001
4597	4951	5297	5647	5981	6317	6689	7013
4603	4957	5303	5651	5987	6323	6691	7019
4621	4967	5309	5653	6007	6329	6701	7027
4637	4969	5323	5657	6011	6337	6703	7039
4639	4973	5333	5659	6029	6343	6709	7043
4643	4987	5347	5669	6037	6353	6719	7057
4649	4993	5351	5683	6043	6359	6733	7069
4651	4999	5381	5689	6047	6361	6737	7079
4657	5003	5387	5693	6053	6367	6761	7103
4663	5009	5393	5701	6067	6373	6763	7109
4673	5011	5399	5711	6073	6379	6779	7121
4679	5021	5407	5717	6079	6389	6781	7127
4691	5023	5413	5737	6089	6397	6791	7129
4703	5039	5417	5741	6091	6421	6793	7151
4721	5051	5419	5743	6101	6427	6803	7159
4723	5059	5431	5749	6113	6449	6823	7177
4729	5077	5437	5779	6121	6451	6827	7187
4733	5081	5441	5783	6131	6469	6829	7193
4751	5087	5443	5791	6133	6473	6833	7207

Table T.2. (*cont.*)

7211	7561	7907	8273	8647	8971	9337	9677
7213	7573	7919	8287	8663	8999	9341	9679
7219	7577	7927	8291	8669	9001	9343	9689
7229	7583	7933	8293	8677	9007	9349	9697
7237	7589	7937	8297	8681	9011	9371	9719
7243	7591	7949	8311	8689	9013	9377	9721
7247	7603	7951	8317	8693	9029	9391	9733
7253	7607	7963	8329	8699	9041	9397	9739
7283	7621	7993	8353	8707	9043	9403	9743
7297	7639	8009	8363	8713	9049	9413	9749
7307	7643	8011	8369	8719	9059	9419	9767
7309	7649	8017	8377	8731	9067	9421	9769
7321	7669	8039	8387	8737	9091	9431	9781
7331	7673	8053	8389	8741	9103	9433	9787
7333	7681	8059	8419	8747	9109	9437	9791
7349	7687	8069	8423	8753	9127	9439	9803
7351	7691	8081	8429	8761	9133	9461	9811
7369	7699	8087	8431	8779	9137	9463	9817
7393	7703	8089	8443	8783	9151	9467	9829
7411	7717	8093	8447	8803	9157	9473	9833
7417	7723	8101	8461	8807	9161	9479	9839
7433	7727	8111	8467	8819	9173	9491	9851
7451	7741	8117	8501	8821	9181	9497	9857
7457	7753	8123	8513	8831	9187	9511	9859
7459	7757	8147	8521	8837	9199	9521	9871
7477	7759	8161	8527	8839	9203	9533	9883
7481	7789	8167	8537	8849	9209	9539	9887
7487	7793	8171	8539	8861	9221	9547	9901
7489	7817	8179	8543	8863	9227	9551	9907
7499	7823	8191	8563	8867	9239	9587	9923
7507	7829	8209	8573	8887	9241	9601	9929
7517	7841	8219	8581	8893	9257	9613	9931
7523	7853	8221	8597	8923	9277	9619	9941
7529	7867	8231	8599	8929	9281	9623	9949
7537	7873	8233	8609	8933	9283	9629	9967
7541	7877	8237	8623	8941	9293	9631	9973
7547	7879	8243	8627	8951	9311	9643	
7549	7883	8263	8629	8963	9319	9649	
7559	7901	8269	8641	8969	9323	9661	

Table T.3. *The values of* $\tau(n)$, $\sigma(n)$, $\varphi(n)$, $\mu(n)$, $\omega(n)$, *and* $\Omega(n)$ *for natural numbers less than or equal to 100.*

n	$\tau(n)$	$\sigma(n)$	$\varphi(n)$	$\mu(n)$	$\omega(n)$	$\Omega(n)$
1	1	1	1	1	1	1
2	2	3	1	−1	1	1
3	2	4	2	−1	1	1
4	3	7	2	0	1	2
5	2	6	4	−1	1	1
6	4	12	2	1	2	2
7	2	8	6	−1	1	1
8	4	15	4	0	1	3
9	3	13	6	0	1	2
10	4	18	4	1	2	2
11	2	12	10	−1	1	1
12	6	28	4	0	2	3
13	2	14	12	−1	1	1
14	4	24	6	1	2	2
15	4	24	8	1	2	2
16	5	31	8	0	1	4
17	2	18	16	−1	1	1
18	6	39	6	0	2	3
19	2	20	18	−1	1	1
20	6	42	8	0	2	3
21	4	32	12	1	2	2
22	4	36	10	1	2	2
23	2	24	22	−1	1	1
24	8	60	8	0	2	4
25	3	31	20	0	1	2
26	4	42	12	1	2	2
27	4	40	18	0	1	3
28	6	56	12	0	2	3
29	2	30	28	−1	1	1
30	8	72	8	−1	3	3
31	2	32	30	−1	1	1
32	6	63	16	0	1	5
33	4	48	20	1	2	2
34	4	54	16	1	2	2
35	4	48	24	1	2	2
36	9	91	12	0	2	4
37	2	38	36	−1	1	1
38	4	60	18	1	2	2
39	4	56	24	1	2	2
40	8	90	16	0	2	4
41	2	42	40	−1	1	1
42	8	96	12	−1	3	3
43	2	44	42	−1	1	1
44	6	84	20	0	2	3
45	6	78	24	0	2	3
46	4	72	22	1	2	2

Table T.3. (*cont.*)

n	$\tau(n)$	$\sigma(n)$	$\varphi(n)$	$\mu(n)$	$\omega(n)$	$\Omega(n)$
47	2	48	46	−1	1	1
48	10	124	16	0	2	5
49	3	57	42	0	1	2
50	6	93	20	0	2	3
51	4	72	32	1	2	2
52	6	98	24	0	2	3
53	2	54	52	−1	1	1
54	8	120	18	0	2	4
55	4	72	40	1	2	2
56	8	120	24	0	2	4
57	4	80	36	1	2	2
58	4	90	28	1	2	2
59	2	60	58	−1	1	1
60	12	168	16	0	3	4
61	2	62	60	−1	1	1
62	4	96	30	1	2	2
63	6	104	36	0	2	3
64	7	127	32	0	1	6
65	4	84	48	1	2	2
66	8	144	20	−1	3	3
67	2	68	66	−1	1	1
68	6	126	32	0	2	3
69	4	96	44	1	2	2
70	8	144	24	−1	3	3
71	2	72	70	−1	1	1
72	12	195	24	0	2	5
73	2	74	72	−1	1	1
74	4	114	36	1	2	2
75	6	124	40	0	2	3
76	6	140	36	0	2	3
77	4	96	60	1	2	2
78	8	168	24	−1	3	3
79	2	80	78	−1	1	1
80	10	186	32	0	2	5
81	5	121	54	0	1	4
82	4	126	40	1	2	2
83	2	84	82	−1	1	1
84	12	224	24	0	3	4
85	4	108	64	1	2	2
86	4	132	42	1	2	2
87	4	120	56	1	2	2
88	8	180	40	0	2	4
89	2	90	88	−1	1	1
90	12	234	24	0	3	4
91	4	112	72	1	2	2
92	6	168	44	0	2	3
93	4	128	60	1	2	2
94	4	144	46	1	2	2

Table T.3. (*cont.*)

n	$\tau(n)$	$\sigma(n)$	$\varphi(n)$	$\mu(n)$	$\omega(n)$	$\Omega(n)$
95	4	120	72	1	2	2
96	12	252	32	0	2	6
97	2	98	96	−1	1	1
98	6	171	42	0	2	3
99	6	156	60	0	2	3
100	9	217	40	0	2	4

Answers to selected exercises

Exercises 1.1

1. Even $\pm$ even $= 2n \pm 2m = 2\ (m \pm n)$ which is even. Odd $\pm$ even $= (2n+1) \pm 2m = 2(n \pm m) + 1$ which is odd. Odd + odd $= (2n + 1) + (2m+1) = 2\ (m+n+1)$ which is even. Odd $-$ odd $= (2n + 1) - (2m+1) = 2\ (m-n)$ which is even.
2. 8, 10, 14, 15, 16, 18, 21, 22, 24, 26.
3. $t_n + t_{n+1} = \dfrac{n(n+1)}{2} + \dfrac{(n+1)(n+2)}{2} = (n+1)^2 = s_{n+1}$.
4. $9t_n + 1 = t_{3n+1}$; $25t_n + 3 = t_{5n+2}$; $49t_n + 6 = t_{7n+3}$; $(2m+1)^{2t_n} + t_m = t_{\{(2m+1)n\}} + m$.
5. $$(t_n)^2 - (t_{n-1})^2 = \left[\frac{n(n+1)}{2}\right]^2 - \left[\frac{(n-1)n}{2}\right]^2 = \frac{n^2}{4}[(n+1)^2 - (n-1)^2] = n^3.$$
6. $t_{608} = 185\,136 = 56 \cdot 57 \cdot 58$.
7. $n(n+1)(n+2)(n+3) + 1 = n^4 + 6n^3 + 11n^2 + 6n + 1 = (n^2 + 3n + 1)^2$.
8. $*_n = s_n + 4t_{n-1} = n^2 + 4\left[\dfrac{(n-1)n}{2}\right] = 3n^2 - 2n$.
9. $8k + 3 = 8(t_n + t_m + t_r) + 3 = 4n(n+1) + 4m(m+1) + 4r(r + 1) + 3 = (2n+1)^2 + (2m+1)^2 + (2r+1)^2$.
10. In odd rows the middle term, $(2n+1)^2$, is flanked on the left by $[(2n+1)^2 - 2n]$, $[(2n+1)^2 - 2n + 2]$, $\ldots$, $[(2n+1)^2 - 2]$ and on the right by $[(2n+1)^2 + 2n]$, $[(2n+1)^2 + 2n - 2]$, $\ldots$, $[(2n+1)^2 + 2]$. Therefore, the sum of the $2n+1$ terms on that row is given by $(2n+1) \cdot (2n+1)^2 = (2n+1)^3$. In even rows, the terms on the left side are $[(2n)^2 - (2n-1)]$, $[(2n)^2 - (2n-3)]$, $\ldots$, $[(2n)^2 - 1]$ and on the right side are $[(2n)^2 + (2n-1)]$, $[(2n)^2 + (2n-3)]$, $\ldots$, $[(2n)^2 + 1]$. Therefore, $2n$ terms sum to $2n \cdot (2n)^2 = (2n)^3$.

11. (a) $s_{2n+1} = (2n+1)^2 = 4n^2 + 4n + 1 = n^2 + (n+1)^2 + 2n(n+1) = s_n + s_{n+1} + 2o_n$.
    (b) $s_{2n} = (2n)^2 = 4n^2 = (n-1)n + n(n+1) + 2n^2 = o_{n-1} + o_n + 2s_n$.

12. $s_n + t_{n-1} = n^2 + \dfrac{(n-1)n}{2} = \dfrac{n(3n-1)}{2} = p^5{}_n$.

13. $p^5{}_n = \dfrac{n(3n-1)}{2} = 3\left[\dfrac{(n-1)n}{2}\right] + n = 3t_{n-1} + n$.

14. $3 \cdot p^5{}_n = 3 \cdot \dfrac{n(3n-1)}{2} = \dfrac{(3n-1)(3n)}{2} = t_{3n-1}$.

15. $n = 24$.

16.
$$\begin{aligned} t_{9n+4} - t_{3n+1} &= \frac{(9n+4)(9n+5)}{2} - \frac{(3n+1)(3n+2)}{2} \\ &= \frac{72n^2 + 72n + 18}{2} = (6n+3)^2 = [3(2n+1)]^2. \end{aligned}$$

17. Both sides equal $(n+1)(n+2)(n^2+3n+4)/8$.

18.
$$\begin{aligned} t_{2mn+m} &= \frac{(2mn+m)(2mn+m+1)}{2} = \frac{4m^2 n(n+1)}{2} \\ &\quad + \frac{m(m+1)}{2} + mn = 4m^2 t_n + t_m + mn. \end{aligned}$$

19. 2, 8, 20, 40, 70, 112, 168, 240, 330, 440.

20. $p^6{}_n = 2n^2 - n$.

21. $40\,755 = t_{285} = p^5{}_{165} = p^6{}_{143}$.

22. We have
$$\begin{aligned} p^m{}_n &= \binom{n-1}{0} + (m-1)\binom{n-1}{1} + (m-2)\binom{n-1}{2} \\ &= 1 + (m-1)(n-1) + \frac{(m-2)(n-1)(n-2)}{2} \\ &= \frac{(m-2)n^2}{2} - \frac{(m-4)n}{2}. \end{aligned}$$

23.
$$\begin{aligned} p^m{}_n + p^3{}_{n-1} &= \left[\frac{(m-2)n^2}{2} - \frac{(m-4)n}{2}\right] + \frac{(n-1)n}{2} \\ &= \frac{(m-1)n^2}{2} - \frac{(m-3)n}{2} \\ &= p^{m+1}{}_n. \end{aligned}$$

24.
$$\begin{aligned} p^m{}_n + p^m{}_r + nr(m-2) &= \left[\frac{(m-2)n^2}{2} - \frac{(m-4)n}{2}\right] \\ &\quad + \left[\frac{(m-2)r^2}{2} - \frac{(m-4)r}{2}\right] + nr(m-2) \end{aligned}$$

$$= \frac{(m-2)(n^2+2nr+r^2)}{2} - \frac{(m-4)(n+r)}{2} = p^m{}_{n+r}.$$

25. $$p^m{}_n = \frac{(m-2)n^2}{2} - \frac{(m-4)n}{2} = \frac{1}{2}(mn^2 - 2n^2 - nm + 4n)$$
$$= \frac{1}{2}(n^2 + n + mn^2 - mn - 3n^2 + 3n)$$
$$= \frac{1}{2}[n(n+1) + (m-3)(n^2 - n)]$$
$$= \frac{n(n+1)}{2} + \frac{(m-3)(n-1)n}{2} = p^3{}_n + (m-3)p^3{}_{n-1}.$$

26. $240 = 3 \cdot 80 = 8 \cdot 30 = 15 \cdot 16$. Hence, $120 = p^{41}{}_3 = p^6{}_8 = p^3{}_{15}$.
27. $8 \cdot 225 \cdot 6 + 4^2 = 104^2$.
28. If x is an m-gonal number, then for some positive integers m and n, $x = (m-2)n^2/2 - (m-4)n/2$. Hence, $2x - (m-2)n^2 = -(m-4)n$, so $4x^2 - 4x(m-2)n^2 + (m-2)^2n^4 = (m-4)^2n^2$. Therefore $8x(m-2) + (m-4)^2 = (2x/n + (m-2)n)^2$.

29. $$P^3{}_n = \binom{n-1}{0} + 3\binom{n-1}{1} + 3\binom{n-1}{2} + \binom{n-1}{3}$$
$$= 1 + 3(n-1) + 3\frac{(n-1)(n-2)}{2} + \frac{(n-1)(n-2)(n-3)}{6}$$
$$= \frac{n^3}{6} + \frac{n^2}{2} + \frac{n}{3} = \frac{n(n+1)(n+2)}{6}.$$

30. $1540 = t_{55} = P^3{}_{20}$; $7140 = t_{119} = P^3{}_{34}$.

31. $$P^3{}_{n-1} + P^3{}_n = \frac{1}{6}(n-1)n(n+1) + \frac{1}{6}n(n+1)(n+2)$$
$$= \frac{1}{6}n(n+1)(2n+1) = P^4{}_n.$$

32. $$P^5{}_n = P^4{}_{n-1} + P^4{}_n = \frac{1}{6}(n-1)n(2n-1) + \frac{1}{6}n(n+1)(2n+1)$$
$$= \frac{1}{3}n(2n^2+1).$$

33. Since $P^m{}_n = p^m{}_1 + p^m{}_2 + \cdots + p^m{}_n$ and

$p^m{}_n = \dfrac{(m-2)n^2}{2} - \dfrac{(m-4)n}{2}$, it follows that

$$P^m{}_n = \left(\frac{m-2}{2}\right)(1^2 + 2^2 + \cdots + n^2) - \left(\frac{m-4}{2}\right)(1 + 2 + \cdots + n)$$
$$= \left(\frac{m-2}{2}\right)\left(\frac{n(n+1)(2n+1)}{6}\right) - \left(\frac{m-4}{2}\right)\left(\frac{n(n+1)}{2}\right)$$
$$= \left(\frac{n+1}{6}\right)\left[\frac{2(m-2)}{2}n^2 - \frac{2(m-4)}{2}n + n\right] = \left(\frac{n+1}{6}\right)(2p^m{}_n + n).$$

34. Since $O_n = P^4_n + P^4_{n-1} = n(2n^2 + 1)/3$, the first 10 octahedral numbers are given by 1, 6, 19, 44, 85, 146, 231, 344, 489, 670.

35. $f^2{}_n = \binom{n+2-1}{2} = \binom{n+1}{2} = \frac{n(n+1)}{2} = t_n.$

$$f^3{}_n = \binom{n+3-1}{3} = \binom{n+2}{3} = \frac{(n+2)(n+1)n}{3}$$
$$= \frac{(n-1)^3 - (n-1)}{6} = F^4{}_n.$$

36. $f^3{}_{n-1} + f^3{}_n = \binom{n+1}{3} + \binom{n+2}{3}$

$$= \frac{(n+1)n(n-1)}{6} + \frac{(n+2)(n+1)n}{6} = \frac{n(n+1)(2n+1)}{6}.$$

37. $n \cdot f^r{}_{n+1} = n\binom{n+r}{r} = \frac{n(n+r)!}{r!n!} = \frac{(r+1)(n+r)!}{(r+1)!(n-1)!}$

$= (n+r)f^{r+1}{}_n.$

38. $xy + x + y = (n^2 + n + 1)^2$; $yz + y + z = (2n^2 + 3n + 3)^2$; $xz + x + z = (2n^2 + n + 2)^2$; $xy + z = (n^2 + n + 2)^2$; $yx + x = (2n^2 + 3n + 2)^2$; $zx + y = (2n^2 + n + 1)^2$.

39. If x, y, A, h denote respectively the legs, area, and hypotenuse of the right triangle, then $x^2, y^2 = \frac{h^2 \pm \sqrt{h^4 - 16A^2}}{2}$.

40. 1, 1, 3, 7, 19, 51, 141, 393, 1107, 3139, 8953.

Exercises 1.2

1. 13 112 221, 1 113 213 211, 31 131 211 131 221.
2. If a four occurs in a look and say sequence then either a four appeared in the previous term or there were four consecutive repeated digits in the previous term. However, because of the linguistic nature of the sequence four or more consecutive repeated digits cannot occur except as the first term. Working backwards, we find that there must be a four in the first or second term. A similar argument applies for the digits 5 through 9.

3. 1, 5, 14, 16, 41, 43, 47, 49, 122, 124.
4. Only (a).
5. 101, 501, 505. Rule: add 4 and reverse the digits.
6. 84, 59, 17. Rule to generate a_n: add $n+2$ and reverse the digits.
7. (a) Happy: 392, 94, 97, 130, 10, 1.
   (b) Happy: 193, 91, 82, 68, 100, 1.
   (c) Sad: 269, 121, 6, 36, 45, 41, 17, 50, 25, 29, 85, 89, 145, 42, 20, 4.
   (d) Sad: 285, 93, 90, 81, 65, 61, 37, 58, 89.
   (e) Sad: 521, 30, 9, 81.
8. There are:
   five of order 1: 1, 153, 370, 371, and 407
   two of order 2: 1459, 919, and 136, 244
   two of order 3: 133, 55, 250, and 217, 352, 160.
9. They are: 1; 1634; 8208; 9747; 6514, 2178; and 13 139, 6725, 4338, 4514, 1138, 4179, 9219.
10. (a) 6, 7, 3, 0, 3, 3, 6, 9, 5, 4, 9, 3, 2, 5, 7, 2, 9, 1, 0, 1, 1, 2, 3, 5, 8, 3, 1, 4, 5, 9, 4, 3, 7, 0, 7, 7, 4, 1, 5, 6, 1, 7, 8, 5, 3, 8, 1, 9, 0, 9, 9, 8, 7, 5, 2, 7, 9, 6, 5, 1 (60); (b) 2, 0, 2, 2, 4, 6, 0, 6, 6, 2, 8, 0, 8, 8, 6, 4, 0, 4, 4, 8 (20); (c) 1, 8, 9, 7, 6, 3, 9, 2, 1, 3, 4, 7 (12); (d) 2, 6, 8, 4 (4); (e) 5, 5, 0 (3); (f) 0 (1). The sum of the periods is 100.
11. (a) 8, 8, 4, 2, 8, 6,
    (b) 4, 8, 2, 6, 2, 2,
    (c) 3, 9, 7, 3, 1, 3,
    (d) 1, 7, 7, 9, 3, 7,
    (e) 6, 4, 4,
    (f) 9, 9, 1,
    (g) 6,
    (h) 5.
12. From the recursive definition of Fibonacci numbers, it follows that u_{3n} is divisible by 2, for any natural number n.
13. Let the sequence be given by a, b, $a+b$, $a+2b$, $2a+3b$, $3a+5b$, $5a+8b$, $8a+13b$, $13a+21b$, $21a+34b$. The sum of the terms equals $55a+88b = 11(5a+8b)$.
14. Set the expression equal to x. Square both sides to obtain $x^2 = x+1$, whose root is τ.
15. Let $|AB| = a$, then $|AD|^2 = |AB|^2 + |BD|^2 = a^2 + (a/2)^2 = (\frac{5}{4})a^2$.

$$|AC| = |AE| = |AD| - |ED| = |AD| - |BD|$$
$$= a\frac{\sqrt{5}}{2} - \left(\frac{a}{2}\right) = a\left(\frac{\sqrt{5}-1}{2}\right).$$

$$\frac{|AB|}{|AC|} = \frac{a}{a\left(\dfrac{\sqrt{5}-1}{2}\right)} = \frac{2}{\sqrt{5}-1} = \tau.$$

16. $|EF|^2 = |EB|^2 = |BC|^2 + |EC|^2 = a^2 + (a/2)^2 = (\frac{5}{4})a^2.$

$$|AG| = |DF| = |DE| + |EF| = \frac{a}{2} + \frac{\sqrt{5}}{2}a = \left(\frac{1+\sqrt{5}}{2}\right)a.$$

    Hence, $|AG|/|AD| = \tau$.

17. Since $\tau^{-1} = \sigma$, multiplying each term by τ gives the desired result.
18. 1, 3, 4, 7, 11, 18, 29, 47, 76, 123.
19. $5778 = t_{107} = \upsilon_{18}$.
20. Let $b_n = \tau^n + \sigma^n$, then $b_{n+2} = \tau^{n+2} + \sigma^{n+2} = (\tau^{n+1} + \tau^n) + (\sigma^{n+1} + \sigma^n) = (\tau^{n+1} + \sigma^{n+1}) + (\tau^n + \sigma^n) = b_{n+1} + b_n$, with $b_1 = \tau + \sigma = 1$ and $b_2 = \tau^2 + \sigma^2 = 3$. Therefore, $b_n = \tau^n + \sigma^n = \upsilon_n$.
21. 1, 1, 2, 4, 7, 13, 24, 44, 81, 149, 274, 504, 927, 1705, 3136, 5768, 10 609, 19 513, 35 890, 66 012. (Tribonacci)
22. 1, 1, 2, 4, 8, 15, 29, 56, 108, 208, 401, 773, 1490, 2872, 5536, 10 671, 20 569, 39 648, 76 424, 147 312. (Tetranacci)
23. (a) 9, 28, 14, 7, 22, 11, 34, 17, 52, 26, 13, 40, 20, 10, 5, 16, 8, 4, 2, 1.
    (b) 50, 25, 76, 38, 19, 58, 29, 88, 44, 22, 11, 34, 17, 52, 26, 13, 40, 20, 10, 5, 16, 8, 4, 2, 1.
    (c) 121, 364, 182, 91, 274, 37, 412, 206, 103, 310, 155, 466, 233, 700, 350, 175, 526, 263, 790, 395, 1186, 593, 1780, 890, 445, 1336, 668, 334, 167, 502, 251, 754, 377, 7132, 566, 283, 850, 425, 1276, 638, 319, 958, 479, 1438, 719, 2158, 1079, 3238, 1619, 4858, 2429, 7288, 3644, 1822, 911, 2734, 1367, 4102, 2051, 6154, 3077, 9232, 4616, 2308, 1154, 577, 1723, 866, 433, 1300, 640, 325, 976, 488, 244, 122, 61, 184, 92, 46, 23, 70, 35, 106, 53, 160, 80, 40, 20, 10, 5, 16, 8, 4, 2, 1.
24. 1, 2; 7, 20, 10, 5, 14; and 17, 50, 25, 74, 37, 110, 55, 164, 82, 41, 122, 61, 182, 91, 272, 136, 68, 34.
25. (a) (b)

$$\begin{array}{rrr}
\text{(a)}\ 9963 & \text{(b)}\ 9421 & 9963 \\
\underline{3699} & \underline{1249} & \underline{3699} \\
6264 & 8172 & 6264 \\
 & & \\
6642 & 8721 & 6642 \\
\underline{2466} & \underline{1278} & \underline{2466} \\
4176 & 7443 & 4176 \\
 & & \\
7641 & 7443 & 7641 \\
\underline{1467} & \underline{3447} & \underline{1467} \\
6174 & 3996 & 6174
\end{array}$$

26. 495.
27.

(a)	(b)	(c)
936	991	864
639	199	468
297	792	396
792	297	(c) 693
1089	1089	1089

28. 2538.
29. 1 2 4 8 7 5, 3 6, and 9.
30. When 9 is added to a natural number the 10s digit is increased by 1 and the units digit decreased by 1 leaving a net change of zero.
31. 220, 224, 232, 239, 253, 263, 274, 287, 304, 311, 316, 326, 337, 350, 358, 374, 388, 407, 418;
    284, 298, 317, 328, 341, 349, 365, 379, 398, 418.
32. Pair up the numbers as follows:

0	– 999 999	sum of digits = 54
1	– 999 998	sum of digits = 54
2	– 999 997	sum of digits = 54
. . . .	. . . .	
499 999	– 500 000	sum of digits = 54
1 000 000		sum of digits = 1

total sum of digits $= 500\,000 \cdot 54 + 1 = 27\,000\,001$.

33. I: K_2, K_4, K_5, K_7, K_8, K_{10}; II: K_3, K_6; III: K_9.
34. 1, 3, 5, 7, 9, 20, 31, 42, 53, 64, 75, 86, 97.
35. If all the digits are less than 5, then double the number will have the sum of its digits equal to 20. If two of the digits are 5, and all the rest zero, double the number will have the sum 2. In all other cases, the sum of the digits of twice the number will be $20 - 9 = 11$.
36. (a) Two: 543, 60, 0.
    (b) Four: 6989, 3888, 1536, 90, 0.
    (c) Seven: 86 898, 27 648, 2688, 768, 336, 54, 20, 0.
    (d) Three: 68 889 789, 13 934 592, 29 160, 0.
    (e) Ten: 3 778 888 999, 438 939 648, 4 478 976, 338 688, 27 648, 2688, 768, 336, 54, 20, 0.
37. 39; 77; 679.
38. 1, 3, 4, 5, 6, 8, 10, 12, 17, 21, 23, 28, 32, 34, 39
39. 2, 3, 5, 7, 8, 9, 13, 14, 18, 19, 24, 25, 29, 30, 35
40. 2, 5, 7, 9, 11, 12, 13, 15, 19, 23, 27, 29, 35, 37, 41, 43, 45, 49, 51, 55, 59, 63, 65, 69, 75, 77, 79, 87, 91, 93
41. 2, 3, 5, 7, 8, 9, 11, 13, 19, 22, 25, 27, 28, 37, 39
42. 1, 2, 4, 7, 10, 13, 16, 19, 22, 25, 28, 31, 34, 37, 40

43. 1, 3, 5, 7, 9, 11, 13, 15, 17, 19, 21, 23, 25, 27, 29
44. 3, 4, 6, 9, 10, 17, 18, 25, 30, 32, 37, 44, 45, 46, 58
45. 2, 3, 6, 12, 18, 24, 36, 48

Exercises 1.3

1. $P(1)$: $1 = 1$.
$P(k+1)$: $[1^2 + 2^2 + \cdots + k^2] + (k+1)^2$

$$= \frac{k(k+1)(2k+1)}{6} + (k+1)^2$$

$$= \left[\frac{k+1}{6}\right][k(2k+1) + 6(k+1)]$$

$$= \frac{(k+1)(k+2)(2k+3)}{6}.$$

2. $P(1)$: $1 = 1$.

$P(k+1)$: $[1^2 + 3^2 + \cdots + (2k-1)^2] + (2k+1)^2$

$$= \left[\frac{(4k^3 - k)}{3}\right] + (2k+1)^2$$

$$= \frac{4k^3 - k + 3(2k+1)^2}{3}$$

$$= \frac{4k^3 + 12k^2 + 11k + 3}{3}$$

$$= \frac{4(k+1)^3 - (k+1)}{3}.$$

3. $P(1)$: $\frac{1}{2} = \frac{1}{2}$.

$P(k+1)$: $\left[\frac{1}{2} + \frac{1}{6} + \frac{1}{12} + \cdots + \frac{1}{k(k+1)}\right] + \frac{1}{(k+1)(k+2)}$

$$= \frac{k}{k+1} + \frac{1}{(k+1)(k+2)}$$

$$= \left[\frac{1}{k+1}\right]\left[k + \frac{1}{k+2}\right]$$

$$= \left[\frac{1}{k+1}\right]\left[\frac{k^2+2k+1}{k+2}\right]$$

$$= \frac{k+1}{k+2}.$$

4. $P(1)$: $1 = 1$.

   $P(k+1)$: $[t_1 + t_2 + \cdots + t_k] + t_{k+1}$

$$= \frac{k(k+1)(k+2)}{6} + \frac{(k+1)(k+2)}{2}$$

$$= \left[\frac{(k+1)(k+2)}{6}\right](k+3).$$

5. $P(1)$: $1 = 1$.

   $P(k+1)$: $[1^3 + 2^3 + 3^3 + \cdots + k^3] + (k+1)^3$

$$= {t_k}^2 + (k+1)^3$$

$$= \frac{k^2(k+1)^2}{4} + (k+1)^3$$

$$= \left[\frac{(k+1)^2}{4}\right][k^2 + 4(k+1)]$$

$$= \frac{(k+1)^2(k+2)^2}{4}$$

$$= (t_{k+1})^2.$$

6. $P(1)$: $1 + a \geqslant 1 + a$.
   $P(k+1)$: $(1+a)^{k+1} > (1+ka)(1+a) = 1 + (k+1)a + ka^2 > 1 + (k+1)a$.
7. $P(4)$: $4! = 24 > 16 = 4^2$. Suppose that $k! > k^2$. Now $(k+1)! > k(k!) > k(k^2) > 3(k^2) = 2k^2 + k^2 > k^2 + 2k + 1 = (k+1)^2$.

8. $P(1)$: $u_1 = 1 = u_2$.
   $P(k+1)$: $[u_1 + u_3 + \cdots + u_{2k-1}] + u_{2k+1} = u_{2k} + u_{2k+1} = u_{2k+2}$.
9. $P(1)$: $u_1^2 = 1 = u_1 \cdot u_2$.
   $$\begin{aligned}[u_1^2 + u_2^2 + \cdots + u_k^2] + u_{k+1}^2 &= u_k u_{k+1} + u_{k+1}^2 \\ &= u_{k+1}(u_k + u_{k+1}) \\ &= u_{k+1}u_{k+2}.\end{aligned}$$
10. $P(1)$: $u_2 = u_3 - 1$.
    $$\begin{aligned}P(k+1)\text{: } [u_2 + u_4 + u_6 + \cdots + u_{2k}] + u_{2k+2} &= u_{2k+1} - 1 + u_{2k+2} \\ &= u_{2k+3} - 1.\end{aligned}$$
11. Since $u_1 = 1 > 1/\tau$ and $u_2 = 1 \geqslant \tau^0$, let $u_k \geqslant \tau^{k-2}$, for $1 \leqslant k < n$; hence, $u_{n-2} \geqslant \tau^{n-4}$ and $u_{n-1} \geqslant \tau^{n-3}$. In addition, $u_n = u_{n-2} + u_{n-1} \geqslant \tau^{n-3} + \tau^{n-4} = (1+\tau)\tau^{n-4} = \tau^2\tau^{n-4} = \tau^{n-2}$ and the result follows from the principle of mathematical induction.
12. For any positive integer n, $u_{n+1}^2 - u_n^2 = (u_{n+1} + u_n)(u_{n+1} - u_n) = u_{n+2}u_{n-1}$.
13. $P(1)$: $u_1 + v_1 = 1 + 1 = 2 = 2u_2$. Suppose the formula is true when $1 \leqslant n \leqslant k$. We have $u_{k+1} + v_{k+1} = (u_k + u_{k-1}) + (v_k + v_{k-1}) = (u_k + v_k) + (u_{k-1} + v_{k-1}) = 2u_{k+1} + 2u_k = 2u_{k+2}$.
14. $P(2)$: $v_1 + v_3 = 1 + 4 = 5 = 5u_1$. Suppose the formula is true for $1 \leqslant n \leqslant k$. We have
    $5u_{k+1} = 5u_k + 5u_{k-1} = (v_{k-1} + v_{k+1}) + (v_{k-2} + v_k) = (v_{k-1} + v_{k-2}) + (v_k + v_{k+1}) = v_k + v_{k+2}$.
15. $P(2)$: $v_2 = 3 = 2 + 1 = u_3 + u_1$. $P(3)$: $v_3 = 4 = 3 + 1 = u_4 + u_2$. Suppose the formula is true for k and $k-1$. We have $v_{k+1} = v_k + v_{k-1} = (u_{k-1} + u_{k+1}) + (u_{k-2} + u_k) = (u_{k-1} + u_{k-2}) + (u_k + u_{k+1}) = u_k + u_{k+2}$.
16. Since $u_n = (\tau^n - \sigma^n)/(\tau - \sigma)$ and $v_n = \tau^n + \sigma^n$, for any positive integer n,
    $$u_n \cdot v_n = \frac{\tau^n - \sigma^n}{\tau - \sigma} \cdot (\tau^n + \sigma^n) = \frac{\tau^{2n} - \sigma^{2n}}{\tau - \sigma} = u_{2n}.$$
17. From the two previous exercises, for any positive integer k, $u_{2k+2} = u_{k+1}v_{k+1} = (u_{k+2} + u_k)(u_{k+2} - u_k) = u_{k+2}^2 - u_k^2$.
18. $P(2)$: $u_3 \cdot u_1 + (-1)^3 = 2 - 1 = 1 = u_2^2$. Suppose that $(-1)^{k+1} = u_k^2 - u_{k+1}u_{k-1}$; then
    $$\begin{aligned}(-1)^{k+2} &= -u_k^2 + u_{k+1}u_{k-1} \\ &= -u_k u_k + u_{k+1}(u_{k+1} - u_k) \\ &= u_{k+1}^2 - u_k(u_{k+1} + u_k) \\ &= u_{k+1}^2 - u_k u_{k+2}.\end{aligned}$$

19. S being infinite leads to a contradiction of the well-ordering principle.
20. Suppose that a is an integer such that $0 < a < 1$. Then $1 > a > a^2 > a^3 > \cdots$ making $\{a, a^2, a^3, \ldots\}$ an infinite set of positive integers having no least element, contradicting the well-ordering principle.
21. The result follows by induction or from the fact that

$$\binom{r}{r} + \binom{r+1}{r} + \binom{r+2}{r} + \cdots + \binom{n+r-1}{r} = \binom{n+r}{r+1}.$$

Exercises 1.4

1. $1\,533\,776\,805 = p^3{}_{55\,385} = p^5{}_{31\,977} = p^6{}_{27\,693}$.
2. 55, 66, 171, 595, 666.
3. $(798\,644)^2 = 637\,832\,238\,736$; $(1\,270\,869)^2 = 1\,615\,108\,015\,161$.
4. $(54\,918)^2 = 3\,015\,986\,724$; $(84\,648)^2 = 7\,165\,283\,904$.
5. No, $11\,826^2 = 139\,854\,276$, '0' is not represented. In Hill's defense, it should be noted that at the time many did not consider 0 to be a digit.
6. $90 \cdot 16\,583\,742 = 1\,492\,536\,780$.
7. $428\,571 = 3 \cdot 142\,857$.
8. $(76)^2 = 57\mathbf{76}$; $(625)^2 = 390\,\mathbf{625}$.
9. $325 = t_{25}$, $195\,625 = t_{625}$, $43\,959\,376 = t_{9376}$.
10. $297^2 = 88\,209$ and $88 + 209 = 297$.
    $142\,857^2 = 20\,408\,122\,449$ and $20\,408 + 122\,449 = 142\,857$.
    $1\,111\,111\,111^2 = 1\,234\,567\,900\,987\,654\,321$ and
    $123\,456\,790 + 0\,987\,654\,321 = 1\,111\,111\,111$.
11. $153 = 1^3 + 5^3 + 3^3$; $371 = 3^3 + 7^3 + 1^3$.
12. $165\,033 = (16)^3 + (50)^3 + (33)^3$.
13. (a) $43 = 4^2 + 3^3$,
    (b) $63 = 6^2 + 3^3$,
    (c) $89 = 8^1 + 9^2$,
    (d) $132 = 1^1 + 3^1 + 2^7$.
14. $2^5 9^2 = 2592$.
15. $4! + 0! + 5! + 8! + 5! = 40\,585$.
16. $21\,978 \times 4 = 87\,912$, $219\,978 \times 4 = 879\,912$, $10\,989 \times 9 = 98\,901$.
17. The sum equals $t_1 + t_2 + \cdots + t_{12}$ or 364 days.
18. The answer represents the month and day that you were born.
19. The result follows from the fact that $7 \cdot 143 = 1001$.
20. Solve $(10x + y)/(10y + z) = \dfrac{x}{z}$ or $9xz = y(10x - z)$ to obtain

$$\frac{16}{64}, \frac{19}{95}, \frac{26}{65}, \frac{49}{98}, \frac{11}{11}, \frac{22}{22}, \ldots, \frac{99}{99}.$$

21. $1 = (9+9-9)/9,$
$2 = (9/9)+(9/9),$
$3 = (9+9+9)/9,$
$4 = (9/\sqrt{9})+(9/9),$
$5 = 9-(9/9)-\sqrt{9},$
$6 = 9+9-9-\sqrt{9},$
$7 = 9+(9/9)-\sqrt{9},$
$8 = (\sqrt{9})(\sqrt{9})-(9/9),$
$9 = (9+9+9)/\sqrt{9},$
$10 = (99-9)/9,$
$11 = 9+\sqrt{9}-9/9,$
$12 = (99+9)/9,$
$13 = 9+\sqrt{9}+9/9,$
$14 = 99/9+\sqrt{9},$
$15 = 9+9-9/\sqrt{9},$
$16 = 9+9-\sqrt{9}+.\overline{9},$
$17 = 9+9-9/9,$
$18 = 9+9+9-9,$
$19 = 9+9+9/9,$
$20 = 99/9+9,$
$21 = 9+9+9/\sqrt{9},$
$22 = 9+9+\sqrt{9}+.\overline{9},$
$23 = 9\sqrt{9}-\sqrt{9}-.\overline{9},$
$24 = 9+9+9-\sqrt{9},$
$25 = 9\sqrt{9}-.\overline{9}-.\overline{9}.$

22. Possible answers include the following.

$$\begin{aligned}
0 &= (9-9)+(9-9) = (\sqrt{9}-\sqrt{9})+(\sqrt{9}-\sqrt{9}) \\
&= (9-9)-(9-9) = (\sqrt{9}-\sqrt{9})-(\sqrt{9}-\sqrt{9}) \\
&= (9+9)-(9+9) = (\sqrt{9}+\sqrt{9})-(\sqrt{9}+\sqrt{9}) \\
&= (9-9)+(\sqrt{9}-\sqrt{9}) = (9/9)-(9/9) = (\sqrt{9}/\sqrt{9})-(9/9) \\
&= (\sqrt{9}/\sqrt{9})-(\sqrt{9}/\sqrt{9}) = 9\cdot 9-9\cdot 9 \\
&= \sqrt{9}\cdot\sqrt{9}-\sqrt{9}\cdot\sqrt{9} = (9\sqrt{9}/9)-\sqrt{9} = 9-9(9/9) \\
&= 9-9(\sqrt{9}/\sqrt{9}) = 9\cdot\sqrt{9}-9\cdot\sqrt{9} \\
&= 9-\sqrt{9}-\sqrt{9}-\sqrt{9} = (9\sqrt{9}/\sqrt{9})-9 \\
&= (.\overline{9}+.\overline{9})-(.\overline{9}+.\overline{9}) = (.\overline{9}-.\overline{9})-(.\overline{9}-.\overline{9}) \\
&= (.\overline{9}-.\overline{9})+(.\overline{9}-.\overline{9}) = (.\overline{9}/.\overline{9})-(.\overline{9}/.\overline{9}) = (9-9)+(.\overline{9}-.\overline{9}) \\
&= (.\overline{9}-.\overline{9})+(\sqrt{9}-\sqrt{9}) = (9-9)-(.\overline{9}-.\overline{9}).
\end{aligned}$$

23. $1 = (4+4-4)/4,$
$2 = (4/4)+(4/4),$
$3 = (4+4+4)/4,$
$4 = (4/\sqrt{4})+(4/\sqrt{4}),$
$5 = 4+(\sqrt{4}+\sqrt{4})/4,$
$6 = 4+(4+4)/4,$
$7 = 4+4-4/4,$
$8 = 4+4+(4-4),$
$9 = 4+4+(4/4),$
$10 = 4+4+(4/\sqrt{4}),$
$11 = (4!-(4/\sqrt{4})/\sqrt{4},$
$12 = 4\cdot(4-(4/4)),$
$13 = (44)/4+\sqrt{4},$
$14 = 4\cdot 4-(4/\sqrt{4}),$
$15 = 4\cdot 4-(4/4),$
$16 = 4+4+4+4,$
$17 = 4\cdot 4+(4/4),$
$18 = 4\cdot 4+(4/\sqrt{4}),$
$19 = 4!-4-(4/4),$
$20 = 4\cdot(4+(4/4)),$
$21 = 4!-4+4/4,$
$22 = ((44)/4)\cdot\sqrt{4},$
$23 = 4!-(\sqrt{4}+\sqrt{4})/4,$
$24 = 4\cdot 4+4+4,$
$25 = 4!+(\sqrt{4}+\sqrt{4})/4.$

24. $$\lim_{n\to\infty} \frac{u_{n+1}}{u_n} = \lim_{n\to\infty} \frac{\tau^{n+1} - \sigma^{n+1}}{\tau^n - \sigma^n} = \tau,$$

since

$$\lim_{n\to\infty}\sigma^{n+1} = \lim_{n\to\infty}\sigma^n = 0,$$

while

$$\lim_{n\to\infty}\tau^{n+1} = \lim_{n\to\infty}\tau^n = \infty.$$

25. If $x_1 = x_2 = \cdots = x_{n-2} = 1$, $x_{n-1} = 2$, and $x_n = n$, $x_1 + x_2 + \cdots + x_n = x_1 \cdot x_2 \cdots x_n = 2n$.

26. $$\sum_{n=1}^{\infty} \frac{1}{t_n} = \sum_{n=1}^{\infty} \frac{2}{n(n+1)} = 2\sum_{n=1}^{\infty}\left(\frac{1}{n} - \frac{1}{n+1}\right) = 2.$$

27. $n = 84$.
29. Thursday
31. June 25, 1963
32. October 2, 1917
33. Tuesday
34. (a) Tuesday; (b) Wednesday; (c) Saturday.
35. 10
36.

1–8	10–15	19–30	24–25	37–44
2–7	11–14	20–29	33–48	38–43
3–6	12–13	21–28	34–47	39–42
4–5	17–32	22–27	35–46	40–41
9–16	18–31	23–26	36–45	

37. 8, 14, 16, 18, and every even integer greater than 22.
38. There are 16 fifth order zigzag numbers, namely 24 351, 25 341, 34 251, 35 241, 45 231, 14 352, 15 342, 34 152, 35 142, 45 132, 14 253, 15 243, 24 153, 25 143, 13 254, and 23 154.
39. $B_6 = \frac{1}{42}$, $B_8 = -\frac{1}{30}$, and $B_{10} = \frac{5}{66}$.
40. 2, 4, 17, 48, 122, 323. No.
41. $A^1 = \begin{pmatrix} 1 & 1 \\ 1 & 0 \end{pmatrix} = \begin{pmatrix} u_2 & u_1 \\ u_1 & u_0 \end{pmatrix}$,

$$A^{n+1} = A^n \cdot A = \begin{pmatrix} u_{n+1} & u_n \\ u_n & u_{n-1} \end{pmatrix} \cdot \begin{pmatrix} 1 & 1 \\ 1 & 0 \end{pmatrix} = \begin{pmatrix} (u_{n+1} + u_n) & u_{n+1} \\ (u_n + u_{n-1}) & u_n \end{pmatrix}$$
$$= \begin{pmatrix} u_{n+2} & u_{n+1} \\ u_{n+1} & u_n \end{pmatrix}.$$

42. $\det(A^n) = (-1)^n$.
43. 2
44. $(a+1)(b+1)(c+1) + (a-1)(b-1)(c-1) = abc + ac + bc + c + ab + a + b + 1 + abc - ac - bc + c - ab + a + b - 1 = 2(a +$

$b + c + abc)$.

45. $[a^4 - (ab + bc + ac)^2] + [b^4 - (ab + bc + ac)^2] = (a^2 - ab - bc - ac)(a^2 + ab + bc + ac) + (b^2 - ab - bc - ac)(b^2 + ab + bc + ac) = (a^2 - ab - bc - ac)[a(a + b + c) + bc] + (b^2 - ab - bc - ac)[b(a + b + c) + ac] = (a^2 - ab - bc - ac)(bc) + (b^2 - ab - bc - ac)(ac) = (-c^2)(a^2 + 2ab + b^2) = c^4$.
46. The formula works for $n = 1, 2, \ldots, 8$, but not for $n > 8$.
47. (a) (1) Great Pyramid of Khufu
    (2) Hanging Gardens of Babylon
    (3) Mausoleum at Halicarnassus
    (4) Artemision at Ephesus
    (5) Colossus of Rhodes
    (6) Olympian Zeus
    (7) Pharos at Alexandria

    (b) (1) Thales of Miletus (natural philosopher)
    (2) Solon of Athens (politician and poet)
    (3) Bias of Priene (philosopher)
    (4) Chilon of Sparta (philosopher)
    (5) Cleobulus of Rhodes (tyrant)
    (6) Periander of Corinth (tyrant)
    (7) Pittacus of Mitylene (statesman and lawyer)

    (c) There are two versions:

    (A) (1) Arsinoe II (Egyptian queen)
    (2) Sappho of Lesbos (poet)
    (3) Corinna (poet)
    (4) Antiochis of Lycia (physician)
    (5) Flavia Publica Nicomachis of Phoecia (politician)
    (6) Apollonia (philosopher)
    (7) Iaia Marcus Varro (artist)

    (B) (1) Arete of Cyrene (philosopher)
    (2) Apasia of Miletus (philosopher)
    (3) Diotima of Mantinea (philosopher)
    (4) Hypatia of Alexandria (mathematician and philosopher)
    (5) Leontium of Athens (philosopher)
    (6) Theano (philosopher and physician)
    (7) Themistoclea (philosopher)

Exercise 2.1

1. Since $d|a$ and $d|b$ there exist integers x and y such that $dx = a$ and $dy = b$. Hence, $c = a - b = dx - dy = d(x - y)$. Since c is a multiple of d, d divides c.
2. Since $a|b$ and $b|c$, there exist integers r and s such that $ar = b$ and $bs = c$. Hence, $c = bs = (ar)s = a(rs)$. Therefore, $a|c$.
3. If $a|b$ and $b|a$ there exist integers r and s such that $ar = b$ and $bs = a$. Hence, $a = bs = (ar)s = a(rs)$ implying that $rs = 1$. Since a and b are positive $r = s = 1$ and $a = b$.
4. Since $a|b$, there exists an integer r such that $ar = b$. Hence, $a + a + \cdots + a = a + (r - 1)a = b$. The sum contains r terms. Since $r - 1 \geqslant 1$, $a \leqslant b$.
5. Let $a|b$ and $c = a + b$. Then $ax = b$ and $c = a + ax = a(1 + x)$. Therefore, $a|c$.
6. Let $ax = b$ and $cy = d$. Thus, $bd = ax \cdot cy = ac \cdot xy$. Therefore, $ac|bd$.
7. (a) False, $6|2 \cdot 3$ yet $6 \nmid 2$ and $6 \nmid 3$.
   (b) False, $6|(3 + 3)$ yet $6 \nmid 3$.
   (c) False, $8^2|4^3$, but $8 \nmid 4$,
   (d) False, $2^2|36$, $3^2|36$, and $2^2 \leqslant 3^2$ yet $2 \nmid 3$.
   (e) True.
8. When $p < q$ and q is divided by p, there are only q possible remainders. Hence, the resulting decimal expansion must repeat after at most q divisions.
9. If $n = 0.123\,123\,123\ldots$, then $1000n = 123.123\,123\ldots$ and $999n = 1000n - n = 123$. Therefore, $n = \frac{123}{999}$.
10. Every integer is of the form $3k$, $3k + 1$ or $3k + 2$. The square of any integer of the form $3k$ is of the form $3m$ and the square of any integer of the form $3k + 1$ or $3k + 2$ is of the form $3m + 1$, where k and m are integers. Suppose that $\sqrt{3} = p/q$, where p and q are integers in lowest form. Hence, $p = \sqrt{3}q$. It follows that $p^2 = 3q^2$, thus 3 divides p^2. Hence, p is divisible by 3 and $p = 3r$. In addition, $3q^2 = p^2 = 9r^2$, thus, $q^2 = 3r^2$ implying that 3 divides q. Thus, p and q have a common factor, contradicting the assumption that p/q was in lowest form.
11. (a) The result follows since either n or $n + 1$ must be even. (b) The result follows since one of n, $n + 1$, or $n + 2$ must be divisible by 3.
12. Since $2|n(n + 1)$, if $3 \nmid n$ and $3 \nmid (n + 1)$, then $n = 3k + 1$, implying that $3|(2n + 1)$. Therefore, $6|n(n + 1)(2n + 1)$.

13. $(2n+1)^2 + (2m+1)^2 = 4(n^2 + m^2 + n + m) + 2 = 4k + 2$ which can never be square by Theorem 2.3.
14. $(n+1)^3 - n^3 = 3n(n+1) + 1 = 6k + 1$, which is always odd.
15. If $n = 2k+1$, then $n^2 - 1 = 4k(k+1) = 8m$.
16. If $3 \nmid n$, then $3|(n-1)$ or $3|(n+1)$. From the previous exercise, $8|(n^2-1)$. Hence, $24|(n^2-1)$.
17. Since 3 divides $(2 \cdot 1^2 + 7) = 9$, suppose that $k(2k^2+7) = 3x$. We have $(k+1)(2(k+1)^2+7) = 3x + 3(2k^2+2k+3) = 3y$ and the result is established by induction.
18. Since 8 divides $5^2 + 7 = 32$, suppose that $5^{2k} + 7 = 8x$. We have $5^{2(k+1)} + 7 = 25(5^{2k}) + 7 = 24(5^{2k}) + 5^{2k} + 7 = 8(3 \cdot 5^{2k} + x) = 8y$ and the result is established by induction.
19. Since 7 divides $3^3 + 2^3$, suppose that $3^{2k+1} + 2^{k+2} = 7x$. We have $3^{2k+3} + 2^{k+3} = 9(3^{2k+1}) + 2(2^{k+2}) = 7(3^{2k+1}) + 2 \cdot 7x = 7y$ and the result follows by induction.
20. Since 5 divides $3^4 + 2^2$, suppose that $3^{3k+1} + 2^{k+1} = 5x$. We have $3^{3k+4} + 2^{k+2} = 27(3^{3k+1}) + 2(2^{k+1}) = 25(3^{3k+1}) + 2(3^{3k+1}) + 2(2^{k+1}) = 5[5 \cdot 3^{3k+1} + 2x] = 5y$ and the result follows by induction.
21. If $n = 2x$ then $n^2 + 2 = 4x^2 + 2$ and if $n = 2k+1$ then $n^2 + 2 = 4(k^2 + k) + 3$. In either case $4 \nmid (n^2+2)$.
22. If an integer is not a perfect square then its divisors can be grouped into distinct pairs.
23. $6k + 5 = 3(2k+1) + 2 = 3m + 2$. However, $8 = 3 \cdot 2 + 2$ and there does not exist a k such that $8 = 6k + 5$.
24. Every integer is of the form $3k$, $3k+1$, or $3k+2$, and $(3k)^2 = 3(3k^2) = 3M$, $(3k+1)^2 = 3(3k^2+2k) + 1 = 3N + 1$, $(3k+2)^2 = 3(3k^2+4k+1) + 1 = 3R + 1$.
25. If $n = 3k$ then $(3k)^3 = 9(3k^3) = 9M$. If $n = 3k+1$ then $(3k+1)^3 = 9(3k^3 + 3k^2 + k) + 1 = 9N + 1$. If $n = 3k+2$ then $(3k+2)^3 = 9(3k^3 + 16k^2 + 4k + 1) - 1 = 9R - 1$.
26. If $n = 5k$ then $(5k)^2 = 5(5k^2) = 5M$. If $n = 5k+1$ then $(5k+1)^2 = 5(5k^2+2k) + 1 = 5N + 1$. If $n = 5k+2$ then $(5k+2)^2 = 5(5k^2+4k) + 4 = 5R + 4$. If $n = 5k+3$ then $(5k+3)^2 = 5(5k^2+6k+1) + 4 = 5S + 4$. If $n = 5k+4$ then $(5k+4)^2 = 5(5k^2+8k+3) + 1 = 5m + 1$. When squared again the results will each be of the form $5m$ or $5m+1$.
27. Since x^2, y^2, and z^2 must be of the form $8m$, $8m+1$, or $8m+4$, $x^2 + y^2 + z^2$ can only be of the form $8k$, $8k+1$, $8k+2$, $8k+3$, $8k+4$, $8k+5$, or $8k+6$.
28. From Exercise 2.1.26 5 divides $n^5 - n$. In addition, $n^5 - n = (n -$

$1)n(n+1)(n^2+1)$. Since $(n-1)n(n+1)$ is the product of three consecutive integers it is divisible by 6. Thus, n^5-n is divisible by $5 \cdot 6 = 30$. If $n = 2m+1$, then $n^5 - n = 8(2m+1)m(m+1)(2m^2 + 2m + 1)$ and is divisible by 16. Therefore, $240 = 15 \cdot 16$ divides $n^5 - n$.

29. Any square must be of the form $3k$ or $3k+1$. If $n = 2k$ then $3n^2 - 1 = 3(4k^2 - 1) + 2 = 3m + 2$. If $n = 2k+1$ then $3n^2 - 1 = 3(4k^2 + 4k) + 2 = 3m + 2$.
30. $11 = 4 \cdot 2 + 3$, $111 = 4 \cdot 27 + 3$. In general, $111 \ldots 1 = 4 \cdot 277 \ldots 7 + 3$ where the integer on the left contains n ones and the second integer in the product on the right contains $n-2$ sevens. Hence, an integer whose digits are all ones is of the form $4m+3$ and thus cannot be square.
31. Suppose $ax = b$ with $a > b/2$. Therefore, $2a > b = ax$. Hence, $2 > x$, which implies that $x = 1$ and $a = b$, contradicting the fact that $a \neq b$.
32. If $a > \sqrt{n}$ and $b > \sqrt{n}$, then $ab > \sqrt{n} \cdot \sqrt{n} = n$, a contradiction.
33. If $n = ab$, with $a \geqslant b$, let $s = (a+b)/2$ and $t = (a-b)/2$; then $n = s^2 - t^2$. Conversely, if $n = s^2 - t^2$ let $a = s + t$ and $b = s - t$.
34. $40 = 101\,000_2$; $40 = 1111_3$; $173 = 10\,101\,101_2$; $173 = 20\,102_3$; $5437 = 1\,010\,100\,111\,101_2$; $5437 = 21\,110\,101_3$.
35. $101\,011_2 = 43$ and $201\,102_3 = 524$.
36. 1 is triangular. Suppose that $(11 \ldots 1)_9$ with k ones is triangular, say $11 \ldots 1_9 = n(n+1)/2$. Consider $11 \ldots 1_9^*$ with $k+1$ ones. We have $11 \ldots 1_9^* = 9 \cdot 11 \ldots 1_9 + 1$. Since $9 \cdot 11 \ldots 1_9 + 1 = 9 \cdot n(n+1)/2 + 1 = (3n+1)(3n+2)/2$, $11 \ldots 1_9^*$ is triangular and the result is established by induction.
37. The weights $1, 2, 2^2, \ldots, 2^{n-1}$ will weigh any integral weight up to $2^n - 1$ and no other set of so few weights is equivalently effective. Any positive integral weight up to 2^{n-1} can be expressed uniquely as $\sum_{k=0}^{n-1} a_k 2^k$, where $a_k = 0$ or 1. One answer is given by 1, 2, 4, 8, 16, 32.
38. Niven numbers: 1, 2, 3, 4, 5, 6, 7, 8, 9, 10, 12, 18, 20, 21, 24, 27, 30, 36, 40, 42, 45, 48, 50, 54, 60.
39. $s_d(7, 2) = s_d(13, 2) = s_d(15, 2) = 3$.
40. When n is even.
41. $n = 2^{15} \cdot 3^{10} \cdot 5^6 = 30\,233\,088\,000\,000$:

$$\left(\frac{n}{2}\right)^{\frac{1}{2}} = 3\,888\,000, \ \left(\frac{n}{3}\right)^{\frac{1}{3}} = 21\,600, \ \left(\frac{n}{5}\right)^{\frac{1}{5}} = 360.$$

Exercises 2.2

1. Since $ax = bc$ and $au + bv = 1$, we have that $auc + bvc = c$. By substitution $c = auc + axv = a(uc + xv)$. Hence, $a|c$.
2. Since $(-1)n + (1)(n+1) = 1$, the result follows from Theorem 2.7.
3. Since $3(22n+7) + (-2)(33n+10) = 1$, the result follows from Theorem 2.7.
4. If $3x = a$ and $3y = b$ then $3(x+y) = 65$, which is impossible since $3 \nmid 65$.
5. If $5x = a$ and $5y = b$, then $5(x+y) = 65$ or $x + y = 13$, which has infinitely many pairs of integers as solutions.
6. If $d|u_{n+1}$ and $d|u_n$ then $d|(u_{n+1} - u_n)$, hence, $d|u_{n-1}$. Continuing this process, it follows that $d|u_1$; hence, $d = 1$.
7. We have $d = au + bv = xdu + ydv$, so $1 = xu + yv$. From Theorem 2.7, $\gcd(x, y) = 1$.
8. Suppose $ax + by = 1$ and $au + cv = 1$, then $(ax + by)(au + cv) = 1 \cdot 1 = 1$. Hence, $a(xau + xcv + byu) + bc(yv) = 1$. By Theorem 2.7, $\gcd(a, bc) = 1$.
9. Let $\gcd(a, b) = 1$. From Exercise 8, $\gcd(a, b^2) = 1$. Suppose $\gcd(a, b^k) = 1$ for some positive integer k. From Exercise 8, $\gcd(a, b^{k+1}) = 1$ and the general result follows from an inductive argument.
10. If $d = \gcd(a, b)$ then $d|a$ and $d|b$, hence, $d|(a+b)$ and $d|(a-b)$. From the definition of gcd, it follows that $d|\gcd(a+b, a-b)$.
11. If $d|(a + ab)$ and $d|b$ then it follows that $d|a$, but since $\gcd(a, b) = 1$ we must have $d = 1$.
12. Let $d = \gcd(a+b, a-b)$. Since $d|(a+b)$ and $d|(a-b)$, $d|[(a+b) \pm (a-b)]$. Thus, $d|2a$ and $d|2b$. Since $\gcd(a, b) = 1$, $d|2$. Therefore, $d = 1$ or 2.
13. When a and b are of different parity.
14. Let $D = \gcd(ac, bc)$ and $d = \gcd(a, b)$. Since $d|a$ and $d|b$, $cd|ca$ and $cd|cb$, so $cd|D$. Conversely, there exist integers x and y such that $d = ax + by$, hence, $cd = acx + bcy$. Hence, $D|cd$. Thus, $cd = D$.
15. Let $d = \gcd(a, a+b)$ so $d|a$ and $d|(a+b)$. Hence, d divides $(-1)a + (a+b)$, that is, $d|b$.
16. Since $\gcd(a, 4) = 2$ and $\gcd(b, 4) = 2$, $a = 2(2n+1)$ and $b = 2(2m+1)$. Thus, $a + b = 4(m+n+1)$. Therefore, $\gcd(a+b, 4) = 4$.
17. From Theorem 2.9, $\operatorname{lcm}(ac, bc) = |ac \cdot bc/\gcd(ac, bc)| =$

$c|ab/\gcd(a, b)| = c\,\mathrm{lcm}(a, b)$.

18. $\gcd(a, b) = |a|$ and $\mathrm{lcm}(a, b) = |b|$.
19. If $a|b$, $\gcd(a, b) = |a|$. From Theorem 2.9 $\mathrm{lcm}(a, b) = |b|$. If $\mathrm{lcm}(a, b) = |b|$, from Theorem 2.9, $|a \cdot b| = \gcd(a, b) \cdot |b|$ implying that $|a| = \gcd(a, b)$, that is, b is a multiple of a or equivalently $a|b$. If $a = b = 0$, the result follows immediately.
20. From Theorem 2.9 and the fact that $\gcd(n, n+1) = 1$, we find that $\mathrm{lcm}(n, n+1) = n(n+1)$.
21. From Theorem 2.9 and the fact that $2(9n+8) + (-3)(6n+5) = 1$, we find that $\mathrm{lcm}(9n+8, 6n+5) = (9n+8)(6n+5) = 54n^2 + 93n + 40$.
22. $\gcd(2, 3, 6) - \mathrm{lcm}(2, 3, 6) = 1 \cdot 6 = 36 = 2 \cdot 3 \cdot 6$.
23. $a = 50$, $b = 20$; $a = 100$, $b = 10$.
24. The largest value for the product of two numbers that sum to 5432 is 7 376 656. In addition, from Theorem 2.9, $a \cdot b = 223\,020 \cdot \gcd(a, b)$. Since $5432 = 7 \cdot 8 \cdot 97$, the only possible values for $\gcd(a, b)$ are 1, 2, 4, 8, 7, 14, and 28. If $\gcd(a, b) = 28$, then we find that $a = 1652$ and $b = 3780$ is a solution.
25. $\{210, 330, 462, 770, 1155\}$.

Exercises 2.3

1.

	gcd	lcm	Lamé	Dixon	Actual
(a)	3	1581	10	6	5
(b)	13	11 063	15	7	6
(c)	2	1 590 446	20	8	7
(d)	1	3 810 183	20	8	5
(e)	77	113 344	20	9	6

2. (a) $3 = 11 \cdot 51 - 6 \cdot 93$.
   (b) $13 = 5 \cdot 481 - 8 \cdot 299$.
   (c) $2 = 413 \cdot 1742 - 394 \cdot 1826$.
   (d) $1 = 803 \cdot 1941 - 794 \cdot 1963$.
   (e) $77 = 9 \cdot 4928 - 25 \cdot 1771$.

Exercises 2.4

1. $(2n^2 + 2n)^2 + (2n+1)^2 = 4n^4 + 8n^3 + 8n^2 + 4n + 1 = (2n^2 + 2n + 1)^2$.
2. $(2n)^2 + (n^2 - 1)^2 = 4n^2 + n^4 - 2n^2 + 1 = n^4 + 2n^2 + 1 = (n^2 + 1)^2$.
3. $(ax - by)^2 + (ay + bx)^2 = (ax)^2 + (by)^2 + (ay)^2 + (bx)^2 = (a^2 +$

$b^2)(x^2 + y^2) = (cz)^2$.

4. $a^2 + (a + d)^2 = (a + 2d)^2$, hence, $(a + d)(a - 3d) = 0$. Thus, $a = 3d$ and the triple is $(3d, 4d, 5d)$.
5. One of s and t must be even.
6. If $s = 2n$ and $t = 2m + 1$, then $s^2 + 2st - t^2 = 4n(n + 1) - 4m(m + 1) - 1 = 8R - 1$. If $s = 2n + 1$ and $t = 2m$, then $s^2 + 2st - t^2 = 4n(n + 1) - 4m(m + 1) + 1 = 8S + 1$.
7. The even numbers occur as the side $x = 2st$. Odd numbers occur when $s = n + 1$ and $t = n$ so $y = s^2 - t^2 = 2n + 1$.
8. $x_n^2 + y_n^2 = (a_{n+1}^4 - 2a_n^2 a_{n+1}^2 + a_n^4) + (4a_n^2 a_{n+1}^2) = a_{n+1}^4 + 2a_n^2 a_{n+1}^2 + a_n^4 - z_n^2$. $x_1 = a_2^2 - a_1^2 = 4 - 1 = 2a_2 a_1 - 1 = y_1 - 1$. Suppose, for some integer k, $x_k = y_k - 1$. That is, $a_{k+1}^2 - a_k^2 = 2a_k a_{k+1} - 1$. Since $a_{k+2} = 2a_{k+1} + a_k$,

$$\begin{aligned} x_{k+1} &= a_{k+2}^2 - a_{k+1}^2 = (a_{k+2} - a_{k+1})^2 - 2a_{k+1}^2 + 2a_{k+2}a_{k+1} \\ &= (a_{k+1} + a_k)^2 - 2a_{k+1}^2 + 2a_{k+2}a_{k+1} \\ &= -a_{k+1}^2 + 2a_k a_{k+1} + a_k^2 + 2a_{k+2}a_{k+1} \\ &= -2a_k a_{k+1} + 1 + 2a_k a_{k+1} + 2a_{k+2}a_{k+1} \\ &= 2a_{k+2}a_{k+1} + 1 = y_{k+1}, \end{aligned}$$

and the result follows by induction.
9. Let $x = 2st$, $y = s^2 - t^2$, $z = s^2 + t^2$, $X = 2(2s + t)s = 4s^2 + 2st$, $Y = (2s + t)^2 - s^2 = 3s^2 + 4st + t^2$, and $Z = 5s^2 + 4st + t^2$. We have $|x - y| = |X - Y|$ and $X^2 + Y^2 = 25s^4 + 40s^3 t + 26s^2 t^2 + 8st^3 + t^4 = Z^2$.
10. (6, 8, 10) and (12, 5, 13).
11. Their perimeters are 120 and their areas 600, 540, and 480.
12. Given the product of three consecutive numbers, say $(2n - 1)(2n)(2n + 1)$, let $s = 2n$ and $t = 1$ to obtain $(s - t)st(s + t)$.
13. Let $s = 2n$, then we have $st(s + t)(s - t) = 2nt(2n + t)(2n - t)$, which is divisible by 2. Since $2n + t$ and $2n - t$ are odd and at equal distances from $2n$ one of the three must be divisible by 3. Thus, $st(s + t)(s - t)$ is divisible by 6.
14. If $st(s^2 - t^2) = w^2$ then, since s and t are coprime, s, t, and $s^2 - t^2$ must be squares, implying that the equation $a^4 - b^4 = c^2$ has a solution, a contradiction.
15. Numbers of the form $2mn + m^2$, for $m = 1, 2, 3, \ldots$.
16. Since one of s and t is even 4 divides $2st(s^2 - t^2)(s^2 + t^2)$. If one of s or t is divisible by 3 then so is xyz. If neither is divisible by 3, say $s = 3u + 1$ and $t = 3v + 1$. We have $s^2 = 3S + 1$ and $t^2 = 3T + 1$ and 3 divides

$y = s^2 - t^2$. Other cases follow similarly. Hence, 12 divides xyz.

17. If s or t is divisible by 5 then so is xyz. If not, go through cases to show that 5 divides xyz, hence 60 divides xyz.
18. Use the following Pythagorean triples:

3	4	5
5	12	13
7	24	25
8	15	17
9	40	41
11	60	61

Hence,

$P_1 = (0, 0)$, $P_i = (x_i, 0)$, for $2 \leqslant i \leqslant 8$, where $x_2 = 3 \cdot 5 \cdot 7 \cdot 8 \cdot 9 \cdot 11$ $= 83160$, $x_3 = 4 \cdot 5 \cdot 7 \cdot 8 \cdot 9 \cdot 11 = 110\,880$, $x_4 = 3 \cdot 12 \cdot 7 \cdot 8 \cdot 9 \cdot 11$ $= 199\,584$, $x_5 = 3 \cdot 5 \cdot 24 \cdot 8 \cdot 9 \cdot 11 = 285\,120$, $x_6 = 3 \cdot 5 \cdot 7 \cdot 15 \cdot 9 \cdot 11$ $= 155\,925$, $x_7 = 3 \cdot 5 \cdot 7 \cdot 8 \cdot 40 \cdot 11 = 369\,600$, $x_8 = 3 \cdot 5 \cdot 7 \cdot 8 \cdot 9 \cdot 60$ $= 453\,600$.

19. There are 16 corresponding to $(s, t) = (2, 1)$, $(3, 2)$, $(4, 1)$, $(4, 3)$, $(5, 2)$, $(5, 4)$, $(6, 1)$, $(6, 5)$, $(7, 2)$, $(7, 4)$, $(7, 6)$, $(8, 1)$, $(8, 3)$, $(8, 5)$, $(9, 2)$, and $(9, 4)$. Lehmer's rule predicts 15.9.

Exercises 2.5

1. It was possible because $\frac{1}{2} + \frac{1}{3} + \frac{1}{9} \neq 1$.
2. $\frac{1}{2} + \frac{1}{4} + \frac{1}{6} = \frac{11}{12}$; $\frac{1}{2} + \frac{1}{4} + \frac{1}{8} = \frac{7}{8}$; $\frac{1}{2} + \frac{1}{3} + \frac{1}{12} = \frac{11}{12}$; $\frac{1}{2} + \frac{1}{3} + \frac{1}{9} = \frac{17}{18}$; $\frac{1}{2} + \frac{1}{4} + \frac{1}{5} = \frac{19}{20}$; $\frac{1}{2} + \frac{1}{3} + \frac{1}{10} = \frac{14}{15}$; $\frac{1}{2} + \frac{1}{3} + \frac{1}{8} = \frac{23}{24}$; $\frac{1}{2} + \frac{1}{3} + \frac{1}{7} = \frac{41}{42}$.
4. The sum of $3k - 2$, $3k - 1$ and $3k$ is $9k - 3$. The repeated sum of the digits of any multiple of 9 is eventually 9, hence, the repeated sum of the digits of $9k - 3$ is eventually 6.
5. The weights $1, 3, 3^2, \ldots, 3^{n-1}$ will weigh any weight up to $(3^n - 1)/2$ when the weights are placed in either pan and no other set is equally effective. Any positive integer up to $3^n - 1$ inclusive can be expressed as $\sum_{k=0}^{n-1} a_k 3^k$, $a_k = 0, 1,$ or 2. Subtracting $1 + 3 + 3^2 + \cdots + 3^{n-1} = (3^n - 1)/2$. Thus every positive or negative integer between $-(3^n - 1)/2$ and $(3^n - 1)/2$ inclusive can be expressed uniquely in the form $\sum_{k=0}^{n-1} b_k 3^k$, where $b_k = -1, 0,$ or 1. One answer is given by 1, 3, 9, 27.
6. The method relies on the binary representation of a number, for example, $83 \cdot 154 = (1 + 2 + 2^4 + 2^6) \cdot 154$. Terms that have already been accounted for in the sum are eliminated.

7. We show that $3 \cdot 10^{3n+1} + 7$ and $7 \cdot 10^{3n+2} + 3$ are both divisible by 37 for any positive integer n. Since 37 divides 30 007 and 37 divides 700 003 both propositions are true when $n = 1$. Suppose that $3 \cdot 10^{3k+1} + 7 = 37r$ for some positive integer k. We have $3 \cdot 10^{3k+4} + 7 = 3 \cdot 10^{3k+1}10^3 + 7 = (37r - 7)10^3 + 7 = 37r10^3 - 6993 = 37(10^3 r - 189)$. If $7 \cdot 10^{3k+2} + 3 = 37s$ for some positive integer k, then $7 \cdot 10^{3k+5} + 3 = 7 \cdot 10^{3k+2}10^3 + 3 = 37s10^3 - 2997 = 37(10^3 s - 81)$. The conclusions follows from the principle of mathematical induction.
8. $f(4) = 12; \frac{1}{2} + \frac{1}{4} + \frac{1}{6} + \frac{1}{12}$.
9. If $(a + b)/ab$ is an integer then $ab|(a + b)$, $a|(a + b)$ and $b|(a + b)$. Hence, $a|b$ and $b|a$ so $a = b$. Hence, $a|2$. Therefore, $a = 1$ or 2.
10. In any subset of $n + 1$ integers selected from $\{1, 2, \ldots, 2n\}$ there must exist two consecutive integers, which are coprime.
11. Such products have the form

$$\frac{n(n-1)(n-2)\cdots(n-k+1)}{k!} = \binom{n}{k},$$

which is an integer.
12. Suppose we have the sequence n, $n + 1$, $n + 2$, $n + 3$, $n + 4$. If n is even and $n + 1$ is not a multiple of 3 then $n + 1$ is coprime to the other four integers. If n is even and $n + 3$ is not divisible by 3 then $n + 3$ is coprime to the rest. If n is odd then $n + 2$ is coprime to the other integers.
13. If n is odd, say $n = 2m + 1$, then $n = m + (m + 1)$. If n is even, say $n = 2^t m$, where m is odd, say $m = 2s + 1$, then $n = (2s + 1)2^t = (2^t - s) + (2^t - s - 1) + \cdots + 2^t + \cdots + (2^t + s - 1) + (2^t + s)$. Otherwise, suppose that $n = 2^t$ and $2^t = m + (m + 1) + \cdots + (m + k) = (k + 1)m + k(k + 1)/2 = (2m + k)(k + 1)/2$. Hence, $2^{t+1} = (2m + k)(k + 1)$. If k is even, say $k = 2r$, then $2^{t+1} = (2r + 1)(2m + 2r)$ or $2^t = (2r + 1)(m + r)$, which is impossible since $2r + 1$ is odd. Suppose that k is odd, say $k = 2r + 1$, then $2^{t+1} = (2r + 2)(2m + 2r + 1)$ or $2^t = (r + 1)[2(m + r) + 1]$, which is impossible since $2(m + r) + 1$ is odd.
14. $s = (2k + 1)^3$, $s - a = 2k^2(2k^2 - 1)^2$, $s - b = 8k^2(2k^2 + 1)$, $s - c = (2k^2 - 1)^2$, hence, $s(s - a)(s - b)(s - c) = 16k^4(2k^2 - 1)^4(2k^2 + 1)^4$.
15. If n is divisible by 9 then its digital root is 9. The assertion is valid for $n < 27$. If n is greater than 27, and not divisible by 9, then, by the division algorithm, $n = 27k + r$, where $0 \leqslant r < 27$. In addition, $\rho(n)$

$= \rho(\rho(27)\rho(k) + \rho(r)) = \rho(9\rho(k) + \rho(r)) = \rho(\rho(9\rho(k)) + \rho(r)) = \rho(\rho(9k) + \rho(r)) = \rho(9 + \rho(r)) = \rho(r)$.

16. The first case in the induction arguments is established by Theorem 2.8. Suppose that if $c_1, c_2, \ldots, c_k$ are pairwise coprime integers and $c_i | n$, for $i = 1, 2, \ldots, k$, and $m = \prod_{i=1}^{k} c_i$, then $m|n$. Let $c_1, c_2, \ldots, c_{k+1}$ be pairwise coprime integers where $c_i|n$, for $i = 1, 2, \ldots, k+1$, and $m = \prod_{i=1}^{k} c_i$. From Theorem 2.8, m and c_{k+1} are coprime, hence, $m \cdot c_{k+1} = \prod_{i=1}^{k+1} c_i$ divides n and the proof is established by induction.
17. For $n = 1$, $\frac{1}{5} + \frac{1}{3} + \frac{7}{15} = 1$. If $n^5/5 + n^3/3 + 7n/15 = m$, then
$$\frac{(n+1)^5}{5} + \frac{(n+1)^3}{3} + \frac{7(n+1)}{15} = m + n^4 + 2n^3 + 3n^2 + 2n + 1,$$
an integer, and the result is established by induction.
18. $S(4) = 4$, $S(5) = 5$, $S(6) = 3$, $S(7) = 7$, $S(8) = 4$, $S(9) = 6$, $S(10) = 5$.
19. $h(7) = h(11) = 6$.
20. $(a_n a_{n+3})^2 + (2a_{n+1}a_{n+2})^2 = [a_n(2a_{n+1} + a_n)]^2 + [2a_{n+1}(a_{n+1} + a_n)]^2$
$= (2a_n a_{n+1} + a_n^2)^2 + (2a_{n+1}^2 + 2a_{n+1}a_n)^2$
$= 4a_{n+1}^2 a_n^2 + 4a_{n+1}a_n^3 + a_n^4 + 4a_{n+1}^4 + 8a_{n+1}^3 a_n + 4a_{n+1}^2 a_n^2$
$= (2a_{n+1}^2 + 2a_{n+1}a_n + a_n^2)^2 = (2a_{n+1}[a_{n+1} + a_n] + a_n^2)^2$
$= (2a_{n+1}a_{n+2} + a_n^2)^2$.
21. $u_0 = 0$, and for n a natural number $u_{-2n} = -u_{2n}$, $u_{-(2n+1)} = u_{2n+1}$.

Exercises 3.1

1. 101, 103, 107, 109, 113, 127, 131, 137, 139, 149, 151, 157, 163, 167, 173, 179, 181, 191, 193, 197, 199, 211, 223, 227, 229, 233, 239, 241.
2. $6(20) - 1 = 119 = 7 \cdot 17$ and $6(20) + 1 = 121 = 11^2$.
3. Every positive integer can be expressed in the form $6k$, $6k+1$, $6k+2$, $6k+3$, $6k+4$, or $6k+5$. For $k \geqslant 1$, $6k$, $6k+2$, $6k+3$, and $6k+4$ are composite. Thus, all primes except 2 and 3 must be of the form $6k+1$ or $6k+5$. Therefore, they are of the form $6n \pm 1$.
4. If $k = 2r+1$ then $3k+1 = 3(2r+1)+1 = 6r+4$ is not prime since it is divisible by 2. Hence, $k = 2r$ and $3k+1 = 3(2r) + 1 = 6r+1$.
5. $1 + \cdots + 128 = 255 = 5 \cdot 51$ and $1 + \cdots + 128 + 256 = 511 = 7 \cdot 73$ are composite.
6. The next numbers are 39, 46, and 49. It is the increasing sequence of

positive integers having exactly two prime factors.

7. (a) $\gcd(m, n) = 2 \cdot 3 \cdot 5 = 30$, $\text{lcm}(m, n) = 2^2 \cdot 3^3 \cdot 5^4 = 67\,500$;
   (b) $\gcd(m, n) = 2 \cdot 5 \cdot 11^2 = 1210$,
   $\text{lcm}(m, n) = 2^3 \cdot 3^2 \cdot 5^2 \cdot 7 \cdot 11^3 = 16\,770\,600$.
8. If n is squarefree, all primes in the canonical representation of n have exponent 1.
9. If $n = p_1^{\alpha_1} p_2^{\alpha_2} \cdots p_r^{\alpha_r} q_1^{\beta_1} q_2^{\beta_2} \cdots q_s^{\beta_s}$, where p_i, q_j are prime, $\alpha_i = 2\gamma_i$ and $\beta_j = 2\delta_i + 1$, for $1 \leqslant i \leqslant r$, $1 \leqslant j \leqslant s$ then $n = (p_1^{\gamma_1} p_2^{\gamma_2} \cdots p_r^{\gamma_r} q_1^{\delta_1} q_2^{\delta_2} \cdots q_s^{\delta_s})^2 (q_1 q_2 \cdots q_s)$, the product of a square and a squarefree number.
10. Since 4 divides every fourth number, the length is 3.
11. $Q(100) = 13$. They are 1, 4, 8, 9, 16, 25, 27, 32, 36, 49, 64, 72, 81.
12. No, since $2 \cdot 15$ is irreducible in E and $(2 \cdot 15)$ divides $(2 \cdot 3)\,(2 \cdot 5)$, but $2 \cdot 15$ divides neither $2 \cdot 3$ nor $2 \cdot 5$.
13. The first 25 Hilbert primes are 5, 9, 13, 17, 21, 29, 33, 37, 41, 49, 53, 57, 61, 69, 73, 77, 89, 93, 97, 101, 109, 113, 121, 129, 133.
14. $4\,937\,775 = 3 \cdot 5 \cdot 5 \cdot 65\,837$ and the sum of the digits on each side of the equality sign is 42.
15. $s_2(4, 2) = 2$, $s_2(6, 2) = 4$, $s_2(8, 2) = 3$, $s_2(9, 2) = 6$, $s_2(12, 2) = 4$, $s_2(14, 2) = 4$, $s_2(15, 2) = 4$, $s_2(16, 2) = 8$.
16. $104 = 2^3 \cdot 13$ and $s_{\text{p}}(104) = 2 + 2 + 2 + 1 + 3 = 10 = 2(1 + 0 + 4) = 2 \cdot s_{\text{d}}(104)$.
17. For any Smith number r, $s_{\text{d}}(r) - s_{\text{p}}(r) = 0$.
18. Suppose $m|n$. If x belongs to M_n then $n|[s_{\text{d}}(x) - s_{\text{p}}(x)]$. Hence, $m|[s_{\text{d}}(x) - s_{\text{p}}(x)]$ and x belongs to M_m.
19. Suppose that x is a k-Smith number. Hence, $s_{\text{p}}(x) = k \cdot s_{\text{d}}(x)$ and $[s_{\text{d}}(x) - s_{\text{p}}(x)] = -(k - 1)s_{\text{d}}(x)$. Thus, $(k - 1)|[s_{\text{d}}(x) - s_{\text{p}}(x)]$. Hence, x is in M_{k-1}.
20. $s_{\text{d}}(1) = 1$, $s_{\text{p}}(10) = 7$, 10 is in M_6, since $6|(1 - 7)$, but 10 is not in S_6 since $6 \nmid 1$.
21. If $17p + 1 = x^2$, then $17p = x^2 - 1 = (x + 1)(x - 1)$. Since $x + 1 = 17$ implies that 15 is prime, $x - 1 = 17$. Therefore, $p = 19$.
22. If all prime factors of the number were of the form $4m + 1$ then the number would be of the form $4k + 1$.
23. Yes, $33 = 4 \cdot 8 + 1 = (4 \cdot 0 + 3)(4 \cdot 2 + 3) = 3 \cdot 11$.
24. For $n > 1$, $n^4 - 1 = (n^2 - 1)(n^2 + 1)$.
25. If $n > 4$ is composite then its factors are included in $(n - 1)!$
26. There are two cycles, $\overline{692\,307}$ and $\overline{153\,846}$, each of length 6.
27. The result follows since $n + 1 = 5^4 \cdot 7$ and $m + 1 = 2^2 \cdot 3^2$.

28. Let $m = p^\alpha x$, $n = p^\beta y$, with $\gcd(p, x) = \gcd(p, y) = 1$, then $mn = (p^\alpha x)(p^\beta y) = (p^\alpha p^\beta)(xy) = (p^{\alpha+\beta})(xy)$. However, $\gcd(xy, p) = 1$, hence, $p^{\alpha+\beta} \| mn$.
29. $2\|10$ and $2\|6$, but $2 \nmid\!\!\| 16$.
30. Suppose $\sqrt[m]{n} = a/b$, in lowest form. Then $a^m = nb^m$. Unless $b = 1$, any prime factor of b is also a prime factor of a, contrary to our assumption that a and b had no comon factor.

Exercises 3.2

1. For any integer n there is a one-to-one correspondence between the set of all divisors d of n and the set of all quotients n/d.
2. (a) $\tau(122) = \tau(2 \cdot 61) = 4$ and $\sigma(122) = 186$.
   (b) $\tau(1424) = \tau(2^4 \cdot 89) = 10$ and $\sigma(1424) = 2790$.
   (c) $\tau(736) = \tau(2^5 \cdot 23) = 12$ and $\sigma(736) = 1512$.
   (d) $\tau(31) = 2$ and $\sigma(31) = 32$.
   (e) $\tau(2^3 \cdot 3^5 \cdot 7^2 \cdot 11) = 144$ and $\sigma(2^3 \cdot 3^5 \cdot 7^2 \cdot 11) = 3\,734\,640$.
3. $\tau(242) = \tau(243) = \tau(244) = \tau(245) = 6$.
4. $\tau(40\,311) = \cdots = \tau(40\,315) = 8$.
5. $\tau(p_1 \cdots p_r) - 1 = 2^r - 1 = (2^r - 1)/(2 - 1) = 1 + \cdots + 2^{r-1}$.
6. If n is squarefree then $2^{\omega(n)} = \tau(n)$, otherwise $2^{\omega(n)} \leqslant \tau(n)$. If n is nonsquare its divisors come in pairs and one of the numbers must be less than $\sqrt{n}$, hence, $\tau(n) \leqslant 2\sqrt{n}$. The inequality is strict since $(n-1) \nmid n$ for $n > 2$. Similarly if n is a square number.
7. If n is not square then the divisors of n pair up, their product is n, and there are $\tau(n)/2$ such pairs.
8. n must be of the form p^3 or pq, where p and q are distinct primes.
9. n must be of the form p^2q or p^5, where p and q are distinct primes.
10. $\tau(10^6) < (10^6)^{2/3} = 10^4$, whereas $\tau(10^6) = \tau(2^6 5^6) = 7 \cdot 7 = 49$. $\sigma(10^6) < 6 \cdot (10^6)^{3/2}/\pi^2 < 607\,927\,101$, whereas $\sigma(10^6) = 2\,480\,437$.
11. $\sigma(10^6) < \frac{1}{6}[7 \cdot 10^6 \cdot \omega(10^6) + 10 \cdot 10^6] = 24 \cdot 10^6/6 = 4 \cdot 10^6$.
12. $\left(\frac{1}{25}\right) \sum_{k=1}^{25} \tau(k) = 3.48$ compared with Dirichlet's 3.37.

    $\left(\frac{1}{50}\right) \sum_{k=1}^{50} \tau(k) = 4.14$ compared with Dirichlet's 4.06.

    $\left(\frac{1}{100}\right) \sum_{k=1}^{100} \tau(k) = 4.84$ compared with Dirichlet's 4.76.

13. $H_1 = 1$, $H_2 = \frac{3}{2}$, $H_3 = \frac{11}{6}$, $H_4 = \frac{50}{24}$, $H_5 = \frac{137}{60}$.
14. $H_1 = 1 = 2(\frac{3}{2} - 1) = 2(H_2 - 1)$. Suppose that for some positive integer k, $H_1 + H_2 + \cdots + H_k = (k+1)(H_{k+1} - 1)$. Then,

$$\begin{aligned} H_1 + H_2 + \cdots + H_k + H_{k+1} &= (k+1)(H_{k+1} - 1) + H_{k+1} \\ &= (k+2)H_{k+1} - k - 1 \\ &= (k+2)\left(H_{k+1} + \frac{1}{k+2}\right) - k - 2 \\ &= (k+2)H_{k+2} - k - 2 \\ &= (k+2)(H_{k+2} - 1), \end{aligned}$$

and the result follows from the principle of mathematical induction.
15. $16! = 20\,922\,789\,888\,000$, Stirling's formula gives $20\,813\,807\,482\,100$.
17. 5040 since $\tau(5040) = \tau(2^4 \cdot 3^2 \cdot 5 \cdot 7) = 60$.
18. $D(8) = 24$; $D(16) = 120$; $D(24) = 360$; $D(32) = 840$.
19. $E(512) = E(2^9) = 1$; $E(24\,137\,569) = E(17^6) = 7$; $E(750) = E(2 \cdot 3 \cdot 5^3) = 0$; $E(2401) = E(7^4) = 1$. $E(19) - E(18) - E(16) + E(13) + E(9) - E(4) = E(19) - 1 - 1 + 2 + 1 - 1 = 0$. Thus, $E(19) = 0$.
20. $\frac{1}{25}\sum_{n=1}^{25} E(n) = \frac{20}{25} = 0.8$; $\pi/4 = 0.785$.
21. For $n = 7$, we have $(1+2)^2 = (1^3 + 2^3) = 9$. For $n = 12$, we have $(1+2+2+3+4+6)^2 = (1^3 + 2^3 + 2^3 + 3^3 + 4^3 + 6^3) = 324$. For $n = 24$, we have $(1+2+2+3+4+4+6+8)^2 = (1^3 + 2^3 + 2^3 + 3^3 + 4^3 + 4^3 + 6^3 + 8^3) = 900$.
22. $\sigma(14) = \sigma(15) = \sigma(23) = 24$.
23.
$$\begin{aligned} \sigma(36) &= \sigma(35) + \sigma(34) - \sigma(31) - \sigma(29) + \sigma(24) + \sigma(21) \\ &\quad - \sigma(14) - \sigma(10) + \sigma(1) + \sigma(-4) \\ &= 48 + 54 - 32 - 30 + 60 + 32 - 24 - 18 + 1 + 0 = 91. \end{aligned}$$
24. When p is odd, $(p^{\alpha+1} - 1)/(p-1) = 1 + p + \cdots + p^\alpha$ must be the sum of an odd number of terms, hence, α must be even. The power of 2 is not a factor in the problem since if $n = 2^k m$, with m odd, then n and m have the same odd divisors which pair up with an even sum if m is not a square.
25. $(1/25)\sum_{k=1}^{25}\sigma(k) = 20.88$ as compared to Dirichlet's 20.56.
$(1/50)\sum_{k=1}^{50}\sigma(k) = 39.78$ as compared to Dirichlet's 41.12.
$(1/100)\sum_{k=1}^{100}\sigma(k) = 83.16$ as compared to Dirichlet's 82.25.
26. 276, 396, 696, 1104, 1872, 3770, 3790, 3050, 2716, 2772, 5964, 28 596.
27. (1, 36), (2, 36), (3, 36), (4, 9), (4, 18), (4, 36), (6, 9), (6, 36), (9, 4), (9, 6), (9, 12), (9, 36), (12, 9), (12, 18), (12, 36), (18, 4), (18, 12),

(18, 36), (36, 1), (36, 2), (36, 3), (36, 4), (36, 6), (36, 9), (36, 12), (36, 18), (36, 36).

28. 12 496 $(2^4 \cdot 11 \cdot 71)$, 14 288 $(2^4 \cdot 19 \cdot 47)$, 15 472 $(2^4 \cdot 967)$, 14 536 $(2^3 \cdot 23 \cdot 79)$, 14 264 $(2^3 \cdot 1783)$.
29. 2 115 324, 3 317 740, 3 649 556, 2 792 612.
30. $\sigma^*(48) = \sigma(48) - 48 - 1 = 124 - 48 - 1 = 75$,
$\sigma^*(75) = \sigma(75) - 75 - 1 = 124 - 75 - 1 = 48$.
31. 36, 54, 65, 18, 20, 21, 10, 7, 1.
32. $(1/50)\sum_{n=1}^{50}\Omega(n) = 2.2$; $\ln(\ln(50)) + 1.0346 = 2.3986$.
33. $\sigma(3^9 \cdot 5^3 \cdot 11^3 \cdot 13^3 \cdot 41^3 \cdot 47^3)$
$= (2^8 \cdot 3^2 \cdot 5 \cdot 7 \cdot 11 \cdot 13 \cdot 17 \cdot 29 \cdot 61)^2$.
34. $\sigma(2^4 \cdot 5^2 \cdot 7^2 \cdot 11^2 \cdot 37^2 \cdot 67^2 \cdot 163^2 \cdot 191^2 \cdot 263^2 \cdot 439^2 \cdot 499^2)$
$= (3^2 \cdot 7^3 \cdot 13 \cdot 19 \cdot 31^2 \cdot 67 \cdot 109)^3$.
35. Both equal 187 131.
36. $s_d(17^3) = s_d(4913) = 17$,
$s_d(18^3) = s_d(5832) = 18$,
$s_d(26^3) = s_d(17\,576) = 26$,
$s_d(27^3) = s_d(19\,683) = 27$.
37. $s_d(22^4) = s_d(234\,256) = 22$,
$s_d(25^4) = s_d(390\,625) = 25$,
$s_d(28^4) = s_d(614\,656) = 28$,
$s_d(36^4) = s_d(1\,679\,616) = 36$.
38. If $n = 2^\alpha p_1^{\alpha_1} p_2^{\alpha_2} \cdots p_r^{\alpha_r}$, where the p_i, for $1 \leqslant i \leqslant r$, are odd, is the canonical representation for n, then $m = p_1^{\alpha_1} p_2^{\alpha_2} \cdots p_r^{\alpha_r}$ and $\sigma(n) - \tau(m) = (2^\alpha - 1)(1 + p_1 + p_1^2 + \cdots + p_1^{\alpha_1}) \cdots (1 + p_r + p_r^2 + \cdots + p_r^{\alpha_r}) - (\alpha_1 + 1) \cdots (\alpha_r + 1)$. The result follows since $2^\alpha - 1$ is odd and $1 + p_i + p_i^2 + \cdots + p_i^{\alpha_i}$ is odd whenever α_i is even and even whenever α_i is odd.
39. The result follows from the fact that if r and s are coprime then $\sigma_k(rs) = \sigma_k(r) \cdot \sigma_k(s)$ and if $n = p^\alpha$,
$$\sigma_k(p^\alpha) = 1 + p^k + p^{2k} + \cdots + p^{\alpha k} = \frac{p^{k(\alpha+1)} - 1}{p - 1}.$$
40. The result follows since
$$\sum_{d|n}\left(\frac{1}{d^2}\right) = \frac{1}{d_1^2} + \frac{1}{d_2^2} + \cdots + \frac{1}{d_r^2} = \frac{d_1^2 + d_2^2 + \cdots + d_r^2}{n^2} = \frac{\sigma_2(n)}{n^2}.$$
41. If $\gcd(m, n) = 1$, $\omega(m) = r$, and $\omega(n) = s$, then $\omega(mn) = r + s = \omega(m) + \omega(n)$, since m and n have no common prime divisor.
42. Let $\Omega(m) = r$ and $\Omega(n) = s$, $\Omega(mn) = r + s = \Omega(m) + \Omega(n)$, since,

in such a product, exponents with common bases are added.

43 $\omega(p^\alpha) = 1 = \omega(p)$ for p a prime and $\alpha > 0$.

44. $\tau(n) = 14$ implies that $n = p^6 q$ or $n = p^{13}$ where p and q are prime. Hence $n = 2^6 \cdot 3$.

45. If $n = p^\alpha > 6$, then $p \geqslant 2$ and $\alpha \geqslant 3$. Hence, $\psi(n) > 6$. If $n = p^\alpha q^\beta > 8$, $p \geqslant 2$, $q \geqslant 2$, $\alpha \geqslant 1$, and $\beta \geqslant 1$. Hence, $\psi(n) > 6$. If $n = \prod_{i=1}^r p_i^{\alpha_i} \geqslant 8$, then $\sum_{i=1}^r \alpha_i p_i + 3 \leqslant \sum_{i=1}^r p_i^{\alpha_i}$.

Exercises 3.3

1. Let $m/n = r$, so $m = n \cdot r$. Since f is completely multiplicative $f(m) = f(n \cdot r) = f(n)f(r)$. Therefore,

$$f\left(\frac{m}{n}\right) = f(r) = \frac{f(m)}{f(n)}.$$

2. $f(r \cdot s) = (r \cdot s)^k = r^k \cdot s^k = f(r)f(s)$.
3. $f(mn) = c^{g(mn)} = c^{g(m)+g(n)} = f(m)f(n)$.
4. Suppose that $\omega(m) = r$ and $\omega(n) = s$. Hence $\omega(m \cdot n) = r + s$. Since $\gcd(m, n) = 1$, $f(m \cdot n) = k^{\omega(m \cdot n)} = k^{r+s} = k^r \cdot k^s = f(m)f(n)$. Conversely, $f(60) = k^3$, but $f(6) \cdot f(10) = k^2 \cdot k^2 = k^4$.
5. If $\gcd(m, n) = 1$, and m, $n > 1$, then $\lambda(mn) = (-1)^{\Omega(mn)} = (-1)^{\Omega(m)+\Omega(n)} = (-1)^{\Omega(m)}(-1)^{\Omega(n)} = \lambda(m)\lambda(n)$.
6. $\lambda(p^\alpha) = (-1)^\alpha$. Hence, $F(p^\alpha) = \lambda(1) + \lambda(p) + \lambda(p^2) + \cdots + \lambda(p^\alpha)$ $= 1 + (-1) + 1 + (-1) + \cdots + (-1)^\alpha = (1 + (-1)^\alpha)/2$. Hence, $F(n) = 1$ if n is square and 0 otherwise.
7. $F(p^\alpha) = \mu(1)\lambda(1) + \mu(p)\lambda(p) = 2$. Therefore, if $n = \prod_{i=1}^r p_i^{\alpha_i}$ $F(n) = 2^r$.
8. See Table A.1.
9. It would suffice to show that they are not multiplicative. We have $\tau_e(6) = \tau_e(2 \cdot 3) = 2$, but $\tau_e(2) \cdot \tau_e(3) = 1 \cdot 0 = 0$. In addition, $\sigma_e(6) = \sigma_e(2 \cdot 3) = 8$, but $\sigma_e(2) \cdot \sigma_e(3) = 2 \cdot 0 = 0$.
10. Suppose that $\gcd(m, n) = 1$; then

$$\begin{aligned}\tau_o(m)\tau_o(n) &= \sum_{\substack{d_1|m \\ d_1 \text{ odd}}} 1 \cdot \sum_{\substack{d_2|n \\ d_2 \text{ odd}}} 1 = \sum_{\substack{d_1|m \\ d_1 \text{ odd}}} \sum_{\substack{d_2|n \\ d_2 \text{ odd}}} 1 = \sum_{\substack{d|mn \\ d \text{ odd}}} 1 \\ &= \tau_o(m \cdot n).\end{aligned}$$

Also,

Table A.1

n	$\tau_e(n)$	$\tau_o(n)$	$\sigma_e(n)$	$\sigma_o(n)$
1	0	1	0	1
2	1	1	2	1
3	0	2	0	4
4	2	1	6	1
5	0	2	0	6
6	2	2	8	4
7	0	2	0	8
8	3	1	14	1
9	0	3	0	13
10	2	2	12	6

$$\sigma_o(m)\sigma_o(n) = \sum_{\substack{d_1|m \\ d_1 \text{ odd}}} d_1 \cdot \sum_{\substack{d_2|n \\ d_2 \text{ odd}}} d_2 = \sum_{\substack{d_1|m \\ d_1 \text{ odd}}} \sum_{\substack{d_2|n \\ d_2 \text{ odd}}} d_1 d_2$$

$$= \sum_{\substack{d|mn \\ d \text{ odd}}} d = \sigma_o(mn).$$

Neither τ_o nor σ_o is completely multiplicative since $\tau_o(60) = 3$, but $\tau_o(6) \cdot \tau_o(10) = 2 \cdot 2 = 4$, and $\sigma_o(60) = 8$, but $\sigma_o(6) \cdot \sigma_o(10) = 4 \cdot 6 = 24$.

11. Suppose that $\gcd(m, n) = 1$. If either m or n is not squarefree then $\mu(mn) = 0 = \mu(m)\mu(n)$. If $m = p_1 \cdot p_2 \cdots p_r$ and $n = q_1 \cdot q_2 \cdots q_s$, where p_i, q_j are prime for $1 \leqslant i \leqslant r$, $1 \leqslant j \leqslant s$, then $\mu(m) = (-1)^r$ and $\mu(n) = (-1)^s$, hence $\mu(mn) = (-1)^{r+s} = (-1)^r(-1)^s = \mu(m)\mu(n)$.
12. One of any four consecutive numbers is divisible by 4, hence $\mu(n)\mu(n+1)\mu(n+2)\mu(n+3) = 0$.
13. $\sum_{k=1}^{\infty}\mu(k!) = 1 + -1 + 1 + 0 + 0 + \cdots = 1$.
14. $n = 33$.
15. Since μ is multiplicative and $\sum_{d|p^\alpha}|\mu(d)| = |\mu(1)| + |\mu(p)| = 2$, for p a prime. Hence, $\sum_{d|n}|\mu(d)| = 2^{\omega(n)}$.
16. Let $F(n) = \sum_{d|n}\mu(d)\tau(n/d)$. Since μ and τ are multiplicative so is F and $F(p^\alpha) = \mu(1)\tau(p^\alpha) + \mu(p)\tau(p^{\alpha-1}) + \mu(p^2)\tau(p^{\alpha-2}) + \cdots + \mu(p^\alpha)\tau(1) = \alpha + 1 + (-1)\alpha + 0 + \cdots + 0 = 1$ for p a prime. Hence, if $n = \prod_{i=1}^r p_i^{\alpha_i}$, $F(n) = 1$.
17. Let $F(n) = \sum_{d|n}\mu(d)\sigma(d)$. Since μ and σ are multiplicative so is F and $F(2^r p^\alpha) = F(2^r)F(p^\alpha) = 2^r \cdot p^\alpha$ for p a prime. Hence, if $n = \prod_{i=1}^r p_i^{\alpha_i}$, $F(n) = n$.

18. Let $F(n) = \sum_{d|n}\mu(d)\tau(d)$. Since μ and τ are multiplicative so is F. For any prime p, $F(p^\alpha) = \mu(1)\tau(1) + \mu(p)\tau(p) + \cdots + \mu(p^\alpha)\tau(p^\alpha) = 1 \cdot 1 + (-1) \cdot 2 + 0 \cdot 0 + \cdots + 0 \cdot 0 = (-1)$. Therefore, if $n = \prod_{i=1}^r p_i^{\alpha_i}$,

$$F(n) = F\left(\prod_{i=1}^r p_i^{\alpha_i}\right) = \prod_{i=1}^r F(p_i^{\alpha_i}) = \prod_{i=1}^r (-1) = (-1)^r = (-1)^{\omega(n)}.$$

19. According to the Möbius inversion formula with $F(n) = 1/n$,

$$f(n) = \sum_{d|n}\mu(d)F\left(\frac{n}{d}\right) = \sum_{d|n}\mu(d)\left(\frac{d}{n}\right).$$

Hence,

$$f(p^\alpha) = \sum_{d|p^\alpha}\mu(d)\left(\frac{d}{p^\alpha}\right) = \frac{1}{p^\alpha} - \frac{p}{p^\alpha} = \frac{1}{p^\alpha}(1-p).$$

Therefore,

$$f(n) = \frac{1}{n}\prod_{p|n}(1-p).$$

20. If $n = p^\alpha$, $\prod_{d|p^\alpha} d^{\tau(d)\mu(n/d)/2} = (p^{\alpha-1})^{-\alpha/2}(p^\alpha)^{(\alpha+1)/2} = p^\alpha$.
21. If $n = p^\alpha$,

$$\sum_{d|p^\alpha}\frac{\omega(d)}{\tau(p^\alpha)} = \frac{1+1+\cdots+1}{\alpha+1} = \frac{\alpha}{\alpha+1}.$$

22. If $n = \prod_{i=1}^r p_i^{\alpha_i}$, $\sum_{d|n}\Lambda(d) = \sum_{i=1}^r\Lambda(p_i^{\alpha_i}) = \sum_{i=1}^r \ln(p_i^{\alpha_i}) = \ln(\prod_{i=1}^r p_i^{\alpha_i}) = \ln(n)$.
23. Since $\sum_{d|n}\Lambda(d) = \ln(n)$ the Möbius inversion formula implies that $\Lambda(n) = \sum_{d|n}\mu(d)\ln(n/d) = \sum_{d|n}\mu(d)\ln(n) - \sum_{d|n}\mu(d)\ln(d) = 0 - \sum_{d|n}\mu(d)\ln(d)$.

Exercises 3.4

1. $114^2 - 12\,971 = 5^2$, thus $12\,971 = (114+5)(114-5) = 119 \cdot 109$.
2. (a) $493 = 18^2 + 13^2 = 22^2 + 3^2 = 17 \cdot 29$.

   (b) $37\,673 = \dfrac{(14^2+36^2)(360^2+36^2)}{4 \cdot 36^2} = 373 \cdot 101$.
3. $a = \dfrac{kmr+ns}{2}$, $b = \dfrac{ms-nr}{2}$, $c = \dfrac{kmr-ns}{2}$, $d = \dfrac{ms+nr}{2}$,

$$N = a^2 + kb^2 = \frac{k^2m^2r^2 + kn^2r^2 + n^2s^2 + km^2s^2}{4},$$

$N = c^2 + kd^2 = \dfrac{k^2m^2r^2 + kn^2r^2 + km^2s^2 + n^2s^2}{4}$. Hence,

$2N = \dfrac{2(k^2m^2r^2 + kn^2r^2 + km^2s^2 + n^2s^2)}{4}$ and

$N = \dfrac{k^2m^2r^2 + kn^2r^2 + km^2s^2 + n^2s^2}{4} = \dfrac{(km^2 + n^2)(kr^2 + s^2)}{4}$.

4. $a + c = 10 \cdot 30$; $a - c = 14$; $ms = d + b = 70$; $nr = d - b = 6$. Hence, $m = 10$; $n = 2$; $r = 3$; $s = 7$ and $34\,889 = 251 \cdot 139$.
5. If $n = pm$ and $m = ab$, with $a > n^{1/3}$ and $b > n^{1/3}$, then $n = pab > n^{1/3}n^{1/3}n^{1/3} > n$, a contradiction.
6. $2\,027\,651\,281 = (45\,041 + 1020)(45\,041 - 1020) = 46\,061 \cdot 44\,021$.

Exercises 3.5

1. Since $[\![x]\!] \leqslant x < [\![x]\!] + 1$, $[\![x]\!] \leqslant x$ and $x < [\![x]\!] + 1$. Hence, $x - 1 < [\![x]\!]$. Therefore, $x - 1 < [\![x]\!] \leqslant x$.
2. Since $x - 1 < [\![x]\!] \leqslant x$ and $-x - 1 < [\![-x]\!] \leqslant -x$, $-2 < [\![x]\!] + [\![-x]\!] \leqslant 0$. If x is an integer $[\![x]\!] + [\![-x]\!] = 0$; otherwise $[\![x]\!] + [\![-x]\!] = [\![|x|]\!] + [\![-|x|]\!] = [\![|x|]\!] + (-[\![|x|]\!] - 1) = -1$.
3. Let $x = [\![x]\!] + \alpha$ and $y = [\![y]\!] + \beta$, where $0 \leqslant \alpha, \beta \leqslant 1$. Hence, $x + y = [\![x]\!] + [\![y]\!] + \alpha + \beta$, where $0 \leqslant \alpha + \beta < 2$. If $0 \leqslant \alpha + \beta < 1$, $[\![x + y]\!] = [\![x]\!] + [\![y]\!]$. If $1 \leqslant \alpha + \beta < 2$, $[\![x + y]\!] = [\![x]\!] + [\![y]\!] + 1$. Therefore, $[\![x + y]\!] \geqslant [\![x]\!] + [\![y]\!]$.
4. (a) $x = n + \alpha$, with n and integer and α real, $0 \leqslant \alpha < 1/2$;
   (b) x any real number;
   (c) x an integer;
   (d) x real, $1 \leqslant x < \frac{10}{9}$.
5. 529, 263, and 131.
6. $n = 30$.
7. 249
8. 150
9. $[\![10\,000/7]\!] - [\![1000/7]\!] = 1428 - 142 = 1286$.
10. $[\![1000/3]\!] - [\![1000/12]\!] = 333 - 83 = 250$.
11. From the inclusion–exclusion principle, $10\,000 - [\![10\,000/3]\!] - [\![10\,000/5]\!] - [\![10\,000/7]\!] + [\![10\,000/15]\!] + [\![10\,000/21]\!] + [\![10\,000/35]\!] - [\![10\,000/105]\!] = 4571$.
12. 369 693 097 digits.
13. If the divisors of k, for $k = 1, \ldots, n$, are listed, k is counted exactly $[\![n/k]\!]$ times.

14. $\sum_{k\geqslant 1}^{\infty}\left(\left[\!\left[\frac{2n}{p^k}\right]\!\right]-2\left[\!\left[\frac{n}{p^k}\right]\!\right]\right)$

15. The result follows from a generalization of Theorem 3.13 and the fact that $[\![n]\!]=\sum_{k=1}^{n}1$.

Exercises 3.6

1. One number in any set of three consecutive integers must be divisible by 3.
2. No, since for $n>2$ $n^2-1=(n+1)(n-1)$.
3. Suppose that there are only finitely many primes of the form $4k+3$, say $q_1, \ldots, q_r$, and consider $N=4(q_1\cdots q_r)-1=4((q_1\cdots q_r)-1)+3$. The product of primes of the form $4k+1$ is always a number of the form $4m+1$. Thus if N is composite one of its factors must be of the form $4r+3$. However, no prime of the form $4r+3$ divides N, a contradiction.
4. Suppose that there are only finitely many primes of the form $4k+1$, say $q_1, \ldots, q_r$, and consider $N=(q_1\cdots q_r)^2+1$. $N>q_i$, for $1\leqslant i\leqslant r$, hence N cannot be prime. Any number of the form a^2+1 has, except possibly for the factor 2, only prime factors of the form $4m+1$. Since division into N by each prime factor of the form $4k+1$ leaves a remainder 1, N cannot be composite, a contradiction. Hence, the number of primes of the form $4k+1$ must be infinite.
5. No, $333\,333\,331=17\cdot 19\,607\,843$.
6. $A(50)=4.63$; $\ln(50)=3.91$.
7. $\lim_{x\to\infty}\dfrac{\displaystyle\int_2^x\frac{\mathrm{d}t}{\ln(t)}}{\dfrac{x}{\ln(x)}}=\lim_{x\to\infty}\dfrac{\ln(x)}{\ln(x)-1}=1.$
8. According to Euler's formula,
$$\zeta(6)=\frac{2^4\cdot\pi^6\cdot|B_{2n}|}{6!}=\frac{16\cdot\pi^6\cdot\left(\dfrac{1}{42}\right)}{720}=\frac{\pi^6}{945}.$$
9. $\sigma(p+2)=p+2+1=p+1+2=\sigma(p)+2$.
10. $17\cdot 19=323$ and $83\,691\,159\,552\,021=323\cdot 259\,105\,757\,127$.
11. If $p=3k+1$, then $p+2=3(k+1)$ which is not prime. Hence, $p=3k+2$, then $2p+2=6(k+1)$, but p is odd, hence $p+1=3(k+1)$ is even implying that $k+1$ is even, hence, $12|(2p+2)$.

12. Yes, $(2n + 1)(2n - 1) = 4n^2 - 1$.
13. 23, 37, 47, 53, 67, 79, 83, 89, 97, 113.
14. (41, 43, 47).
15. (101, 103, 107, 109) or (191, 193, 197, 199).
16. 76 883, 6883, 883, 83, and 3 are prime.
17. 59 393 339, 5 939 333, 593 933, 59 393, 5939, 593, 59 and 5 are prime.
18. 521, since $125 = 5^3$; 487 since $784 = 28^2$; 691 since $196 = 14^2$; 1297, since $7921 = 89^2$; 1861, since $1681 = 41^2$; 4441, since $1444 = 38^2$; 4483, since $3844 = 62^2$; 5209, since $9025 = 95^2$; 5227, since $7225 = 85^2$; 9049, since $9409 = 97^2$; 806 041, since $140\,608 = 52^3$.
19. 11, 13, 17, 31, 37, 71, 73, 79, 97.
20. The 3-digit palindromic primes are 101, 131, 151, 181, 191, 313, 353, 373, 383, 727, 757, 787, 797, 919, and 929.
21. $1441 = 11 \cdot 131$ and $3443 = 11 \cdot 313$.
22. 113, 131, and 311 are prime.
23. 1423, 2341, 2143, and 4231 are prime; 1847, 8147, 8741, 1487, 7481, 4817, 4871, and 7841 are prime.
25. 1111 is prime and $1111 \cdot 3304 = 3\,670\,744 = 2 \cdot 2 \cdot 2 \cdot 7 \cdot 11 \cdot 59 \cdot 101$, $s_d(3\,670\,744) = 31 = s_d(2 \cdot 2 \cdot 2 \cdot 7 \cdot 11 \cdot 59 \cdot 101)$.
26. 11, 23, 29, 41, 53, 83, 89, 113, 131, 173, 179, 191.
27. 5, 11, 23, 47
28. $2 \cdot 3 \cdot 5 \cdot 7 \cdot 11 \cdot 13 + 1 = 30\,031$ and $30\,047 - 30\,030 = 17 = f_6$. $2 \cdot 3 \cdot 5 \cdot 7 \cdot 11 \cdot 13 \cdot 17 + 1 = 510\,511$ and $510\,529 - 510\,510 = 19 = f_7$. $2 \cdot 3 \cdot 5 \cdot 7 \cdot 11 \cdot 13 \cdot 17 \cdot 19 + 1 = 9\,699\,691$ and $9\,699\,713 - 9\,699\,690 = 23 = f_8$. Yes.
29. 19, 23, 29, 37, 47, 59, 73, 89, 107, 127, 149, 173, 199, 227, 257 are prime, but $f(16) = 289 = 17^2$.
30. 13, 19, 29, 43, 61, 83, 109, 139, 173, and 211 are prime but $f(11) = 253 = 11 \cdot 23$.
31. $f(25) = 251$, $f(30) = 131$, $f(40) = 41$, $f(60) = 461$ are prime but $f(80) = 1681 = 41^2$.
32. $f([(p-1)! + 1]/k,\ p - 1) = p$.
33. $f(n, n) = 2$.
34. 1, 3, 7, 9, 13, 15, 21, 25, 31, 33, 37, 43, 49, 51, 63, 67, 69, 73, 75, 79, 85, 93, 99, 105, 111, 115, 127, 129, 133, 135, 141, 151, 159, 163, 169, 171, 189, 193, 195, 201, 205, 211, 219, 223, 231, 235, 237, 241, 259, 261, 267, 273, 283, 285, 289, 297.

35. $6 = 3 + 3$, $8 = 7 + 1$, $10 = 7 + 3$, $12 = 9 + 3$, $14 = 13 + 1$, $16 = 15 + 1$, $18 = 15 + 3$, $20 = 13 + 7$, $22 = 21 + 1$, $24 = 21 + 3$, $26 = 25 + 1$, $28 = 25 + 3$, $30 = 21 + 9$, $32 = 31 + 1$, $34 = 31 + 3$, $36 = 33 + 3$, $38 = 37 + 1$, $40 = 37 + 3$, $42 = 37 + 5$, $44 = 43 + 1$, $46 = 43 + 3$, $48 = 33 + 15$, $50 = 49 + 1$.

36. Suppose that $2n \geqslant 6$; if $2n - 2 = p + q$, where p and q are prime, then $2n = 2 + p + q$, and $2n + 1 = 3 + p + q$. Hence, every positive integer greater than unity is the sum of three or fewer primes. Conversely if $2n \geqslant 4$ then $2n + 2 = p + q + r$, where p, q, r are prime. Since one of p, q, r is even, say $r = 2$, we have $2n = p + q$.

37. $4 = 2 + 2$, $6 = 3 + 3$, $8 = 5 + 3$, $10 = 7 + 3$, $12 = 7 + 5$, $14 = 7 + 7$, $16 = 11 + 5$, $18 = 13 + 5$, $20 = 17 + 3$, $22 = 11 + 11$, $24 = 13 + 11$, $26 = 13 + 13$, $28 = 23 + 5$, $30 = 23 + 7$, $32 = 29 + 3$, $34 = 31 + 3$, $36 = 31 + 5$, $38 = 31 + 7$, $40 = 37 + 3$, $42 = 37 + 5$, $44 = 37 + 7$, $46 = 41 + 5$, $48 = 41 + 7$, $50 = 47 + 3$.

38. $10 = 3 + 7 = 5 + 5$,
$16 = 13 + 3 = 11 + 5$,
$18 = 13 + 5 = 7 + 11$.

39. $22 = 11 + 11 = 19 + 3 = 17 + 5$,
$24 = 7 + 17 = 19 + 5 = 13 + 11$,
$26 = 23 + 3 = 19 + 7 = 13 + 13$.

40. $7 = 2 \cdot 2 + 3$, $9 = 2 \cdot 2 + 5$, $11 = 2 \cdot 2 + 7$, $13 = 2 \cdot 3 + 7$, $15 = 2 \cdot 5 + 5$, $17 = 2 \cdot 5 + 7$, $19 = 2 \cdot 7 + 5$, $21 = 2 \cdot 5 + 11$, $23 = 2 \cdot 5 + 13$, $25 = 2 \cdot 11 + 3$, $27 = 2 \cdot 11 + 5$, $29 = 2 \cdot 11 + 7$, $31 = 2 \cdot 13 + 5$, $33 = 2 \cdot 13 + 7$, $35 = 2 \cdot 11 + 13$, $37 = 2 \cdot 17 + 3$, $39 = 2 \cdot 17 + 5$, $41 = 2 \cdot 17 + 7$, $43 = 2 \cdot 19 + 5$, $45 = 2 \cdot 19 + 7$, $47 = 2 \cdot 17 + 13$, $49 = 2 \cdot 13 + 23$.

41. $\pi(10^9)/10^9 = 50\,847\,478/10^9 = 0.050\,847\,7$ or about 5%.

42. $(\zeta(s))^2 = \left(\sum_{u=1}^{\infty} \frac{1}{u^s}\right)\left(\sum_{v=1}^{\infty} \frac{1}{v^s}\right) = \sum_{n=1}^{\infty} \frac{\tau(n)}{n^s}$, where $u \cdot v = n$.

43. $$\zeta(s)\cdot\zeta(s-1) = \left(\sum_{u=1}^{\infty}\frac{1}{u^s}\right)\left(\sum_{v=1}^{\infty}\frac{1}{v^{s-1}}\right) = \left(\sum_{u=1}^{\infty}\frac{1}{u^s}\right)\left(\sum_{v=1}^{\infty}\frac{v}{v^s}\right)$$
$$= \sum_{n=1}^{\infty}\frac{\sigma(n)}{n^s}, \text{ where } u\cdot v = n.$$

44. $$\zeta(s)\cdot\zeta(s-k) = \left(\sum_{u=1}^{\infty}\frac{1}{u^s}\right)\left(\sum_{v=1}^{\infty}\frac{1}{v^{s-k}}\right) = \left(\sum_{u=1}^{\infty}\frac{1}{u^s}\right)\left(\sum_{v=1}^{\infty}\frac{v^k}{v^s}\right)$$
$$= \sum_{n=1}^{\infty}\frac{\sigma_k(n)}{n^s}, \text{ where } u\cdot v = n.$$

45. $$\left(\sum_{n=1}^{\infty}\frac{\mu(n)}{n^s}\right)\left(\sum_{n=1}^{\infty}\frac{1}{n^s}\right) = \sum_{n=1}^{\infty}\frac{\left(\sum_{k|n}^{\infty}\mu(k)\right)}{n^s} = \sum_{n=1}^{\infty}\frac{\nu(n)}{n^s} = 1.$$

Exercises 3.7

1. $\gcd(a^m, b^n) = p^{\min\{m, n\}}$.
2. $\gcd(a+b, p^4) = p$ and $\gcd(ab, p^4) = p^3$.
3.

1	2	3	~~4~~	5	6	7	~~8~~	9	10	11	~~12~~	13	14	15	~~16~~	17	18	19	~~20~~
1	3	~~6~~		11	17	~~24~~		33	43	~~54~~		67	81	~~96~~		113	131	~~150~~	
1	~~4~~			15	~~32~~			65	~~108~~			175	~~256~~			369	~~500~~		
1				16				81				256				625			

4.

1 16 81 256 625 1296 2401 4096

15 65 175 369 671 1105 1695

50 110 194 302 434 590

60 84 108 132 156

24 24 24 24

5. If n is in the array, let r and c denote, respectively, the row and column indicating n's position in the array. Since the numbers in each row and column form an arithmetic progression, $n = 4 + 3(c-1) + (2c+1)(r-1)$. Hence, $2n+1 = 2[4 + 3(c-1) + (2c+1)(r-1)] + 1 = (2r+1)(2c+1)$, which is composite and odd. In addition, all odd composite numbers can be obtained in this manner. If p is an odd prime, then $m = (p-1)/2$ is a positive integer that cannot appear in the array.
6. n is a positive integer such that $n+1$ is not an odd prime.
7. The order of the factors counts. Hence, the number of solutions to $xy = n$, $d_2(n)$, equals $\tau(n)$. Similarly, $d_1(n) = 1$.
8. $\tau(24) = 24 - t(23, 1) - t(22, 2) - t(21, 3) - t(20, 4)$
$\quad - t(19, 5) - t(18, 6) - t(17, 7) - t(16, 8)$

$- t(15, 9) - t(14, 10) - t(13, 11) - t(12, 12)$
$= 24 - 1 - 2 - 2 - 3 - 1 - 2 - 1 - 1 - 1 - 1 - 1 - 0 = 8.$

9. $\theta(10) + \theta(9) + \theta(7) + \theta(4) + \theta(0) = \theta(10) + 13 + 8 + (-5) + (-10) = 0$. Hence, $\theta(10) = -6$. $\theta(24) + \theta(23) + \theta(21) + \theta(18) + \theta(14) + \theta(9) + \theta(3) = \theta(24) + 24 + 32 + (-13) + (-8) + 13 + 4 = 0$. Hence, $\theta(24) = -52$.
10. Let $m = 2t + 1$ denote the largest odd divisor of n. For any proper odd divisor $2r + 1$ of m, with $n/(2r + 1) = s$, we have that $(s - r) + \cdots + (s - 1) + s + (s + 1) + \cdots + (s + r) = (2r + 1)s = n$. In addition, if $n = 2^{\alpha} p_1^{\alpha_1} p_2^{\alpha_2} \cdots p_r^{\alpha_r}$, where the p_i, for $1 \leqslant i \leqslant r$, are odd, is the canonical representation for n, we have that $m = t + (t + 1) = p_1^{\alpha_1} p_2^{\alpha_2} \cdots p_r^{\alpha_r}$. Hence, $(t - (2^{\alpha} - 1)) + \cdots + (t - 1) + t + (t + 1) + (t + 2) + \cdots + (t + (2^{\alpha} - 1)) + (t + 2^{\alpha}) = 2^{\alpha+1} \cdot t + 2^{\alpha} = 2^{\alpha}(2t + 1) = 2^{\alpha} m = n$.
11. Suppose $S = 1 + \frac{1}{2} + \frac{1}{3} + \frac{1}{4} + \cdots + 1/n$ is an integer. Let m be the largest integer such that $2^m \leqslant n$ and $P = 1 \cdot 3 \cdot 5 \cdots (2r + 1)$, with $2r + 1 \leqslant n$. Then, each term of the sum $2^{m-1} \cdot P \cdot S$ is an integer except $2^{m-1} \cdot P/2^m$. Hence, S is not an integer.
12. The area of the polygonal region equals $I + B/2 - 1$.
13. The area of the polygonal region remains equal to $I + B/2 - 1$.
15. 1, 2, 4, 5, 7, 8, 9, 11, 12, 14, 15, 16, 18, 19, 21, 22, ... [Beatty sequence]
16. 3, 6, 10, 13, 17, 20, 23, 27, 30, 34, 37, 40, 44, 49, 51, 54, 58, 61, 64, 68.
17. If a_n denotes the nth term of the sequence, the positive integer k first appears in the sequence when

$$n = 1 + 2 + \cdots + (k - 1) + 1 = \left[\!\left[\frac{(k-1)k}{2}\right]\!\right] + 1.$$

Thus

$$a_n = k \text{ for } n = \left[\!\left[\frac{(k-1)k}{2}\right]\!\right] + 1 + r,$$

where $r = 0, 1, \ldots, k - 1$. Hence,

$$0 \leqslant n - \frac{(k-1)k}{2} - 1 \leqslant k - 1,$$

or

$$\frac{k^2 - k + 2}{2} \leqslant n \leqslant \frac{k^2 + k}{2}.$$

Thus,

$$(2k-1)^2+7 \leqslant 8n \leqslant (2k+1)^2-1,$$

or

$$(2k-1)^2 \leqslant 8n-7 \leqslant (2k+1)^2-8<(2k+1)^2.$$

So

$$2k-1 \leqslant (8n-7)^{1/2} \leqslant 2k+1,$$

or

$$k \leqslant \left[\!\left[\frac{1+\sqrt{8n-7}}{2}\right]\!\right] \leqslant k+1.$$

Therefore,

$$a_n = \left[\!\left[\frac{1+\sqrt{8n-7}}{2}\right]\!\right].$$

18. The result follows directly from the previous exercise.
19. The only factors of $p^n!$ divisible by p are $p, 2p, 3p, \ldots, p^{n-2}p$, $(p^{n-2}+1)p, \ldots, 2p^{n-2}p, (2p^{n-2}+1)p, \ldots, 3p^{n-2}p, \ldots, p^n$. The number of these factors is p^n. Since p is prime, after dividing each of these factors by p, there remain only the quotients from the factors $p^{n-1}, 2p^{n-1}, 3p^{n-1}, \ldots, p^n$ still divisible by p, and the number of these is p^{n-1}. Dividing these by p, there remain only the quotients from the factors $p^{n-2}, 2p^{n-2}, 3p^{n-2}, \ldots, p^n$ still divisible by p, and the number of these is p^{n-2}. Continuing this process, eventually there remains only one quotient, namely that from p^n, divisible still by p. Therefore $p^n!$ is divisible by p to the power $p^{n-1}+ p^{n-2}+\cdots + p+1=(p^n-1)/(p-1)$.
20. From the previous exercise, $x = 2^n - 1$.
21. If $n = 2^\alpha \cdot m$, where m is odd, any factor of n which gives an odd quotient must have 2^α as an element, and therefore is of the form $2^\alpha \cdot d$, where d is any odd divisor. Therefore, $A = \sum 2^\alpha \cdot d = 2^\alpha \sum d = 2^\alpha \cdot C$. If r is any factor of n giving an even quotient, then r must contain a power of 2 not greater than $\alpha - 1$. Therefore, r is of the form $2^\beta \cdot d$, where $\beta < \alpha$. Thus, corresponding to any odd factor d, the sum of divisors giving an even quotient is $(1+2+2^2+\cdots+2^{\alpha-1})d = (2^\alpha-1)d$. Hence, $B=(2^\alpha-1)\sum d = (2^\alpha-1)C$ and $A = B + C$.
23. If p is prime,

$$\tau_2(p^\alpha) = \tau(1)+\tau(p)+\cdots+\tau(p^\alpha) = 1+2+\cdots+(\alpha+1)$$
$$= \frac{(\alpha+1)(\alpha+2)}{2} = \binom{\alpha+2}{2}.$$

Since τ_2 is multiplicative the result is established.

24.
$$\frac{\pi^2}{8} = \iint_T \mathrm{d}u\mathrm{d}v = \iint_S (1 - x^2y^2)^{-1}\mathrm{d}x\mathrm{d}y$$
$$= \frac{1}{1^2} + \frac{1}{3^2} + \frac{1}{5^2} + \cdots = \left(1 - \frac{1}{2^2}\right)\zeta(2).$$

The Jacobian of the transformation is $1 - x^2y^2$.

Exercises 4.1

1. No, $10 \cdot 11 = 110$ and $\sigma(110) = \sigma(2 \cdot 5 \cdot 11) = 216 < 220 = 2 \cdot 110$.
2. If $\sigma(n) \geqslant 2n$, then $\sigma(kn) > k \cdot \sigma(n) \geqslant k \cdot (2n) = 2kn$, hence, all multiples of perfect and abundant numbers are abundant.
3. (a) The primes greater than 2 are odd and deficient,
   (b) number of the form $2p$, where $p > 5$ is prime, are even and deficient.
4. Suppose $\sigma(n) = 2n$, $d|n$, $d \neq n$ and $\sigma(d) \geqslant 2d$; then $2n = \sigma(n) = \sigma(d \cdot (n/d)) > 2d(n/d) = 2n$, a contradiction.
5. $6 = 110_2, 28 = 11\,100_2, 496 = 111\,110\,000_2, 8128 = 1\,111\,111\,000\,000_2$. $(2^{p-1}(2^p - 1))$ to base 2 is p ones followed by $p - 1$ zeros.
6. $\rho(137\,438\,691\,328) = 1$.
7. $2^{p-1}(2^p - 1) = t_{2^p-1}$.
8. $2^{p-1}(2^p - 1) = p^6{}_{2^{p-1}}$.
9. $\displaystyle\sum_{d|n} \frac{1}{d} = \frac{\sum_{d|n} d}{n} = \frac{\sigma(n)}{n} = \frac{2n}{n} = 2.$
10. If $n = 2^{p-1}(2^p - 1)$, $\displaystyle\prod_{d|n} d = 2^{0+1+\cdots+(p-1)}(2^p - 1)^p \cdot 2^{0+1+\cdots+(p-1)}$
$$= (2^p - 1)^p[2^{p(p-1)/2}]^2 = (2^p - 1)^p(2^{p-1})^p = n^p.$$
11. $(1\,398\,269)\log(2) + 1 = 420\,921$.
12. If $n = p^\alpha$ and $2p^\alpha = \sigma(p^\alpha) = (p^{\alpha+1} - 1)/(p - 1)$ then $p^{\alpha+1} = 2p^{\alpha+1} - 2p^\alpha + 1$ or $p^{\alpha+1} = 2p^\alpha - 1$. Hence, $p = 1$, a contradiction. If $n = pq$ and $2pq = \sigma(pq) = (p + 1)(q + 1)$ then one of $p + 1$ and $q + 1$ must be even. Thus, without loss of generality, $q + 1 = 2p$ and $p + 1 = q$, by the Fundamental Theorem of Arithmetic. Hence, $p = 2$ and $q = 3$.
13. If $n = p_1 \cdots p_r$ and $\sigma(n) = 2p_1 \cdots p_r = (p_1 + 1) \cdots (p_r + 1)$, with

$p_1 < p_2 < \cdots < p_r$, then we must have $2p_1 = p_r + 1$, $p_2 = p_1 + 1$, $p_3 = p_2 + 1, \cdots, p_r = p_{r-1} + 1$, as in the previous answer, with $n = 6$ as the only solution.

14. From Exercise 1.4.5, $2^{n-1}(2^n - 1) = 1^3 + 3^3 + 5^3 + \cdots + (2^{(n+1)/2} - 1)^3 = 1^3 + \cdots + (2 \cdot 2^{(n-1)/2} - 1)^3$, the sum of the first $2^{(n-1)/2}$ odd cubes.
15. For any positive integer k, the units digit of 2^{4k} is 6, of $2^{4k+1} - 1$ is 1, of 2^{4k+2} is 4, and of $2^{4k+3} - 1$ is 7. Hence, the units digit of $2^{4k}(2^{4k+1} - 1)$ is 6 and that of $2^{4k+2}(2^{4k+3} - 1)$ is 8.
16. If p is an odd prime, $2^{p-1} - 1$ is divisible by 3. Hence, $2^{p-1} = 3k + 1$ for some k. Thus, $2^p = 6k + 2$ and $2^p - 1 = 6k + 1$. Hence, $2^{p-1}(2^p - 1) = (3k + 1)(6k + 1) = 18k^2 + 9k + 1 = 9M + 1$.
17. $\sigma(\sigma(6)) = \sigma(12) = 28$. Suppose that n is even, then $n = 2^{p-1}(2^p - 1)$ and $\sigma(\sigma(n)) = 2^p(2^{p+1} - 1)$ is even and perfect. Hence, $2^{p+1} - 1$ and $p + 1$ are prime. Thus, $p = 2$ and $n = 6$. If n is odd then $\sigma(n) = 2n$ implies that $\sigma(\sigma(n)) = 6n = 2^{p-1}(2^p - 1)$ so $n = 1$.
18. 4, 14, 67, 42, 111, 0.
19. If $\sigma(2 \cdot 3^\alpha) = 2 \cdot 2 \cdot 3^\alpha$, $3(3^{\alpha+1} - 1)/2 = 4 \cdot 3^\alpha$. Hence, $3^{\alpha+1} - 1 = 8 \cdot 3^{\alpha-1}$, which is true only if $\alpha = 1$.
20. 6, 10, 14, 15, 21, 22, 26, 27, 33, 34, 35, 38, 39, 46, 51.
21. n is product perfect if $\tau(n) = 4$.
22. One.

Exercises 4.2

1. The digital roots are, respectively 3, 5, 8, 5, 8, 5.
2. $n = 3, 4, 5, 6, 8, 10, 12, 15, 16, 17, 18, 20, 24, 26$.
3. $F_0 = F_1 - 2$. Suppose that $\prod_{0 \leqslant n \leqslant k} F_n = F_{k+1} - 2$. Hence,
$$\prod_{0 \leqslant n \leqslant k+1} F_n = (F_{k+1} - 2)F_{k+1} = (2^{2^{k+1}} - 1)(2^{2^{k+1}} + 1) = 2^{2^{k+2}} - 1 = F_{k+2} - 2.$$
4. The last digit of $F_2 = 2^{2^2} + 1 = 17$ is 7. If the last digit of $F_n = 2^{2^n} + 1$ is 7, the last digit of 2^{2^n} is 6. Therefore the last digit of $F_{n+1} = 2^{2^{n+1}} + 1 = (2^{2^n})^2 + 1$ is 7.
5. If a prime p divides $\gcd(F_m, F_n)$, where $n = m + k$, then $rp = 2^{2^n} + 1$ and we have $2^{2^n} = 2^{2^{m+k}} = (2^{2^m})^{2^k} = (pr - 1)^{2^k}$, which is not divisible by p.
6. Let $n = m + k$; then $F_{m+k} - 2 = 2^{2^{m+k}} - 1 = (2^{2^m})^{2^k} - 1$. Since

$(x+1)|(x^{2^n}-1)$, $(2^{2^m}+1)|[(2^{2^m})^{2^k}-1]$. Therefore, F_m divides $F_{m+k}-2$.

7. $F_1=5$. If $F_n=2^{2^n}+1=12k+5$, $2^{2^n}=12k+4$. Therefore, $F_{n+1}=2^{2^{n+1}}+1=(2^{2^n})^2+1=(12k+4)^2+1=12m+5$.
8. If $F_n=2^{2^n}+1=(2k\pm 1)^2$, $2^{2^n}=4k(k\pm 1)$, a contradiction since k or $k+1$ is odd.
9. If $F_n=2^{2^n}+1=(2k+1)^3$, $2^{2^n}=2k(4k^2+6k+3)$, a contradiction since $4k^2+6k+3$ is odd.
10. Suppose $2^{2^n}+1=k(k+1)/2$. Multiplying both sides of the equation by 2, we obtain $2^{2^n+1}+2=k(k+1)$. Hence, $2^{2^n+1}=(k+2)(k-1)$, a contradiction, since one of the factors on the right is odd.

Exercises 4.3

1. (a) $220=2^2\cdot 5\cdot 11, 284=2^2\cdot 71$, and $\sigma(220)=504=\sigma(284)$.
   (b) $1184=2^5\cdot 37$, $1210=2\cdot 5\cdot 11^2$, and $\sigma(1184)=2394=\sigma(1210)$.
   (c) $17\,296=2^4\cdot 23\cdot 47$, $18\,416=2^4\cdot 1151$, and $\sigma(17\,296)=35\,712=\sigma(18\,416)$.
   (d) $176\,272=2^4\cdot 23\cdot 479$, $180\,848=2^4\cdot 89\cdot 127$, and $\sigma(176\,272)=357\,120=\sigma(180\,848)$.
2. If $\sum_{d|m} d=\sum_{d|n} d=m+n$ then

$$\sum_{d|m}\frac{1}{d}=\frac{\sum_{d/m} d}{m}=\frac{m+n}{m}$$

and

$$\sum_{d|n}\frac{1}{d}=\frac{\sum_{d|n} d}{n}=\frac{m+n}{n}.$$

Hence,

$$\frac{1}{\sum_{d|m} d}+\frac{1}{\sum_{d|n} d}=\frac{m}{m+n}+\frac{n}{m+n}=\frac{m+n}{m+n}=1.$$

3. (a) The sum of the digits of the pair (63 020, 76 084) is 36;
   (b) the sum of the digits of the pair (652 664, 643 336) is 54.
4. $48=2^4\cdot 3$ and $75=3\cdot 5^2$; $\sigma(48)=124=\sigma(75)$. $140=2^2\cdot 5\cdot 7$ and $195=3\cdot 5\cdot 13$; $\sigma(140)=336=\sigma(195)$. $1575=3^2\cdot 5^2\cdot 7$ and

$1648 = 2^4 \cdot 103$; $\sigma(1575) = 3224 = \sigma(1648)$.

5. $\sigma(2^5 \cdot 3^2 \cdot 47 \cdot 109) = \sigma(2^5 \cdot 3^2 \cdot 7 \cdot 659) = \sigma(2^5 \cdot 3^2 \cdot 5279)$
$= 4\,324\,320 = 1\,475\,424 + 1\,328\,544 + 1\,520\,352$.
6. $\sigma(2^2 \cdot 3^2 \cdot 5 \cdot 11) = \sigma(2^5 \cdot 3^2 \cdot 7) = \sigma(2^2 \cdot 3^2 \cdot 71) = 6552$.
7. $s(123\,228\,768) = 103\,340\,640 + 124\,015\,008 = 227\,355\,648$
$s(103\,340\,640) = 123\,228\,768 + 124\,015\,008 = 247\,244\,377$
$s(124\,015\,008) = 123\,228\,768 + 103\,340\,640 = 276\,569\,408$

Exercises 4.4

1. $\sigma(120) = \sigma(2^3 \cdot 3 \cdot 5) = 3 \cdot 120 = 360$.
$\sigma(672) = \sigma(2^5 \cdot 3 \cdot 7) = 3 \cdot 672 = 2016$.
$\sigma(2^9 \cdot 3 \cdot 11 \cdot 31) = 1023 \cdot 4 \cdot 12 \cdot 32 = 3(2^9 \cdot 3 \cdot 11 \cdot 31) = 3 \cdot 523\,776$.
2. Suppose n is squarefree and 3-perfect. Since $n = p_1 \cdots p_r$ and $\sigma(n) = 3n$, $3p_1 \cdots p_r = (p_1 + 1) \cdots (p_r + 1)$, a contradiction, since $2^{r-1} | (p_1 + 1) \cdots (p_r + 1)$, but $2^{r-1} \nmid (3p_1 \cdots p_r)$, unless $r \leqslant 2$, which is easily eliminated.
3. $\sigma(30\,240) = \sigma(2^5 \cdot 3^3 \cdot 5 \cdot 7) = 63 \cdot 6 \cdot 8 \cdot 40 = 120\,960 = 4 \cdot 30\,240$.
4. $\sigma(14\,182\,439\,040) = 255 \cdot 121 \cdot 6 \cdot 8 \cdot 133 \cdot 18 \cdot 20 = 70\,912\,195\,200 = 5 \cdot 14\,182\,439\,040$.
5. If n is k-perfect then $\sigma(n) = kn$. Hence, $\dfrac{\sigma(n) - n}{n} = \dfrac{kn - n}{n} = k - 1$.
6. $2 \cdot \sigma(21) = 2 \cdot 32 = 3 \cdot 21 + 1 = 64$.
$2133 = 3^3 \cdot 79$ and $2 \cdot \sigma(2133) = 2 \cdot 3200 = 6400 = 3 \cdot 2133 + 1$.
$19\,521 = 3^4 \cdot 241$ and $2 \cdot \sigma(19\,521) = 2 \cdot 29\,282 = 58\,564 = 3 \cdot 19\,521 + 1$.
7. $3 \cdot \sigma(325) = 3 \cdot 434 = 1302 = 4 \cdot 325 + 2$.
8. $36 = 6 + 12 + 18$; $40 = 10 + 20 + 2 + 8$;
$770 = 35 + 5 + 385 + 154 + 110 + 70 + 11$;
$945 = 3 + 7 + 135 + 105 + 189 + 315 + 21 + 27 + 63 + 45 + 35$.
9. 770 and 945 are semiperfect and none of their divisors are semiperfect.
10. $\sigma(70) = 144 > 140 = 2 \cdot 70$.
11. $2^{161\,038} - 2 = 2(2^{161\,037} - 1) = (2^9)^{29\cdot 617} - 1^{29\cdot 167} = (2^9 - 1)(\ldots) = 511(\ldots) = 7 \cdot 73(\ldots)$ and $2^{161\,037} - 1 = (2^{29})^{9\cdot 617} - 1^{9\cdot 617} = (2^{29} - 1)(\ldots) = 1103 \cdot 486\,737(\ldots)$. Hence, the primes 73 and 1103 both divide $2^{161\,037} - 1$.

12.
$$\begin{aligned} 1 &= 1, & 13 &= 12 + 1,\\ 2 &= 2, & 14 &= 12 + 2,\\ 3 &= 3, & 15 &= 12 + 3,\\ 4 &= 4, & 16 &= 12 + 4,\\ 5 &= 4 + 1, & 17 &= 12 + 3 + 2,\\ 6 &= 4 + 2, & 18 &= 12 + 6,\\ 7 &= 4 + 3, & 19 &= 12 + 6 + 1,\\ 8 &= 6 + 2, & 20 &= 12 + 8,\\ 9 &= 6 + 3, & 21 &= 12 + 8 + 1,\\ 10 &= 6 + 4, & 22 &= 12 + 8 + 2,\\ 11 &= 8 + 3, & 23 &= 12 + 8 + 3.\\ 12 &= 8 + 4, \end{aligned}$$

13. The result follows since it is possible using the binary system to represent any integer from 1 to $2^p - 1$ as a sum of 1, 2, ..., 2^{p-1}.
14. $23 = 8 + 6 + 4 + 3 + 2 = 12 + 8 + 3$.
15. If n is perfect then $\sigma(n) = 2n$ and $\sigma(n) - n - 1 = n - 1$.
16. $140 \cdot \tau(140)/\sigma(140) = 5$.
17. If n is perfect $\sigma(n) = 2n$ and $\tau(n)$ is even. Hence,
$$\frac{n \cdot \tau(n)}{\sigma(n)} = \frac{n \cdot 2r}{2n} = r.$$
18. $\alpha(60) = 168 - 120 = 48$; $\delta(26) = 52 - 42 = 10$.
19. $A(p^\alpha) = \dfrac{\sigma(p^\alpha)}{\tau(p^\alpha)} = \dfrac{p^{\alpha+1}}{(p-1)(\alpha+1)}$.
20. 1, 3, 5, 6, 7, 11, 13, 14, 15, 17.
21. $H(p^\alpha) = \dfrac{(p-1)(\alpha+1)}{p^{\alpha+1} - 1}\, p^\alpha$.
22. $H(1) = 1$, $H(4) = \frac{12}{7}$, $H(6) = 2$, and $H(p) = 2p/(p+1) < 2$, for p a prime.
23. $H(2^{n-1}(2^n - 1)) = n$.
24. $G(p^\alpha) = (1 \cdot p \cdot p^2 \cdots p^\alpha)^{\alpha+1} = p^{\alpha(\alpha+1)^2/2}$.
25. $A(n)$ and $H(n)$ are multiplicative because σ and τ are multiplicative. However, $G(6) = 68 \neq 64 = G(2) \cdot G(3)$.
26. $\sigma(2^n) + 1 = (2^{n+1} - 1)/(2 - 1) + 1 = 2^{n+1} = 2(2^n)$.
27. $\sigma(\sigma(16)) = \sigma(31) = 32 = 2 \cdot 16$.
28. $90 = 2 \cdot 3^2 \cdot 5$ and $\sigma^*(90) = 180$;
$87\,360 = 2^6 \cdot 3 \cdot 5 \cdot 7 \cdot 13$ and $\sigma^*(87\,360) = 174\,720$.
29. The result follows since σ^* is multiplicative and $\sigma^*(p^\alpha) = p^\alpha + 1$.
30. $n = 32$.
31. $\sigma^*(114) = \sigma^*(126) = 114 + 126 = 240$.

Exercises 5.1

1. If $a \equiv b \pmod{m}$, then $a = b + km$ for some integer k. Hence, $a \pm c = b \pm c + km$ or $a \pm c \equiv b \pm c \pmod{m}$. Similarly, $a \cdot c = b \cdot c + ckm$ or $a \cdot c \equiv b \cdot c \pmod{m}$. The third property follows since $a^n = (b + km)^n \equiv b^n \pmod{m}$.
2. For $i = 1, \ldots, n$, suppose that $a_i \equiv b_i \pmod{m}$. There exist k_i such that $a_i = b_i + k_i m$. The additive case may be handled without using induction since
$$\sum_{i=1}^{n} a_i = \sum_{i=1}^{n} (b_i + k_i m) = \sum_{i=1}^{n} b_i + \left(\sum_{i=1}^{n} k_i\right) m.$$
Hence, $\sum_{i=1}^{n} a_i \equiv \sum_{i=1}^{n} b_i \pmod{m}$. We have already shown that if $a \equiv b \pmod{m}$ and $c \equiv d \pmod{m}$, then $ac \equiv bd \pmod{m}$. Suppose that $\prod_{i=1}^{n} a_i = \prod_{i=1}^{n} b_i$ and $a_{n+1} \equiv b_{n+1} \pmod{m}$. The result follows since
$$\prod_{i=1}^{n+1} a_i \equiv \left(\prod_{i=1}^{n} a_i\right) a_{n+1} \equiv \left(\prod_{i=1}^{n} b_i\right) b_{n+1} \equiv \prod_{i=1}^{n+1} b_i \pmod{m}.$$
3. If $a \equiv b \pmod{m_1}$, $a \equiv b \pmod{m_2}$, and $\gcd(m_1, m_2) = 1$, then $a - b = rm_1$, $a - b = sm_2$, and $m_1 u + m_2 v = 1$. Multiplying the latter equation by $a - b$, we obtain $(a - b)m_1 u + (a - b)m_2 v = a - b$. Therefore, $a - b = sm_2 m_1 u + rm_1 m_2 v = (su + rv)m_2 m_1$. Thus, $m_1 m_2$ divides $a - b$ or $a \equiv b \pmod{m_1 m_2}$.
4. Suppose $a \equiv b \pmod{m}$ and d divides m, where $d > 0$. There are integers s and t such that $a = b + cm$ and $dt = m$. Hence $a = b + c(dt) = b + (ct)d$. Therefore, $a \equiv b \pmod{d}$.
5. The result follows directly from the previous exercise.
6. If $a \equiv b \pmod{m}$ and $c \equiv d \pmod{m}$ there exist integers s and t such that $a = b + sm$ and $c = d + tm$. Hence $ax = bx + sxm$ and $cy = dy + tym$. Thus, $ax + cy = (bx + sxm) + (dy + tym) = bx + dy + (sx + ty)m$, implying that $ax + cy \equiv bx + dy \pmod{m}$.
7. If $a \equiv b \pmod{m}$ then there is an integer k such that $a = b + km$ or $b = a - km$. Hence $\gcd(a, m)$ divides $\gcd(b, m)$ and $\gcd(b, m)$ divides $\gcd(a, m)$. Thus, $\gcd(a, m) = \gcd(b, m)$.
8. If $a^2 \equiv b^2 \pmod{p}$, where p is prime, then there exists an integer k such that $a^2 - b^2 = (a + b)(a - b) = kp$. Hence, since p is prime, from Euclid's Lemma, either $p|(a + b)$ or $p|(a - b)$.
9. $47 \equiv 5$, $86 \equiv 2$, $22 \equiv 1$, $-14 \equiv 0$, $32 \equiv 4$, $20 \equiv 6$, and $143 \equiv 3 \pmod{7}$.

10. $-88, -69, -50, -31, -12, 7, 26, 45, 64$, and 83.
11. $0 \equiv 7 \cdot 0$, $1 \equiv 7 \cdot 8$, $2 \equiv 7 \cdot 5$, $3 \equiv 7 \cdot 2$, $4 \equiv 7 \cdot 10$, $5 \equiv 7 \cdot 7$, $6 \equiv 7 \cdot 4, 7 \equiv 7 \cdot 1, 8 \equiv 7 \cdot 9, 9 \equiv 7 \cdot 6$, and $10 \equiv 7 \cdot 3 \pmod{11}$.
12. If m is even then $m \equiv 2m \equiv 0 \pmod{m}$ and the integers 2, 4, 6, $\ldots, 2m$ are not all distinct. If m is odd, $\gcd(2, m) = 1$. Since $2r \equiv 2s \pmod{m}$ for $1 \leqslant r, s \leqslant m$ implies $r \equiv s \pmod{m}$, it follows $2, 4, \ldots, 2m$ are distinct.
13. If $m > 2$, then $(m-1)^2 \equiv 1 \pmod{m}$. Hence $\{1^2, 2^2, 3^2, \ldots, m^2\}$ does not contain m distinct elements modulo m.
14. $1941 \equiv 2 \pmod{7}$, $1941^3 \equiv 1 \pmod{7}$. Hence, $1941^{1963} \equiv 1941^{3 \cdot 654+1} \equiv 1941 \equiv 2 \pmod{7}$. Similarly, $1963 \equiv 3 \pmod{7}$, $1963^6 \equiv 1 \pmod{7}$. Hence, $1963^{1991} = 1963^{6 \cdot 331+5} \equiv 1963^5 \equiv 3^5 \equiv 243 \equiv 5 \pmod{7}$. Therefore, $1941^{1963} + 1963^{1991} \equiv 2 + 5 \equiv 0 \pmod{7}$.
15. $9^{10} \equiv 1 \pmod{100}$. Hence, $9^{9^9} = 9^{387\,420\,489} = 9^{10 \cdot 38\,742\,048+9} \equiv 1^{38\,742\,048} \cdot 9^9 \equiv 1 \cdot 387\,420\,489 \equiv 89 \pmod{100}$. Therefore, the last two digits of 9^{9^9} are 89.
16. $53^{103} + 103^{53} \equiv 53 \cdot (53^2)^{51} + 103 \cdot (103^2)^{26} \equiv 53(1)^{51} + 103(1)^{26} \equiv 53 + 103 \equiv 156 \equiv 0 \pmod{39}$.
17. $111^{333} + 333^{111} \equiv (-1)^{333} + 333 \cdot (333^2)^{55} \equiv -1 + 4 \cdot 2^{55} \equiv -1 + 4 \cdot (2^3)^{18}\, 2 \equiv -1 + 8 \equiv 7 \equiv 0 \pmod{7}$.
18. $19^2 \equiv 19^{32} \equiv 20 \pmod{31}$,
    $19^4 \equiv 19^{64} \equiv 28 \pmod{31}$,
    $19^8 \equiv 19^{128} \equiv 9 \pmod{31}$,
    $19^{16} \equiv 19^{256} \equiv 19 \pmod{31}$.
    Therfore, $19^{385} = 19^{256+128+1} \equiv 19 \cdot 9 \cdot 19 \equiv 25 \pmod{31}$.
19. $3^{97} = (3^4)^{24} \cdot 3 \equiv 1^{24} \cdot 3 \equiv 3 \pmod{10}$. Hence, the last digit is 3.
20. $3^{1000} \equiv (3^{40})^{25} \equiv 1^{25} \equiv 1 \pmod{100}$. Hence, the last two digits are 01.
21. $1! + 2! + \cdots + 100! \equiv 1 + 2 + 6 + 9 + 0 + \cdots + 0 \equiv 3 \pmod{15}$.
22. $1^5 + 2^5 + \cdots + 100^5 \equiv 1 + 0 + 3 + 0 + 1 + \cdots + 3 + 0 \equiv (1 + 0 + 3 + 0) \cdot 25 \equiv 0 \pmod{4}$.
23. $63! - 61! = (63 \cdot 62 - 1)61! = 71 \cdot 55 \cdot 61! \equiv 0 \pmod{71}$.
24. $5^{2n} + 3 \cdot 2^{5n-2} \equiv (5^2)^n + 3 \cdot (2^5)^n \cdot 2^{-2} \equiv 4^n + 3 \cdot 4^n \cdot 2 \equiv 7 \cdot 4^n \equiv 0 \pmod{7}$.
25. $3^{n+2} + 4^{2n+1} \equiv 9 \cdot 3^n + (16)^n \cdot 4 \equiv 9 \cdot 3^n + 4 \cdot 3^n \equiv 13 \cdot 3^n \equiv 0 \pmod{13}$.
26. If $n = 2k + 1$, then $n^2 - 1 = (2k+1)^2 - 1 = 4k^2 + 4k = 4k(k+1) = 8m$ since either k or $k + 1$ is even. Therefore, $n^2 - 1 \equiv 0 \pmod{8}$.
27. $a = 0$, $b = 5$, $c = 16$, $d = 28$, and $e = 4$. Therefore, Easter fell on

April 23, 1916.

29. If $x \equiv 0 \pmod{12}$, then $x \equiv 0 \pmod 2$.
    If $x \equiv 1 \pmod{12}$, then $x \equiv 1 \pmod 4$.
    If $x \equiv 2 \pmod{12}$, then $x \equiv 0 \pmod 2$.
    If $x \equiv 3 \pmod{12}$, then $x \equiv 0 \pmod 3$.
    If $x \equiv 4 \pmod{12}$, then $x \equiv 0 \pmod 2$.
    If $x \equiv 5 \pmod{12}$, then $x \equiv 1 \pmod 4$.
    If $x \equiv 6 \pmod{12}$, then $x \equiv 0 \pmod 2$.
    If $x \equiv 7 \pmod{12}$, then $x \equiv 1 \pmod 6$.
    If $x \equiv 8 \pmod{12}$, then $x \equiv 0 \pmod 2$.
    If $x \equiv 9 \pmod{12}$, then $x \equiv 0 \pmod 3$.
    If $x \equiv 10 \pmod{12}$, then $x \equiv 0 \pmod 2$.
    If $x \equiv 11 \pmod{12}$, then $x \equiv 11 \pmod{12}$.
30. $(3n)^3 \equiv 0 \pmod 9$, $(3n+1)^3 \equiv 1 \pmod 9$, and $(3n+2)^3 \equiv 8 \pmod 9$.
31. The result follows immediately from the previous exercise.
32. If $0 < c_k < b$, $0 \leqslant c_i < b$, for $i = 1, 2, \ldots, k-1$, and $b > 1$ is a positive integer, then $c_k b^k + \cdots + c_1 b + c_0 \equiv c_0 + \cdots + c_k \pmod{b-1}$
33. Suppose that there exist integers u and v such that $n = r + mu$ and $n = s + (m+1)v$. Hence, $n(m+1) = r(m+1) + m(m+1)u$ and $nm^2 = sm^2 + m^2(m+1)v$. Combining and simplifying, we obtain $n = r(m+1) + m^2 s + m(m+1)(u + vm^2 - n)$. Therefore, $n \equiv (m+1)r + m^2 s \pmod{m(m+1)}$.

Exercises 5.2

1. If 7 divides $(2a+b)$ then $2a + b = 7k$. Hence, $100a + b = 98a + 7k = 7(14a + k)$. Conversely, if 7 divides $(100a + b)$ then $7s = 100a + b = 14a(7) + 2a + b$. Hence, $2a + b = 7(s - 14a)$.
2. From the proof of Theorem 5.8, $10 \equiv 1 \pmod 9$, hence $f(10) \equiv f(1) \pmod 9$ so $a \equiv s \pmod 9$. Therefore, $a - s \equiv 0 \pmod 9$.
3. (a) $x = 2$, (b) $x = 5$, (c) $x = 4$.
4. From Theorem 5.8, $9|R_n$ if and only if the number of ones in R_n is a multiple of 9. That is, if and only if $9|n$.
5. From Theorem 5.8, 11 divides R_n if and only if the number of ones in R_n is even. That is, if and only if n is even.
6. 691 504 249 989, 13 830 085 087, 276 601 787, 5 532 121, 110 663, 2275, 119, 21, which is divisible by 7. Therefore, 691 504 249 989 is

divisible by 7.

7. 67 911 603 138 353, 6 791 160 313 847, 679 116 031 412, 67 911 603 149, 6 791 160 314, 679 116 035, 67 911 623, 6 791 174, 679 133, 67 925, 6 812, 639, 104, 26, which is divisible by 13. Therefore, 67 911 603 138 353 is divisible by 13.

8. 2 ╳ 8 (5 above, 7 below), $5 \neq 7$. Therefore, a mistake has been made.

9. Drop the units digit from the number and subtract 5 times it from what remains. The result is divisible by 17 if and only if the original number is divisible by 17.

10. Let $n = 7 \cdot 5^{41}$. Since $\log(n) = \log(7) + 41 \cdot \log(5) = 29.5$, n has 30 digits. The only 30-digit numbers not having four repeated digits are those in which each digit occurs exactly three times. However, each of these is divisible by 3. Since $n \equiv 1 \cdot 2^{41} \equiv 2 \pmod 3$, $3 \nmid n$. Therefore, in the decimal representation of n at least one digit appears at least four times.

Exercises 5.3

1. (a) $\phi(406) = \phi(2 \cdot 7 \cdot 29) = 168$.
   (b) $\phi(756) = \phi(2^2 \cdot 3^3 \cdot 7) = 216$.
   (c) $\phi(1228) = \phi(2^2 \cdot 307) = 612$.
   (d) $\phi(7642) = \phi(2 \cdot 3821) = 3820$.
2. $\{1, 5, 7, 11, 13, 17\}$
3. $\phi(25\,930) = \phi(2 \cdot 5 \cdot 2593) = 10\,368$.
   $\phi(25\,935) = \phi(3 \cdot 5 \cdot 7 \cdot 13 \cdot 19) = 10\,368$.
   $\phi(25\,940) = \phi(2 \cdot 2 \cdot 5 \cdot 1297) = 10\,368$.
   $\phi(25\,942) = \phi(2 \cdot 7 \cdot 17 \cdot 109) = 10\,368$.
4. $\phi(p+2) = p+1 = p-1+2 = \phi(p)+2$.
5. If n is prime, $(\phi(n)\sigma(n)+1)/n = n$.
6. $1 + \phi(p) + \cdots + \phi(p^n) = 1 + (p-1) + (p^2 - p) + \cdots + (p^n - p^{n-1}) = p^n$.
7. $f(p^k) = \phi(p^k)/p^k = (p-1)/p = \phi(p)/p = f(p)$.
8. (a) If $n > 2$ then there will always be a factor of the form p or $p-1$ that is even, hence, $\phi(n)$ is even, thus, $n = 1$ or 2;
   (b) n is prime;
   (c) $n = 1$, 2^r, or $2^r 3^s$, where r and s are positive integers;
   (d) n has at least two district odd prime factors, or one prime factor of the form $4k+1$, or is divisible by 4, except 4 itself;

(e) $n = 2^{k+1}, 3 \cdot 2^k$, or $5 \cdot 2^{k-1}$;
(f) $n = 2^k$;
(g) there are none;
(h) power of 2 dividing n plus number of distinct prime factors of the form $4r + 3$ plus twice number of distinct prime factors of the form $4r + 1$ is at least k if n is odd or $k + 1$ if n is even.

9. $$\phi(n^2) = n^2 \prod_{p|n^2}\left(1 - \frac{1}{p}\right) = n \cdot n \prod_{p|n}\left(1 - \frac{1}{p}\right) = n\phi(n).$$
10. $\phi(11^k \cdot p) = 10 \cdot 11^{k-1} \cdot (p - 1)$.
11. $\phi(2^{2k+1}) = (2^k)^2$.
12. $\phi(125) = 100$. Hence, $a^{100} \equiv 1 \pmod{125}$ if $5 \nmid a$, and $a^{100} \equiv 0 \pmod{125}$ if $5|a$.
13. $5 < \phi(100) \leqslant 36.7$; $15.8 < \phi(1000) \leqslant 81.3$.
14. The average is 30.34; $6n/\pi^2 = 60.79$.
15. The numbers k which are less than n and coprime to n occur in pairs $(k, p - k)$ whose sum is p and there are $\phi(n)/2$ such pairs.
16. If n is nonsquare its divisors pair up and one of them is less than $\sqrt{n}$. Thus the divisors d of n and their pairs would be less than but not be coprime to n. If n is square only $\sqrt{n}, 2\sqrt{n}, \ldots, (\sqrt{n} - 1)\sqrt{n}$ are less than n and not coprime to n.
17. The result follows from Theorem 5.12 and the fact that if $k \leqslant n$, then $\phi(k)$ occurs as often as there are multiples of k that are less than n.
18. 36.
19. If $n = p^\alpha$, $\phi(p^\alpha) + \sigma(p^\alpha) = p^{\alpha-1}(p - 1) + (1 + p + \cdots + p^{\alpha-1} + p^\alpha) = 2p^\alpha + 1 + p + \cdots + p^{\alpha-2} \geqslant 2p^\alpha$.
20. Whenever n is 1 or a prime.
21. Since $\gcd(m, n) = 1$,
$$f(mn) = \frac{\sigma(mn)\phi(mn)}{(mn)^2} = \frac{\sigma(m)\sigma(n)\phi(m)\phi(n)}{(mn)^2} = \frac{\sigma(m)\phi(m)}{m^2}\frac{\sigma(n)\phi(n)}{n^2}$$
$$= f(m)f(n).$$
22. $$\binom{p}{k} = \frac{p!}{k!(p - k)!}$$
is an integer and none of the factors in the denominator divides the p in the numerator.
23. (a) $1^{p-1} + \cdots + (p - 1)^{p-1} \equiv 1 + \cdots + 1 \equiv p - 1 \equiv -1 \pmod{p}$.
(b) $1^p + \cdots + (p - 1)^p \equiv 1 + 2 + \cdots + (p - 1) \equiv p(p - 1)/2 \equiv 0 \pmod{p}$.
24. If $\gcd(m, n) = 1$, $m^{\phi(n)} \equiv 1 \pmod{n}$ and $n^{\phi(m)} \equiv 1 \pmod{m}$. Hence,

there exist integers r and s such that $m^{\phi(n)} - 1 = ns$ and $n^{\phi(m)} - 1 = ms$. Multiplying, we obtain $m^{\phi(n)}n^{\phi(m)} - m^{\phi(n)} - n^{\phi(m)} + 1 = rsmn$. Thus, $m^{\phi(n)} + n^{\phi(m)} - 1 = nm(-rs + m^{\phi(n)-1} n^{\phi(m)-1})$. Therefore, $m^{\phi(n)} + n^{\phi(m)} \equiv 1 \pmod{mn}$.

25. Since $\phi(62) = 30$, multiplying both sides of the congruence by 41^{29} yields $41^{30}x \equiv 41^{29} \cdot 53 \pmod{62}$. Therefore, $x \equiv 41^{29} \cdot 53 \equiv (41)^{2\cdot 14+1} \cdot 53 \equiv (7)^{14} \cdot 41 \cdot 53 \equiv 9 \cdot 41 \cdot 53 \equiv 27 \pmod{62}$.
26. $6601 = 7 \cdot 23 \cdot 41$, and 6, 22, and 40 each divide 6600. Hence, if $\gcd(a, 6601) = 1$, a^{6600} is congruent to 1 modulo 7, 23, and 41. Therefore, $a^{6600} \equiv 1 \pmod{6601}$
27. $1 \cdot \mu(105) + 3 \cdot \mu(35) + 5 \cdot \mu(21) + 15 \cdot \mu(7) = -1 + 3 + 5 - 15$

$$= -8 = \frac{(-1) \cdot 48}{6} = \frac{\mu(7) \cdot \phi(105)}{\phi(7)} = \frac{\mu\left(\frac{105}{15}\right) \cdot \phi(105)}{\phi\left(\frac{105}{15}\right)}.$$

28. Since ϕ and τ are multiplicative let $n = p^\alpha$,

$$\sum_{d \mid p^\alpha} \phi(d) \cdot \tau\left(\frac{p^\alpha}{d}\right) = 1 \cdot (\alpha + 1) + (p-1)\alpha + p(p-1)\alpha + \cdots$$
$$+ p^{\alpha-1}(p-1) \cdot 1 = 1 + p + \cdots + p^\alpha = \phi(p^\alpha).$$

29. Since ϕ and σ are multiplicative let $n = p^\alpha$,

$$\sum_{d \mid p^\alpha} \phi(d) \cdot \sigma\left(\frac{p^\alpha}{d}\right) = 1 \cdot (1 + p + \cdots + p^{\alpha-1})$$
$$+ (p-1)(1 + p + \cdots + p^{\alpha-2}) + \cdots + p^{\alpha-1}(p-1) \cdot 1$$
$$= (1 + p + \cdots + p^{\alpha-1}) + (p + p^2 + \cdots + p^{\alpha-1})$$
$$- (1 + p + \cdots + p^{\alpha-2}) + p^\alpha - p^{\alpha-1} = np^\alpha.$$

30. If n is prime, $\sigma(n) = n + 1$, $\phi(n) = n - 1$, and $\tau(n) = 2$. Hence, $\sigma(n) + \phi(n) = n \cdot \tau(n)$. Suppose $\sigma(n) + \phi(n) = n \cdot \tau(n)$ and $n > 1$ is not prime. Thus, $\sigma(n) < n$, $\tau(n) = k \geqslant 3$, and there exists a divisor d^* of n such that $kd^* < n$ and $n - d^* \geqslant 1$. Therefore, $n \cdot \tau(n) - \sigma(n) = kn - \sum_{d \mid n} d = \sum_{d \mid n}(n - d) \geqslant (n-1) + (n - d^*) + 0 \geqslant n - 1 + 1 \geqslant n > \phi(n)$, a contradiction.
31. $12 = 6 + 4 + 3 - 1 = \tau(12) + \phi(12) + \xi(12) - 1$.
32. Both equal 4.
33. Both equal 3.
34. Let $n = p^\alpha$.

$$\sum_{d|p^\alpha}\frac{\mu^2(d)}{\phi(d)}=\frac{\mu^2(1)}{\phi(1)}+\frac{\mu^2(p)}{\phi(p)}+\cdots+\frac{\mu^2(p^\alpha)}{\phi(p^\alpha)}=1+\frac{1}{p-1}$$

$$=\frac{p^\alpha}{p^\alpha-p^{\alpha-1}}=\frac{p^\alpha}{\phi(p^\alpha)}.$$

35. $\frac{1}{2}\sum_{k=1}^{10}\phi(k)=15$; $3(10/\pi)^2=30.4$.
36. $\mathscr{F}_7=\{\frac{0}{1},\frac{1}{7},\frac{1}{6},\frac{1}{5},\frac{1}{4},\frac{2}{7},\frac{1}{3},\frac{2}{5},\frac{3}{7},\frac{1}{2},\frac{4}{7},\frac{3}{5},\frac{2}{3},\frac{5}{7},\frac{3}{4},\frac{4}{5},\frac{5}{6},\frac{6}{7},\frac{1}{1}\}$.
37. It is true for the first row. Suppose it is true for the $(n-1)$st row. Any consecutive fractions on the nth row will be of the form

$$\frac{a}{b},\frac{c}{d},$$
$$\frac{a}{b},\frac{a+c}{b+d},$$

or

$$\frac{a+c}{b+d},\frac{c}{d},$$

where a/b and c/d are consecutive fractions on the $(n-1)$st row, hence, $ad-bc=1$. In the second case, $ab+ad-ba-bc=1$. In the third case, $ad+cd-bc-dc=1$.
38. If $a/b<c/d$, then $ad<bc$. Hence, $ab+ad<ba+bc$ and $ad+cd<bc+cd$. Therefore,

$$\frac{a}{b}<\frac{a+c}{b+d}<\frac{c}{d}.$$

39. $\dfrac{c}{d}-\dfrac{a}{b}=\dfrac{ad-bc}{bd}=\dfrac{1}{bd}$.
40. $\dfrac{x}{y}=\dfrac{ma+nc}{mb+nd}$.

Exercises 5.4

1. (a) $x\equiv 18 \pmod{29}$.
   (b) $x=4+16t$, for $t=0, 1, 2, 3$.
   Hence, $x\equiv 4, 20, 36, 52 \pmod{64}$.
   (c) $x\equiv 56 \pmod{77}$.
   (d) No solution.
   (e) $x\equiv 14 \pmod{29}$.
2. $x=-36-51t$; $y=3+4t$.
3. $x=2-3t$; $y=2t$.
4. $h=9+21t$; $c=71-31t$. Hence, $(h, c)=(51, 9)$, $(30, 40)$ or $(9, 71)$.
5. $17p+15a=143$ or $17p\equiv 143 \pmod{15}$, implying that $p\equiv 4$

(mod 15). Therefore, $a = 5$ and $p = 4$.

6. $x + y = 100$, $x \equiv 0 \pmod{7}$, $x \equiv 0 \pmod{11}$. Hence, $7S + 11t = 100$. Thus, $S \equiv 8 \pmod{11}$ and $t \equiv 4 \pmod{7}$. Therefore, $x = 44$ and $q = 56$ is a solution.
7. $x \equiv 49 \pmod{61}$.
8. $$x + y + z = 100,$$ $$3x + 2y + \frac{z}{2} = 100$$

   or

   $$x + y + z = 100,$$ $$6x + 4y + z = 200.$$

   Therefore,

   $$x = 2 + 3t,$$ $$y = 30 - 5t,$$ $$z = 68 + 2t.$$

   Solutions (m, w, c) are given by (2, 30, 68), (5, 25, 70), (8, 20, 72), (11, 15, 74), (14, 10, 76), (17, 5, 78), and (20, 0, 80).
9. $$x + y + z = 100,$$ $$5x + y = \frac{z}{20} = 100,$$

   or

   $$x + y + z = 100,$$ $$100x + 20y + z = 2000.$$

   Therefore, buying 100 chickens is a solution.
10. We seek solutions to

    $$x + y + z = 41 \text{ and } 4x + 3y + \frac{1}{3}z = 40$$

    or equivalently to

    $$x + y + z = 41 \text{ and } 12x + 9y + z = 120.$$

    Subtracting, we obtain $11x + 8y = 79$. Hence, $8y \equiv 79 \equiv 2 \pmod{11}$, implying that $y \equiv 3 \pmod{11}$. Thus, $y = 3 + 11t$, $x = 5 - 8t$, $z = 33 - 3t$. Therefore, there were 5 men, 3 women, and 33 children.
11. No integral solutions.
12. $x = 18$, $y = 0$, $z = 12$.
13. 5.
14. 59.
15. 1103.
16. $x \equiv 1982 \equiv 2 \pmod{6}$;
    $x \equiv 1978 \equiv 4 \pmod{7}$;

$x \equiv 32 \pmod{42}$. Therefore $x = 2006$.

17. $x \equiv 3 \pmod{17}$,
    $x \equiv 10 \pmod{16}$,
    $x \equiv 0 \pmod{15}$.
    Therefore, $x \equiv 3930 \pmod{4080}$.
18. $x \equiv 6 \pmod{23}$,
    $x \equiv 7 \pmod{28}$,
    $x \equiv 8 \pmod{33}$.
    Therefore, $x \equiv 17\,003 \pmod{21\,252}$ or 46+ years.
19. $x = -7$, $y = 2$, $z = 3$.
20. $x = 7$, $y = 1$, $z = 1$.
21. $x = 114$, $y = 87$, $z = 39$.
22. We have $n^2 \equiv n \pmod{2^5 \cdot 5^5}$. We solve $n(n-1) \equiv 0 \pmod{32}$ and $n(n-1) \equiv 0 \pmod{3125}$ and use the Chinese Remainder Theorem to obtain $n \equiv 8\,212\,890625 \equiv 90\,625 \pmod{100\,000}$.
23. None.

Exercises 5.5

1. Let the integers be $a_1, \ldots, a_n$ and consider $a_1 + a_2$, $a_1 + a_3$, ..., $a_1 + a_n$. If one of these is divisible by n then we are done. If two of them, say $a_1 + a_i$ and $a_1 + a_j$, have the same remainder modulo n then $(a_1 + a_i) - (a_1 + a_j) = a_i - a_j$ is divisible by n. Otherwise the remainders 1, 2, ..., $n - 1$ must be counted once each when dividing the numbers by n, so one of them must have the same remainder as $a_1 - a_2$, say it is $a_1 + a_k$. Hence, $(a_1 + a_k) - (a_1 - a_2) = a_k + a_2$ is divisible by n.
2. Let the numbers be $a_1, a_2, \ldots, a_n$ and consider the numbers $a_1 + a_2$, $a_1 + a_2 + a_3$, ..., $a_1 + a_2 + \cdots + a_n$. When divided by n each of the numbers must leave a remainder from 0 to $n - 1$. So either one gives a remainder 0, and hence is divisible by n, or two have the same remainder and subtracting the smaller from the larger gives the desired sum.
3. Suppose $\gcd(na_i + mb_j, mn) = d$ and p is a prime such that $p|d$. Since $p|mn$, $p|m$ or $p|n$. If $p|m$ then $p \not| n$ since $\gcd(m, n) = 1$. We have $p|(na_i + mb_j)$, hence, $p|na_i$ implying thay $p|a_i$. A contradiction since $\gcd(a_i, m) = 1$. Therefore, $\gcd(na_j + mb_j, mn) = 1$.
4. No two elements in T can be congruent since $\gcd(a_i, m) = 1$ and $\gcd(b_j, n) = 1$. Thus, every integer coprime to mn is counted exactly once and $\phi(m)\phi(n) = \phi(mn)$.

5. Let $n = 2^k p_1^{\alpha_1} p_2^{\alpha_2} \cdots p_r^{\alpha_r}$ and $m = p_1^{\alpha_1} p_2^{\alpha_2} \cdots p_r^{\alpha_r}$, where p_i, for $i = 1, 2, \ldots, r$, are odd primes.

$$\begin{aligned}
\phi(2n) &= \phi(2^{k+1} p_1^{\alpha_1} p_2^{\alpha_2} \cdots p_r^{\alpha_r}) \\
&= \phi(2^{k+1})\phi(p_1^{\alpha_1} p_2^{\alpha_2} \cdots p_r^{\alpha_r}) \\
&= 2^k \phi(p_1^{\alpha_1} p_2^{\alpha_2} \cdots p_r^{\alpha_r}) \\
&= 2 \cdot 2^{k-1} \phi(p_1^{\alpha_1} p_2^{\alpha_2} \cdots p_r^{\alpha_r}) \\
&= 2 \cdot \phi(2^k)\phi(p_1^{\alpha_1} p_2^{\alpha_2} \cdots p_r^{\alpha_r}) \\
&= 2 \cdot \phi(2^k p_1^{\alpha_1} p_2^{\alpha_2} \cdots p_r^{\alpha_r}) \\
&= 2\phi(n). \\
\phi(2m) &= \phi(2 p_1^{\alpha_1} p_2^{\alpha_2} \cdots p_r^{\alpha_r}) \\
&= \phi(2)\phi(p_1^{\alpha_1} p_2^{\alpha_2} \cdots p_r^{\alpha_r}) \\
&= \phi(p_1^{\alpha_1} p_2^{\alpha_2} \cdots p_r^{\alpha_r}) \\
&= \phi(m).
\end{aligned}$$

6. Let $n = 3^k p_1^{\alpha_1} p_2^{\alpha_2} \cdots p_r^{\alpha_r}$ and $m = p_1^{\alpha_1} p_2^{\alpha_2} \cdots p_r^{\alpha_r}$, where p_i, for $i = 1, 2, \ldots, r$, are primes with none equal to 3.

$$\begin{aligned}
\phi(3n) &= \phi(3^{k+1} p_1^{\alpha_1} p_2^{\alpha_2} \cdots p_r^{\alpha_r}) \\
&= \phi(3^{k+1})\phi(p_1^{\alpha_1} p_2^{\alpha_2} \cdots p_r^{\alpha_r}) \\
&= 2 \cdot 3^k \cdot \phi(p_1^{\alpha_1} p_2^{\alpha_2} \cdots p_r^{\alpha_r}) \\
&= 3 \cdot \phi(3^k)\phi(p_1^{\alpha_1} p_2^{\alpha_2} \cdots p_r^{\alpha_r}) \\
&= 3 \cdot \phi(3^k p_1^{\alpha_1} p_2^{\alpha_2} \cdots p_r^{\alpha_r}) \\
&= 3\phi(n). \\
\phi(3m) &= \phi(3 p_1^{\alpha_1} p_2^{\alpha_2} \cdots p_r^{\alpha_r}) \\
&= \phi(3)\phi(p_1^{\alpha_1} p_2^{\alpha_2} \cdots p_r^{\alpha_r}) \\
&= 2 \cdot \phi(p_1^{\alpha_1} p_2^{\alpha_2} \cdots p_r^{\alpha_r}) \\
&= 2\phi(m).
\end{aligned}$$

7. (a) $\Lambda_c(24) = 4$, (b) $\Lambda_c(81) = 54$, (c) $\Lambda_c(341) = 36$, (d) $\Lambda_c(561) = 16$, (e) $\Lambda_c(2^6 \cdot 3^4 \cdot 5^2 \cdot 7 \cdot 19) = \text{lcm}(32, 54, 20, 6, 18) = 2480$.
8. $x = 2$, $y = 3$, $z = 4$, $w = 5$.
9. $77w \equiv 707 \pmod 3$. Therefore, a solution is given by $w = 1 + 3s$, $z = t$, $y = 6 + 2s + 6t + 9u$, $x = 16 - 11s - 9t - 11u$.
10. $x = 5 + 8t$, $y = 3 - 11t$, $z = 33 - 3t$, or $x = 20 - t - 6s$, $y = 2t$, $z = -t + 5s$.
11. $x + y + z = 100$ and $x/2 + 3y + 10z = 100$, or $x + y + z = 100$ and

$x + 6y + 20z = 200$. Thus, $x = 80 + 14t$, $y = 20 - 19t$, $z = 5t$.

12. For any integer n, Fermat's Little Theorem implies that 7 divides $n^7 - n$. If $n = 3k$, $3k + 3$, or $3k + 2$ or $n = 2k$ or $2k + 1$, 6 divides $n^7 - n = n(n^3 - 1)(n^3 + 1)$. Hence, 42 divides $n^7 - n$.
13. If $n = p^\alpha$,

$$\begin{aligned}\sum_{d|p^\alpha} d \cdot \phi(d) \cdot \sigma\left(\frac{p^\alpha}{d}\right) &= 1 \cdot 1 \cdot (1 + p + \cdots + p^\alpha) \\ &\quad + p(p-1)(1 + p + \cdots + p^{\alpha-1}) \\ &\quad + p^2(p^2 - p)(1 + p + \cdots + p^{\alpha-2}) + \cdots \\ &\quad + p^\alpha(p^\alpha - p^{\alpha-1}) \cdot 1 \\ &= 1 + p^2 + p^4 + \cdots + p^{2\alpha} = \sum_{d|p^\alpha} d^2.\end{aligned}$$

14. If $n = p^\alpha$, $\sum_{d|p^\alpha} \mu(d) \cdot \phi(d) = \mu(1)\phi(1) + \mu(p)\phi(p) = 1 + (-1)p(p-1) = 2 - p$.
15. $2^{64} + 1 \equiv 0 \pmod{1071 \cdot 2^8 + 1}$. Suppose that $(-1071)^n + 2^{64-8n} \equiv 0 \pmod{1071 \cdot 2^8 + 1}$. It follows that $(-1071)^{n+1} + 2^{64-8(n+1)} \equiv (-1071)^{n+1} + 2^{64-8(n+1)} - 2^{64-8(n+1)}(1071 \cdot 2^8 + 1) \equiv (-1071) \times [(-1071)^n + 2^{64-8n}] \equiv 0 \pmod{1071 \cdot 2^8 + 1}$.
16. For $0 \leqslant r \leqslant 9$, $-r \equiv 10 - r$.
17. $1^{-1} = 1$, $2^{-1} = 6$, $3^{-1} = 4$, $4^{-1} = 3$, $5^{-1} = 9$, $6^{-1} = 2$, $7^{-1} = 8$, $8^{-1} = 7$, $9^{-1} = 5$.
18. $1^{-1} = 1$, $5^{-1} = 5$, $7^{-1} = 7$, $11^{-1} = 11$.
19. $aa^{-1} = e$, $ea^{-1} = a^{-1}$, and $a(b^{-1})^{-1} = ab$ are in H. Elements in H are associative because they are elements of G.
20. The multiples of r, where $0 \leqslant r \leqslant m - 1$.
21. 0, 2, 3, 4, have no multiplicative inverses in Z_6.
22. $1^{-1} = 1$, $2^{-1} = 4$, $3^{-1} = 5$, $4^{-1} = 2$, $5^{-1} = 3$, $6^{-1} = 6$.
23. Let $1 \leqslant r \leqslant m$, and c be such that $ac \equiv 1 \pmod m$. If $x = (r - b)c$, then $ax = r - b$, or $ax + b = r$.

Exercises 6.1

1. (a) $x \equiv 0, 1, 2, 3, 4 \pmod 5$,
   (b) no solution.
2. (a) $x \equiv 2, 3 \pmod 5$ and $x \equiv 2, 4, 5 \pmod 7$. Hence, $x \equiv 2, 32, 12, 23, 18, 33 \pmod{35}$.
   (b) $x \equiv 1, 3 \pmod 5$ and $x \equiv 1, 2, 6 \pmod 9$. Hence, $x \equiv 1, 6, 11, 28,$

33, 38 (mod 45).

(c) $x \equiv 1, 3, 5 \pmod 7$ and $x \equiv 1, 3, 5 \pmod{11}$. Hence, $x \equiv 1, 3, 5, 12, 36, 38, 45, 47, 71 \pmod{77}$.

3. (a) $x_1 \equiv 5, 6 \pmod{11}$, hence, $x \equiv 38$ and $83 \pmod{121}$.
   (b) $x_1 \equiv 5 \pmod 7$ and $x_2 \equiv 40 \pmod{49}$. Hence, $x \equiv 89 \pmod{343}$.
   (c) No solution.
4. $x \equiv 1, 3 \pmod 6$ and $x \equiv 5, 12 \pmod{17}$. Hence, $x \equiv 73, 97, 39, 63 \pmod{102}$.
5. $16! = (16)(15 \cdot 8)(14 \cdot 11)(13 \cdot 4)(12 \cdot 10)(9 \cdot 2)(7 \cdot 5)(6 \cdot 3)(1) \equiv (-1) \cdot (1) \cdots (1) \equiv -1 \pmod{17}$.
6. 17 is prime, hence, $16! = 16 \cdot 15! \equiv (-1) \cdot 15! \equiv -1 \pmod{17}$. Therefore, $15! \equiv 1 \pmod{17}$.
7. $437 = 19 \cdot 23$. Since 23 is prime, $-1 \equiv 22! \equiv 22 \cdot 21 \cdot 20 \cdot 19 \cdot 18! \equiv (-1)(-2)(-3)(-4) \cdot (18!) \equiv 18! \pmod{23}$. From Wilson's Theorem, $18! \equiv -1 \pmod{19}$. Hence, $18! \equiv -1 \pmod{437}$.
8. Since $(p - k) + k \equiv 0 \pmod p$, $(p - k) \equiv -k \pmod p$. From Wilson's Theorem $(p - 1)! \equiv -1 \pmod p$. Substituting, we obtain $1^2 \cdot 3^2 \cdots (p-2)^2 \equiv (-1)^{(p+1)/2} \pmod p$ and $2^2 \cdot 4^2 \cdots (p-1)^2 \equiv (-1)^{(p+1)/2} \pmod p$.
9. The two incongruent solutions are 1 and $p - 1$.
10. $(x^{99} + x^{98} + x^{97} + \cdots + x + 1) \cdot x(x-1) = x^{101} - x \equiv 0 \pmod{101}$. Hence, $x^{99} + x^{98} + x^{97} + \cdots + x + 1$ has 99 solutions modulo 101.
11. In Z_p^*, $p - 1$ is its own inverse. Every other element has a distinct inverse. Therefore, $(p-1)! \equiv (p-1) \cdot 1 \cdot 1 \cdots 1 \equiv p - 1 \equiv -1 \pmod p$.
12. From Wilson's Theorem $(p-1)! \equiv -1 \pmod p$, $1 \cdot 2 \cdot 3 \cdots (p-2) \equiv 1 \pmod p$, and $1 + 2 + \cdots + (k-1) + (k+1) + \cdots + (p-1) \equiv -k \pmod p$ for $k = 2, \ldots, p-2$. If each fraction in the sum is replaced by an equivalent fraction with denominator $(p-1)!$ and the fractions added together the numerator will be congruent modulo p to $1 + 2 + \cdots + (p-1)$ which is congruent to 0 modulo p. Therefore,
$$1 + \frac{1}{2} + \frac{1}{3} + \cdots + \frac{1}{p-1} \equiv 0 \pmod p.$$
13. $k = 1$, when $p = 2$ or 3; never.

Exercises 6.2

1. 1, 4, 5, 6, 7, 9, 13, 16, 20, 22, 23, 24, 25, 28.

2. (a) -1, (b) 1, (c) 1, (d) -1, (e) 1, (f) -1, (g) -1, (h) 1.
3. (b), (c), (e) and (h).
4. (a) Yes, since $\left(\dfrac{9}{19}\right) = 1$; $x \equiv 5$ and 17 (mod 19).

   (b) Yes, since $\left(\dfrac{16}{17}\right) = 1$; $x \equiv 13$ and 16 (mod 17).

   (c) No, since $\left(\dfrac{6}{61}\right) = -1$.

5. (a) $$\left(\frac{21}{221}\right) = \left(\frac{3}{221}\right)\left(\frac{7}{221}\right) = \left(\frac{3}{17}\right)\left(\frac{3}{13}\right)\left(\frac{7}{17}\right)\left(\frac{7}{13}\right)$$
   $$= (-1)(1)(-1)(-1) = -1.$$

   (b) $$\left(\frac{215}{253}\right) = \left(\frac{43}{23}\right)\left(\frac{43}{11}\right)\left(\frac{5}{23}\right)\left(\frac{5}{11}\right) = (-1)(-1)(-1)(1) = -1.$$

   (c) $$\left(\frac{631}{1099}\right) = \left(\frac{631}{157}\right)\left(\frac{631}{7}\right) = \left(\frac{3}{157}\right)\left(\frac{1}{7}\right) = \left(\frac{1}{3}\right) = 1.$$

   (d) $$\left(\frac{1050}{1573}\right) = \left(\frac{2}{11}\right)^2\left(\frac{2}{13}\right)\left(\frac{525}{11}\right)\left(\frac{525}{13}\right)$$
   $$= (1)(-1)(1)\left(\frac{5}{13}\right) = 1.$$

   (e) $$\left(\frac{89}{197}\right) = \left(\frac{197}{89}\right) = \left(\frac{19}{89}\right) = \left(\frac{89}{19}\right) = \left(\frac{13}{19}\right) = \left(\frac{19}{13}\right)$$
   $$= \left(\frac{6}{13}\right) = \left(\frac{2}{13}\right)\left(\frac{3}{13}\right) = (1)(-1) = -1.$$

6. Half the values of $(\frac{a}{p})$ equal 1 and the other half equal -1. Hence, their sum is zero.
7. $\gcd(a,\ p) = \gcd(b,\ p) = 1$ implies $\gcd(ab,\ p) = 1$. Thus,
   $$\left(\frac{ab}{p}\right) = \left(\frac{a}{p}\right)\left(\frac{b}{p}\right).$$

   The only possibilities are

   QR = QR · QR, QR = QNR · QNR, QNR = QNR · QR, and QNR = QR · QNR.

8. If $p = 1 + 4k$, then
$$\left(\frac{-1}{p}\right) = (-1)^{(p-1)/2} = (-1)^{2k} = 1.$$

9. $$\left(\frac{p}{q}\right)\left(\frac{q}{p}\right) = \left(\frac{p}{q}\right)\left(\frac{2q-1}{p}\right) = \left(\frac{p}{q}\right)\left(\frac{-1}{p}\right) = (-1)^{\frac{1}{2}(p-1)\frac{1}{2}(2p-2)}$$
$$= (-1)^{(p-1)^2/2} = 1.$$

Hence,
$$\left(\frac{p}{q}\right) = \left(\frac{-1}{p}\right).$$

10. $$\left(\frac{p}{q}\right) = \left(\frac{q}{p}\right)(-1)^{\frac{1}{2}(3+4t-1)\frac{1}{2}(3+4s-1)} = -\left(\frac{q}{p}\right).$$

11. If $4n^2 + 4 \equiv 0 \pmod{19}$ then $4n^2 \equiv -4 \pmod{19}$. Hence, $n^2 \equiv -1 \pmod{19}$ which is impossible.
12. If $0 \leqslant k \leqslant p$, then $(p - k) \equiv -k \pmod{p}$. Hence, $(p-1)(p-2) \cdots (p-k) \equiv (-1)^k(k)! \pmod{p}$. If $h = p - k - 1$, then $h! = (p - k - 1)!$ and $(p-1)! \equiv (-1)^k(k!)(h!) \pmod{p}$. Therefore, $h!k! \equiv (-1)^k(p-1)! \equiv (-1)^{k+1} \pmod{p}$.
13. If $p \equiv 1 \pmod{4}$, then $p = 1 + 4r$ for some integer r. If $h = k = 2r$, then $h + k = 4r$ and $[(2r)!]^2 \equiv (-1)^{2k+1} \equiv -1 \pmod{p}$.

Exercises 6.3

1. Since $\phi(\phi(m)) = 1$, $m = 2, 3, 4$, or 6.
2. $F_3 = 257$ and $3^{(257-1)/2} \equiv 3^{128} \equiv (3^{20})^6 \cdot 3^8 \equiv 123^6 \cdot 136 \equiv 17 \cdot 136 \equiv -1 \pmod{257}$.
3. See Table A.2.
4. $5^{14} \equiv 1 \pmod{29}$.
5. They are 2^1, 2^3, 2^5, 2^9, 2^{11}, 2^{13}, 2^{15}, 2^{17}, 2^{19}, 2^{23}, 2^{25}, and 2^{27}, or 2, 3, 8, 10, 11, 14, 15, 18, 19, 21, 26, 27.
6. See Table A.3.
7. (a) gcd(4, 28) = 4. Hence, the fourth power residues are 2^4, 2^8, 2^{12}, 2^{16}, 2^{20}, 2^{24}, and 2^{28}, or 1, 7, 16, 20, 23, 24 and 25.
   (b) gcd(7, 28) = 7. Hence, the seventh power residues are 2^7, 2^{14}, 2^{21}, and 2^{28}, or 1, 12, 17, and 28.
8. $x^7 \equiv 12 \pmod{29}$ or $7I(x) \equiv 7 \pmod{28}$, or $I(x) \equiv 1 \pmod{4}$. Hence, $I(x) \equiv 1, 5, 9, 13, 17, 21, 25$, and $x \equiv 2^1, 2^5, 2^9, 2^{13}, 2^{17}, 2^{21}$, and 2^{25}, or 2, 3, 11, 14, 17, 19 and 21.

Table A.2.

k	$2k$	2^k
1	2	2
2	4	4
3	6	8
4	8	16
5	10	3
6	12	6
7	14	12
8	16	24
9	18	19
10	20	9
11	22	18
12	24	7
13	26	14
14	28	28
15	1	27
16	3	25
17	5	21
18	7	13
19	9	26
20	11	23
21	13	17
22	15	5
23	17	10
24	19	20
25	21	11
26	23	22
27	25	15
28	27	1

Table A.3.

k	$I(k)$
1	28
2	1
3	5
4	2
5	22
6	6
7	12
8	3
9	10
10	23
11	25
12	7
13	18
14	13
15	27
16	4
17	21
18	11
19	9
20	24
21	17
22	26
23	20
24	8
25	16
26	19
27	15
28	14

9. $9 \cdot I(x) \equiv 7 \pmod{28}$. Hence, $-I(x) \equiv 21 \pmod{28}$, implying that $I(x) \equiv -21 \equiv 7 \pmod{28}$. Therefore, $x \equiv 12 \pmod{29}$.
10. (a) $x \equiv 8 \pmod{17}$, (b) $x \equiv 10 \pmod{17}$, (c) no solution.
11. See Table A.4. (a) $x \equiv 7 \pmod{11}$, (b) $x \equiv 5, 6 \pmod{11}$, (c) no solution.
12. $I(x) \equiv I(3^{24} \cdot 5^{13}) \equiv 24 \cdot I(3) + 13 \cdot I(5) \equiv 24 + 65 \equiv 89 \equiv 9 \pmod{16}$. Hence, $x \equiv 14 \pmod{17}$.
13. $x \equiv 4 \pmod{29}$.
14. If q is a primitive root modulo p, then q^b is also a primitive root if and only if $\gcd(b, \phi(n)) = 1$. Hence,

Table A.4.

k	1	2	3	4	5	6	7	8	9	10
$I(k)$	2	4	8	5	10	9	7	3	6	1

$$\sum_{\gcd(b,\phi(n))=1} q^b \equiv q^{\sum_{\gcd(b,\phi(n))=1} b} \equiv q^{k(p-1)} \equiv (q^{(p-1)})^k \equiv 1^k \equiv 1 \pmod{p}.$$

15. $\left(\frac{7}{p}\right) = \left(\frac{p}{7}\right)(-1)^{6(p-1)/2} = \left(\frac{28k+3}{7}\right)(-1)^{6(7k+1)} = \left(\frac{3}{7}\right) = 1.$

16. $$\left(\frac{3}{p}\right) = \left(\frac{p}{3}\right)(-1)^{(p-1)/2}.$$

If $p = 1 + 12k$ or $p = 11 + 12k$, $\left(\frac{p}{3}\right) = 1$. If $p = 5 + 12k$ or $p = 7 + 12k$, $\left(\frac{p}{3}\right) = -1$.

17. $$\left(\frac{5}{p}\right) = \left(\frac{p}{5}\right)(-1)^{(p-1)/2} = \left(\frac{p}{5}\right).$$

If $p = 1 + 10k$ or $p = 9 + 10^k$, $\left(\frac{5}{p}\right) = 1$. If $p = 3 + 10k$ or $p = 7 + 10k$, $\left(\frac{5}{p}\right) = -1$.

Exercises 6.4

1. Using indices modulo 13, the equation $3^n \equiv 12 \pmod{13}$ leads to the equation $3n \equiv 7 \pmod{12}$ which has no solutions. Using indices modulo 29, the least solution to the equation $3^n \equiv 28 \pmod{29}$ is $n = 14$.
2. If d divides $p - 1$, $x^{p-1} - 1 \equiv (x^d - 1)(x^{d(k-1)} + \cdots + x^d + 1)$ and the expression in the second set of partentheses on the right has at most $d(k-1)$ solutions. Thus $x^d - 1$ has at least $(p-1) - d(k-1) = d$ solutions. By Lagrange's Theorem it has at most d solutions. Hence, it has exactly d solutions.
3. If q is a primitive root of the odd prime p, it follows from Theorem 6.3 and Theorem 6.20 that $q^{((p-1)/d)k}$, for $k = 1, 2, \ldots, d$, are d incongruent solutions to $x^d - 1 \equiv 0 \pmod{p}$.
4. From Theorem 6.10, if p is of the form $8k + 3$, 2 is a QNR of p.

Every primitive root of p is a QNR of p. In addition, there are $(p-1)/2 = q$ QNRs of p and $\phi(p-1) = q-1$ primitive roots of p. From Theorem 6.7, $p-1$ is a QNR of p and is not a primitive root of p since it has order 2. Hence, all other QNRs, including 2, are primitive roots of p.

5. As in the previous exercise, we need only show that -2 is a QNR of p. However, from Theorem 6.10, 2 is a QR of p and -1 is a QNR of p. Hence, their product $2 \cdot (-1)$ is a QNR of p.
6. The order of 3 must be a divisor of $\phi(p-1) = 4q$. However, 3 is a QNR of p. Since $3^{2q} \equiv -1 \pmod{p}$, the order of 3 cannot be 1, 2, 4, or $2q$. In addition, p does not divide $3^4 - 1$. Thus the order of 3 cannot be 4. Therefore, the order of 3 is $4q$ and 3 is a primitive root of p.
7. $\gcd(k, p-1) = 1$ if and only if $\gcd((p-1)-k, p-1) = 1$. Since $q^{(p-1)-k}q^k \equiv q^{p-1} \equiv 1 \pmod{p}$, $q^{(p-1)-k} \equiv -q^k \pmod{p}$. Therefore, the sum of all primitive roots is 0.
8. See Table A.5.
9. Z_p^* is generated by any primitive root of p.
10. 2 is a primitive root of 13. Hence, the primitive roots of 13 are $2^1(2)$, $2^5(6)$, $2^7(11)$, and $2^{11}(7)$. Therefore, the generators of Z_{13}^* are 2, 6, 7, and 11.
11. The subgroups of Z_{13}^* are $\{1\}$, $\{1, 2^6\}$, $\{1, 2^4, 2^8\}$, $\{1, 2^2, 2^4, 2^6, 2^8, 2^{10}\}$, and Z_{13}^*. That is, they are $\{1\}$, $\{1, 12\}$, $\{1, 3, 9\}$, $\{1, 3, 4, 9, 10, 12\}$, and Z_{13}^*.

Table A.5.

	p								
q	3	5	7	11	13	17	19	23	29
3		−1	1	−1	1	−1	1	−1	−1
5	−1		−1	1	−1	−1	1	−1	1
7	−1	−1		1	−1	−1	−1	1	1
11	1	1	−1		−1	−1	−1	1	−1
13	1	−1	−1	−1		1	−1	1	1
17	−1	−1	−1	−1	1		1	−1	−1
19	−1	1	1	1	−1	1		1	−1
23	1	−1	−1	−1	1	−1	−1		1
29	−1	1	1	−1	1	−1	−1	1	

Exercises 7.1

1. 33 34 32 11 33 24 43 11 33 24 43 31 11 33 14
2. ITS GREEK TO ME
3. (a) L KDYH D VHFUHW,
   (b) VLF VHPSHU WBUDQQLV,
   (c) VHQG KHOS.
4. (a) ALL MEN ARE MORTAL,
   (b) PERICULUM IN MORA (He who hesitates is lost),
   (c) INVITO PATRE SIDERA VERSO (Against my father's will, I study the stars).
5. There are 27XS and 23MS. If we assume E in the plaintext became X in the ciphertext then $k = 19$ and we obtain:
   WE HOLD THESE TRUTHS TO BE SELF EVIDENT THAT ALL MEN ARE CREATED EQUAL THAT THEY ARE ENDOWED BY THEIR CREATOR WITH CERTAIN UNALIENABLE RIGHTS THAT AMONG THESE ARE LIFE LIBERTY AND THE PURSIUT OF HAPPINESS
6. HBGTG IAEKY DGIRH BGYNN ISGXX
7. STUDY HARD FOR THE FINAL EXAM
8. $k = 14$.
9. Assuming E was enciphered as P, the most common letter in the ciphertext, $k = 11$. Hence, $P \equiv C + 15$. The plaintext message reads NUMBER THEORY IS USEFUL FOR ENCIPHERING MESSAGES.
10. We have $21 \equiv 4a + b \pmod{26}$ and $0 \equiv 19a + b \pmod{26}$. Therefore, $a = 9$, and $b = 11$.
11. If E and T are enciphered as L and U, respectively, $4a + b \equiv 11 \pmod{26}$ and $19a + b \equiv 20 \pmod{26}$. Hence, $a = 11$ and $b = 19$. Therefore, $P \equiv 19(C - 19) \equiv 19C + 3 \pmod{26}$.
    WHEN THE ONE GREAT SCORER COMES TO MARK AGAINST YOUR NAME THE MARK IS NOT FOR WHETHER YOU WON OR LOST BUT HOW YOU PLAYED THE GAME
12. DRESSED TO THE NINES
13. TWENTY THREE SKIDDOO
14. NEVER WAS SO MUCH OWED BY SO MANY TO SO FEW – WSC

Exercises 7.2

1. IDUCORFPHOPFBHG
2. SURRENDER AT ONCE
3. MBMH QD JHG

4. HERE WE ARE NOT AFRAID TO FOLLOW THE TRUTH WHEREVER IT MAY LEAD NOR TOLERATE ERROR AS LONG AS REASON IS FREE TO COMBAT IT—Jefferson
5. MVMOQ HXFPQ XVFAT DQYRL KYV
6. ARE YOU LOST
7. IL IMPORTE DE CHERCHER TOUJOURS LA VERITE.
8. HOW DO I LOVE THEE, LET ME COUNT THE WAYS.
9. APRIL IS THE CRUELEST MONTH.
10. FHX DZALX UAZE RHJPS.
11. DE MORTUIS NIHIL NISI BONUMXX; (say) nothing but good about the dead.
12. KVC GIF KZG XKD ERV
13. GOOD LUCK
14. See Table A.6.
    HQBASDGLTPLQ.
15. IF I SHOULD DIE THINK ONLY THIS OF ME THAT THERE IS SOME CORNER OF A FOREIGN FIELD THAT IS FOREVER ENGLAND. — Rupert Brooke.

Exercises 7.3

1. 12 635 8645
2. ICU TOO
3. $\begin{pmatrix} 7 & 12 \\ 8 & 15 \end{pmatrix}^{-1} \equiv \left(\frac{1}{9}\right)\begin{pmatrix} 15 & -12 \\ -8 & 7 \end{pmatrix} \equiv 3 \cdot \begin{pmatrix} 15 & 14 \\ 18 & 7 \end{pmatrix} \equiv \begin{pmatrix} 19 & 16 \\ 2 & 21 \end{pmatrix}$
   (mod 26).
4. ZOLWN WCOKR OIHPA PPEOI HPVIX.
5. THE HOUSTON EULERS.
6. $A \cdot \begin{pmatrix} 19 & 7 \\ 7 & 4 \end{pmatrix} \equiv \begin{pmatrix} 25 & 20 \\ 8 & 6 \end{pmatrix}$ (mod 26),
   $A \equiv \begin{pmatrix} 25 & 20 \\ 8 & 6 \end{pmatrix}\begin{pmatrix} 4 & 19 \\ 19 & 19 \end{pmatrix} \equiv \begin{pmatrix} 12 & 23 \\ 16 & 6 \end{pmatrix}$ (mod 26).

Table A.6.

K	E	L	V	IJ
N	A	B	C	D
F	G	H	M	O
P	Q	R	S	T
U	W	X	Y	Z

7. $A \equiv \begin{pmatrix} 0 & 5 & 7 \\ 22 & 12 & 23 \\ 6 & 3 & 9 \end{pmatrix} \begin{pmatrix} 19 & 6 & 4 \\ 7 & 13 & 17 \\ 4 & 3 & 4 \end{pmatrix}^{-1} \equiv$
$\begin{pmatrix} 0 & 5 & 7 \\ 22 & 12 & 23 \\ 6 & 3 & 9 \end{pmatrix} \begin{pmatrix} -1 & -12 & 0 \\ 12 & -8 & 9 \\ 5 & 5 & 13 \end{pmatrix} \equiv \begin{pmatrix} 15 & 19 & 6 \\ 3 & 15 & 17 \\ -3 & 1 & 14 \end{pmatrix} \pmod{26}.$

Exercises 7.4

1. 0793 0082 0003 2251 0815 0481
2. 0569 1608 0044 0927 1946 2766 0244 2766 2437 2131 1539
3. 6505 4891 3049 0532
4. $e^{-1} \equiv 71 \pmod{3372}$;
   THE END IS NEAR.
5. $e^{-1} \equiv 33 \pmod{2590}$;
   WAHOO WAH.
6. $e^{-1} \equiv 109 \pmod{2670}$;
   MEET ME TONIGHT AT THE HAT AND FEATHERS.
7. $p = 3019$, $q = 1453$, $t = 3\,505\,709$.
8. $p = 2153$, $q = 1867$, $t = 708\,641$.
9. $k = 1817 \equiv 61^{17 \cdot 31} \pmod{8461}$.

Exercises 8.1

1. $x^2 + y^2 = \dfrac{4a^2m^2}{(m^2+1)^2} + \dfrac{a^2(m^2-1)^2}{(m^2+1)^2} = \dfrac{a^2(m^2+1)^2}{(m^2+1)^2} = a^2.$
2. $8650 = 89^2 + 27^2 = 93^2 + 1^2$.
3. See Table A.7.
4. Suppose that $x = 2n + 1$. If $y = 2m$ then $n = x^2 + y^2$ is of the form $4k + 1$. If $y = 2m + 1$, then $n = x^2 + y^2$ is of the form $4k + 2$. In neither case is n a multiple of 4.
5. If $n = 12 + 16k = x^2 + y^2$, then both x and y are even, say $x = 2r$ and $y = 2s$. It follows that $3 + 4k = r^2 + s^2$, contradicting Theorem 8.1.
6. Suppose that $n = 8k + 6 = x^2 + y^2$. If $x = 2r$ and $y = 2s$, then $6 = 4(r^2 + s^2) - 8k$, implying that 4 divides 6. If $x = 2r + 1$ and $y = 2s$, then $8k + 6 = 4r^2 + 4r + 1 + 4s^2$ so $8k + 5 = 4(r^2 + s^2 + r)$, implying that 4 divides 5. If $x = 2r + 1$ and $y = 2s + 1$, then $8k + 6 = 4r^2 + 4r + 1 + 4s^2 + 4s + 1$, or $8k + 4 = 4r^2 + 4s^2 + 4r + 4s$, or $2k + 1 = r(r + 1) + s(s + 1)$, an even number, a contra-

Table A.7.

n	$h(n)$	$f(n)$	n	$h(n)$	$f(n)$	n	$h(n)$	$f(n)$	n	$h(n)$	$f(n)$
101	1	8	126	0	0	151	0	0	176	0	0
102	0	0	127	0	0	152	0	0	177	0	0
103	0	0	128	1	4	153	1	8	178	1	8
104	1	8	129	0	0	154	0	0	179	0	0
105	0	0	130	1	16	155	0	0	180	1	8
106	1	8	131	0	0	156	0	0	181	1	8
107	0	0	132	0	0	157	1	8	182	0	0
108	0	0	133	0	0	158	0	0	183	0	0
109	1	8	134	0	0	159	0	0	184	0	0
110	0	0	135	0	0	160	1	8	185	1	16
111	0	0	136	1	8	161	0	0	186	0	0
112	0	0	137	1	8	162	1	4	187	0	0
113	1	8	138	0	0	163	0	0	188	0	0
114	0	0	139	0	0	164	1	8	189	0	0
115	0	0	140	0	0	165	0	0	190	0	0
116	1	8	141	0	0	166	0	0	191	0	0
117	1	8	142	0	0	167	0	0	192	0	0
118	0	0	143	0	0	168	0	0	193	1	8
119	0	0	144	1	4	169	1	12	194	1	8
120	0	0	145	1	16	170	1	16	195	0	0
121	1	4	146	1	8	171	0	0	196	1	4
122	1	8	147	0	0	172	0	0	197	1	8
123	0	0	148	1	8	173	1	8	198	0	0
124	0	0	149	1	8	174	0	0	199	0	0
125	1	16	150	0	0	175	0	0	200	1	12

diction. Therefore, n cannot be written as a sum of squares.

7. Suppose that $n = 8k + 7 = x^2 + y^2$. If $x = 2r$ and $y = 2s$, then $7 = 4(r^2 = s^2) - 8k$, implying that 4 divides 7. If $x = 2r + 1$ and $y = 2s$, then $8k + 7 = 4r^2 + 4r + 1 + 4s^2$ so $8k + 6 = 4(r^2 + s^2 + r)$, implying that 4 divides 6. If $x = 2r + 1$ and $y = 2s + 1$, then $8k + 7 = 4r^2 + 4r + 1 + 4s^2 + 4s + 1$, or $8k + 5 = 4r^2 + 4s^2 + 4r + 4s$, implying that 4 divides 5. Therefore, n cannot be written as a sum of two squares.
8. Suppose that $6n = x^2 + y^2$. Clearly, x and y must be of the form $3k$, $3k + 1$, or $3k + 2$. The only case not leading to a divisibility contradiction is the case where x and y are both multiples of 3.
9. If $n = x^2 + y^2$, then $2n = (x + y)^2 + (x - y)^2$.
10. 50 is the smallest such number; $50 = 1^2 + 7^2 = 5^2 + 5^2$.
11. $425 = (9^2 + 2^2)(2^2 + 1^2) = 20^2 + 5^2 = 19^2 + 8^2 = 16^2 + 13^2$.
12. $2^{2k+1} = (2^k)^2 + (2^k)^2$.

13. If $2^{2k} = x^2 + y^2$, and x and y are both even, both odd or one is even and the other odd, a contradiction arises. Hence, one of x and y must be 0 and the other 2^k.
14. (a) $3185 = 5 \cdot 7^2 \cdot 13$, $\tau(1, 3185) = 7$, $\tau(3, 3185) = 5$. Therefore, $f(3185) = 8$.
    (b) $7735 = 5 \cdot 7 \cdot 13 \cdot 17$, $\tau(1, 7735) = \tau(3, 7735) = 8$. Therefore, $f(7735) = 0$.
    (c) $72\,581 = 181 \cdot 401$, $\tau(1, 72\,581) = 4$, $\tau(3, 72\,581) = 0$. Therefore, $f(72\,581) = 16$.
    (d) $226\,067 = 23 \cdot 9829$, $\tau(1, 226\,067) = 2$, $\tau(3, 226\,067) = 2$. Therefore, $f(226\,067) = 0$.
15. $6525 = 78^2 + 21^2$. Thus, from Theorem 2.13, $s = 78$, $t = 21$, $y = 78^2 - 21^2 = 5643$, and $x = 2 \cdot 21 \cdot 78 = 3276$. Therefore, (3276, 5643, 6525) is a primitive Pythagorean triple.
16. $6370 = 77^2 + 21^2$. Thus, from Theorem 2.13, $s = 77$, $t = 21$, $y = 77^2 - 21^2 = 5488$, and $x = 2 \cdot 21 \cdot 77 = 3234$. Therefore, (3234, 5488, 6370) is a Pythagorean triple.
17. If n cannot be expressed as the sum of three squares then $n = 4^m(8k + 7)$. We have $2n = 2 \cdot 4^m(8k + 7) = 4^m(8r + 6)$. Hence, $2n$ can be expressed as the sum of three integral squares.
18. $1729 = 1^3 + 12^3 = 9^3 + 10^3$.
19. $40\,033 = 16^3 + 33^3 = 9^3 + 34^3$.
20. (a) none, $16\,120 = 2^3 \cdot 5 \cdot 13 \cdot 31$.
    (b) none; $56\,144 = 2^4 \cdot 11 \cdot 319$.
21. $870 = 12^2 + 1^2 + 14^2 + 23^2$.
22. $3^3 + 4^3 + 5^3 = 6^3$.
23. $a = b = c = 18$, $d = 7$.
24. $n = 2$ produces $10^2 + 11^2 + 12^2 = 365 = 13^2 + 14^2$; $n = 4$ produces $36^2 + 37^2 + 38^2 + 39^2 + 40^2 = 7230 = 41^2 + 42^2 + 43^2 + 44^2$.
25. Since x^2 and y^2 are congruent to 0 or 1 modulo 4, it follows that $x^2 - y^2$ is congruent to 0, 1, or 3 modulo 4. If n is congruent to 1 or 3 modulo 4, then

    $$n = \left(\frac{n+1}{2}\right)^2 - \left(\frac{n-1}{2}\right)^2.$$

    If n is congruent to 0 modulo 4, then

    $$n = \left(\frac{n}{4}+1\right)^2 = \left(\frac{n}{4}-1\right)^2.$$

26. $2^{2^n} + 1 = (2^{2^{n-1}} + 1)^2 - (2^{2^{n-1}})^2$.

27. If p is an odd prime, then
$$p = \left(\frac{p+1}{2}\right)^2 - \left(\frac{p-1}{2}\right)^2.$$
28. $113 = 7^2 + 8^2$; $181 = 9^2 + 10^2$; $313 = 12^2 + 13^2$.
29. $509 = 12^2 + 13^2 + 14^2$; $677 = 14^2 + 15^2 + 16^2$; $1877 = 24^2 + 25^2 + 26^2$.
30. $459 = 15^2 + 15^2 + 3^2$.
32. Suppose that $3n = a^2 + b^2 + c^2 + d^2$. Since $x^2 \equiv 0$ or $1 \pmod 3$, at least one of a, b, c, d is congruent to 0 modulo 3. Let a be divisible by 3. Hence, $s = 3r$ for some integer r. Since $b^2 + c^2 + d^2 \equiv 0 \pmod 3$, where b, c, d may be negative, $b \equiv c \equiv d \pmod 3$. Therefore,
$$n = \left(\frac{b+c+d}{3}\right)^2 + \left(\frac{a+c-d}{3}\right)^2 + \left(\frac{a-c+d}{3}\right)^2 + \left(\frac{a+b-d}{3}\right)^2.$$
33. If $n = 192$, $8n + 3 = 1539 = 37^2 + 11^2 + 7^2$. Hence, $192 = t_{18} + t_5 + t_3$.
34. If $p \nmid xyz$, then $\gcd(x, p) = \gcd(y, p) = \gcd(z, p) = 1$ and $x^{p-1} \equiv y^{p-1} \equiv z^{p-1} \equiv 1 \pmod p$. Hence, $x^{p-1} + y^{p-1} \equiv 1 + 1 \equiv 2 \not\equiv 1 \equiv z^{p-1} \pmod p$ and $x^{p-1} + y^{p-1} \neq z^{p-1}$.
35. If $\gcd(x, p) = \gcd(y, p) = \gcd(z, p) = 1$ then $x^p \equiv x$, $y^p \equiv y$, $z^p \equiv z \pmod p$. Hence, $x^p + y^p - z^p \equiv x + y - z \equiv 0 \pmod p$.
36. No, 1999 is a prime of the form $4k + 3$.
37. No, $5\,941\,232 = 4^2(8 \cdot 46\,415 + 7)$.
39. $4 = 5^3 - 11^2$; $5 = 3^2 - 2^2$; $7 = 2^4 - 3^2$; $8 = 2^4 - 2^3$; $9 = 5^2 - 4^2$; $10 = 13^3 - 3^7$; $11 = 6^2 - 5^2$; $12 = 2^4 - 2^2$; $13 = 2^8 - 3^5$.

Exercises 8.2

1. $s_{6930} = t_{9800} = 48\,024\,900$.
2. (a) (7, 4); (b) (161, 72); (c) (49, 20).
3. $x^2 = (3y^2 - y)/2$ implies that $3y^2 - y = 2x^2$. Hence, $36y^2 - 12y = 24x^2$, or $36y^2 - 12y + 1 = 24x^2 + 1$. Hence, $(6y - 1)^2 = 24x^2 + 1$, or $z^2 = 24x^2 + 1$.
4. Two solutions (x, z) are given by (1, 5) and (99, 485). Hence 1 and 9801 are square–pentagonal numbers.
5. 1 and 210.
6. 48 024 900 and 1 631 432 881.
7. If $d = n^2$, the equation $y^2 - (nx)^2 = 1$ would have no solutions.
8. Clearly, x and y cannot be of opposite parity. Suppose that they are

both even, say $x = 2r$ and $y = 2s$. We obtain $4s^2 = 8r^3 + 2$ or $4(s^2 + 2r^3) = 2$, a contradiction since $4 \nmid 2$. Hence, both x and y must be odd.

9. If (a, b) is a solution, then $3a^2 + 2 = b^2$. Hence, $b^2 \equiv 2 \pmod 3$, a contradiction since 2 is not a quadratic residue of 3.

Exercises 8.3

1. It is reflexive since under the identity transformation $f \sim f$. It is symmetric. If $f \sim g$ under the transformation $x = au + bv$, $y = cu + dv$, then $g \sim f$ under the transformation
$$u = \left(\frac{d}{\Delta}\right)x + \left(\frac{b}{-\Delta}\right)y,\ v = \left(\frac{c}{-\Delta}\right)x + \left(\frac{a}{\Delta}\right)y,$$
where $\Delta = ad - bc = \pm 1$ and
$$\frac{d}{\Delta}\cdot\frac{a}{\Delta} - \frac{b}{-\Delta}\cdot\frac{c}{-\Delta} = \frac{\Delta}{\Delta^2} = \pm 1.$$
It is transitive. If $f \sim g$ under the transformation $x = au + bv$, $y = cu + dv$, and $g \sim h$ under the transformation $u = qw + rz$, $v = sw + tz$, where $ad - bc = qt - rs = \pm 1$, then $f \sim h$ under the transformation $x = (aq + bs)w + (ar + bt)z$, $y = (cq + ds)w + (cr + dt)z$, with $(aq + bs)(cr + dt) - (ar + bt)(cq + ds) = (ad - bc)(qt - rs) = \pm 1$.
2. $f(x, y) = -x^2 + 2y^2 = -(2u + v)^2 + 2(3u + 2v)^2 = 14u^2 + 20uv + 7v^2 = g(u, v)$ since $2 \cdot 2 - 1 \cdot 3 = 1$.
3. $u = 4$, $v = -5$.
4. Use the transformation $x = 3u + 2v$, $y = 4u + 3v$.
5. Use the transformation $x = 3u - 2v$, $y = 2u - v$.
6. $2x^2 + 5xy - y^2 = 2(5u + 2v)^2 + 5(5u + 2v)(7u + 3v) - (7u + 3v)^2$ $= 176u^2 + 143uv + 29v^2$.
7. Suppose that $f(x, y) = ax^2 + bxy + cy^2$, $x = Au + Bv$, and $y = Cu + Dv$, with $AD - BC = \pm 1$. Then $f(x, y) = a(Au + Bv)^2 + b(Au + Bv)(Cu + Dv) + c(Cu + Dv)^2 = (aA^2 + bAC + cC^2)u^2 + (2aAB + b(AD + BC) + 2cCD)uv + (aB^2 + bBD + cD^2)v^2$. The discriminant equals $(2aAB + b(AD + BC) + 2cCD)^2 - 4(aA^2 + bAC + cC^2)(aB^2 + bBD + cD^2) = (b^2 + 4ac)(AD - BC)^2 = b^2 - 4ac$.
8. If b is even, $d \equiv 0 \pmod 4$. If b is odd, $d \equiv 1 \pmod 4$.
9. $f(x, y) = x^2 + 4xy + y^2$.
10. Only f_4.
11. No; the discriminant of f is -15 and that of g is -4.

12. Yes, $31 = 5^2 + 6 \cdot 1^2$; yes, $415 = 11^2 + 6 \cdot 7^2$.
13. If $b^2 - 4ac < 0$, the only critical point, where $\partial f/\partial x = \partial f/\partial y = 0$, is at (0, 0) which is, from the second derivative test, a relative minimum. If $b^2 - 4ac = 0$ then the critical points lie on the lines $2ax + by = 0$ and $bx + 2cy = 0$; however, the second derivative test fails. Plugging the critical points into $f(x, y)$ we obtain cy^2 and $(4ac - b^2)y^2$, respectively. In either case, $f(x, y) \geqslant 0$.
14. $1 = x^2 + 3y^2 = (2u + v)^2 + 3(u + v)^2 = 7u^2 + 10uv + 4y^2$.

Exercises 8.4

1. $\dfrac{33}{23}$
2. $\dfrac{1393}{972}$
3. If $x = [a_1, a_2, \ldots, a_n]$,
$$\frac{1}{x} = 0 + \frac{1}{[a_1, \ldots, a_n]} = [0, a_1, a_2, \ldots, a_n].$$
4. $0, \frac{1}{1}, \frac{3}{4}, \frac{19}{25}, \frac{79}{104}, \frac{177}{233}$.
5. The equation $ax + by = c$ has solution $x = (-1)^n c y_{n-1}$, $y = (-1)^{n+1} c x_{n-1}$.
6. If $c_i = x_i/y_i$, then $x_i/x_{i-1} = [a_i, a_{i-1}, \ldots, a_2, a_1] = y_i/y_{i-1}$. Hence, $x_n/x_{n-1} = [a_n, a_{n-1}, \ldots, a_2, a_1] = x_n/y_n$ and the condition is $x_{n-1} = y_n$.
7. The formulas for obtaining convergents in Theorem 8.14 are the rules for multiplication of matrices given in the exercise.

Exercises 8.5

1. (a) $\sqrt{3} = [1, \overline{1, 2}]$, (b) $\sqrt{5} = [2, \overline{4}]$, (c) $\sqrt{7} = [2, \overline{1, 1, 1, 4}]$, (d) $\sqrt{10} = [3, \overline{6}]$.
2. [2, 1, 2, 1, 1, 4, 1, 1, 6, 1, 1, 8].
3. [2, 6, 10, 14, 18, 22, 30, 1, 1, 5, 1, 1]
4. See Table A.8
   Hence, $(x, y) = (2, 1) = (7, 4) = (26, 15) = (97, 56) = (362, 209)$.
5. If $x = [\overline{n}]$, then $x = n + 1/[\overline{n}] = n + 1/x$. Hence,

Table A.8.

		1	1	2	1	2	1	2	1	2	1
0	1	1	2	5	7	19	26	71	97	265	362
1	0	1	1	3	4	11	15	41	56	153	209

$$x = \frac{n + \sqrt{n^2 + 4}}{2}.$$

6. $$\sqrt{13} = \sqrt{3^2 + 4} = 3 + \cfrac{4}{6 + \cfrac{4}{6 + \cfrac{4}{6 + \dots}}}$$

$$\sqrt{18} = \sqrt{4^2 + 2} = 4 + \cfrac{2}{8 + \cfrac{2}{8 + \cfrac{2}{8 + \dots}}}.$$

7. Since, γ is irrational neither $|\gamma - a/b| = 1/2b^2$ nor $|\gamma - c/d| = 1/2d^2$. If $|\gamma - a/b| > 1/2b^2$ and $|\gamma - c/d| > 1/2d^2$ then
$$\frac{1}{bd} = \frac{bc - ad}{bd} = \frac{c}{d} - \frac{a}{b} = \left|\gamma - \frac{a}{b}\right| + \left|\gamma - \frac{c}{d}\right| > \frac{1}{2b^2} + \frac{1}{2d^2}.$$
Hence, $2bd > b^2 + d^2$, implying that $(b - d)^2 < 0$, a contradiction. Therefore, either $|\gamma - a/b| < 1/2b^2$ or $|\gamma - c/d| < 1/2d^2$. Hence, by Theorem 8.21 a/b or c/d is a convergent of γ.

Exercises 8.6

1. The absolute value function is not non-Archimedean since $|1 + 2| = 3 > 2 = \max\{|1|, |2|\}$. Therefore, it is Archimedean.
2. (a) If $v(e) \neq 0$, then $v(e) = v(e \cdot e) = v(e)v(e)$ implies that $v(e) = 1$.
   (b) $1 = v(e) = v((-e)(-e)) = v(-e)v(-e)$.
   (c) $v(a) = 1 \cdot v(a) = v(-e)v(a) = v(-e \cdot a) = v(-a)$.
3. If $x = 0$, $|x|_0 = 0$. If $x \neq 0$, $|x|_0 = 1$. In either case, $|x|_0 \geqslant 0$. A case

Table A.9.

x	y	$v(xy)$ $\lvert xy\rvert_0$	$v(x) \cdot v(y)$ $\lvert x\rvert_0 \cdot \lvert y\rvert_0$	$v(x + y)$ $\lvert x + y\rvert_0$	$v(x) + v(y)$ $\lvert x\rvert_0 + \lvert y\rvert_0$
$\neq 0$	$\neq 0$	1	1	0 or 1	2
$\neq 0$	$= 0$	0	0	1	1
$= 0$	$= 0$	0	0	0	0

by case approach (see Table A.9) shows that $|xy|_0 = |x|_0 \cdot |y|_0$, as well as $|x + y|_0 \leqslant |x|_0 + |y|_0$.

4. The maximum of $v(x)$ and $v(y)$ is less than or equal to $v(x) + v(y)$.
5. $|600|_2 = 1/2^3$, $|600|_3 = 1/3$, $|600|_5 = 1/5^2$, $|600|_p = 1$ for any other prime p.
6. If $p = q$, $|q^k|_p = 1/q^k$. If $p \neq q$, $|q^k|_p = 1$.
7. Let $p^\alpha \| x$ and $p^\beta \| y$ so $x = (a/b)p^\alpha$ and $y = (c/d)p^\beta$, where $\gcd(a, b) = \gcd(c, d) = 1$. Suppose further that $\alpha \geqslant \beta$; hence, $|x|_p = 1/p^\alpha \leqslant 1/p^\beta = |y|_p$, thus $\max\{|x|_p, |y|_p\} = |y|_p = 1/p^\beta$.
$$|x + y|_p = \left|\left(\frac{a}{b}\right)p^\alpha + \left(\frac{c}{d}\right)p^\beta\right|_p = \left|\left(\frac{p^\beta}{bd}\right)(adp^{\alpha-\beta} + bc)\right|_p$$
$$= \left|\frac{p^\beta}{bd}\right|_p \cdot |adp^{\alpha-\beta} + bc|_p \leqslant p^{-\beta};$$
the latter inequality follows since $adp^{\beta-\alpha} + bc$ is an integer. Hence, $|abp^{\beta-\alpha} + bc|_p \leqslant 1$.
8. (1) If $q \neq 0$, then $0 < |q|_p \leqslant 1$ and $|0|_p = 0$.
(2) Let $p^\alpha \| x$ and $p^\beta \| y$ so $x = (a/b)p^\alpha$ and $y = (c/d)p^\beta$, where $\gcd(a, b) = \gcd(c, d) = 1$.
$$|x|_p \cdot |y|_p = \frac{1}{p^\alpha} \cdot \frac{1}{p^\beta} = \frac{1}{p^{\alpha+\beta}}$$
and
$$|xy|_p = \left|\left(\frac{ac}{bd}\right)p^\alpha p^\beta\right| = \frac{1}{p^{\alpha+\beta}}$$
since $p^{\alpha+\beta} \| xy$.
(3) Follows from the previous exercise.
9. Suppose $r = \prod_{i=1}^k p_i^{\alpha_i}$ and $s = \prod_{i=1}^k p_i^{\beta_i}$, where α_i and β_i nonnegative for $1 \leqslant i \leqslant k$. If $p = p_i$, for $1 \leqslant i \leqslant k$, then r divides s if and only if $\alpha_i \leqslant \beta_j$ if and only if $1/p^\beta \leqslant 1/p^\alpha$ if and only if $|s|_p \leqslant |r|_p$. If $p \neq p_i$, for $1 \leqslant i \leqslant k$, then $|s|_p = |r|_p = 0$.
10. If $x = 1 + 2 + 2^2 + 2^3 + \cdots$, then $x + 1 = 0$. Hence, $x = -1$.
11. If $x = 5 + 2 \cdot 3 + 2 \cdot 3^2 + 2 \cdot 3^3 + \cdots$, then $3x = 6$. Hence, $x = 2$.
12. $\dfrac{5}{6} = 9 + 7 + 7^2 + 7^3 + \cdots$.
13. Suppose $x = 6 + 6 \cdot 7 + 6 \cdot 7^2 + 6 \cdot 7^3 + 6 \cdot 7^4 + \cdots$. Adding 1 to both sides of the equation, we obtain $7 + 6 \cdot 7 + 6 \cdot 7^2 + 6 \cdot 7^3 + 6 \cdot 7^4 + \cdots = 7 \cdot 7 + 6 \cdot 7^2 + 6 \cdot 7^3 + 6 \cdot 7^4 + \cdots = 7 \cdot 7^2 + 6 \cdot 7^3 + 6 \cdot 7^4 + \cdots = 0$. Hence, $x = -1$.

14. $|48 - 36|_p = |12|_p = \begin{cases} \frac{1}{4} & \text{if } p = 2 \\ \frac{1}{3} & \text{if } p = 3, \text{ and} \\ 1 & \text{if } p \neq 2, 3. \end{cases}$
15. $3 + 4 \cdot 7 + 4 \cdot 7^2 + 4 \cdot 7^3 + \cdots$.
16. 98, 784, 5586, 39 200, ...
17. (a) the Cartesian plane,
    (b) the closed circular disk of radius 1 centered at the origin,
    (c) a square centered at the origin with vertices at $(\pm 1, 0)$ and $(0, \pm 1)$,
    (d) a rhombus centered at the origin with vertices at $(\pm 1, 0)$ and $(0, \pm 1)$,
    (e) $\mathbb{Q} \times \mathbb{Q}$.
18. Let x be a point in the interior of $D(a; r)$ and z be a point such that $d(a, z) = r$. Since $d(x, a) < r$ and the two longest sides of every triangle in a non-Archimedean geometry are equal in length, $d(x, z) = r$. Hence, x can be considered as being at the center.

Exercises 9.1

1. 1, 4, 10, 20, 35, 56, 84, 120, 165, 220, 286, . . . , the tetrahedral numbers.
2. 1, 5, 15, 35, 70, 126, 210, 330, 495, . . . , the fourth order figurate numbers.
3. The nth order figurate numbers.
4. 0, 1, 4, 10, 20, 35, 56, 84, 120, 165, 220, 286, ...
5. 0, 1, 5, 15, 35, 70, 126, 210, 330, 495, ...
6. 0 followed by the nth order figurate numbers.
7. 0, 0, 1, 2, 3, 4, ...
8. 1, 3, 5, 7, 9, . . . , the odd positive integers.
9. 0, 1, 4, 9, 16, 25, 36, . . . , squares of the nonnegative integers.
10. 0, 1, 8, 27, 64, 125, 216, 343, . . . , cubes of the nonnegative integers.
11. $G(x) = \dfrac{x(x^3 + 11x^2 + 11x + 1)}{(1 - x)^5}$.
12. $G(x) = \displaystyle\sum_{n=1}^{\infty} \frac{n^k x^n}{1 - x^n}$ is the generating function for σ_k.
13. $G(x) = \dfrac{1 + 2x}{1 - x - x^2}$.

14. $G(x) = \dfrac{x}{1 - 3x + 7x^2}$.

Exercises 9.2

1. 4, $3+1$, $1+3$, $2+2$ $2+1+1$, $1+2+1$, $1+1+2$, and $1+1+1+1$. 5, $4+1$, $1+4$, $3+2$, $2+3$, $3+1+1$, $1+3+1$, $1+1+3$, $2+2+1$, $2+1+2$, $1+2+2$, $2+1+1+1$, $1+2+1+1$, $1+1+2+1$, $1+1+1+2$, $1+1+1+1+1$.
2. $p(8) = 22$

   8
   7 + 1
   6 + 2
   6 + 1 + 1
   5 + 3
   5 + 2 + 1
   5 + 1 + 1 + 1
   4 + 4
   4 + 3 + 1
   4 + 2 + 2
   4 + 2 + 1 + 1
   4 + 1 + 1 + 1 + 1
   3 + 3 + 2
   3 + 3 + 1 + 1
   3 + 2 + 2 + 1
   3 + 2 + 1 + 1 + 1
   3 + 1 + 1 + 1 + 1 + 1
   2 + 2 + 2 + 2
   2 + 2 + 2 + 1 + 1
   2 + 2 + 1 + 1 + 1 + 1
   2 + 1 + 1 + 1 + 1 + 1 + 1
   1 + 1 + 1 + 1 + 1 + 1 + 1 + 1

   $p(9) = 30$

   9
   8 + 1
   7 + 2
   7 + 1 + 1
   6 + 3
   6 + 2 + 1
   6 + 1 + 1 + 1
   5 + 4
   5 + 3 + 1
   5 + 2 + 2
   5 + 2 + 1 + 1
   5 + 1 + 1 + 1 + 1
   4 + 4 + 1
   4 + 3 + 2
   4 + 3 + 1 + 1
   4 + 2 + 2 + 1
   4 + 2 + 1 + 1 + 1
   4 + 1 + 1 + 1 + 1 + 1
   3 + 3 + 3
   3 + 3 + 2 + 1
   3 + 3 + 1 + 1 + 1
   3 + 2 + 2 + 2
   3 + 2 + 2 + 1 + 1
   3 + 2 + 1 + 1 + 1 + 1
   3 + 1 + 1 + 1 + 1 + 1 + 1
   2 + 2 + 2 + 2 + 1
   2 + 2 + 2 + 1 + 1 + 1
   2 + 2 + 1 + 1 + 1 + 1 + 1
   2 + 1 + 1 + 1 + 1 + 1 + 1 + 1
   1 + 1 + 1 + 1 + 1 + 1 + 1 + 1 + 1

3. The coefficient of $x^n z^m$ represents the number of different ways n can be written as a sum of m distinct terms of the sequence, $a, b, c, d, e \ldots$.
4. Expanding we obtain $(1 + x^a z + x^{2a} z^2 + x^{3a} z^3 + \cdots)(1 + x^b z + x^{2b} z^2 + x^{3b} z^3 + \cdots) \cdots$. Hence, the coefficient of $x^n z^m$ represents the number of ways that n can be written as a sum of m, not necessarily distinct, terms from the sequence $a, b, c, d, e, \ldots$.
5. $\prod_{n=1}^{\infty}(1 + x^{n^3})$.
6. $\prod_{n=1}^{\infty}(1 + x^{n(n+1)/2})$.
7. $\prod_{p}(1 + x^p)$, where p runs through all primes.
8. $\prod_{n=1}^{\infty}\frac{1}{1 - x^{n^3}}$.
9. $\prod_{n=1}^{\infty}\frac{1}{(1 - x^{n(n+1)/2})}$.
10. $\prod_{p}\frac{1}{1 - x^p}$, where p runs through all primes.
11. $\prod_{p}\frac{1}{1 - x^p}$, where p runs through all primes greater than 7.
12. $\prod_{n=6}^{\infty}\frac{1}{1 - x^{2n+1}}$.
13. $\prod_{n=3}^{10}\frac{1}{1 - x^{2n}}$.
14. $(1 + x)(1 + x^2)(1 + x^4)(1 + x^8)(1 + x^{16}) \cdots = 1 + x + x^2 + x^3 + x^4 + x^5 + x^6 + x^7 + x^8 + x^9 + \cdots$.
15. 9, $7 + 1 + 1$, $5 + 3 + 1$, $5 + 1 + 1 + 1 + 1$, $3 + 3 + 3$, $3 + 3 + 1 + 1 + 1$, $3 + 1 + 1 + 1 + 1 + 1 + 1$, and $1 + 1 + 1 + 1 + 1 + 1 + 1 + 1 + 1$. 9, $8 + 1$, $7 + 2$, $6 + 3$, $6 + 2 + 1$, $5 + 4$, $5 + 3 + 1$, $4 + 3 + 2$.
16. 10, $8 + 2$, $6 + 4$, $6 + 2 + 2$, $4 + 4 + 2$, $4 + 2 + 2 + 2$, $2 + 2 + 2 + 2 + 2$.
17. $6 + 4$, $7 + 3$, $5 + 5$, $4 + 3 + 3$.
18. The partitions of n into at most two parts are n and $(n - k) + k$, for $k = 1, \ldots, [\![n/2]\!] - 1$.
19. See Table A.10.

Table A.10.

n	$p(n)$	$p_e(n)$	$p_o(n)$	$p_d(n)$	$p_{ed}(n)$	$p_{od}(n)$	$p_1(n)$
1	1	0	1	1	0	1	1
2	2	1	1	1	0	1	2
3	3	0	2	2	1	1	1
4	5	2	2	2	1	1	7
5	7	0	3	3	2	1	12
6	11	3	4	4	3	3	45
7	15	0	5	5	4	4	87
8	22	5	6	6	3	3	45
9	30	0	8	8	4	4	87

Exercises 9.3

1. 1, 8, 24, 39, 47, 44, 38, 29, 22, 15, 11, 7, 5, 3, 2, 1, 1.
2. See Table A.11.
   The only selfconjugate partition of 7 is $4 + 1 + 1 + 1$.
4. $1 = 1$
   $2 = 2$
   $3 = 2 + 1$
   $4 = 2 + 2$
   $5 = 2 + 2 + 1$
   $6 = 2 + 2 + 2$
5.
$$\prod_{n=1}^{\infty}(1 - x^{2n})(1 + x^{2n-1}z)(1 + x^{2n-1}z^{-1})$$
$$= \prod_{n=1}^{\infty}(1 - u^{n^3})(1 - u^{3n-3/2+1/2})(1 - u^{3n-3/2-1/2})$$
$$= \prod_{n=1}^{\infty}(1 - u^{3n-2})(1 - u^{3n-1})(1 - u^{3n}) = \prod_{k=1}^{\infty}(1 - u^k),$$
   and
$$\sum_{n=-\infty}^{\infty} x^{n^2} z^n = \sum_{n=-\infty}^{\infty} u^{3n^2/2}(-u^{1/2})^n = \sum_{-\infty}^{\infty}(-1)^n u^{n(3n+1)/2}.$$
6. The largest part of the conjugate is the number of parts of the partition and vice versa.
7. The sum is 0. See Table A.12.
8. The sum is 0. See Table A.13.
9. If $n = 4 \pmod 5$ arrange the partitions of n into five classes such that the ranks of the partitions in each class have the same residue modulo

Table A.11.

Partitions	Number of distinct parts in each partition
7	1
6 + 1	2
5 + 2	2
5 + 1 + 1	2
4 + 3	2
4 + 2 + 1	3
4 + 1 + 1 + 1	2
3 + 3 + 1	2
3 + 2 + 2	2
3 + 2 + 1 + 1	3
3 + 1 + 1 + 1 + 1	2
2 + 2 + 2 + 1	2
2 + 2 + 1 + 1 + 1	2
2 + 1 + 1 + 1 + 1 + 1	2
1 + 1 + 1 + 1 + 1 + 1 + 1	1
Total	$30 = p_1(7)$

5. There will be the same number of partitions in each class. The result follows since $0 + 1 + \cdots + (n - 1) \equiv 0 \pmod 5$.

10.

```
1   2  4   6  9      1   3  5   9  11
3   5  7  10         2   4  6  10
8  11                7  12
12                   8
```

11.

```
1 2 3   1 2 3   1 2 4   1 2 4   1 2 5   1 2 5   1 2 6
4 5     4 6     3 5     3 6     3 4     3 6     3 4
6       5       6       5       6       4       5

1 2 6   1 3 4   1 3 4   1 3 5   1 3 5   1 3 6   1 3 6
3 5     2 5     2 6     2 4     2 6     2 4     2 5
4       6       5       6       4       5       4

1 4 6   1 4 5
2 5     2 6
3       3
```

Table A.12.

Partition	Rank	Modulo 5
4	3	3
3 + 1	1	1
2 + 2	0	0
2 + 1 + 1	−1	4
1 + 1 + 1 + 1	−3	2

Table A.13.

Partition	Rank	Modulo 5
9	8	3
8 + 1	6	1
7 + 2	5	0
7 + 1 + 1	4	4
6 + 3	4	4
6 + 2 + 1	3	3
6 + 1 + 1 + 1	2	2
5 + 4	3	3
5 + 3 + 1	2	2
5 + 2 + 2	2	2
5 + 2 + 1 + 1	1	1
5 + 1 + 1 + 1 + 1	0	0
4 + 4 + 1	1	1
4 + 3 + 2	1	1
4 + 3 + 1 + 1	0	0
4 + 2 + 2 + 1	0	0
4 + 2 + 1 + 1 + 1	−1	4
4 + 1 + 1 + 1 + 1 + 1	−2	3
3 + 3 + 3	0	0
3 + 3 + 2 + 1	−1	4
3 + 3 + 1 + 1 + 1	−2	3
3 + 2 + 2 + 2	−1	4
3 + 2 + 2 + 1 + 1	−2	3
3 + 2 + 1 + 1 + 1 + 1	−3	0
3 + 1 + 1 + 1 + 1 + 1 + 1	−4	1
2 + 2 + 2 + 2 + 1	−3	2
2 + 2 + 2 + 1 + 1 + 1	−4	1
2 + 2 + 1 + 1 + 1 + 1 + 1	−5	0
2 + 1 + 1 + 1 + 1 + 1 + 1 + 1	−6	4
1 + 1 + 1 + 1 + 1 + 1 + 1 + 1 + 1	−8	2

Bibliography

Mathematics (general)

Adams, W.W. and Goldstein, L.J., *Introduction to Number Theory*, Prentice-Hall, Englewood Cliffs, NJ, 1976.

Adler, A. and Coury, J.E., *The Theory of Numbers: a Text and Source Book of Problems*, Jones and Bartlett, Boston, Mass., 1995.

Andrews, G.E., *Number Theory*, W.B. Saunders, Philadelphia, 1971.

Apostol, T.M., *Introduction to Analytic Number Theory*, Springer-Verlag, New York, 1976.

Ball, W.W.R. and Coxeter, H.S.M., *Mathematical Recreations and Essays*, Dover, New York, 1987.

Barnett, I.A., *Some Ideas about Number Theory*, National Council of Teachers of Mathematics, Reston, VA, 1961.

Barnett, I.A., The fascination of whole numbers, *Mathematics Teacher* **64** (1971), 103–8.

Bennett, M., *et al.*, eds., *Surveys in Number Theory: Papers from the Millennial Conference on Number Theory*, A.K. Peters, Natick, 2003.

Burton, D.M., *Elementary Number Theory*, 2nd ed., Wm. C. Brown, Dubuque, IA, 1989.

Davenport, H., *The Higher Arithmetic*, 5th ed., Cambridge University Press, Cambridge, 1982.

Dickson, L.E., *Modern Elementary Theory of Numbers*, University of Chicago Press, Chicago, 1939.

Dudley, U., *Elementary Number Theory*, W.H. Freeman, San Francisco, 1969.

Erdős, P., Some unconventional problems in number theory, *Mathematics Magazine* **52** (1979), 67–70.

Erdős, P. and Dudley U., Some remarks and problems in number theory related to the work of Euler, *Mathematics Magazine* **56** (1983), 292–8.

Euclid, *The Thirteen Books of the Elements*, 3 vols., 2nd ed., trans. Thomas Heath, Dover, New York, 1956.

Fermat, P. *Varia opera mathematica*, J. Pech, Toulouse, 1679.

Grosswald, E., *Topics from the Theory of Numbers*, Macmillan, New York, 1966.

Guy, R.K., *Unsolved Problems in Number Theory*, 3rd ed., Springer-Verlag, New York, 1982.

Guy, R.K., *Unsolved Problems in Number Theory*, 3rd ed., Springer, New York, 2004
Hardy, G.H. and Wright, E.M., *An Introduction to the Theory of Numbers*, 4th ed., Oxford University Press, Oxford, 1960.
Hurwitz, A. and Kritikos, N., *Lectures on Number Theory*, Springer-Verlag, New York, 1986.
Legendre, A.M., *Théorie des nombres*, 4th ed., Librairie Scientifique et Technique, Paris, 1955.
LeVeque, W.J., *Reviews in Number Theory*, 6 vols., American Mathematical Society, Providence, RI, 1974.
LeVeque, W.J. *Fundamentals of Number Theory*, Addison-Wesley, Reading, MA., 1977.
Ogilvy, C.S. and Anderson, J.T., *Excursions in Number Theory*, Oxford University Press, New York, 1966.
Ore, O., *An Invitation to Number Theory*, Mathematical Association of America, Washington, DC, 1967.
Ribenbaum, O., *My Numbers, My Friends*, Spinger-Verlag, New York, 2000.
Roberts, J., *Elementary Number Theory*, MIT Press, Cambridge, MA., 1967.
Roberts, J., *Lure of the Integers*, Mathematical Association of America, Washington, DC, 1992.
Rose, H.E., *A Course in Number Theory*, Clarendon, Oxford, 1988.
Rosen, K.H., *Elementary Number Theory and its Applications*, 3rd ed., Addison-Wesley, Reading, Mass., 1993.
Scharlau, W. and Opolka, H., *From Fermat to Minkowski: Lectures on the Theory of Numbers and its Historical Development*, Springer-Verlag, New York, 1985.
Shapiro, H.N., *Introduction to the Theory of Numbers*, Wiley, New York, 1983.
Sierpiński, W., *Elementary Theory of Numbers*, North-Holland, Amsterdam, 1988.
Silverman, J.H., *A Friendly Introduction to Number Theory*, Prentice-Hall, Upper Saddle River, NJ, 1997.
Stark, Harold M., *An Introduction to Number Theory*, MIT Press, Cambridge, MA., 1987.
Stewart, B.M., *Theory of Numbers*, 2nd ed., Macmillan, New York, 1964.
Stopple, J., *A Primer of Analytic Number Theory: From Pythagoras to Riemann*, Cambridge University Press, Cambridge, 2003.
Uspensky, J.V. and Heaslett, M.A., *Elementary Number Theory*, McGraw-Hill, New York, 1939.

History (general)

Berndt, B. and Rankin, R., *Ramanujan: Letters and Commentary*, Hist. of Math. vol. 9, American Mathematical Society and London Mathematical Society, Providence, RI, 1995.
Bühler, W.K., *Gauss: a Biographical Study*, Springer, New York, 1981.
Bunt, N.H., Jones, P. S. and Bedient, J.D., *The Historical Roots of Elementary Mathematics*, Prentice-Hall, Englewood Cliffs, NJ, 1976.
Canfora, Luciano, *The Vanished Library*, University of California Press, Berkeley, 1987.

Dickson, L.E., *History of the Theory of Numbers*, 3 vols., Carnegie Institute, Washington, DC, 1919, reprinted Chelsea, New York, 1952.

Dunham, W., *Euler, The Master of Us All*, Mathematical Association of America, Washington, D.C., 1999.

Dzielska, M., *Hypatia of Alexandria*, Harvard University Press, Cambridge, MA., 1995.

Hardy, G.H., *A Mathematician's Apology*, Cambridge University Press, Cambridge, 1967.

Heath, T.L., *Diophantos of Alexandria*, Cambridge University Press, Cambridge, 1885.

Hoffman, P., *The Man Who Loved Only Numbers: the Story of Paul Erdős and the Search for Mathematical Truth*, Hyperion, New York, 1998.

Kanigel, R., *The Man Who Knew Infinity*, Charles Scribner's, New York, 1991.

Katz, V., *A History of Mathematics*, 2nd ed., Addison-Wesley Longman, Reading, MA., 1998.

Kendall, D.G., The scale of perfection, *Journal of Applied Probability* **19A** (1982), 125–38.

Littlewood, J.E., *A Mathematic Miscellany*, revised ed. Cambridge University Press, Cambridge, 1986.

Mahoney, M.S., *The Mathematical Career of Pierre de Fermat 1601–1665*, 2nd ed., Princeton University Press, Princeton, NJ, 1994.

Mordell, L.J., Reminiscences of an octogenarian mathematician, *American Mathematical Monthly* **78** (1971), 952–61.

Reid, C., *Hilbert*, Springer-Verlag, New York, 1970.

Reid, C., *The Search for E.T. Bell: Also Known as John Taine*, Mathematical Association of America, Washington, D.C., 1993.

Reid, C., *Julia: a Life in Mathematics*, Mathematical Association of America, Washington, D.C., 1996.

Schecter, B., *My Brain Is Open: the Mathematical Journeys of Paul Erdős*, Simon and Schuster, New York, 1998.

Weil, A., *Number Theory: an Approach Through History from Hammurapi to Legendre*, Birkhäuser, Boston, Mass., 1984.

Yan, L. and Shiran, D., *Chinese Mathematics: a Concise History*. Clarendon, Oxford, 1987.

Chapter 1

Aaboe, Asger, *Episodes from the Early History of Mathematics*, New Mathematics Library vol. 13, Mathematical Association of America, Washington, D.C., 1964.

Bachet, C.-G., *Problèmes plaisants et délectables qui se font par les nombres*, 5th ed. A. Labosne, reprinted Librairie Scientifique et Technique, Paris, 1959.

Caffrey, J., Variations on Trigg's Constant, *Journal of Recreational Mathematics* **23** (1991), 192–3.

Conway, J.H. and Guy, R.K., *The Book of Numbers*, Springer-Verlag, New York, 1996.

DeTemple, D.W., The triangle of smallest perimeter which circumscribes a given semicircle, *Fibonacci Quarterly* **30** (1992), 274.

Duncan, D.C., Happy integers, *Mathematics Teacher* **65** (1972), 627–9.

Edwards, A.W.F., *Pascal's Arithmetic Triangle*, Oxford University Press, New York, 1987.

El-Sedy, E. and Siksek, S., On happy integers, *Rocky Mountain Journal of Mathematics* **30** (2000), 565–70

Ewell, J.A., On representations of numbers by sums of two triangular numbers, *Fibonacci Quarterly* **30** (1992), 175–8.

Ewell, J.A., On sums of triangular numbers and sums of squares, *American Mathematical Monthly* **99** (1992), 752–7.

Fibonacci, *The Book of Squares: an Annotated Translation into Modern English by L.E. Sigler*, Academic Press, Boston, MA., 1987.

Gagan, T., Automorphic numbers, *Australian Mathematical Society Gazette* **20** (1993), 37–44.

Haggard, P.W. and Sadler, B.L, A generalization of triangular numbers, *International Journal of Mathematical Education in Science and Technology* **25** (1994), 195–202.

Hemmerly, H.O., Polyhedral numbers, *Mathematics Teacher* **66** (1973), 356–62.

Hutton, C., *Ozanam's Recreational Mathematics and Natural Philosophy*, E. Riddle, London, 1803.

Juzuk, D., Curiosa 56, *Scripta Mathematica* **6** (1939), 218.

Kaprekar, D.R., Another solitaire game, *Scripta Mathematica* **15** (1949), 244–5.

Kay, D.C., On a generalizaton of the Collatz algorithm, *Pi Mu Epsilon Journal* **5** (1972), 338.

Koshy, T., *Fibonacci and Lucas Numbers with Applications*. Wiley-Interscience, New York, 2001.

Kullman, D.E., Sums of powers of digits, *Journal of Recreational Mathematics* **14** (1981–82), 4–10.

Kumar, V.S., On the digital root series, *Journal of Recreational Mathematics* **12** (1979–80), 267–70.

Larison, C.B., Sidney's series, *Fibonacci Quarterly* **24** (1986), 313–15.

Livio, M., *The Golden Ratio*, Broadway Books, New York, 2002.

Long, C.T., Gregory interpolation: a trick for problem solvers out of the past, *Mathematics Teacher* **81** (1983), 323–5.

Lucas, E., Théorie des fonctions numériques simplement périodiques, *American Journal of Mathematics* **1** (1878), 184–240, 289–321.

Lucas, E., *Récréations mathématiques*, 4 vols., Gauthier-Villars et fils, Paris, 1891–6.

MacKay, J.S., A puzzle of Dr. Whewell's, *Mathematical Notes of the Edinburgh Mathematical Society* **11** (1912), 121–5.

May, K.O., Galileo sequences: a good dangling problem, *American Mathematical Monthly* **79** (1972), 67–9.

Matsuoka, Y., On a problem in number theory, *Science Reports from Kagoshima University* **14** (1965), 7–11.

Mohanty, S.P., Which triangular numbers are products of three consecutive integers? *Acta Mathematica Hungarica* **58** (1991), 31–6.

Naranam, S. An elephantine equation, *American Mathematical Monthly* **66** (1973), 276–8.

Nicomachus, *Introductio arithmeticae*, trans. M.L. D'Ooge, Macmillan, London, 1926, reprinted Johnson, New York, 1972.

Porges, A., A set of eight numbers, *American Mathematical Monthly* **52** (1945), 379–82.

Porges, A., The Devil and Simon Flagg, *The Magazine of Fantasy and Science Fiction*, 1954; reprinted in *Fantasia mathematica*, ed. Clifton Fadiman, Simon and Schuster, New York, 1958.

Schwartzman, J. and Shultz, H.S., Square dance numbers, *Mathematics Teacher* **82** (1989), 380–2.

Sloane, N.J.A., The persistence of a number, *Journal of Recreational Mathematics* **6** (1973), 97–8.

Sloane, N.J.A. and Plouffe, S., *Encyclopedia of Integer Sequences*, Academic Press, San Diego, 1995.

Stewart, B.M., Sums of functions of digits, *Canadian Journal of Mathematics* **12** (1960), 374–89.

Theon of Smyrna, *Mathematics for Useful Understanding of Plato*, trans. Robert and Deborah Lawlor, Wizard's Bookshelf, San Diego, 1979.

Trigg, C.W., A new routine leads to a new constant, 2538, *Journal of Recreational Mathematics*, **22** (1990), 34–6.

Vajda, S., *Fibonacci & Lucas Numbers, and the Golden Section: Theory and Application*, Ellis Horwood, Chichester, 1989.

Vovob'ev, N.N., *Fibonacci Numbers*, Blaisdell, New York, 1961.

Zeckendorf, E., Répresentation des nombres naturels par une somme de nombres de Fibonacci ou de nombres Lucas, *Bulletin de la Sociètè Royale des Sciences de Liege* **41** (1972) 179–82.

Chapter 2

Bouton, C.L., Nim, a game with a complete mathematical theory, *Annals of Mathematics* **3** (1902), 35–9.

Collison, M.J., The unique factorization theorem: from Euclid to Gauss, *Mathematics Magazine* **53** (1980), 96–100.

Cooper, C. and Kennedy, R.E., On consecutive Niven numbers, *Fibonacci Quarterly* **31** (1993), 146–51.

Dixon, J., The number of steps in the Euclidean algorithm, *Journal of Number Theory* **2** (1970), 414–22.

Horadam, A.F., A generalized Fibonacci sequence, *American Mathematical Monthly* **68** (1961), 455–9.

Kalman, D., Fractions with cycling digit property, *The College of Mathematics Journal* **27** (1996) 109–15.

Kennedy, R., Digital sums, Niven numbers, and natural density, *Crux Mathematicorum* **8** (1982) 131–5.

Kennedy, R., Goodman, T., and Best, C., Mathematical discovery and Niven numbers, *Mathematics for Two-Year Colleges Journal* **14** (1980) 21–5.

Mohanty, S.P., Pythagorean numbers, *Fibonacci Quarterly* **28** (1990), 31–42.

Moore, E.H., A generalization of the game called nim, *Annals of Mathematics* **11** (1910), 93–4.

Neugebauer, O. and Sachs, A., *Mathematical Cuneiform Texts*, American Oriental Society and the American School of Oriental Research, New Haven, CT., 1945, pp. 38–41.

Saunderson, N., *The Elements of Algebra in Ten Books*, Cambridge University Press, Cambridge, 1740.

Schroeder, M.R., *Number Theory in Science and Communication*, Springer-Verlag, New York, 1984.

Tattersall, J.J., Nicholas Saunderson: the blind Lucasian Professor, *Historia Mathematica* **19** (1992), 356–70.

Chapter 3

Apostol, T., Some properties of completely multiplicative arithmetic functions, *American Mathematical Monthly* **78** (1971), 266–71.

Apostol, T., A characteristic property of the Möbius function, *American Mathematical Monthly* **72** (1965), 279–82.

Bressoud, D.M., *Factorizaion and Primality Testing*, Springer-Verlag, New York, 1989.

Crandall, R. and Pomerance, C., *Prime Numbers*, Springer-Verlag, New York, 2001.

Dirichlet, P.G.L., *Vorlesung über Zahlentheorie*, F. Vieweg, Braunschweig, 1894, reprinted Chelsea, New York, 1968.

Dixon, J., Factoring and primality tests, *American Mathematical Monthly* **91** (1984), 333–51.

Dudley, U., Formulas for primes, *Mathematics Magazine* **56** (1983), 17–22.

Dudley, U., Smith numbers, *Mathematics Magazine* **67** (1994), 62–5.

Duncan, R.L., Note on the divisors of a number, *American Mathematical Monthly* **68** (1961), 356–9.

Friedlander, R.J., Factoring factorials, *Two-Year College Mathematics Journal* **12** (1981), 12–20.

Gardiner, V., Lazarus, R., Metropolis, N., and Ulam, S., On certain sequences of integers defined by sieves, *Mathematics Magazine* **29** (1956), 117–22.

Gardner, M., The remarkable lore of the prime number, *Scientific American* **210** (March 1964), 120–8.

Goldstein, L., A history of the prime number theorem, *American Mathematical Monthly* **80** (1973), 599–615.

Long, C.T., On the Moessner theorem on integral powers, *American Mathematical Monthly* **73** (1966), 846–51.

Long, C.T., A note on Moessner's theorem, *Fibonacci Quarterly* **24** (1986), 349–55.

McDaniel, W.L., The existence of infinitely many k-Smith numbers, *Fibonacci Quarterly* **25** (1987), 76–80.

Pollard, J.M., Theorems of factorization and primality testing, *Proceedings of the Cambridge Philosophical Society* **76** (1974), 521–8.

Pomerance, C., The search for prime numbers, *Scientific American* **247** (Dec. 1982), 122–30.

Ramanujan, S., Highly composite numbers, *Proceedings of the London Mathematical Society* (2) **14** (1915), 347–409.
Ribenboim, P., *The Book of Prime Number Records*, 2nd ed., Springer-Verlag, New York, 1989.
Ribenboim, P., *The New Book of Prime Number Records*, Springer-Verlag, New York, 1996.
Ribenboim, P., *The Little Book of Big Primes*, Springer-Verlag, New York, 1991.
Smith, M., Cousins of Smith numbers: Monica and Suzanne sets, *Fibonacci Quarterly* **34** (1996), 102–4.
Solovay, R. and Strassen, V., A fast Monte Carlo test for primality, *SIAM Journal for Computation* **6** (1977), 84–5, erratum, **7** (1978), 118.
Ulam, S., *Problems in Modern Mathematics*, Interscience, New York, 1964, ix.
Wagon, S., Primality testing, *Mathematical Intelligencer* **8** (1986), 58–61.
Wallis, J., *Treatise of Algebra Both Historical and Practical*, J. Playford, London, 1685.
Willansky, A., Smith numbers, *Two-Year College Mathematics Journal* **13** (1982), 21.

Chapter 4

Bezuska, S., *Perfect Numbers*, Boston College Press, Boston, MA., 1980.
Brooke, M. On the digital roots of perfect numbers, *Mathematics Magazine* **34** (1960–61), 100, 124.
Brown, A.L., Multiperfect numbers, *Scripta Mathematica* **20** (1954), 103–6.
Carmichael, R.D., A table of multiperfect numbers, *Bulletin of the American Mathematical Society* **13** (1907), 383–6.
Chernick, J., On Fermat's simple theorem. *Bulletin of the American Mathematical Society* **45** (1939), 269–74.
Escott, E.B., Amicable numbers, *Scripta Mathematica* **12** (1946), 61–72.
Franqui, B. and García, M., Some new multiperfect numbers, *American Mathematical Monthly* **60** (1953), 459–62.
Franqui, B. and García, M., 57 new multiperfect numbers, *Scripta Mathematica* **20** (1954), 169–71.
García, M., New amicable pairs, *Scripta Mathematica* **21** (1957), 167–71.
Lee, E.J. and Madachy, J., The history and discovery of amicable numbers, *Journal of Recreational Mathematics* **5** (1972), 77–93, 153–73, 231–49.
Minoli, D. and Bear, R., Hyperperfect numbers, *Pi Mu Epsilon Journal* **6** (1974), 153–7.
Pinch, R.G.E., The Carmichael numbers up to 10^{15}, *Mathematics of Computation* **61** (1993), 703–22.
Poulet, P., *La Chasse aux nombres*, pp. 9–27, Stevens Frères, Bruxelles, 1929.
Poulet, P., 43 new couples of amicable numbers, *Scripta Mathematica* **14** (1948), 77.
Prielipp, R.W., Digital sums of perfect numbers and triangular numbers, *Mathematics Teacher* **62** (1969), 170–82.
Prielipp, R.W., Perfect numbers, abundant numbers, and deficient numbers, *Mathematics Teacher* **63** (1970) 692–6.
Shoemaker, R.W., *Perfect Numbers*, National Council of Teachers of Mathematics, Reston, VA, 1973.

Wall, C.R., *Selected Topics in Elementary Number Theory*, p. 68, University of South Carolina Press, 1974.

Chapters 5 and 6

Brown, E., The first proof of the quadratic reciprocity law, revisited, *American Mathematical Monthly* **88** (1981), 257–64.

Cox, D., Quadratic reciprocity: its conjecture and application, *American Mathematical Monthly* **95** (1988), 442–8.

Edwards, H., Euler and quadratic reciprocity, *Mathematics Magazine* **56** (1983), 285–91.

Gauss, K.F., *Disquisitiones arithmeticae*, trans. A.A. Clarke, revised W.C. Waterhouse, Springer-Verlag, New York, 1966.

Legendre, A.M. *Essai sur la théorie des nombres*, Duprat, Paris 1808.

Murty, M.R., Artin's conjecture for primitive roots, *Mathematical Intelligencer* **10** (1988), 59–67.

Sylvester, J.J., On the problem of the virgins, and the general theory of compound partitions, *Philosophical Magazine* **(4) 16** (1858), 371–6.

Tattersall, J.J. and McMurran, S.L., Hertha Ayrton: a persistent experimenter, *Journal of Women's History* **7** (1995), 86–112.

Western, A.E. and Miller, J.C.P., *Tables of Indices and Primitive Roots*, Royal Society Mathematical Tables, Vol. 9, Cambridge University Press, London, 1968.

Chapter 7

Beutelspacher, A., *Cryptology*, Mathematical Association of America, Washington, DC, 1994.

Diffie, W. and Hellman, M.E., New directions in cryptography, *IEEE Transactions on Information Theory* **IT-22** (1976), 644–54.

Herodotus, *The History*, trans. David Grene, University of Chicago Press, Chicago, 1987.

Hill, L.S., Concerning certain linear transformations apparatus of cryptography, *American Mathematical Monthly* **38** (1931), 135–54.

Kahn, D., *The Codebreakers: the Story of Secret Writing*, Macmillan, New York, 1967.

Koblitz, N., *A Course in Number Theory and Cryptography*, Springer-Verlag, New York, 1987.

Luciano, D. and Prichett, G., Cryptology: from Caesar ciphers to public-key cryptosystems, *College Mathematics Journal* **18** (1987), 2–17.

Merkle, R.C., Secure communications over insecure channels, *Communications of the ACM* **21** (1978), 294–9.

Rivist, R., Shamir, A., and Adleman, L., A method for obtaining digital signatures and public-key cryptosystems, *Communications of the ACM* **21** (1978), 120–6.

US Army, Signal Communications Field Manual M-94. Superseded by Field Army Signal Communications FM 11–125, December 1969.

Chapter 8

Bachman, G., *Introduction to p-adic Numbers and Valuation Theory*, Academic Press, New York, 1964.

Baker, A., *A Concise Introduction to the Theory of Numbers*, Cambridge University Press, Cambridge, 1984.

Bell, E.T., *The Last Problem*, Mathematical Association of America, Washington, DC, 1990.

Brezinski, C., *History of Continued Fractions and Padé Approximations*, Springer-Verlag, New York, 1991.

Cassels, J.W.S., *Rational Quadratic Forms*, Academic Press, London, 1978.

Conway, J.H., *The Sensual Quadratic Form*, Mathematical Association of America, Washington, DC, 1997.

Edgar, H.M., *A First Course in Number Theory*, Wadsworth, Belmont, NY, 1988.

Edwards, H., *Fermat's Last Theorem*, Springer-Verlag, New York, 1977.

Elkies, N., On $A^4 + B^4 + C^4 = D^4$, *Mathematics of Computation* **51** (1988), 825–35.

Ellison, W.J., Waring's problem, *American Mathematical Monthly* **78** (1981) 10–36.

Gouvêa, F.G., *p-adic Numbers: an Introduction*, Springer, New York, 1991.

Lander, L.J. and Parkin, T.R., Counterexamples to Euler's conjectures on sums of like powers, *Bulletin of the American Mathematical Society* **72** (1966), 1079.

MacDuffie, C.C. The p-adic numbers of Hensel, *American Mathematical Monthly* **45** (1938) 500–8; reprinted in *Selected Papers in Algebra*, Mathematical Association of America, Washington, DC, 1977, 241–9.

Nelson, H.L., A solution to Archimedes' cattle problem, *Journal of Recreational Mathematics* **13** (1980–81), 164–76.

Niven, I., Zuckerman, H.S., and Montgomery, H.L., *An Introduction to the Theory of Numbers*, 5th ed., Wiley, New York, 1991.

Olds, C.D., *Continued Fractions*, Random House, New York, 1963.

Paton, W.R. (trans.) *The Greek Anthology*, 5 vols. vol. 5, 93–5, Heinemann, London, 1918.

Ribenboim P., *13 Lectures on Fermat's Last Theorem*, Springer-Verlag, New York, 1979.

Ribenboim P., *Catalan's Conjecture*, Academic Press, Boston, MA., 1994.

Robbins, N., *Beginning Number Theory*, Wm. C. Brown, Dubuque, IA, 1993.

Singh, S., *Fermat's Enigma*, Walker, New York, 1997.

Stanton, R.G., A representation problem, *Mathematics Magazine* **43** (1970), 130–7.

Steiner, R.P., On Mordell's equation $y^2 - k = x^3$, *Mathematics of Computation* **46** (1986), 703–14.

Strayer, J.K., *Elementary Number Theory*, PWS, Boston, Mass., 1994.

Tausky, O., Sums of squares, *American Mathematical Monthly* **77** (1970), 805–30.

Van der Poorten, A., *Notes on Fermat's Last Theorem*, Wiley-Interscience, New York, 1996.

Vardi, Ilan, Archimedes' Cattle Problem, *American Mathematical Monthly* **105** (1998), 305–19.

Chapter 9

Alder, H.L., Partition identities – from Euler to the present, *American Mathematical Monthly* **76** (1969), 733–46.

Alder, H.L., The use of generating functions to discuss and prove partition identities, *Two-Year College Mathematics Journal* **10** (1979), 318–29.

Andrews, G.E., *The Theory of Partitions*, Addison-Wesley, Reading, Mass., 1976.

Euler, L. *Introductio in analysin infinitorum*, M. Bousquet, Lausanne, 1748. *Introduction to the Analysis of the Infinite*, 2 vols., trans. John Blanton, Springer-Verlag, New York, 1988, 1990.

Hardy, G.H. and Ramanujan, S., Asymptotic formulae in combinatorial analysis, *Proceedings of the London Mathematical Society* (2) **17** (1918), 75–118.

Guy, R.K., Two theorems on partitions, *Mathematics Gazette* **42** (1958), 84–6.

Rademacher, H., On the partition function $p(n)$, *Proceedings of the London Mathematical Society* (2) **43** (1937), 241–57.

Ramanujan, S., Proof of certain identities in combinatorial analysis, *Proceedings of the Cambridge Philosophical Society* **19** (1919), 214–16.

Index